T4-AUB-524

SOURCEBOOK OF ATM AND IP INTERNETWORKING

Books of Related Interest from IEEE Press

TELECOMMUNICATIONS NETWORK MANAGEMENT INTO THE 21st CENTURY: Techniques, Standards, Technologies, and Applications
Edited by Salah Aidarous and Thomas Plevyak
1994 Hardcover 448 pp. IEEE Order No. PC3624 ISBN 0-7803-1013-6

TELECOMMUNICATIONS NETWORK MANAGEMENT: Techniques and Implementations
Edited by Salah Aidarous and Thomas Plevyak
1997 Hardcover 352 pp. IEEE Order No. PC5711 ISBN 0-7803-3454-X

FUNDAMENTALS OF TELECOMMUNICATIONS NETWORK MANAGEMENT
Lakshmi G. Raman
1999 Hardcover 368 pp. IEEE Order No. PC5723 ISBN 0-7803-3466-3

SECURITY FOR TELECOMMUNICATIONS MANAGEMENT NETWORKS
Moshe Rozenblit
2000 Hardcover 320 pp. IEEE Order No. PC5793 ISBN 0-7803-3490-6

INTEGRATED TELECOMMUNICATIONS MANAGEMENT SOLUTIONS
Graaham Chen and Qinzheng Kong
2001 Hardcover 261 pp. IEEE Order No. PC5795 ISBN 0-7803-5353-6

SOURCEBOOK OF ATM AND IP INTERNETWORKING

Khalid Ahmad

Salah Aidarous and Thomas Plevyak, *Series Editors*

IEEE Press

JOHN WILEY & SONS, INC.

This text is printed on acid-free paper. ♾

For ordering and customer service, call 1-800-CALL-WILEY.

Library of Congress Cataloging in Publication Data is available.

ISBN 0-471-20815-9

Printed in the United States of America.

10 9 8 7 6 5 4 3 2 1

Dedicated to the memory of my parents—
Dr. Nazir Ahmad and Begum Razia Nazir Ahmad

Contents

Preface

Internetworking technologies such as ATM, frame relay, and IP, extensively developed and deployed over the last decade, have so transformed the traditional landscape of telecommunications and data communications that catch phrases such as "information highway" and "information revolution" have become commonplace in the popular consciousness. The pace of technology development poses a challenge even to telecommunications and datacom professionals, who must attempt to make sense of the plethora of networking protocols, already deployed or being developed, in designing communications systems and networks for increasingly demanding applications.

In surveying these technological developments, it is often easy to lose sight of the broad symbiotic relationships that underlie networking technologies in the struggle to understand details necessary to utilize the protocols. The evolution of conventional circuit switched voice telephony networks toward integrated voice and data communications based on a variety of fast packet technologies inevitably results in the internetworking of packet switching mechanisms from different origins. Consequently, a detailed understanding of the functional similarities and differences between the various approaches is an invaluable requisite to the network or system designer, who must forge a networking solution derived from either ATM-, FR-, or IP-based technologies.

This book is intended to provide such an understanding. By focusing on fundamental architectural precepts and essential functional descriptions of the protocols, highlighting commonalities as well as differences between the major internetworking protocols, the intent is to assist the reader in finding his or her way through the labyrinth of concepts, architecture, and specifications that appear with increasing frequency in the literature. Although the account is intended to be reasonably comprehensive, the stress is on describing the essential functionality necessary for robust internetworking, rather than attempting a cursory coverage of every specialized feature that any be desirable for any given network application.

Thus, no attempt is made to describe potentially vast areas of study such as access network or wireless ATM applications, each of which would require a separate treatise. It is also not necessary to describe in rigorous detail well-established technologies such as frame relay and TCP/IP, which have already generated a copious literature of their own. On the other hand, in order to understand the interworking of these technologies with ATM in some detail, it was felt to be useful to summarize the essential aspects of IP and FR protocols from the perspective of internetworking with ATM networks.

This work grew out of a series of courses on ATM and IP internetworking I gave to Nortel Networks R&D engineers as well as network operators in various parts of the globe. In presenting these courses I realized that though several good books on ATM were available, most did not fulfill a need for a detailed as well as comprehensive functional description that builds on the unifying protocol architecture and relationships with other packet technologies. Having been involved in the development of most aspects of ATM technology from its beginnings in the ITU-T and ATM Forum, as well as frame relay and more recently the MPLS evolution of IP, I felt a book stressing the underlying simplicity and unity of the ATM vision would prove useful to network planners and researchers seeking to use the technology in real world networking applications.

It will be evident that the development of ATM has profoundly influenced, and in turn been influenced by, other networking technologies such as IP and FR. Useful concepts from one approach are incorporated in another, a feat that can often be readily accomplished by simply changing the terminology. This cross-fertilization of ideas, while fascinating in itself, is also beneficial in that interworking between the different approaches is facilitated.

However, it can also be confusing for network designers in that proponents of one camp or the other may focus on differences rather than stressing the underlying functional commonalities. It is inevitable that each architectural approach will stress its advantages relative to the competing alternatives from the perspective of performance or cost aspects. In evaluating these alternatives for any given internetworking solution, it is important for the system and network designer to compare them on a detailed function-by-function basis. This book is intended to assist the reader in making such evaluations, by providing the necessary function detail while highlighting the essential conceptual similarities or differences between the various internetworking architectures.

ACKNOWLEDGMENTS

The technology developments described in this book are the result of the extensive collective work carried out by many researchers in the fields of networking technologies over the years, whose efforts in this endeavor must be acknowledged. In particular, I have been fortunate in being able to learn about ATM and related technologies directly from many distinguished colleagues, for which I remain deeply grateful.

For this, I wish to thank my many colleagues in the International Telecommunications Union (ITU-T) and ATM Forum standardization endeavor for ATM for their vision and patience. Although too numerous to list here, these notably include: Drs. Luc Le Beller, Jean-Pierre Coudreause, Laurent Hue, Pierre Boyer, Annie Gravey, Michel Bonnifait, Nicholas Lozach, Gilles Joncour, J. C. Sapanel, Dominique Delisle and many others from the Centre National d'Études de Telecommunications (CNET) Laboratories, France Telecom; Hiroshi Ohta, N. Morita, K. Asatani, Y. Maeda, N. Kawarasaki, and others from NTT Laboratories; Dr. Jon Andersen from Lucent Technologies; Dipl. Eng. Werner Hug from Alcatel; and Dipl. Eng. Joachim Klink from Siemens.

I am also deeply grateful to numerous colleagues in Nortel Networks, who over the years have provided me with insights into the intricacies of ATM, FR, and IP technologies. Notable among these include Drs. Osama Aboul-Magd, Sameh Rabie, Rungroj Kositpaiboon, Gordon Fitzpatrick, Ghassem Koleyni, Riad Hartani, Ernst Munter, Marek Wernik, Stephen Shew, Russ Pretty, Peter Ashwood-Smith, and the many others who helped me.

It should be acknowledged that initial developments related to ATM were promoted under the auspices of the ITU Telecommunications standardization sector, in which I have been fortunate to participate. I wish to acknowledge the generous help provided by the staff and counselors of the ITU-T secretariat, notably Fabio Bigi, Paolo Rosa, and Arshey Oderdra, in allowing me to reference the documentation of the ITU-T.

I also gratefully acknowledge the support of the ATM Forum, whose initiatives have greatly contributed to the widespread acceptance of ATM as the technology of choice for high-performance networking, for allowing me to reference their ATM specifications.

I would like to gratefully acknowledge the many helpful comments and suggestions of the IEEE Press and review teams for their help. Notable among these include Drs. Salah Aidarous, Tom Robertazzi, Mostafa Sherif, M. Baker, Jianjun Han, D. New, and others of the review teams, and Catherine Feduska, Linda Matarazzo, and Karen Hawkins of the IEEE Press editorial team.

This book could never been written were it not for the painstaking work of Dipl. Ing. Ana Szpaizer, who undertook the onerous task of preparing and editing the entire manuscript and artwork for publication. I wish to acknowledge my grateful appreciation for the enormous amount of work Ana Szpaizer accomplished, as well as the resourcefulness and encouragement she demonstrated in completing this massive task.

KHALID AHMAD

Nortel Networks
Ottawa, Canada
November 2001

CHAPTER 1

Introduction

Asynchronous transfer mode (ATM) technology has emerged in recent years as the most promising switching and transport technology for enabling the integration and enhancement of traditional telecommunications and data networks. The widespread use (and abuse) of the catchphrase "The Information Age" to capture an essential element of today's working environment essentially implies a symbiotic combination of electronic data processing with traditional voice telephony and video image transport and processing. A unifying networking technology that allows for the relatively elegant integration of voice, video, and data information sources for transport and switching functions is clearly an essential element for the implementation of telecommunication networks that seek to handle all forms of electronic information sources.

ATM has been designed to provide such a networking (implying both transport and switching in the sense of ubiquitous connectivity) technology, essentially by combining the advantages of traditional voice telephony circuit switching with those of packet-based data networking. This results in a combined switching and transport technology capable of very high performance and scalability, ideally suited to the widespread deployment of modern digital switching exchanges and optical fiber transmission systems.

Although the ATM protocols described here have been specifically designed for the integrated switching and transport of all electronic information sources, it must be stressed at the outset that ATM is not the only technology currently in use that is capable of the integrated handling of voice, video, and data sources. There are several other networking technologies also being developed to provide the capability of integrating voice, video, and data sources toward the same ends as ATM. Notable examples of these alternative switching and transport mechanisms include the internetworking protocol (IP) based global internet, and frame relay services.

A more detailed discussion of these protocols may be found in later chapters, essentially in the context of understanding how ATM can interwork with these alternative networking technologies. Despite the presence of such alternative protocols, and as evidenced by the enormous and growing global investment in ATM technology, the potential flexibility and performance advantages of ATM make it the "technology of choice" for next generation telecommunications networks.

However, in the foreseeable future, ATM based networks will likely coexist with other wide area networking (WAN) technologies such as the TCP/IP Internet and frame relay bearer services (FRBS), necessitating interworking capabilities with these and other networks to enable the provision of seamless services to end users of any of the different networks. The interworking and mutual interaction of ATM with other important networking technologies is a major area of study and development, and is discussed in later chapters.

ATM-based networks are presently either in service or being actively evaluated by numerous network operators in all parts of the developed world. Since ATM is essentially intended to provide the underlying switching and transport infrastructure, a wide variety of applications and telecommunication services are being offered or undergoing preservice trials. However, even at

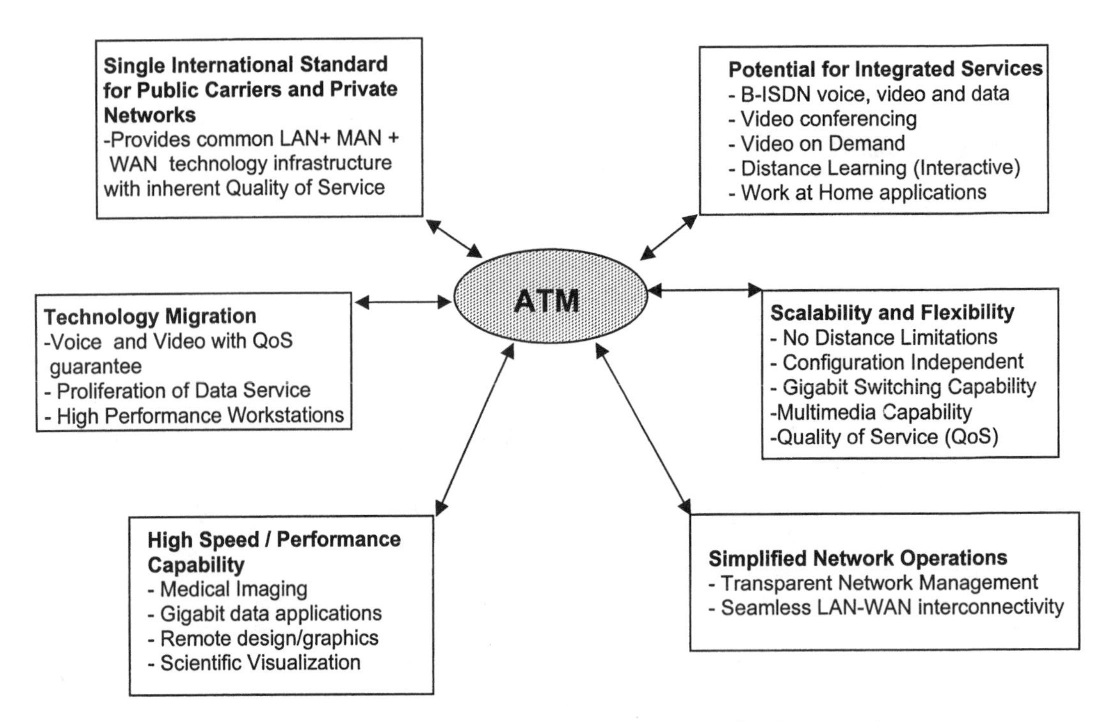

Figure 1.1. Advantages of ATM and some network application examples.

this stage, important examples of the use of ATM can be mentioned. These include the transport and switching of the majority of data (e.g., Internet) traffic in carrier networks, as well as the use of ATM in mission critical applications where quality of service (QoS) and reliability are key requirements. It is likely that the number of ATM applications and services will grow rapidly as the inherent flexibility and high performance capabilities of ATM technology are adapted by enterprising service providers to their specific commercial ends. Consequently, it is not useful at this stage to provide a comprehensive catalog of ATM applications and services since these are evolving so rapidly as to render any list obsolete before long.

The main objective of this text is to emphasize the fundamental principles and protocols utilized in ATM technology, a clear understanding of which will enable the reader to readily follow any new applications and developments in a rapidly evolving technology.

Figure 1.1 summarizes some of the typical rationales used for ATM technology deployment, as well as some potential applications. This is not intended to be comprehensive, but simply to provide some answers to a frequently asked question: Why ATM?

CHAPTER 2

ATM Principles and Basic Definitions

The basic principles of ATM (asynchronous transfer mode) can be characterized as follows:

1. ATM is a specific packet-oriented transfer mode in which the digitized information from any source (e.g., voice samples, data packets, video image frames, etc.) is organized into blocks of fixed size called "cells." An ATM cell is simply a "packet" of digital information of fixed size.
2. The ATM cells consist of an "information field" 48 octets long, appended to an ATM "header" 5 octets long. Consequently, all ATM cells have a fixed size of 53 octets. The ATM cell structure is shown in Fig. 2-1.
3. The ATM cells from either different or the same source are multiplexed together in a cell stream using the asynchronous time division multiplexing technique, which is essentially similar to conventional packet-based multiplexing. The term "asynchronous time division" is used in the sense that there is no fixed time relationship between the arrival of the ATM cells from a given source in the cell stream. That is, the ATM cells from a given source may be emitted with different intervals between them in general, and are then multiplexed with the cells from other sources, resulting in a cell stream in which cells belonging to the different sources are asynchronously spaced, although always of fixed size of 53 octets long.
4. The ATM header in each cell contains the (logical) connection identifier, which enables the cells belonging to the same "virtual channel" to be uniquely identified within the asynchronously multiplexed flow of cells from the various sources. The concept of a virtual channel (or virtual connection) is essentially the same as for any packet-based information transfer mechanism. The term "virtual" in this context implies that the capacity (bandwidth) of the channel (or connection) is only utilized when a packet is present. If a given source has no cells at a specific time interval, the capacity may be used by another source, which may have information to send at that specific instance. This important fundamental concept is discussed further below.
5. The sequence of the ATM cells belonging to any individual virtual channel must be preserved end-to-end for the connection. This fundamental requirement for ATM is termed "cell sequence integrity," and is an essential attribute contributing to the very high performance of ATM networks. Since cells transmitted in sequence from a source arrive in sequence at the receiver, there is no need for resequencing of the cells in reassembling a data packet, with consequent saving in processing and time. From an engineering standpoint, maintaining ATM cell sequence integrity implies a "first in, first out" (FIFO) queuing discipline at any connecting point along the ATM connection.
6. The information transfer capacity (in loose terms, the effective bandwidth) assigned to the ATM virtual channels in the asynchronously multiplexed cell flow is based on a negotiat-

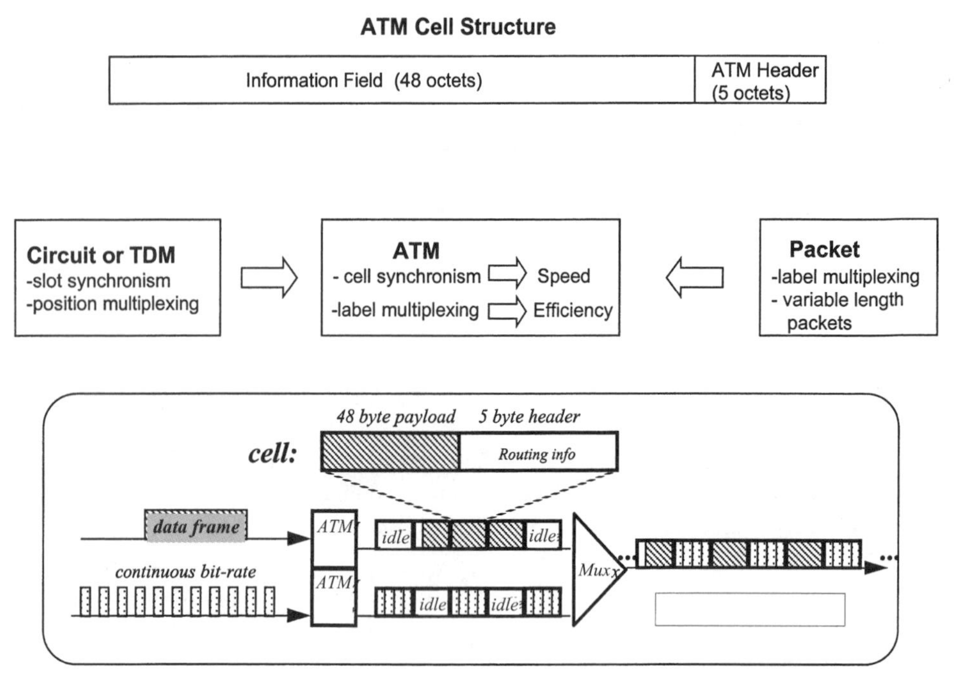

Figure 2-1. ATM cell structure and relationship with STM and packet mode.

ed value derived from the bandwidth required by the source and the available capacity in the ATM network. This concept is related to the notion of a "traffic contract" between the ATM cell traffic source (be it data, voice, or video) and the ATM network. The regulation of ATM traffic and resource management is an essential aspect of ATM technology, and is dealt with in greater detail in subsequent chapters.

7. ATM is essentially a "connection-oriented" packet-transfer mode. This implies that prior to the transport of any actual data from an ATM traffic source, an end-to-end ATM virtual channel connection must be established through the ATM network elements (NE) either by some form of signaling procedures, or by network management actions initiated by a request for service. In this respect, ATM technology is exactly analogous to conventional voice telephony, where the connection is initiated by signaling (dial-up) before a voice conversation can proceed. The signaling and management procedures for ATM connection setup are discussed in further detail below. Only after an end-to-end ATM (virtual) connection is established can the end user send data in the ATM cells. As indicated above, the ATM cells contain the connection identifiers that associate any given cell to its connection.

The connection-oriented nature of ATM cell transport should be contrasted with the so-called "connectionless" packet transport mechanisms used in other networking technologies such as, for example, the IP-based internet. The essential feature of connectionless packet switching (e.g., IP) is that each individual packet (typically of variable size in octets) contains a "globally unique" address identifying its destination. The switching elements in the connectionless networks "read" the address in the packet (e.g., an IP address) and route the packet towards its destination along some optimally computed path. This path may vary from packet to packet, depending on network conditions.

The relative advantages and disadvantages of connection-oriented versus connectionless packet switching have been discussed extensively in the literature since the advent of data networking and need not concern us unduly here. Suffice it to say that for applications where high performance (e.g., low delay and latency as required for voice and video image) and high speeds (say, transfer rates in excess of tens of megabits/second) are a prerequisite, it is generally recognized that a connection-oriented packet-switching mode is to be preferred, despite the initial burden of connection establishment and subsequent release.

In addition, it is now increasingly recognized that it may be possible to combine the convenience of connectionless packet transfer with the potential high performance of ATM in a number of ways, typically by utilizing the ATM mechanism as an underlying connection-oriented high-speed transport for the connectionless data packets. From this perspective, the connectionless packet service is simply a specific application of the underlying high-speed ATM network.

The concept of a "virtual channel," also referred to as a "virtual circuit," alluded to earlier is fundamental to ATM and needs further clarification in the ATM context.

In conventional telephony, the (digitized) voice or data information samples are assigned to specific "time slots" in the multiplexed information stream. The time relationship of these time slots is synchronized to a reference time slot in the multiplexed bit stream. The position of a given time slot with respect to this reference identifies the connection, or circuit, to which it belongs in the multiplexed stream. This mechanism is generally termed "position multiplexing" and is the basis of the so-called synchronous time division multiplexing, or synchronous transfer mode (STM) techniques used in existing circuit switched telephony networks.

In STM, as described above, when a source has no data (or voice sample) to transmit, its assigned time slot must however be transmitted in any case, since its position in the multiplexed bit stream needs to be maintained as long as the connection is held. The "empty" time slots in the multiplexed bit stream consequently represent wasted capacity (bandwidth), and cannot be used by other connections, which have been assigned different time slots. The synchronized time slots represent the circuits, or channels, which have to be maintained as long as the end-to-end connection exists.

For the case of packet networks, and for ATM specifically, the ATM cells belonging to a specific connection are only inserted into the multiplexed cell stream when the source has information to send. During time intervals when a source has no information to transmit, no cell need be generated, thereby enabling other connections to share the available capacity of the multiplexed stream. As noted above, the ATM cells of a given connection are distinguished by the connection identifier or "label" in the cell header, and not by position relative to any reference time slot. Consequently, the cells being emitted by a given source into the multiplexed cell stream need bear no time relationship to the previous cells emitted by the source, in contrast to the case of STM position multiplexing. For the ATM case, "label multiplexing" is employed by the inclusion of the logical connection identifier in the cell header.

Since the cells from a given source are not present when the source has no information to transmit, although a "connection" has been set up, connection-oriented packet transport mechanisms are said to effectively be utilizing a virtual circuit or virtual channel. In effect, for virtual channels, the capacity of the channel (or circuit) is only being utilized when a cell of the connection is emitted into the multiplexed stream at any time. Otherwise, this capacity could be used by cells of another connection. This notion is also sometimes loosely referred to as "bandwidth-on-demand." It should be noted that the concept of a virtual channel is an inherent attribute of any connection-oriented packet-transport mechanism, and is not specific to ATM technology.

Based on the above considerations, the relationship between ATM, conventional STM (circuit switching), and packet-mode transport can be described as shown in Fig. 2-1. Notice that ATM combines the key features of STM (circuit) and packet-mode transport in a unique way to achieve high-speed capability, coupled with flexibility and bandwidth efficiency.

From conventional STM (circuit mode), ATM adapts the concept of "slot synchronism" by using fixed-size cells (of 53 octets) instead of time slots to obtain the possibility of very high switching speeds not possible with packets of variable size. This is because once the start of a cell is identified (as described below), the fact that every cell contains 53 octets may be used by the cell-switching equipment to relay ATM cells at very high speeds. Throughputs up to several tens of gigabits/second are readily obtained with present technology, and higher speeds are likely in the near future. Thus, the possibility of high switching speeds follows from "cell synchronism" as opposed to slot synchronism.

Note that the notion of cell synchronism as described above has nothing to do with the asynchronous (i.e., time-unrelated) arrival or emission of cells belonging to a given connection, but relates to the multiplexed cell stream comprising ATM cells from different sources.

From packet mode transport, ATM takes the concept of "label" multiplexing to achieve the bandwidth efficiency of packet switching techniques. The "label" or logical connection identifier in the ATM cell header identifies the virtual channel to which the cell belongs. Since the capacity associated with the connection is only used when a cell is transmitted, the potential for efficient sharing of the total available capacity amongst many more connections arises, as inherent in the concept of virtual channel (or circuit). In addition, the flexibility inherent in packet-based transport is taken over in ATM, since the fixed-sized cells may carry any type of data as the payload in its information field, independent of the characteristics of the source, be it voice, video, or data, etc.

2.1 THE B-ISDN PROTOCOL REFERENCE MODEL (PRM)

It was noted above that ATM is the information transfer mode solution for implementing broadband integrated services digital networks (B-ISDNs), as defined by the ITU-T in Recommendations I.150 [2.1] and I.121 [2.2]. The vision of B-ISDNs itself derives from, and is based on, the earlier concept of the 64 kbit/sec or narrowband integrated services digital network (ISDN).

The (narrowband) ISDN sought to enhance the performance and capability of traditional telephony networks to encompass the handling of voice and data traffic using circuit switching techniques (STM), coupled with enhanced network signaling procedures (SS7) [2.3, 2.4]. However, the universal selection of ATM technology as the basis for the B-ISDN concept marked a radical shift away from the synchronous time division multiplexing transport and switching mechanism of ISDN to the packet-oriented approach, adopted in ATM transport and switching.

In addition to the above fundamental shift of transfer mode, the B-ISDN concept also adopts a well-defined protocol reference model (PRM) for ATM, which serves to clarify the relationships between ATM functions associated with the various layers, as well as functions associated with network management and control (signaling).

As shown in Fig. 2.2, the ATM/B-ISDN PRM is generally represented as a three-dimensional cube to facilitate visualization of the relationship between the layered protocol and the associated management and signaling (also called control) functions. The underlying protocol layering concept in ATM is analogous to the generic (seven layer) ISO–OSI protocol model applicable to describing most data network protocols. [2.5, 2.6]. The term analogous should be noted here, since there are arguably differences between the generic ISO–OSI seven layer protocol model and the ATM PRM. These differences of detail need not concern us unduly here in seeking to obtain an understanding of ATM protocols.

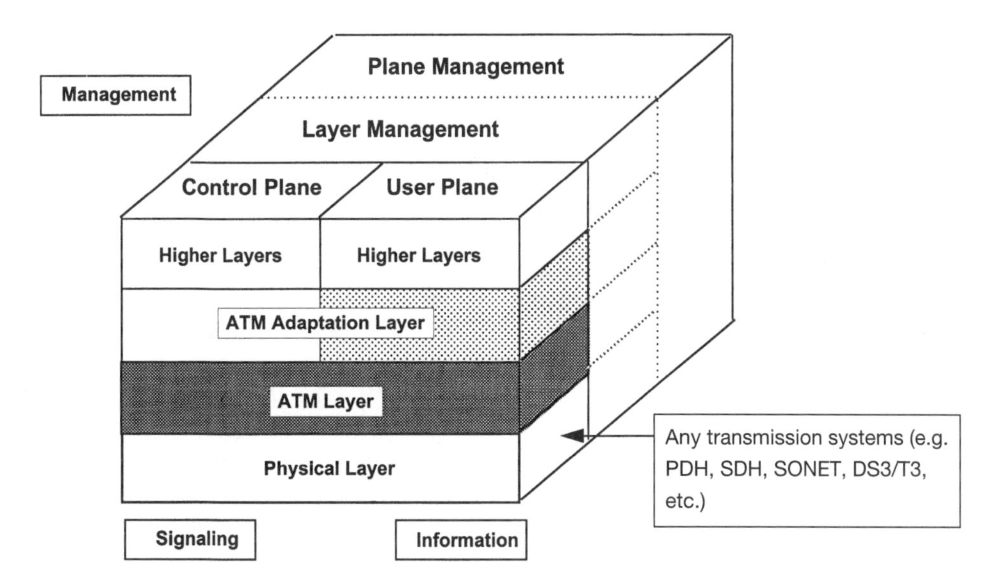

Figure 2.2. B-ISDN protocol reference model (PRM).

The significant point to note from Fig. 2.2 is the broad identification of four main layers:

1. Physical layer
2. ATM layer
3. ATM adaptation layer (AAL)
4. Higher layers (may be several)

The four main layers may be subdivided into so-called "sublayers" for convenience, as discussed below. The other major point to note is the separation of the PRM cube into functional blocks identified with user information transfer, the so-called "user plane" (also referred to as the transfer plane), and management and control (signaling) functional blocks.

The management functional block is further subdivided into those management functions associated within a given layer, termed "layer management" (functions), and the set of functions that may be associated with management of the network or system as a whole. The latter set of management functions comprise the "plane management," also called system management. The plane management functions are not layered, and may interact with any of the layer management functions in this model. Correspondingly, the layer management functions are only responsible for management actions within a given layer, and may only interact with other layers via the plane (or system) management functional block.

Although at first glance such a separation of the ATM/B-ISDN management functions into layer management blocks and plane (or system) management blocks may appear overly subtle, the value of such an approach will become evident when we discuss the management functional interactions in further detail in subsequent chapters. It should, however, be noted that such a distinction of the management functions is unique to the ATM/B-ISDN concept, and is not present in other internetworking protocols, including the 64 kbit/sec based ISDN from which B-ISDN is derived.

It should also be noted that the processing of signaling information (the control functions) are grouped in the so-called "control plane" (C-plane). In addition, the signaling functional block may use the ATM layer functions for the transfer of signaling information together with user information, although on separate virtual channel connections. Underlying this sharing of the ATM layer functions for all types of information, whether management, signaling, or user data (of whatever characteristics—voice, video, or data), is the notion that the ATM layer functions are "service independent." In other words, the ATM cells provide transport and multiplexing functionality common to all types of information, and any service-specific attributes that may be necessary are only visible either in the ATM adaptation layer or at the higher layers.

In effect, the service/application-independent nature of the ATM layer functions results in the need for the ATM adaptation layer (AAL). The AAL provides those functions that enable any specific type of service or application to be mapped into the underlying essentially "service-independent" ATM cell structure. An example of a typical function required in the AAL would be the "segmentation" (at the transmitter end) and "reassembly" (at the receiver end) of data frames longer than 48 octets to be packed into the information fields of ATM cells. Other AAL functions are described in detail in subsequent chapters, but here it is important to note that a number of different AAL protocols have been specified to accommodate different applications, and that the AAL may itself be further sublayered to enable more convenient separation of the AAL functions for protocol modeling purposes.

Although the ATM/B-ISDN PRM as described briefly above provides a very useful conceptual framework for the overall modeling of the ATM protocols, it should be borne in mind that it is not intended to constrain any practical implementation of an ATM system in either hardware or software. Thus, the actual partitioning of the ATM-related functions in the ICs or the associated software architecture of an ATM network element may not bear any relationship to the PRM, and is purely a matter for the ATM system designer. For example, a given ATM VLSI chip may group functions of the AAL together with those of the ATM layer, or those of the physical layer, and so on, depending on design imperatives or technology choices. This does not constitute any violation of the ATM PRM, since its purpose is essentially to describe in abstract terms the relationships between the different functional elements, be they management, signaling, or transfer related, necessary to implement an ATM system.

2.2. TRANSFER PLANE FUNCTIONS AND ATM PROCESS OVERVIEW

Before undertaking a detailed description of the individual ATM functions or functional blocks, it is useful to obtain an overview of the basic ATM processes involved in the transfer of information from any given source, with reference to the overall ATM PRM description above.

The transfer (or user) plane functions are listed in Fig. 2-3, together with the layering specified by the ATM/B-ISDN PRM of Fig. 2.2. Notice that the functions listed in the physical layer are sublayered into those of the "physical media (PM) sublayer" and those of the "transmission convergence (TC) sublayer."

The physical media sublayer consists of the physical media-dependent functions related to the electrical or optical transmission system used to carry the ATM cells. These functions will clearly be heavily transmission-technology-specific, and may include electrooptic conversion, line conditioning, and scrambling, as well as bit timing extraction and generation.

Although ATM was initially conceived with the capabilities of modern high-speed optical

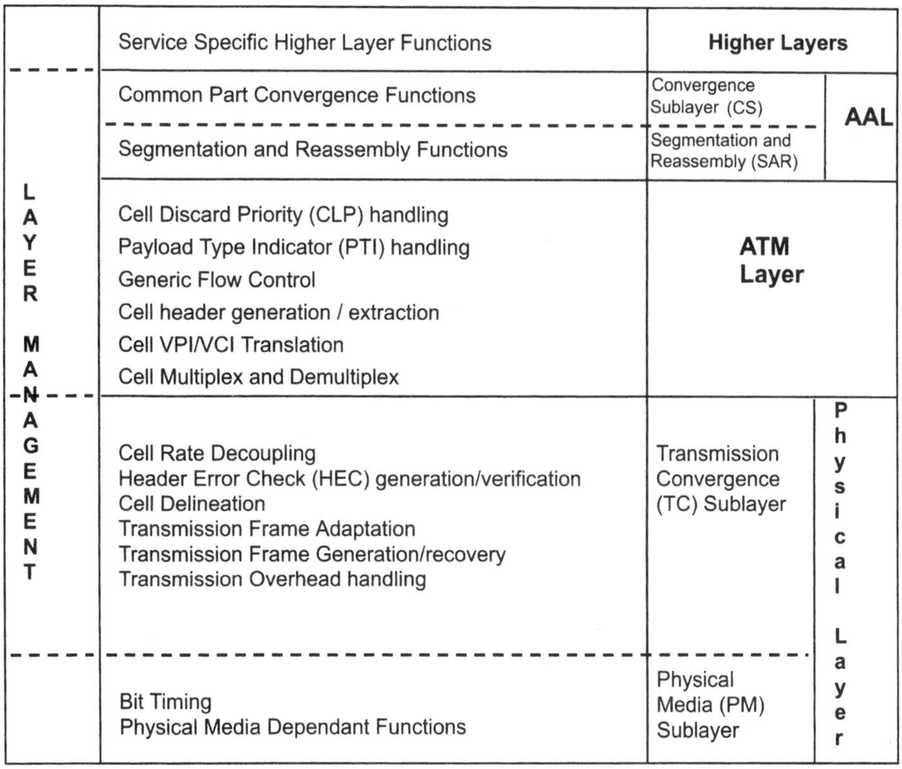

	Functions	Sublayer	Layer
	Service Specific Higher Layer Functions	**Higher Layers**	
LAYER MANAGEMENT	Common Part Convergence Functions	Convergence Sublayer (CS)	**AAL**
	Segmentation and Reassembly Functions	Segmentation and Reassembly (SAR)	
	Cell Discard Priority (CLP) handling Payload Type Indicator (PTI) handling Generic Flow Control Cell header generation / extraction Cell VPI/VCI Translation Cell Multiplex and Demultiplex		**ATM Layer**
	Cell Rate Decoupling Header Error Check (HEC) generation/verification Cell Delineation Transmission Frame Adaptation Transmission Frame Generation/recovery Transmission Overhead handling	Transmission Convergence (TC) Sublayer	**Physical Layer**
	Bit Timing Physical Media Dependant Functions	Physical Media (PM) Sublayer	

Figure 2-3. Transfer plane functions.

fiber transmission systems in mind, its essential independence from the underlying transmission system has allowed ATM to be considered over virtually any transmission media, including wireless recently. Consequently, the physical media-dependent functions may vary widely.

The functions grouped in the transmission convergence (TC) sublayer are also somewhat dependent on the type of underlying transmission system used, particularly with respect to the type of transmission overhead handling, frame generation or recovery, and frame adaptation. However, notice that there are ATM-specific functions in the TC sublayer. These are ATM cell delineation, ATM header error control and generation, as well as cell rate decoupling. These ATM-specific functions, described in detail below, indicate that the TC sublayer is essentially a "bridge" or adaptation between the ATM layer and the underlying ATM-independent transmission media.

The ATM Layer functions include:

1. Cell multiplexing and demultiplexing
2. Processing (translation) of virtual connection identifiers (VPI/VCI)
3. Cell header generation/extraction
4. Generic flow control
5. Processing of payload type
6. Processing of cell discard priority

These functions will be described in detail in the subsequent sections; for the purposes of this overview, it is only sufficient to note that the processing of the virtual channel identifiers denoted as virtual path indicator (VPI) and virtual channel indicator (VCI) allows for the multiplexing and switching of cells belonging to any given virtual channel in the ATM network. The VPI/VCI values constitute the "labels" that serve to identify the individual VCs in the network elements and, as such, their assignment and processing constitute an essential function in ATM-based information transfer. As mentioned earlier, the AAL functions may also be sublayered into the (1) segmentation and reassembly (SAR) sublayer and (2) convergence sublayer (CS). Typically, the AAL convergence sublayer may also be further sublayered into a common part convergence sublayer (CPCS) and a service-specific convergence sublayer (SSCS), but this separation need not concern us unduly in this overview description.

In general terms, the basic ATM process and the corresponding representation from the perspective of the protocol architecture can be broken down into the steps shown schematically in Fig. 2.4. The digitally processed user information is handled by the AAL and ATM layer functions. Note that the underlying assumption in ATM technology is that the analog to digital conversion processing is "external" to the ATM/B-ISDN PRM, which concerns itself primarily with information already in digital format. As summarized in Fig. 2.4, the AAL functions are essentially:

1. Map the higher layer user information flow into an AAL protocol data unit (PDU) format
2. Provide service specific functions such as error detection, timing recovery, etc.
3. Segment (or reassemble) the data into cell sized (i.e. 48 octet) PDUs

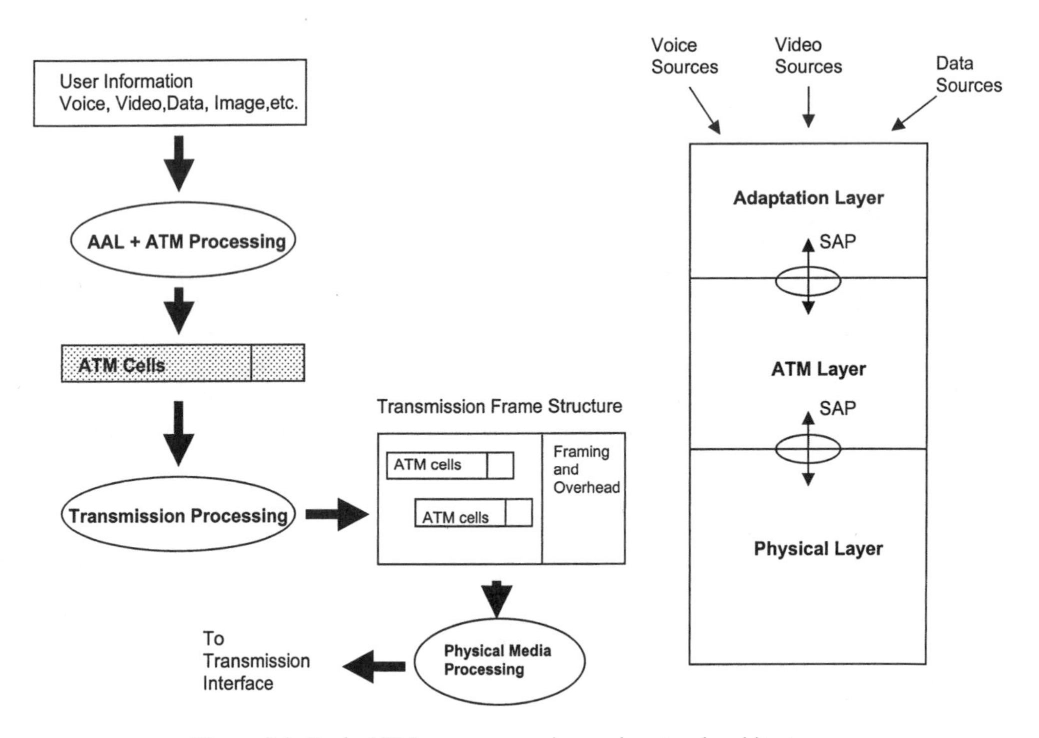

Figure 2.4. Basic ATM process overview and protocol architecture.

The ATM Layer processing:

1. Generates the ATM header
2. Adds/removes the header to 48 octet payload
3. Multiplexes/demultiplexes the cells into a virtual connection identified by the VPI and VCI labels in each NE

The transmission processing adapts the rate at which cells are emitted from various sources to the speed of the transmission system being used, adds (or removes at the receiver end) the transmission overhead, and conditions the resulting bit stream for transmission over any given physical transmission interface. As depicted schematically in Fig. 2.4, the ATM cells are mapped into the payload of the transmission frame structure in a standardized way. The actual transmission frame format and overhead structure depends on whether traditional asynchronous multiplex hierarchy (also loosely referred to as the plesiochronous digital hierarchy (PDH) or optical-fiber based synchronous digital hierarchy (SDH) transmission systems are utilized in the ATM network.

2.3 NETWORK MODELS AND REFERENCE POINTS

Just as it is useful to bear in mind an overall view of the ATM transfer plane processes in general terms, an abstract "model" of a general ATM network is also useful to identify reference points of relevance at which an interface may be defined.

It is important to note the distinction between the concept of a "reference point" and that of an "interface." An interface may or may not be associated with any given reference point. In network terms, a reference point is an abstract point (or plane) in a model of the network or protocol architecture. The reference point essentially serves to partition functions or configurations and so assists in the description of the model, as well as serves as a point of interoperability between different parts of the network. On the other hand, if an interface is defined at any given location in a network, which may or may not correspond to any recognized reference point, this implies the possibility that a "real" partition may be made at that interface, allowing different physical network elements to be connected in some (specified) way across the interface. The interface may also imply some form of administrative or legal separation of domains. In practice, two separate approaches have been taken to modeling ATM networks in general. From the conventional telecommunications perspective, ATM technology is viewed as the transfer mode of choice for implementing B-ISDNs. Consequently, the B-ISDN is modeled as an extension of the existing 64 kbit/sec (or narrowband) ISDN as depicted in Fig. 2.5. The prefix B denotes that a broadband (i.e., ATM) capability is present in the terminal equipment (TE), the terminal adapter (TA) function, the network termination (NT) 2 and NT1, and so on.

The essential functional partitioning between B-TE, B-TA, B-NT2, and B-NT1 is similar to that specified in the conventional 64 kbits ISDN case, and is a generalized abstraction used primarily to define the reference points S, T, etc. for administrative purposes. Thus, the international specification of the user network interface (UNI) is associated with the T_B reference point in telecommunications. According to this definition, the B-NT2 functional block includes customer premises network (CPN) functions such as multiplexing, internal switching within a private network (e.g., PABX), and so on. The B-NT1 functional block broadly includes the carrier (service provider) networks transmission facilities (feeders, cables, and line conditioning equipment, access network multiplexers, etc.). The B-LT and B-ET refer to the local

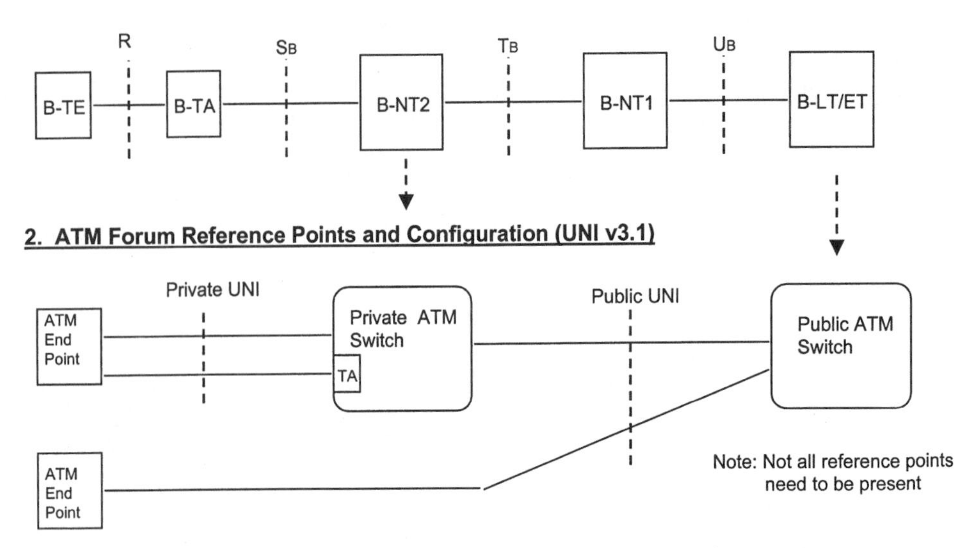

Figure 2.5. B-ISDN reference points/network models.

switching exchange functions (line termination, exchange termination) typically implemented in the public carrier's central office switching exchange.

It should be noted that such a model of a network mirrors the broad existing structure of the voice-oriented telecom industry, and is more useful for the purposes of defining administrative and/or legal domains than for detailed and rigorous engineering-oriented functional modeling. In addition, the functional blocks may not be present in any given implementation. From a somewhat more implementation-oriented perspective, the ATM Forum adopted a slightly different model of the local ATM network, as shown in Fig. 2.5. Here the distinction between private and public carrier ATM network elements is more clearly delineated by the public UNI. It is important to note that the ATM Forum view does not specify a "reference point" at the public (or private) UNI, but an interface, in the sense of an interoperable "physical connection," between two networks. In contrast, the ITU-T model formally defines the UNI as the "interface at the T_B reference point."

Although these two perspectives on modeling an ATM network appear disparate (and have generated some confusion in the ATM industry), the main points to note here are the broad correspondences between the ATM Forum's public UNI and the (ITU-T based) user network interface at the T_B reference point. In addition, the so-called "private UNI" corresponds to the interface at the S_B reference point. In practice, from the perspective of ATM layer functionality, there is generally no difference between the interfaces at the S_B and T_B reference points (private and public UNIs).

In addition to the definition of the UNI, we need to define interfaces between network elements and networks in general to ensure end-to-end interoperability. These are loosely referred to as network node interface and/or network–network interface (NNI). A further distinction is also made between "public" (carrier) NNIs and private NNI (PNNI) in the ATM Forum specifications, although the ITU-T model does not specifically distinguish network node interfaces (NNI) within a customer premises network as NNI.

Whichever of these broadly similar network models a network designer selects as a basis for specifying given ATM network applications, the essential point to bear in mind is the distinction between simple reference points and interfaces associated with the reference point. For the interface definition to have meaning, it is necessary to uniquely specify the peer-to-peer protocols at all layers across the interface. Only then can the interface be considered as a point of interoperability or interconnection between different domains of the overall ATM network.

In addition to such basic and general interfaces as the UNI and NNI, other more specialized interfaces are also being considered for specific applications of ATM, such as for residential broadband networks (RBB) and access network configurations. These specialized interfaces generally derive from the basic UNI or NNI functionality, but may differ in a number of important details by adding or subtracting functions related to the specific network applications. Some examples of such specific interfaces are discussed later, whereas others are still somewhat imprecise at this stage.

CHAPTER 3

Functions of the ATM Layer

In this chapter, we discuss in some detail the functions of the ATM Layer. As noted earlier, the ATM Layer is designed to be "service independent" and hence common to either any type of user data, signaling/control messages, or higher-layer management messages. Consequently, a detailed understanding of all aspects of the ATM layer functions enables us to accommodate the transport and switching of all of the above forms of information.

The structure of the ATM cell header shown in Fig. 3-1 is based on ITU-T Recommendation I.361 [3.4]. It should be immediately noted that there are two distinct formats for the ATM cell header, distinguished by the presence of the generic flow control (GFC) function (field) in the first octet of the cell header at the UNI only. The GFC field is not present at the NNI. The use of the GFC function is discussed later, but it is useful to note that apart from the presence of the GFC field in the UNI cell format, in other respects the cell fields are identical at the UNI and NNI. All the ATM header fields (the protocol control information, PCI) are binary encoded. By convention, the numbering of the octets and individual bits in the ATM cell header is as shown in Fig. 3-1. The highest bit number signifies the most significant bit (MSB), whereas the lowest octet number signifies the first octet transmitted. Thus, the transmission order is the MSB of octet 1 followed by the MSB of octet 2 and so on up to octet 5. The information field (from octet 6 to 53) is then transmitted in continuing order.

3.1 VIRTUAL PATH AND VIRTUAL CHANNEL IDENTIFIERS

As noted earlier, ATM employs the label multiplexing technique used by some other packet mode switching technologies. The label, or more properly the logical channel number (LCN), uniquely associates each cell (packet) with its virtual channel. However, the label multiplexing mechanism in ATM possesses some features that are unique to ATM.

The first unique feature associated with the label multiplexing technique in ATM is that there are two levels of connection hierarchy defined for the virtual channels in ATM networks, as described in ITU-T Recommendation I.311 [3.5]. The virtual path identifier (VPI) value in the VPI field of the cell header identifies the virtual path level of the hierarchy. Connections associated with the virtual path level of the hierarchy are called virtual path connections (VPCs). ATM cells associated with a virtual path connection are identified by the VPI value alone in the ATM cell header. Similarly, the virtual channel identifier (VCI) value in the VCI field identifies the virtual channel level of the hierarchy. Connections associated with the virtual channel level of the hierarchy are called virtual channel connections (VCCs). ATM cells associated with a VCC are identified by the VPI and VCI values in the ATM cell header.

Note that from the ATM cell header structure of Fig. 3-1, the VPI values constitute the more significant bits in the total label space of VPI + VCI. This implies that the VPC level provides a higher logical granularity in the multiplexed cell flow than that provided by the values of the vir-

ATM Cell Header Formats

8	7	6	5	4	3	2	1	Bit / Octet
GFC				VPI				1
VPI				VCI				2
VCI								3
VCI				PT			CLP	4
HEC								5

ATM cell header structure at the UNI

8	7	6	5	4	3	2	1	Bit / Octet
VPI								1
VPI				VCI				2
VCI								3
VCI				PT			CLP	4
HEC								5

ATM cell header structure at the NN

GFC - Generic Flow Control
VPI - Virtual Path Identifier
VCI - Virtual Channel Identifier
PT - Payload Type
CLP - Cell Loss Priority
HEC - Header Error Control

**ATM cell structure / codings and protocols defined in ITU-T Recommendation I.361 [3.4].
Different ATM cell header structure at the User Network Interface (UNI)
and Network Node Interface (NNI) results from inclusion of the Generic Flow Control at UNI.**

Figure 3-1. ATM cell header formats.

tual channel identifiers. In addition, the number of VPI values differs at the UNI, where a total of 256 (2^8) VPI values are available for a given physical transmission path (TP) or physical link. At the NNI, the number of possible VPI values is 4096 (2^{12}) per physical TP. The difference, of course, is due to the presence at the UNI of the GFC field of 4 bits. For each VPI value there are a total of 65,536 (2^{16}) possible VCI values available to identify the individual virtual channel connections (VCCs) in the virtual path connection. The hierarchical relationship between the virtual path connection level and the virtual channel connection level for a single (physical) transmission path (or link) can be visualized schematically as shown in Fig. 3-2.

Considering the UNI case first, it can be seen from the above visualization that a single physical transmission path (e.g., say a 155.54 Mbit/sec optical fiber carrying STM-1 signals with a TP payload of 149.76 Mbit/sec) may support up to 256 VPCs. Each VPC may support up to 65,536 VCCs. Thus, a theoretical maximum of up to 16,777,216 virtual channels may be supported at a given UNI. This is clearly a very large number of individual virtual channels per physical interface. In practice, only a small fraction of the total number of possible virtual channels per physical link may be utilized. Similarly, at the NNI, the number of possible virtual path connections per transmission path is 4096 (12 bits). Each VPC may support up to 65,536 virtual channel connections (VCCs), resulting in a theoretical total of up to 268,435,456 (28 bits) virtual channels per physical interface. Again, this huge number of potential virtual channels is in practice unlikely to be implemented for cost or complexity reasons, and a fraction of the possible number may be sufficient for most applications.

It is important to recognize that although cells associated with a VPC are identified by the VPI value alone as noted above, the cells associated with a given VCC must be identified by *both* the VPI value AND the VCI value. In fact, the unique specification of the VCC implies

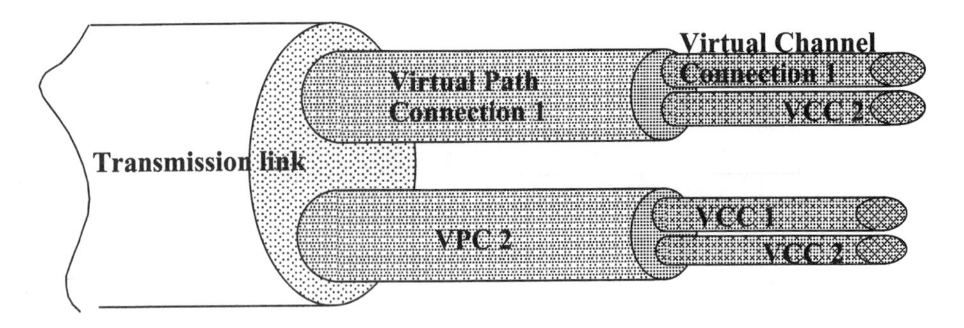

The separation of the Logical Connection Identifier Field into two connection identifiers:
Virtual Path and Virtual Channel enables two levels of virtual connection in any given physical interface

Cells associated with a Virtual Path Connection are identified by the VPI value alone
Cells associated with a Virtual Channel Connection are identified by both VPI and VCI values

In general, ATM connections are "bidirectional"
The VPI and VCI values assigned are the same for both directions of a given VPC or VCC

Figure 3-2. Virtual path and virtual channel connections.

identification of the transmission path (i.e., the actual physical link) as well as the two logical identifiers—the VPI and VCI values. In practice, the identification of any given TP or physical port is generally dependent on the implementation of the ATM network element (e.g., a physical port identifier or MAC address/serial number, etc.), and hence does not need to appear explicitly in the cell header. In other words, the identification of the physical port to which any given cell is switched is internal to the ATM network element itself, and hence need not be explicitly specified as part of the ATM PCI.

3.2 RELATION BETWEEN VPC, VCC AND VPL, VCL

As is the case for some other connection-oriented packet-based technologies, it is important to understand that the values assigned to any VPI or VCI are purely to identify individual logical links between ATM NEs [3.5]. This is usually stated as saying that the VPI (or VCI) value has only "local significance." The concept of local significance implies that a given VPI value is associated with a virtual path link (VPL) between any two adjacent ATM network elements. As shown in Fig. 3-3, the VPI values on either side of the ATM NE may in general be different. Consequently, the ATM switching function in the NE constitutes "translating" the incoming VPI value associated with the input virtual path link (VPL) on any given physical port to the outgoing VPI value associated with the output VPL on any (other) physical port. Since the translation of the VPI values associated with each VPL occurs at each NE in the path of the virtual path connection (VPC), it is evident that a VPC is essentially a concatenation of VPLs, where each VPL is identified by its VPI value.

A similar relationship exists for the case of the virtual channel connection (VCC). As can be inferred from Fig. 3-3, the VCC is a concatenation of virtual channel links (VCLs), where each VCL is identified by the associated values assigned to the VPI + VCI field. As noted earlier, both the VPI and the VCI values are required to uniquely identify the cells belonging to a given VCL in any given transmission path (or physical port).

The fact that a virtual channel connection (or VPC) is made up of a concatenation of VCLs

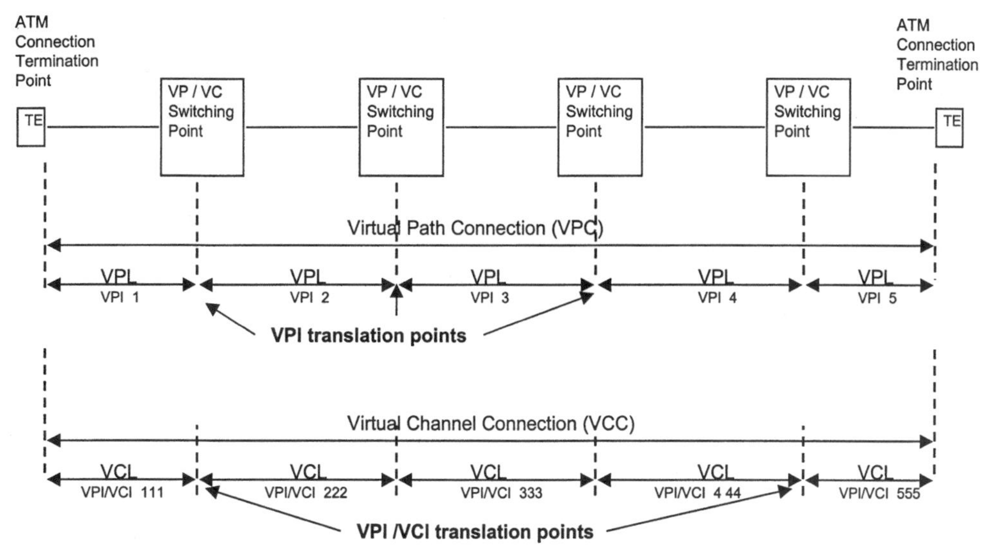

A Virtual Path Connection (VPC) is a concatenation of Virtual Path Links (VPL)—
each VPL is associated with a VPI value. Switching points may change the VPI values

A Virtual Channel Connection (VCC) is a concatenation of Virtual Channel Links (VCL)—
each VCL is associated with a VPI and VCI value. Switching points may change the VPI/VCI values

Figure 3-3. Relation between VPC, VCC, and VPL, VCL.

(or VPLs) is essentially only a consequence of the fundamental connection-oriented nature of ATM. It implies that when an ATM connection (either a VPC or a VCC) is set up, each ATM NE in the path of the connection must assign the appropriate VPI (and VCI for the case of VCC) values to the incoming and outgoing VPLs (or VCLs). The association between the incoming VPI value and the outgoing VPI value is maintained in the connection memory of the ATM NE as long as the connection is maintained. The switching function then essentially involves transferring the cells with the incoming VPI (and VCI) value on the input VPL (or VCL) to the outgoing VPL (or VCL) with the associated VPI (and VCI) values. The ATM switch is said to "translate" the incoming VPI (and VCI) value to the outgoing VPI (and VCI) value. Since the incoming VPI (and VCI) value identifies the incoming VPL (or VCL) on any given physical port (or TP), and the outgoing VPI (and VCI) value identifies the outgoing VPL (or VCL), the cells belonging to a given connection (VPC or VCC) are effectively switched from one VPL (or VCL) to another VPL (or VCL) by this translation mechanism.

The process of cell switching by means of the label translation process described above can be done extremely rapidly by implementing the process in hardware if required. It enables ATM switching (or cross connection) up to very high speeds, in marked contrast to many connectionless packet switching technologies, where it is necessary to translate a global address (e.g., an IP address) contained in the packet header in order to route the connectionless packet to its destination. It is useful to note here that the performance advantages (as related to packet throughputs) claimed by ATM technology result primarily from its fundamental connection-oriented nature, as well as other factors, such as the use of fixed size cells and high-speed physical transmission systems.

As noted earlier, a very large number of possible VPI + VCI values are available at any physical transmission path (physical interface). However, this does not imply that all possible values

TABLE 3-1. VPI/VCI allocation rules

	UNI	NNI
Virtual paths (per transmission path)[a]	256 (8 bits)	4096 (12 bits)
Virtual channels per virtual path[b]	65,536 (16 bits)	65,536 (16 bits)
Maximum number of virtual channels[c]	16,777,216 (24 bits)	268,435,456 (28 bits)

[a]The actual number of virtual channels per transmission path (physical link) may be much less than the theoretical maximums possible (e.g., to reduce implementation costs).

[b]The number of bits available for active connections may be established by mutual negotiation at UNI/NNI (e.g., in North America the total number of virtual channels has been set to 1,048,576: 20 bits).

[c]The potential number of ATM Virtual Channels possible on a given transmission path is extremely large.

need to be accommodated by every ATM NE. In practice, the actual number of VPI + VCI values may be limited in order to simplify design as well as reduce the amount of connection memory required in the switching hardware. In addition, it may also be possible to negotiate the number of bits available for active connections at any given interface (UNI or NNI), thereby giving the network operators some flexibility in the design and configuration of their ATM networks.

Although the number of possible active connections per NE is primarily a design parameter largely dependent on implementation choices, some ATM network standards have established guidelines for maximum values. Thus in the United States, the ANSI standards have limited the total number of virtual channels at any given interface to the equivalent of 20 bits of VPI + VCI (i.e., 1,048.576 virtual channels per transmission path). Even with this limitation, there appears to be a sufficient number of simultaneous active connections for all foreseeable network applications. Table-3-1 summarizes the VPI and VCI allocation rules at the UNI and NNI respectively.

3.3 PREASSIGNED HEADER VALUES

Certain values of the ATM header fields have been assigned for specific functions that are required to enable the ATM layer and its interaction with the physical layer to operate. These so-called "preassigned" header values are divided into two groups consistent with the ATM PRM described earlier. The preassigned header values for the physical layer are given in Table 3-2

The other group of preassigned header values relate to use by the ATM layer functions, and are summarized in Fig. 3-4, based on the description given in Recommendation I.361 [3.4], which defines the basic ATM layer coding.

In addition to these preassigned ATM header values, it is now generally accepted by the ATM standards bodies that the range of VCI values up to and including VCI = 32 are reserved to allow for ongoing (and future) growth of preassigned header values for specific ATM functions that may require them. These reserved VCI header values may be assigned by the standards bod-

TABLE 3-2. Preassigned header values for physical layer use

	Octet 1	Octet 2	Octet 3	Octet 4
Idle cell identification	0000 0000	0000 0000	0000 0000	0000 0001
Physical layer OAM cell identification	0000 0000	0000 0000	0000 0000	0000 1001
Reserved for use of physical layer	PPPP 0000	0000 0000	0000 0000	0000 PPP1

P indicates the bit is available for use by the physical layer. Values assigned to these bits have no meaning with respect to corresponding bit positions in ATM layer.

Use	VPI	VCI	PTI	CLP
Unassigned cell	0000 0000	00000000 00000000	Any	0
Invalid	Any VPI not 0	00000000 00000000	Any	B
Meta-Signaling (*)	XXXX XXXX	00000000 00000001	0AA	C
General broadcast Signaling (*)	XXXX XXXX	00000000 00000010	0AA	C
Point-to-point Signaling	XXXX XXXX	00000000 00000101	0AA	C
Segment OAM F4 flow cell	Any VPI value	00000000 00000011	0A0	A
End-to-end OAM F4 flow cell	Any VPI value	00000000 00000100	0A0	A
VP resource management cell	Any VPI value	00000000 00000110	110	A
Reserved for future VP functions	Any VPI value	00000000 00000111	0AA	A
Reserved for future functions	Any PVI value	00000000 000SSSSS	0AA	A
Reserved for future functions	Any VPI value	00000000 000TTTTT	0AA	A
Segment OAM F5 flow cell (#)	Any VPI value	Any VCI not preassn.	100	A
End-to-end OAM F5 flow cell (#)	Any VPI value	Any VCI not preassn.	101	A
VC resource management cell (#)	Any VPI value	Any VCI not preassn.	110	A
Res. for future VC functions(#)	Any VPI value	Any VCI not preassn.	111	A

Key

A - may be 1 or 0

B - any value

C - set to 1 by orig.

XXXXXXXX - Any value but see notes in Rec. I.361

SSSSS - any value from 01000 to 01111

TTTTT - any value from 10000 to11111

See also other notes in Rec. I.361 (1995)

(*) - Not defined at NNI

(#) - See I.361 for values

Figure 3-4. Preassigned header values based on ITU-T Recommendation I.361.

ies (the ITU-T and the ATM Forum), based on general acceptance of the proposals for additional required functions in future.

Consider first the preassigned header values for physical layer use shown in Table 3-2. At first glance, it may appear paradoxical that we even consider ATM header fields in the physical layer, which after all is intended to specify the transmission-related functions carrying ATM cells as a payload. However, there are two important reasons for specifying the cell structure within the physical layer.

The first reason results from the need to perform "rate adaptation" between the speed of the transmission system (e.g., 155.54 Mbit/sec for the case of an SDH system), and the rate at which ATM cells are emitted across the boundary between the ATM and the physical layer. This transfer of cells across the ATM to the physical layer boundary may be much less than the actual transmission system speed, depending on the service being provided at the higher layers. The difference is made up by so-called "idle cells" generated (or removed) in the physical layer. The rate adaptation mechanism is described later, but here it should be noted that idle cells, which exist *only* in the physical layer, are distinguished by the preassigned header values defined in Table 3-2.

The second reason for specifying ATM cells header values in the physical layer relates to the use of the so-called "ATM cell-based" transmission system. In fact, from its onset, ATM technology was viewed by many as not only a switching (or multiplexing) technology, but also as a transmission technology, whereby the ATM cells are converted directly into the physical medium (optical fiber or copper) without the need for any intervening frame-based transmission system such as SDH. In this scheme, sometimes also called "pure ATM" transmission, the transmission overhead is transported in special cells called physical layer OAM (PL-OAM) cells, distinguished by the preassigned header values in Table 3-2.

The cell-based transmission system is described in more detail later, but it should be noted here that the "pure" cell-based physical layer has been defined in ITU-T standards (Recommendation I.432) [4.6] as an alternative option to the more commonly employed frame-based transmission systems such as SDH or PDH.

In addition to the idle cell and the PL-OAM cell, additional values of the physical layer cells have been reserved for future use of physical layer functions, as shown in Table 3-2. It is important to stress that these physical layer cells only have meaning for the physical layer functions, and are *not* transferred to the ATM layer across the ATM to the physical layer boundary. It should also be noted that only the first four octets of the ATM cell header are shown in Table 3-2. The fifth octet contains the header error control (HEC) function, which operates as normal [i.e., cyclic redundancy check 8 (CRC 8)] over the header, irrespective of whether values are preassigned or not.

As noted earlier in describing the B-ISDN PRM, the physical layer is sublayered into the transmission convergence (TC) sublayer and the physical media (PM) dependant sublayer functions. The relationship of the (preassigned) physical layer cells to this sublayering should now be clear. For the commonly used frame-based based transmission systems such as SDH, the idle cells are generated (or terminated) as part of the TC sublayer functions, and PL-OAM cells are not used, since SDH employs its own overhead structure. However, for the pure ATM cell-based transmission system, either idle cells or PL-OAM cells, which may also carry the transmission overhead, may be used for rate adaptation in the TC sublayer, but for such systems, the sublayering of the physical layer functions has less significance, since the PM sublayer may only involve electrical-to-optical conversion functions.

Figure 3-4 shows that many more preassigned header values have been defined for the ATM layer functions than for the physical layer. Since many of these ATM layer functions are described in detail in subsequent chapters, it is only sufficient to note at this stage that the preassigned values fall into three broad categories, which essentially reflect the fundamental PRM underlying ATM. Thus, it may be seen that preassigned header values have been specified for functions related to the transfer (or user) plane, the management plane (or OAM functions), and the control plane (or signaling functions).

In particular, for ATM signaling it should be noted that VCI = 5 in any VPI (i.e., in any virtual path link) is reserved to carry point-to-point signaling messages for the control of on-demand virtual channel connections, or potentially virtual path connections. In this sense, the B-ISDN model is consistent with the philosophy of "common channel" or "out-of-band" signaling upon which its predecessor, the 64 kbit/sec ISDN, is based. In the ATM case, signaling messages used to perform the functions of the control plane (e.g., setting up or tearing down VCCs identified by sets of different concatenated VCI values) are transported by cells with VCI = 5 in any VPI, which form a common "out-of-band" virtual channel. This signaling VCC is analogous to the "link set(s)" used to transport DSS2 or ISUP signaling messages in conventional ISDN SS7 systems [6.2, 6.8].

In addition to the above VCI value for point-to-point signaling, preassigned header values have also been defined for the so-called "meta-signaling" (VCI = 1) and for general broadcast signaling (VCI = 2), although these are specific only for the UNI. To date, these specialized functions have not been used to any extent, and we shall defer the description of meta-signaling and general broadcasting signaling to the chapters describing signaling in more detail. Here it only needs to be noted that in defining a variety of ways in which signaling messages may be transported by preassigned VCLs, the intention is to enable the support of on-demand connectivity in any foreseeable ATM network configuration.

In this context, it is also worth pointing out that the availability of the preassigned, or default VCL identified by VCI = 5 for point-to-point signaling is not intended to preclude any other dedicated VCC being set up to carry signaling messages across any given interface, if the network operator chooses to do so. VCI = 5 simply represents a convenient "default" VCL, which may be used for signaling in the event that it is not deemed necessary to configure a dedicated VCC for transport of signaling messages.

The other group of preassigned header values in Fig. 3-4 are intended for the various operations and maintenance (OAM) functions defined for ATM layer management and resource (i.e., traffic) management. These include the segment as well as end-to-end OAM flows for both the virtual path (VP) and virtual channel (VC) levels, and the so-called Resource Management (RM) cells, which are intended only for ATM traffic or congestion control purposes.

The ATM layer OAM functions are described in detail in subsequent chapters, but with respect to the preassigned values given above, two key considerations may be recognized from Fig. 3-4. In the first place, it should be noted that the ATM OAM cells have the same VPI/VCI values as the data (or user) cells in the virtual channel (or virtual path) connection. Thus, the OAM information is carried "in-band" or as part of the connection, in contrast to the signaling information, which is carried in an "out-of-band" common signaling virtual channel. This per connection characteristic of the ATM layer OAM cell flow applies to both the "segment" or the "end-to-end" OAM cells, which are simply distinguished by different preassigned header values.

Secondly, it should be noted that for any given VPC, the OAM cells are distinguished from the user information cells of the VPC by preassigned VCI values. These are VCI = 3 for segment OAM cells and VCI = 4 for the end-to-end OAM cells for any value of the VPIs. However, for any VCC, the OAM cells flowing in the connection (the so-called F5 flow) are distinguished from the user cells by specific values of the payload type (PT) field in the ATM header. Thus, PTI = 100 (= 4) identifies a segment OAM cell and PTI = 101 (= 5) identifies an end-to-end OAM cell in any VCC.

The same principles described above for the OAM cells have also been extended to distinguish the resource management (RM) cells, although there is no concept of segment or end-to-end for RM cells flowing in any connection. Thus, the RM cells in any VPC are distinguished from the user cells in that connection by the preassigned value VCI = 6, and the RM cells flowing in any VCC are distinguished from the user cells of that VCC by the specific PTI value of PTI = 110 (= 6).

In addition to the "in-band" or per connection ATM Layer OAM information that can be exchanged between NEs via the predefined OAM cells as described above, management information may also be exchanged between NEs via a separate or default (out-of-band) ATM connection, which can be dedicated to carry only management information. ATM connections (VPCs or VCCs) set up specifically for management information (sometimes loosely referred to as a management interface) may exist between ATM NEs or between the NE and the network operating system function (OSF). Such dedicated management ATM connections need to be distinguished from the in-band, per connection ATM OAM cells, which exchange the ATM layer OAM functions such as fault and performance monitoring. The detailed interactions between these management functions will be described later.

In contrast, since RM cells are only intended for per connection traffic control purposes, as described in detail later, a dedicated out-of-band connection set up specifically for RM information has not been required.

3.4 THE PAYLOAD TYPE FIELD

The primary intent of the payload type (PT) field is to distinguish the user data cells from the various special purpose management cells, as indicated above. However, the PT field is also used to perform some other functions at the ATM layer, so it is important to understand the interpretation of the codepoints defined for the PT field. These are shown in Table 3-3.

As noted earlier, specific PT codepoints identify the segment and end-to-end OAM cells and the RM cell (PTI = 110). However, the normal ATM cells carrying user data may be coded in one of the four ways specified in Table 3-3, to indicate the presence (or absence) of congestion along the connection path. Thus, user data cells that do not experience any traffic congestion along their path have their PTI coded as 000. If, however, a NE along the connection is experiencing traffic congestion, it has the option to set the congestion indication bit (bit 3 of the PT field) to = 1. The threshold at which a NE is considered to be in a state of "congestion" is somewhat arbitrary, and may be settable within a given range by a network operator management system. These aspects of congestion are considered in further detail in subsequent chapters.

Here it should simply be noted that the coding of the PT field enables an explicit indication of traffic congestion along a connection. This function is termed explicit forward congestion indication (EFCI), as it enables the downstream virtual connection end-point to know of the onset of traffic congestion by simply monitoring the PT coding. The use of the EFCI function for potential congestion control is discussed in subsequent chapters.

In addition to the EFCI function, the PT field also includes the so-called "ATM user-to-user indication" function. In its most general interpretation, this function allows the user of the ATM layer service (i.e., the AAL) to indicate a change depending on whether bit 2 of the PT field is set to 0 or 1. In practice, as detailed later, this function is used primarily by the AAL Type 5 protocol to signal the end of an AAL Type 5 frame (or more strictly, PDU), when the ATM user-to-user bit is set to 1 to signify an "end of frame" ATM cell. The detailed use of this function in the AAL Type 5 protocol is discussed in subsequent chapters.

It is useful to notice from Table 3-3 that only the value of bit 4 of the PT field indicates whether the cell is a "user data" cell (i.e., where bit 4 = 0) or "management" cell (where bit 4 is set to = 1). For the user data cells, the value of bit number 3 is used for the EFCI function, with bit number 3 = 0 signifying "congestion not experienced" and bit number 3 set to 1 signifying "congestion experienced." The last bit of the PT field, bit number 2, is used for the ATM user-to-ATM user function, with bit number 2 set to 0 or 1 (when used for the AAL Type 5 protocol, the ATM user-to-ATM user indication = 1 signifies the "end-of-frame" cell, for example). Consequently, 4 codepoints out of the possible 8 codepoints of the PT field are utilized for user data cells.

TABLE 3-3. Payload type field

PTI coding (bits 432)[a]	Interpretation
000	User data cell, congestion not experienced. ATM-user-to-ATM-user indication = 0
001	User data cell, congestion not experienced. ATM-user-to-ATM-user indication = 1
010	User data cell, congestion experienced. ATM-user-to-ATM-user indication = 0
011	User data cell, congestion experienced. ATM-user-to-ATM-user indication = 1
100	OAM F5 segment associated cell
101	OAM F5 end-to-end associated cell
110	Resource management cell
111	Reserved for future VC functions

[a]Any congested ATM network element may modify the PTI as follows: cells received with PTI = 000 or 010 are transmitted with PTI = 010; cells received with PTI = 001 or 011 are transmitted with PTI = 011; noncongested network elements should not change PTI. The ATM-user-to-ATM-user indication is used by AAL Type 5 to delineate end of frame condition. Payload type function is used to distinguish user data cells from OAM cells.

For the management cells designated by bit number 4 set to 1, two codepoints are used to distinguish segment OAM cells (PT value = 100) and end-to-end OAM cells (PT value = 101) for the virtual channel level. The resource management (RM) cell is coded PT = 110, leaving one codepoint (PT = 111) reserved for future extension of functionality. Since the codepoint space in the PT field (and, for that matter, in the ATM cell header) is severely limited, it is evident that any function that competes for the use of the reserved codepoints will have to be strongly justified.

The terminology used for the ATM user-to-ATM user function suggests that this function is intended for connection end point usage, as is the case for its initially intended use in the AAL Type 5 protocol. Intermediate network elements that do not include AAL functions may not interpret the ATM user-to-ATM user bit. However, a number of congestion control mechanisms have been proposed that may utilize this function, as described later. Consequently, it is no longer valid to consider the ATM user-to-ATM user function as strictly pertaining to an end-to-end usage related to AAL Type 5.

3.5 THE CELL LOSS PRIORITY (CLP) FIELD

Arguably, no other function in the ATM cell header fields has generated as much controversy as the cell loss priority (CLP) function. This despite the fact that the concept of cell loss priority is simple: The cell loss priority field is intended to indicate explicitly whether a cell in a given connection (VCC or VPC) has high or low priority. Thus:

CLP = 0 indicates a "high-priority" cell in a connection (VPC or VCC)
CLP = 1 indicates a "low-priority" cell in a connection (VPC or VCC)

The essential idea behind CLP is that in the event of congestion in an ATM NE, the NE may "selectively discard" the low-priority (CLP = 1) cells before discarding the high-priority (CLP = 0) cells to ease the traffic congestion conditions. In this respect, the CLP function is closely related to the ATM traffic control policy and quality of service (QoS) parameters available to a user. However, it is evident that the use of CLP for selective discard congestion control may not be applicable for all the services ATM has been designed to support, and therein lies one of the factors fueling the discussion on the usefulness of the CLP function, given that it does inevitably add to the complexity of an implementation.

One of the original motivations for including a selective discard capability in ATM derived from its potential use in certain applications such as variable bit rate (VBR) layer coded video where it was felt that the more essential video information such as framing synchronizing may be carried within "high-priority" cells, whereas the "background" picture information may be carried by the lower-priority cells. Since the overall video quality may be more drastically affected by a loss of synchronization information than by loss of picture information carried in the low priority "discard eligible" cells in the event of traffic congestion, providing two levels of priority in a virtual connection may be beneficial in such cases. However, to date the use of layered video coding schemes has not generally been widespread, and some workers have questioned the relative benefit of CLP use even for such schemes since as will be described later, the CLP bit may also be used for the purpose of "tagging" cells for congestion control purposes. In this case, under certain conditions, the high-priority (CLP = 0) cells may be overwritten to low priority (CLP = 1). As a result of this uncertainty, some administrations (notably in the ETSI standards) do not support the use of CLP function in their ATM networks. However, in North America, where ANSI standards are followed, CLP capability is generally endorsed.

It may also be recognized by those readers familiar with frame relay technology that the CLP function is exactly analogous to the use of the so-called "discard eligibility" (DE) function in frame relay services (see Chapter 11). Typically, the concept of relative loss priority is widely accepted in data networking services and is not unique to ATM. Note that when the CLP function is utilized, there are two relative cell loss probabilities implied for the given virtual connection (VPC or VCC). The cell loss probabilities are "relative" in the sense that the actual (i.e., numerical) rate of cells lost depends on the QoS provided to a user by the network operator, which in turn depends largely on the traffic engineering policy and dimensioning of the network resources.

It is important to distinguish the above "explicit" relative cell loss priority within a virtual connection based on CLP use from the so-called "implicit" loss priority that is also possible for groups of virtual connections (VPCs or VCCs) by simply partitioning network resources into high- and low-priority connections. Thus to implement implicit relative priority for any virtual connection (or group), the VPC or VCC as a whole may be classed as "high priority" by partitioning the network resources (e.g., buffers in the NE) in such a way that the virtual connection in question experiences a relatively low cell loss compared to other (groups) of virtual connections offered by the network provider. In theory, the concept of implicit relative priority based solely on VPI/VCI values implies that there may be as many "relative" priorities (or QoS) as there are active VPI/VCI values. Of course, in practice this would be virtually impossible to manage within the network, so that typically only a few implicit QoS classes may be offered by partitioning of network resources between the virtual connections supported.

3.6 GENERIC FLOW CONTROL (GFC) FUNCTION

The generic flow control (GFC) mechanism has been designed to allow for the control of ATM cell flows across the UNI (T_B reference point) or within the customer premises networks (CPN or S_B reference points). An example of a configuration where the GFC may be used is shown in Fig. 3-5, based on ITU-T Recommendations I.361 [3.4] and I.150 [2.1]. Here multiple ATM terminal equipment (or, in general, network elements) are connected to a single GFC controller function in an ATM NE across the UNI. Each terminal may have one or more ATM virtual connections (VPCs or VCCs) within the physical transmission path that attempt to transmit cells across the UNI. If the rates at which each terminal transmits cells is comparable to the transmission capacity of the output physical link after the multiplexing function, cells will be lost as each connection contends for the capacity on the output link. The GFC function may be used to control the cell flow from each ATM TE based on a simple "stop–go" protocol to enable sharing of the capacity (bandwidth) on the output physical path.

The GFC protocol may be readily understood by examining the meaning assigned to the codepoints defined for the GFC field of the ATM header, as shown in Fig. 3-5. In the first place, it should be noted that the GFC mechanism is asymmetrical across the interface and that it applies to all virtual path or virtual channel links (VPLs or VCLs) in the transmission path between the "controlled" equipment (e.g., TE) and the "controlling" equipment (e.g., ATM switch). Thus two sets of codepoints are defined as shown in Fig. 3-5.

Two modes of operation are defined for GFC:

1. In the "uncontrolled mode," the GFC procedures are not used. In this case the GFC field is set to 0000.
2. For the "controlled mode," the GFC mechanism applies. Three functions are possible: (a) cyclic halt of traffic on all ATM connections to reduce the total cell flow to the network

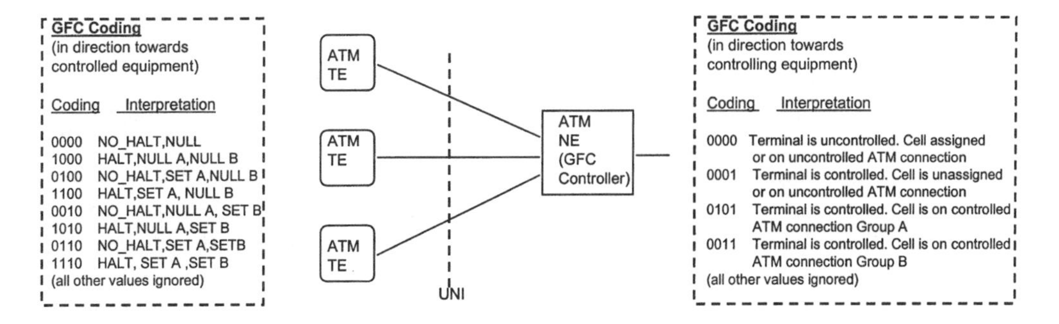

- The GFC mechanism may be used to control ATM cell flow at the S_B or T_B reference points (UNI only)
- Two modes of operation defined in Recommendations I.150 and I.361:
  1) Uncontrolled mode - GFC procedures not used (GFC Field set to 0000)
  2) Controlled mode - GFC procedures apply

- If implemented, GFC enables the following three functions:
  1) Cyclic HALT of traffic (as an option) on all ATM connections to reduce traffic to network across UNI
  2) Access control to the network of traffic on controlled ATM connections
  3) Explicit indication to controlling equipment that traffic is on a controlled ATM connection

- GFC protocol uses both assigned and unassigned cells to carry the GFC codes

- Two groups of controlled ATM connections are supported (Group A and Group B)

Figure 3-5. Generic flow control.

across the interface in question; (b) access control to the network of the traffic on the "controlled" ATM virtual connections; (c) an explicit indication to the controlling equipment that the traffic is on a (GFC) controlled ATM connection.

The GFC procedures allow for two groups of controlled ATM connections, called group A and group B. The grouping of the controlled virtual connections may be used to distinguish between different traffic types, priority or QoS classes, for example. The controlling ATM NE may regulate the traffic by setting the GFC field codepoints in the direction toward the controlled equipment (e.g., the terminal) as shown in Fig. 3-5. Thus the coding 0000 indicates that cells may flow as for an uncontrolled mode of operation. The coding 1000, 1100, 1010, and 1110 signifies no traffic (i.e., "halt condition"). Note that the MSB of the GFC field indicates the "no-halt" (0) or "halt" (1) condition for the cell flow.

For the two groups of controlled virtual connections, the "null" codepoint indicates that group is not supported, whereas the "SET" codepoints implies a particular group is in operation. Thus, the codings 0100 and 1100 signify that traffic is allowed to flow (0100) and not allowed to flow (1100) on the group A virtual connections, whereas group B is not defined. Similarly, the codepoints 0010 and 1010 indicate that traffic is allowed to flow (0010) and not allowed to flow (1010) on the group B virtual connections, and group A is not defined. Finally, the codepoints 0110 and 1110 indicate that both groups A and B are controlled with traffic allowed (0110) or not allowed (1110) to be transmitted toward the GFC controller NE.

The GFC field codepoints in the direction toward the controlling equipment are clearly different, as may be expected given the asymmetrical nature of the GFC mechanism. The asymmetry follows from the requirement that the equipment incorporating the GFC controller function determines the cell flow from the controlled equipment, since it needs to resolve contention for the bandwidth on the (output) network node interface (NNI). Thus, in the direction towards the controlling equipment, the codepoint 0000 indicates that the terminal is operating in "uncontrolled mode," that is, the GFC mechanism is not used. The codepoints 0001, 0101, and 0011 indicate that the terminal is in "controlled mode," with the codings of the middle two bits signifying, as before, whether or not the virtual connections in group A or B are being controlled. The codepoint 0001 indicates that although the terminal is in controlled mode, the cell belongs to an uncontrolled ATM connection (or is unassigned).

The role of the so-called "unassigned" cell (with the preassigned header value described above) in the transfer of the GFC information needs to be noted. As long as user data cells are being generated, the GFC control information can be transported across the interface in the GFC field of the ATM header (at the T or S reference point). The GFC information thus transported is used to control the traffic flow into the network by means of the simple "halt" or "no-halt" commands. However, if no user data cells are available to carry the GFC information at any given instance, the unassigned cells may be inserted to transport the GFC information. This in fact is the reason for the unassigned cells, which are used solely to carry GFC information if a user data cell is not available. The unassigned cells are discarded in the ATM Layer functionality after the GFC information is extracted from them.

There has been considerable confusion in the ATM literature between the role of the unassigned cells and that of the idle cells, which we noted earlier exist only in the physical layer functionality and do not appear in the ATM layer. Since the idle cells are not visible at the ATM layer, they cannot be used to carry GFC information, since GFC is an ATM layer function. In fact, the idle cells are used for rate adaptation purposes only, as described later, whereas the unassigned cells are used (at the ATM layer) for transport of GFC information in the absence of normal assigned (i.e., user data) cells.

It should also be recognized that the GFC controlling function may vary the amount of traffic from the controlled (e.g., terminal) equipment by varying the duration between the "halt" and "no-halt" commands it transmits towards the controlled equipment. If the duration of the "halt" period is relatively long, for example, on average only a small fraction of the traffic from a given controlled virtual connection (or group) may be admitted into the network across a UNI, or vice versa. Although such an "adjustable" GFC protocol may be useful for some configurations on network applications, it is clear that such an implementation will be somewhat more complex than one in which, for example, a simple "round robin" or cyclic discipline is used between the multiple terminals attached to the GFC controller. This aspect is essentially an implementation choice.

In effect, the role of the GFC controller may be likened to that of a traffic policeman at a busy road junction where several streams of traffic are attempting to converge into a main road. By using simple "stop or go" signals, the traffic policeman may cyclically regulate the traffic from each of the incoming streams toward the main road. Moreover, by varying the intervals of the "go" period for some of the incoming traffic streams, the traffic policeman may choose to preferentially regulate one or other of the traffic streams.

It is interesting to reflect that when the GFC concept was originally proposed in the early days of ATM protocol standardization, it was envisaged more as a possible mechanism for providing a form of media access control (MAC) for configurations where ATM terminals were attached to a shared physical media, analogous to legacy LAN configurations. This concept was motivated by the related work on the IEEE 802.6 Metropolitan Area Network (MAN) Standards [3.1] based on the Distributed Queue Dual Bus (DQDB) protocol, which also used a fixed-size cell structure very similar to the ATM cell. It was felt by some workers that incorporation of GFC mechanism in ATM would facilitate interworking between the IEEE 802.6 DQDB protocols and the ATM network, which were envisaged to support common services for data internetworking.

However, after extensive and prolonged discussions and numerous proposals for possible

GFC protocols, it was recognized that a "star" network configuration at the UNI was more suitable for the development of the GFC concept in ATM, and the LAN-like shared media configurations were generally abandoned. Also, the need for possible interworking with IEEE 802.6 DQDB networks decreased as these were only deployed in a few cases.

The GFC mechanism, even as described above, has not been without its share of controversy, and to date has not been widely endorsed in the industry. The ATM Forum's UNI specifications, while recognizing the existence of the GFC protocol, do not mandate its implementation in general. As described in subsequent chapters, more complex per connection flow control mechanisms have been developed [e.g., the so-called available bit rate (ABR) mechanism] that allow more efficient utilization of the transmission path bandwidth, so that many workers regard the simple "stop–go" mechanism of the GFC as inadequate. However, the potential use of the simple GFC procedures for certain CPN configurations or applications cannot be ruled out, and it is likely that future implementations may utilize the GFC protocol more widely.

3.7 PRIMITIVES AND SERVICE ACCESS POINTS

The concept of service access points (SAP) and the associated primitives are widely used in the formal description of data protocol architectures and are not specific to ATM. However, since the ATM literature also employs this description, it is useful to review its application to the ATM protocol architecture here.

In the layered OSI data networking protocol model [2.5, 2.6], the independence between the functions in each layer (when present) is characterized by regarding the lower layer as providing a "layer service" to the adjacent higher layer through a "service access point" (SAP). In effect, the concept of a SAP is an abstract way of describing the "interface" between two layers, and is hence completely independent of any given implementation of the functions in that layer. The information that is exchanged between the layers through the SAP is described in an abstract way by means of so-called "primitives" that are envisaged to pass across the SAP. The primitives are typically termed service data unit (SDU) to reflect the concept of the layer service provided. Although in the general OSI layer model four primitives are defined at the SAP, not all may be defined for specific protocol architectures, and this is the case for ATM.

For the ATM protocol stack, SAPs are defined between the physical layer and ATM layer, and between the ATM layer and the AAL, as shown in Fig. 2-4. SAPs are also defined on top of the AAL, but these are not shown in this figure. Note that although each layer itself may sublayered (for example between VP and VC levels), SAPs, and hence primitives, are not defined between sublayers.

There are two primitives defined at the SAP between the ATM layer and the AAL, as shown in Fig. 2-4. These are:

1. ATM-DATA Request. This primitive is issued *by* the upper layer to "request" transfer of an ATM-SDU to its corresponding (peer) entity over an ATM connection.
2. ATM-DATA Indication. This primitive is issued *to* the upper layer (AAL) to indicate arrival of an ATM-SDU over an ATM connection.

The above definitions of the primitives imply that the ATM-DATA request primitive is passed "down" from the AAL to the ATM layer where it is processed to enable data transfer via the ATM connection to its peer end point. Correspondingly, the ATM-DATA indication is passed "up" from the ATM layer to the AAL to transfer an ATM-SDU arriving from the peer entity (essentially the ATM cell payload) to the AAL for processing.

At the SAP between the physical layer and the ATM layer as well, two primitives have been defined. These are:

1. PHY-DATA Request. This primitive is generated by the ATM layer to request the transfer of a PHY-SDU (ATM cell) from a "local" ATM entity to a peer ATM entity over the PHY connection (e.g., the transmission path).
2. PHY-DATA Indication. This primitive is issued by the physical layer to the ATM layer to indicate arrival of a PHY-SDU from the corresponding PHY entity over the existing PHY connection (e.g., the transmission path).

The above definitions of the primitives between the physical layer and ATM layer imply that the PHY-DATA request primitive is passed down from the ATM layer to the physical layer where it is processed to enable the transfer of data over the physical "connection," namely the transmission path. The SDU passed is essentially the ATM cell. Correspondingly, the PHY-DATA indication primitive is passed "up" from the physical layer to the ATM layer to enable transfer of the PHY-SDU (essentially the ATM cell) to the ATM layer for processing (of the ATM header).

It should be apparent that these formal descriptions of the data transfer and processing are primarily intended to clarify the layered protocol model in general terms, and should not be interpreted in any literal sense for actual implementations in hardware or software. Thus, the partitioning of the physical layer and ATM layer functionality in any particular hardware implementation may not correspond to the layer boundaries in any physical way, and actual information transfer may be partitioned in any number of implementation-specific ways in a VLSI chip or in software code.

The concept of SAPs and the associated primitive exchanging SDUs in the layered OSI protocol model envisages ATM from the perspective of a specific packet mode "data" oriented viewpoint. However, as pointed out earlier, ATM may also be viewed from the perspective of a specific form of "transmission" (or transport) technology, which traditionally has developed its own modeling methodology, particularly as applied to SDH transmission technology [3.2].

Although the transport modeling methodology is also layered, the principles are somewhat different from that of the ISO-OSI data protocol model, and the notion of SAPs and associated primitives have not been incorporated into the transport layer models. This modeling approach to ATM is described later in more detail, but it should be pointed out here that the differences between these two approaches are nowhere more evident than at the boundary between the physical layer (where transport layer modeling has traditionally been used) and the ATM layer (which adopted the OSI modeling approach of traditional data networks).

These different approaches to modeling ATM has led to considerable discussion as to where exactly the boundary should be drawn between "physical layer" functions and "ATM layer" functions, not to mention terminological confusion in describing equivalent functions. This is hardly surprising in view of the fact that the boundary between the ATM and physical layers is where the traditional telecommunications world of synchronous transfer mode (STM) based technology overlaps with the packet-based data-oriented asynchronous transfer mode technology, and both worlds have claimed ATM as their own.

Attempts have been made, particularly by the ITU-T, to reconcile the two different modeling descriptions of ATM, as well as to develop a uniform approach combining the key aspects of both methodologies in a unified model [3.3, 9.2]. However, these efforts have yet to achieve widespread usage in the ATM industry, and the reader should be aware that some confusion in terminology between the different modeling approaches still exists. Fortunately, these differences in the fine points of the abstract modeling of ATM functions do not generally affect actual implementations in hardware or software or interoperability between ATM equipment, provided that the standards have been strictly complied with.

CHAPTER 4

Functions of the Physical Layer

It was noted earlier that the intention of ATM protocols is to be essentially independent of the underlying physical transmission system used to transport the ATM cells of any given virtual connection. With this in mind, the reader may well question why it is even necessary to discuss physical layer functionality in a treatise dealing primarily with ATM. In principle, ATM cells containing user data, signaling information, or management information may be transported over virtually any type of physical transport systems characterized by a wide variety of rates and formats. However, the "mapping" of the ATM cells into the transmission system format needs to be precisely defined and understood, and there is inevitably an interaction between the type of transmission format used and specific ATM related functions.

This interplay between ATM and the strictly transmission related functionality is clearly reflected in the sublayering of the physical layer of the B-ISDN PRM, as depicted in Fig. 2.3. As shown, the physical layer functions are sublayered into the so-called transmission convergence (TC) sublayer and the physical medium (PM) dependent sublayer (also known as PMD sublayer). The functions in the TC sublayer include:

- Cell rate adaptation
- Cell delineation
- HEC processing
- Transmission frame adaptation/generation
- Transmission overhead processing

The functions in the PM sublayer include:

- Bit timing
- Line coding
- Adaptation to physical medium
- Connector related functions

It will be noted from Fig. 2.3 and the above list of functions that the more "ATM-related" functions such as cell delineation, etc. are grouped in the TC sublayer as would be expected, whereas the more "physical" functions such as adaptation to physical medium (e.g., electrical-to-optical conversion) are grouped near the lower end of the PM sublayer. Although the classification of these functions may be somewhat evident, other functions in the proximity of the sublayer boundary are perhaps not so easy to place. Thus, there has been considerable discussion on the exact allocation of functions to TC and PM sublayers, and it may be argued that the allocation depicted in Fig. 2.3 is to some extent somewhat arbitrary. This uncertainty is not surprising in view of the observations made earlier that this boundary represents the "interface" to two markedly different technologies in telecommunications.

It is not within the scope of this work to describe in any detail the vast technology of telecommunications transmission systems. Detailed accounts of the various types of transmission systems hierarchies are given in the extensive literature (for a general overview see, for example, [4.1–4.3]. In this section, we will focus on the ATM-related aspects of the main physical layer functions and interface rates of more general use in the industry. Moreover, the wide range of physical interface rates and formats that have already been defined or are under development at this stage offer a bewildering choice to any prospective system designer or user of ATM equipment. To some extent, this enormous choice of possible interfaces is a reflection of the versatility and universal appeal of the ATM technology, as different applications of ATM seek to optimize and adapt suitable physical interface rates and formats to their own requirements.

The primary function of any telecommunications transport system is the reliable transmission and recovery of user information, whether in analog or digital form, over wide areas, while minimizing errors, loss, or distortion of the information that is being carried by the transmission system. Most telecommunications-grade transmission systems are structured as multiplexed hierarchies to enable the efficient transport of large amounts of information (or equivalently, support a large number of users). An obvious classification may be made in terms of the traditional analog signal transmission technologies and the more modern (and now widespread) digital transmission systems. In the digital domain, another natural distinction may be made between the copper or coaxial (so-called electrical) medium and the more recent optical fiber transmission systems, with their enormous information carriage capacities. It may be of interest to reflect that ATM was initially envisaged as primarily applicable over high-capacity optical fiber transport systems, but its subsequent wide acceptance has resulted in its adoption for use over many lower-speed copper-based transport systems as well.

4.1 FRAME-BASED AND CELL-BASED PHYSICAL INTERFACES

The fact that ATM may also be viewed as an inherent transmission technology as well as a "switching" technology (in the traditional telecommunications sense of the words) led from the outset to an extended debate as to whether ATM really needs to use the conventional "frame-based" transmission technology as typified in the SDH system, for example. It was argued by some researchers, with some justification, that since ATM is inherently an asynchronous packet-based transfer mode, there is really no additional requirement to impose a framing structure. Moreover, any required transmission overhead (e.g., for fault or error detection) may readily be carried by specially dedicated physical layer cells that can be inserted into the ATM cell stream. From this perspective, the imposition of an unnecessary framing structure and additional transmission overhead was seen as redundant, resulting only in more inefficient use of the available transmission bandwidth.

The primary function of the framing structure in a conventional transmission system is to enable timing recovery and synchronization of the time slots at the receiver end, so that demultiplexing of the individual (voice) channels may be performed. A well-known example is the North American T1 system [4.4], in which 24 64 kbit/sec (DS0) voice channels are multiplexed into a bit stream that includes a "framing" bit that serves as a reference to identify the start of the multiplex structure in every transmission frame. The receiver must synchronize with the framing bits in the incoming bit stream, to enable the extraction of the individual time slots in the multiplex hierarchy. Although many different types of transmission frame formats and methods of carrying other overhead information have been devised over the years for transmission systems engineering, the basic principles of the underlying frame-based synchronization remain the same, and have been extensively proven in the field with the development of highly robust and versatile systems. In the case of ATM, however, the cells may be generated at unrelated time intervals, depending on the type of higher-layer application or traffic source. In effect, this is the significance of the term "asynchronous" in ATM. In addition, as will be described below, ATM provides a mechanism whereby the cells can be delineated independently of any framing structure, so that cells belonging to any given virtual connection may be delineated and recovered from the multiplex cell stream without recourse to any underlying transmission-framing mechanism. Consequently, it was argued that the conventional frame-based transmission hierarchies developed primarily for 64 kbit/sec based, time-slot oriented STM technology are simply redundant and inefficient for ATM transmission.

On the other hand, it was recognized that an enormous investment in terms of both actual transmission equipment deployment in the field and technological development globally had already been committed, particularly for the SDH-based transmission systems being deployed at that stage. There was an understandable reluctance on the part of many networks operators and vendors globally to jeopardize their huge investment in proven frame-based transmission technology and adopt an untried and relatively untested cell-based transmission technology for ATM, no matter how elegant it seemed from a theoretical standpoint.

Faced with this seeming impasse between technological elegance and hard commercial reality, it was eventually agreed in the ITU-T (then called the CCITT) that both frame-based and cell-based transmission systems may be standardized for B-ISDN interfaces, with a simple means to interwork between these interfaces if required. Although, on the face of it, this adoption of both types of ATM physical interface in the ITU-T B-ISDN standards may appear as a classic example of a standards "compromise" in the event of a deadlock, there is also some underlying wisdom in this result. Accordingly, it would be possible to utilize the vast installed base of existing frame-based transmission equipment, as well as not endanger the SDH based systems under development based on the ITU-T's extensive and well-defined standardization program. It was envisaged that this course of action would enable earlier deployment of ATM on top of the installed base of transmission equipment. On the other hand, the cell-based transmission system proposed for ATM, which was not well defined at that stage, could be more rigorously tested in the laboratory, and may in future be used for some network applications where the conventional frame-based transmission systems are not suitable.

4.2 THE CELL-BASED PHYSICAL INTERFACE

As noted above, for the cell-based physical interface, no additional framing structure is required. The ATM cells (including those for signaling and OAM as well as user data) from the ATM layer are passed as a PHY-SDU via the SAP at the physical–ATM layer boundary to the TC sublayer of the cell-based interface. The TC functions serve only to add the physical layer cells (which may be either PL-OAM cells or idle cells) to the cell stream, which is then converted directly to the physical media. Thus, the actual structure at the physical interface (UNI or NNI) is a continuous stream of cells, ATM data cells interspersed with physical layer cells. In order to adapt the rate specified as, say, 155 Mbit/sec, it was agreed to limit the maximum spacing between successive physical layer cells to 26 ATM layer cells. This implies that after a maximum of 26 consecutive ATM cells have been passed down, a physical layer cell must be inserted into the cell stream. If there are no ATM cells to be transmitted, physical layer cells are transmitted. If there is no PL-OAM information to be transmitted at any given time, idle cells are transmitted for interface rate adaptation purposes.

The PL-OAM cells are used to carry the OAM information necessary to maintain the transmission path of the cell-based interface. The structure and functions of the PL-OAM cells will be described in more detail when we come to discuss the OAM flows in general. As noted before, the physical layer cells are distinguished by preassigned header patterns, and physical layer cells *must not* be passed to the ATM layer, where they will be discarded as invalid cells. In this respect, the appearance of any physical layer cell (idle cell or PL-OAM cell) in the ATM layer functions defined by coding of the ATM cell header is logged as a "defect" or protocol error. An excessive number of such "invalid" cells may consequently indicate a failure of the equipment.

Since no external framing function is provided for the cell-based transmission system, the provision of timing recovery in the end system or ATM terminal equipment needs to be derived either by a local clock or a separate timing interface for those applications that may require a timing recovery, such as voice signals. In addition, as will be described in subsequent chapters, it is possible to derive such timing information for these applications from the AAL Type 1 protocol if used for the application. Consequently, the provision of timing recovery for ATM cell-based transmission systems is not viewed as an essential requirement, as it is for frame-based transmission systems.

Since the PL-OAM cells are intended to carry the OAM information pertinent to the transmission system, the rate at which the PL-OAM cells are generated depends on the OAM requirements for any given system. As noted above, the maximum rate has been specified as 1 PL-OAM cell every 26 ATM cells in order to match the cell-based transmission rate to the 155.52 Mbit/sec STM-1 SDH rate. It has also been specified that the minimum rate of PL-OAM cell insertion should not be less than one PL-OAM cell every 513 cells. This minimum rate is intended to ensure that a certain minimum OAM information is sustained for system integrity. However, since to date cell-based ATM transmission systems have not been widely implemented for large scale deployment, it has not been verified that the range of PL-OAM cell rates specified above achieves the desired objectives.

More recently, there has been a resurgence of interest in the ATM forum in cell-based transmission for ATM, in particular for campus or private ATM networks, where deployment of SDH transmission systems may not make economic sense. If this interest results in the availability of commercial cell-based ATM transmission systems in the future, it is possible that further refinement of the PL-OAM functions will also result. In principle, there appears to be no inherent reason why cell-based transmission systems should not be able to provide as much capability as the corresponding frame-based SDH transmission systems. Moreover, the ability to vary the insertion rate of the PL-OAM cells into the overall ATM cell stream over a wide range allows for more efficient transport of the OAM information if required (i.e., less bandwidth for the OAM overhead).

4.3 THE FRAME-BASED PHYSICAL INTERFACE

ATM was originally intended to be carried primarily by the SDH frame-based transmission hierarchy, which was also being extensively developed and deployed for optical fiber networks at the time. Since the underlying SDH physical transmission system is essentially designed to be oblivious to the type of traffic being transported in its payload, the ATM cells can simply be "mapped" into the payload of the SDH transmission path at the point where the payload is being assembled in the transmission equipment. The cells can then be extracted from the transmission payload at the termination of the transmission path. The formats and mechanisms to map the ATM cells into the SDH STM-1 and STM-4 virtual container structure has been defined in detail by the ITU-T and has been extensively implemented in hardware by many vendors. The most common rates and formats to date are the 155.52 Mbit/sec STM-1 and the 622.08 Mbit/sec STM-4 SDH transmission interfaces used at both UNI and NNI.

The SDH System is also closely related to the North American "Synchronous Optical Network" (SONET) hierarchy, from which it was generalized so that mapping of ATM cells into SONET payloads essentially follows the same principles. The mapping of ATM into SONET transmission systems has been standardized by the ATM Forum as well as the American National Standards Institute (ANSI), Committees T1.S1 and T1.X1.

However, it was soon recognized by most network operators that although the SDH system will increasingly be used for optical fiber transport globally, the existing installed base of the so-called plesiochronous digital hierarchy (PDH) transmission systems may also be used for the transport of ATM cells. This would have the commercial advantages of utilizing the vast installed base of existing transmission systems based on the North American T1 (or 1.54 Mbit/sec DS-1 and DS-3) multiplex hierarchy and the European E1 or 2.048 Mbit/sec multiplex hierarchies. In addition, since the T1 and E1 systems offer transmission at lower speeds than the SDH systems, the cost of offering ATM services to users would be lower and would encourage earlier deployment of ATM. Consequently, the standards defining the formats and mechanism to map ATM cells into the T1 and E1 PDH systems were rapidly developed and deployed by vendors and operators anxious to exploit the existing transmission infrastructure.

For either the SDH or PDH transmission systems, the principles underlying the mapping of ATM cells into the transmission payload are essentially the same. At the transmitter side, ATM cells generated by the ATM layer functions are inserted (or, more commonly, "mapped") into the available payload of the transmission system, with any spare capacity taken up by the idle cells generated by the physical layer hardware. This is shown schematically in Fig. 4-1. The idle cells are therefore inserted to compensate, or adapt, for any differential in rate between the ATM cells being generated by the ATM layer functions and the actual physical layer transmission rate of the particular physical interface. Conversely, at the receiver side, the ATM cells are extracted from the payload of the transmission system and the idle cells are discarded by the physical layer functions to reconstruct the original ATM cell rate passed up to the ATM layer functions.

Since the ATM cells are simply "packed" end-to-end to form the transmission path payload, an important function that must be performed at the termination of the transmission path is delineation of the ATM cell boundaries. It is only after delineation of the ATM cells that individ-

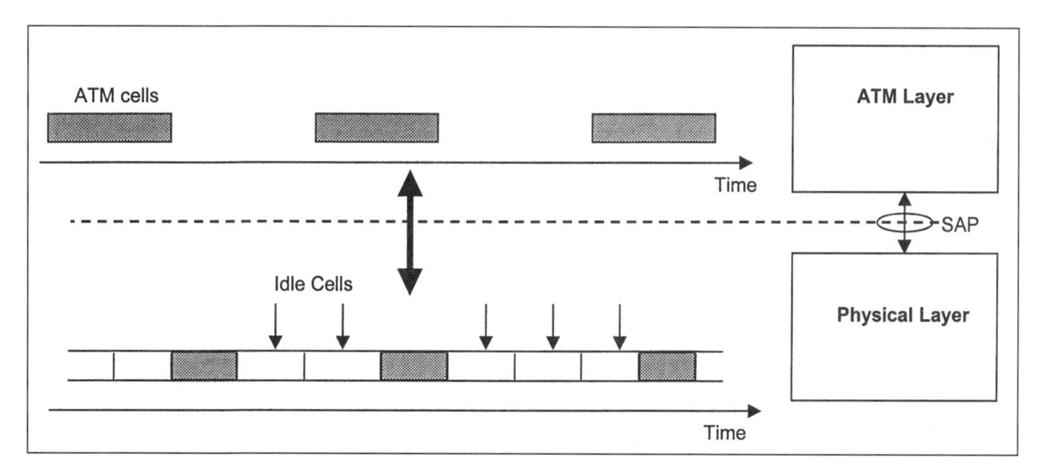

Figure 4-1. Cell rate adaptation function. Cell rate adaptation uses insertion/extraction of idle cells in the physical layer. For cell-based interfaces, both idle cells and PL-OAM cells may be used.

ual ATM cells can be "passed" up to the ATM Layer functions via the SAP at the ATM layer–physical layer boundary. The detailed mechanism by which ATM cell delineation is performed using the header error control (HEC) function is described below. Here we consider primarily the mapping structure used to transport the aggregate ATM cell stream in the transmission path.

The mapping of ATM cells into the 155.52 Mbit/sec STM-1 SDH frame structure is depicted in Fig. 4-2. The ATM cell stream (including any idle cells inserted for rate adaptation purposes) are packed into the VC-4 container together with the VC-4 path overhead (POH). The STM-1 section overhead bytes are then added to generate the standard STM-1 (9 bytes × 270 bytes) frame structure, which is transmitted every 125 μsec. It is not the intention here to discuss in any detail the structure and handling of the path and section overhead bytes that constitute the SDH overhead structure, since this belongs more appropriately to the realm of transmission technology. A detailed description may be found in ITU-T Recommendations G.707 [4.5] and I.432 [4.6]. However, Fig. 4-3 indicates the section and path layer overhead structure to convey some indication of the inherent complexity of the transmission system protocols as well as a measure of the potential capabilities possible with the SDH system. As a result of the extensive overhead provided in the SDH system, it should be noted that the actual net payload capacity for ATM cells is reduced to 149.76 Mbits/sec for the STM-1 rate of 155.52 Mbits/sec. In addition, for many transmission system applications, not all of the overhead bytes may be used, which results in wasted bandwidth. These aspects of SDH-based physical layer for ATM transport have been criticized by proponents of ATM cell-based trans-

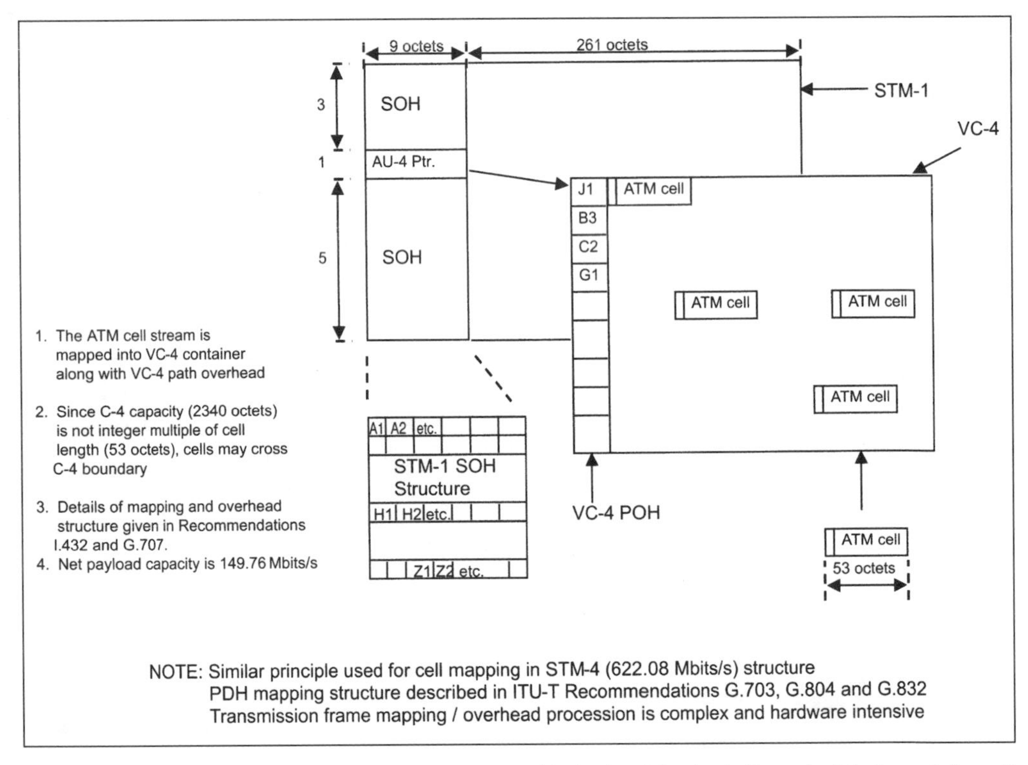

Figure 4-2. Mapping ATM cells into SDH (155.52 Mbits/sec). *Note:* A similar principle is used for cell mapping in STM-4 (622.08 Mbits/s) structure. PDH mapping structure described in ITU-T Recommendations G.703, G.804, and G.832 Transmission frame mapping/overhead procession is complex and hardware-intensive.

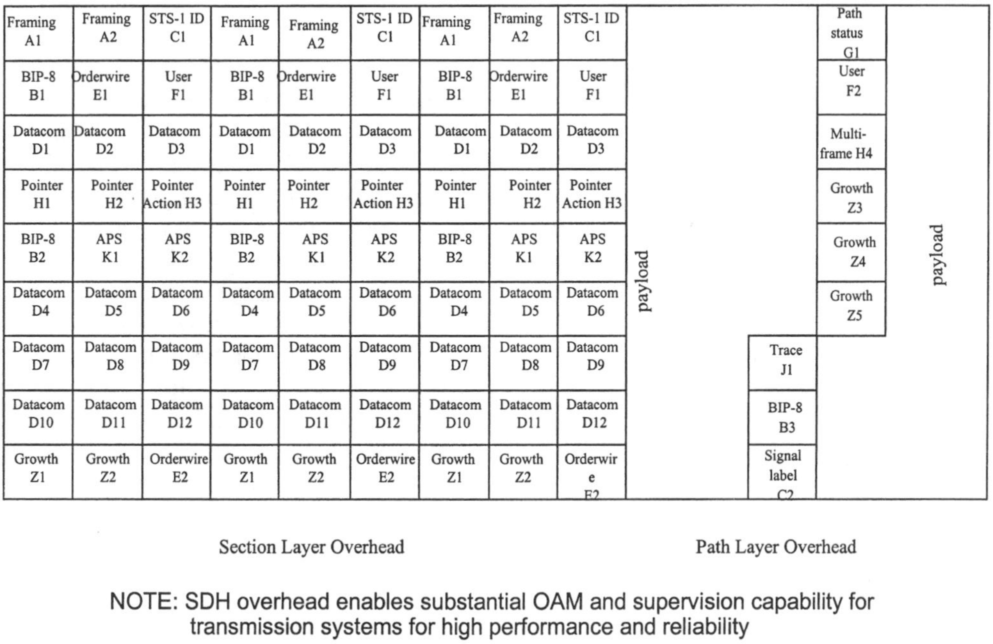

Framing A1	Framing A2	STS-1 ID C1	Framing A1	Framing A2	STS-1 ID C1	Framing A1	Framing A2	STS-1 ID C1	payload	Path status G1
BIP-8 B1	Orderwire E1	User F1	BIP-8 B1	Orderwire E1	User F1	BIP-8 B1	Orderwire E1	User F1		User F2
Datacom D1	Datacom D2	Datacom D3	Datacom D1	Datacom D2	Datacom D3	Datacom D1	Datacom D2	Datacom D3		Multi-frame H4
Pointer H1	Pointer H2	Pointer Action H3	Pointer H1	Pointer H2	Pointer Action H3	Pointer H1	Pointer H2	Pointer Action H3		Growth Z3
BIP-8 B2	APS K1	APS K2	BIP-8 B2	APS K1	APS K2	BIP-8 B2	APS K1	APS K2		Growth Z4
Datacom D4	Datacom D5	Datacom D6	Datacom D4	Datacom D5	Datacom D6	Datacom D4	Datacom D5	Datacom D6		Growth Z5
Datacom D7	Datacom D8	Datacom D9	Datacom D7	Datacom D8	Datacom D9	Datacom D7	Datacom D8	Datacom D9		Trace J1
Datacom D10	Datacom D11	Datacom D12	Datacom D10	Datacom D11	Datacom D12	Datacom D10	Datacom D11	Datacom D12		BIP-8 B3
Growth Z1	Growth Z2	Orderwire E2	Growth Z1	Growth Z2	Orderwire E2	Growth Z1	Growth Z2	Orderwire E2		Signal label C2

payload

Section Layer Overhead

Path Layer Overhead

NOTE: SDH overhead enables substantial OAM and supervision capability for transmission systems for high performance and reliability

Figure 4-3. SDH overhead structure. *Note:* SDH overhead enables substantial OAM and supervision capability for transmission systems for high performance and reliability.

mission from the perspective of transmission efficiency, although, clearly, the robustness and flexibility of the SDH system is not in question.

It is important to note that since the C-4 (payload) capacity of 260 × 9 (2340) octets is *not* an integer multiple of the ATM cell length (53 octets), the ATM cells may overlap the SDH frame boundaries. Conceptually, this key aspect of ATM cell mapping may be envisaged by regarding the ATM cell stream as essentially independent of the underlying transmission frame structure. The ATM cells packed into the transmission path payload simply cross the frame boundaries seamlessly as if they did not exist. One way to imagine this is to view the ATM cell stream as floating independently of the repetitive underlying frame structure. This aspect is also true of the STM-4 (622.08 Mbits/sec) interface and the so-called "direct mapped" DS-3 (PDH) interface. However, the situation is more complex in the PDH case, since two different mapping structures have been standardized, with the earlier "PLCP based" mapping employing a different mapping procedure to the currently more widely used direct mapping structure. The so-called physical layer convergence protocol (PLCP), originally developed for the IEEE 802.6 distributed queue dual bus (DQDB) protocol, packs a given number of ATM cells into the DS-3 payload and fills the remaining empty octets with padding to complete the DS-3 frame structure. In this respect, the PLCP mapping procedure effectively "ties" the ATM cell stream to the DS-3 frame structure, unlike the other direct mapping mechanism, which allows cell overlap across adjacent transmission frames.

To an extent, the independence of the ATM cell stream boundaries to the transmission frame boundaries is a direct consequence of the layered architecture of the B-ISDN PRM. Each layer should be able to operate independently of the adjacent layers in this model, with interactions between layers occurring via primitives across the SAPs, or via indirect communication between layer management functions and plane management functions, as indicated earlier. In practice, this theoretical requirement of the layer model is also assisted by the mechanisms defined to delineate ATM cell boundaries, which operate independently of the transmission frame boundaries.

The most commonly used cell delineation mechanism utilizes the properties of the HEC function, as noted earlier and described in more detail below. However, for the SDH transmission system, an additional mechanism has also been defined and may be used in some implementations. This alternative mechanism uses the so-called H4 byte of the path overhead (POH) as a "pointer" to indicate the start of the first ATM cell in the VC-4 container (payload). In this technique, the H4 byte is coded at the transmitter side with the offset (i.e., distance in bytes) of the first ATM cell boundary from the start of the SDH payload. At the receiver side, the H4 pointer offset value is used to locate the first octet of the ATM cell boundary in the SDH payload. Since each cell has a fixed size of 53 octets, subsequent cell boundaries in the frame can readily be delineated if the first cell in the frame is delineated. In addition, since each frame will include the H4 "pointer" value, the cell delineation can be verified with each frame to minimize the occurrence of a loss of cell delineation (LCD) condition at the receiver.

In addition to the possible use of the H4 byte for cell delineation purposes, the C2 byte in the POH also has relevance for ATM payloads. The so-called "signal label" C2 byte may be used to denote the nature of the payload to the transmission path termination equipment. Thus, the C2 byte codepoint value of hexadecimal 13 denotes an ATM payload as defined in ITU-T Recommendation G.707 [4.5]. It should be noted that the possible use of the H4 and C2 bytes in the SDH POH implies a degree of interdependence between the SDH physical layer and the ATM payload being transported, which is not strictly in accord with the layered PRM. Interestingly, for the case of direct mapped PDH-based transmission systems, no such interdependence is possible since the equivalent of H4 and C2 byte functions do not exist in this case.

4.4 CELL DELINEATION USING THE HEC

It was noted earlier that the header error control (HEC) mechanism in octet 5 of the ATM cell header provides for two important functions in ATM. The first function is error detection and/or correction over the first four octets of the cell header, based in a cyclic redundancy check 8 (CRC 8) algorithm encoded in the HEC field. The second function is to enable ATM cell boundary delineation by using the inherent properties of cyclic codes [4.7].

To enable the ATM header error control function, the transmitter side performs a CRC 8 calculation over the entire cell header using a standardized HEC algorithm defined in ITU-T Recommendation I.432 [4.6] as follows: The HEC field is encoded with an 8-bit sequence, which is the remainder of a modulo 2 division by the CRC generator polynomial

$$x^8 + x^2 + x + 1$$

of the product of x^8 multiplied by the content of the first four octets of the ATM cell header. To improve robustness of the CRC 8 mechanism, the HEC algorithm also requires that the eight-bit pattern 01010101 is added modulo 2 to the remainder derived as above, before insertion into the HEC field.

At the receiver side, the above HEC mechanism is capable of either (1) single bit error correction, or (2) multiple bit error detection, depending on whether the receiver is configured in "correction" or "detection" mode. The relationship between these modes and resulting actions is shown in the simplified receiver state diagram of Fig. 4-4. It should be noted that the default mode is single-bit error correction. To achieve this, the receiver side performs the CRC 8 based

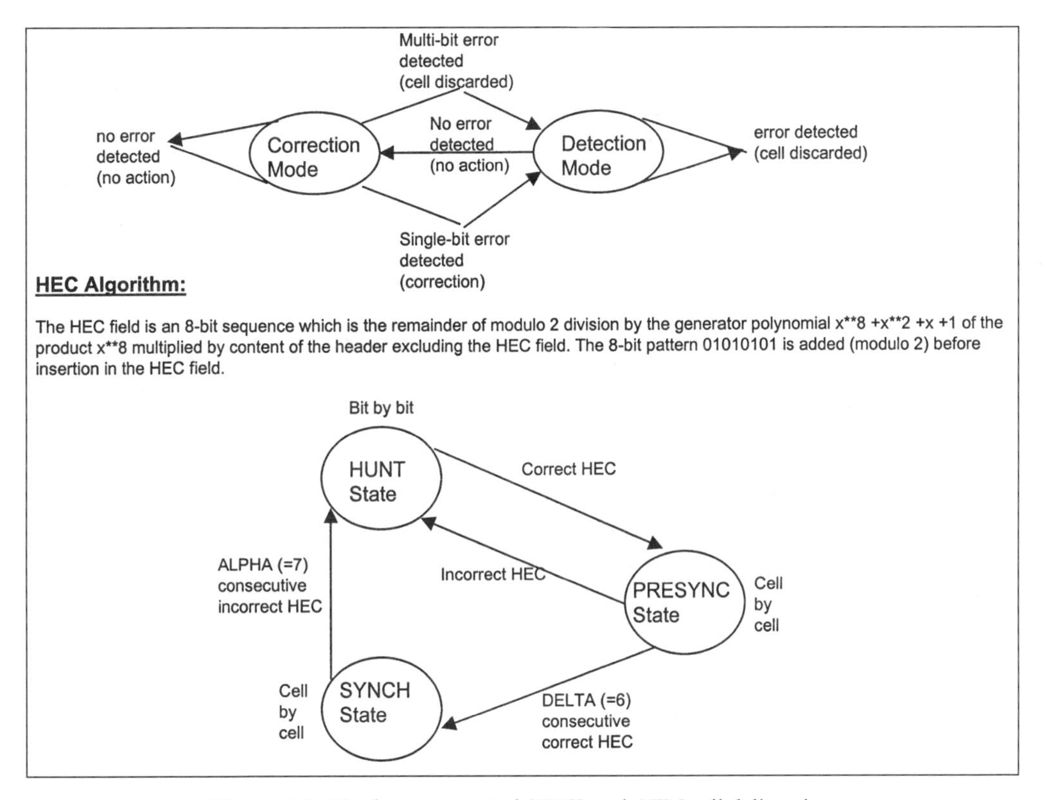

Figure 4-4. Header error control (HEC) and ATM cell delineation.

algorithm described above on a cell-by-cell basis and compares it with the encoded value inserted in the HEC field by the transmitter side. Any discrepancy in the sequences implies errors engendered by the intervening transmission system and may result in the cell being discarded if the error cannot be corrected. The mathematical properties of CRC 8 sequence over 32 bits allows for the correction of a single bit error or the detection only of multiple bit errors, caused, for example, by an error burst in the transmission system.

The use of error detection codes for the protection of information transmitted in packet-based systems has been extensively studied and is well established. For ATM, the importance of detecting errors in the cell header is obvious. Since the switching of ATM cells is performed by translation of the VPI and VCI values in every NE along the connection, any errors in the VPI/VCI values may result in undesirable misrouting or misinsertion of cells. Clearly, this should be avoided for robust operation of ATM-based systems. However, it has been demonstrated in a number of studies that for practically achievable transmission system performance, the ATM HEC mechanism enables robust error detection performance to be obtained in practice [4.5]. Typically, for random transmission bit error rate (BER) of better than 10^{-6} the probability of cells being discarded is better than 10^{-9}. In practice, well-designed optical fiber transmission systems can achieve BERs of 10^{-9} or better, so the probability of cell losses in this case is negligible.

However, it has been observed that typical transmission system impairments result from the presence of burst errors rather than random single bit errors. The probability distributions of such error bursts and their duration are much more complex to model and are typically highly media-dependent. Clearly, in such cases, the probability of ATM cell losses can be much higher than for the case of random single bit errors, since the single bit error correcting property of the CRC 8 HEC cannot be effective in the presence of error bursts. Nonetheless, experience with practical ATM systems has demonstrated that for well-engineered transmission systems, cell losses even in the presence of error bursts can be maintained well within acceptable levels, even for demanding applications such as broadcast-quality video signal transport.

The HEC algorithm also provides an elegant way to perform cell delineation at the receiver side. The cell delineation is performed by using the correlation between the 32 (four octet) error-protected header bits and the eight "control" bits inserted in the HEC field by the transmitter side. It will be recognized that the CRC 8 based algorithm described above introduces a direct correlation between the pattern in the four error-protected octets of the header and the pattern in the HEC field as a result of the mathematical properties of cyclic codes. By detecting this correlation between the patterns, the receiver side can determine the initial bit position of the ATM cell header in the received bit stream. The correlation can then be verified by repeating the process on a cell-by-cell basis.

The simplified state diagram for the HEC-based cell delineation mechanism shown in Fig. 4-4 illustrates how the correlation introduced by the CRC 8 algorithm is used by the receive side. In the HUNT state, the delineation process is performed by checking the incoming bit stream bit by bit for the correct HEC value for an "assumed" cell header field. Thus, the receiver assumes any initial five octet sequence of bits to be a cell header and performs the HEC algorithm over the first four octets to compare the resulting value with the pattern in octet 5. If the value is correct (i.e., the difference or so-called syndrome is zero), the receiver cell delineation mechanism enters the PRESYNCH state. If the resulting value is not correct, then the mechanism returns to the HUNT state and repeats the process starting from the next consecutive bit and so on. It should be noted that if the octet boundaries are available on processing the transmission path payload, as with an SDH based system, the delineation process may be performed on an octet by octet basis rather than bit by bit to improve the synchronization process.

Once the correlation between the 32 protected bits and the HEC pattern is established, the receiver mechanism enters the PRESYNCH state, in which the delineation is verified by checking the pattern correlation cell by cell until the correct HEC pattern has been confirmed DELTA times consecutively. The value of DELTA has been chosen to ensure the process is sufficiently robust while minimizing the delay required to achieve the SYNCH state. Values have been standardized as DELTA = 6 for SDH-based transmission systems and DELTA = 8 for cell-based systems. It should also be noted that for the cell-based physical layer, only the last six bits of the HEC field are used for the cell delineation mechanism, whereas for the SDH-based physical layer all eight bits of the HEC algorithm are specified for use in the cell delineation process.

After the correct HEC pattern has been confirmed DELTA times, as described above, the receiver cell delineation mechanism enters the SYNCH state, as shown in Fig. 4-4. In the SYNCH state, the receiver continues to check the pattern correlation between the HEC value and the first four octets of the header on a cell-by-cell basis. However, if an incorrect HEC value is detected ALPHA times consecutively, it is assumed that cell delineation has been lost, and the receiver returns to the HUNT state. The value of ALPHA = 7 has been specified for both SDH- and cell-based physical layers, as before, to minimize loss of synchronization delay while maintaining robustness of the overall cell delineation mechanism.

Although at first glance the cell delineation mechanism deriving from the HEC algorithm appears somewhat complex, especially when compared with the conceptually simpler H4 byte "offset pointer" based mechanism described earlier for the SDH based system, it has proven itself to be technically superior to the earlier H4-based pointer mechanism in terms of robustness and performance. Consequently, most of the present implementations of ATM equipment have adopted the HEC-based technique for cell delineation, although some earlier equipment that used the H4 pointer mechanism may still be available in the field. It should be pointed out that interoperability between the two different cell delineation mechanisms will only be a problem if the more recent ATM equipment implements only the HEC-based mechanism and does not process the H4 byte in the POH, e.g., to reduce hardware complexity.

In order to improve randomization of the data for transmission, the physical layer also includes a scrambler function. For the SDH-based transmission system, a so-called self-synchronizing scrambler based on the polynomial $x^{43} + 1$ is used. This particular polynomial has been selected in order to minimize the error multiplication effect (by a factor of 2) which is introduced by using a self-synchronous scrambler to randomize the bit stream. The operation of the scrambler is arranged such that the scrambler algorithm only randomizes the ATM cell 43 octet information field and not the five octet header. Thus, the scrambler mechanism at the transmitter side is disabled over the ATM cell header but enabled over the payload fields. Similarly, in order for the HEC-based cell delineation mechanism to work, the receiver side also disables the descrambler mechanism over all fields in the HUNT state and over the ATM header fields in the PRESYNCH and SYNCH states.

For the cell-based physical layer, a different scrambler algorithm has been selected. In this case the so-called distributed sample scrambler to the 31st order is used based on the generator polynomial $x^{31} + x^{28} + 1$. The distributed sample scrambler randomizes the transmitted data by the modulo 2 addition of a pseudorandom binary sequence (PRBS) to the bit stream based on the specified generator polynomial. At the receiver side, the data is descrambled by modulo 2 addition of a local generated PRBS based on the same generator polynomial function. This mechanism requires phase synchronization between the scrambling and descrambling pseudorandom sequences to be maintained. For the distributed sample scrambler, the phase synchronization technique requires the use of the two most significant bits (MSB) of the HEC field and is described in detail in Recommendation I.432 [4.6]. This is the reason that for the cell-based physical layer, only six bits of the HEC field are used for the cell delineation functions, as noted before. Although the detailed operation of the cell-based distributed sample scrambler mechanism need not be considered here, it should be noted that the technique, although somewhat more complex than the self-synchronizing scrambler mechanism utilized for the SDH frame-based physical layer, nonetheless allows the cell delineation process to proceed as described earlier while enabling suitable randomizing of the transmitted bit stream for robustness.

CHAPTER 5

ATM Traffic Control

5.1 THE NEED FOR TRAFFIC CONTROL IN ATM

In common with other packet-switched technologies, ATM networks will also need to be engineered and dimensioned in such a way that a suitable level of performance in terms of cell losses and cell delays can be maintained under the anticipated user traffic loads, potentially from a large number of sources emitting cells at different rates. Consequently, although the general problem of network traffic control is not unique to ATM technology, this aspect of the overall network design may be made significantly more complex in ATM since the technology is intended for multimedia traffic sources as well as high rates of information transfer for some of these sources. It is therefore not surprising that an enormous amount of work has been done in recent years to study the behavior of both real and simulated ATM networks from the perspective of traffic control and resource management.

It is not our purpose here to review the extensive literature on ATM traffic performance resulting from this work, nor treat the subject with the mathematical rigor that has been applied to it in some studies. For the reader interested in delving deeper into the intricacies of this perennially absorbing field, a good starting point can be the overviews given in [5.1–5.5]. Here we will describe the essential results from this work, and focus on the mechanisms that have been incorporated into ATM technology to ensure that the QoS requirements can be met with suitable network design.

It is useful to note at the outset that partially as a result of the extensive empirical and simulation studies of ATM traffic performance characteristics, and partially as a result of the varied applications (traffic sources) intended for use with ATM, a relatively large number of traffic control mechanisms have been proposed and many incorporated into ATM specifications and standards. This has unfortunately resulted in the public perception that traffic control or resource management of ATM networks is necessarily complex or somehow inherently problematical. This perception, which is also conveniently often promoted by proponents of competing packet-mode technologies, is, however, far from the truth. In fact, there is nothing inherently more complex in designing an ATM network from a traffic control perspective than any other packet-mode based network to achieve a given level of traffic performance and quality of service. More often, the issue in ATM is to select which of the many available traffic control mechanisms to use, since no unique solution exists to achieve the required performance level for a given class of network applications. In a sense, for ATM the network designer is "spoiled for choice," since a range of traffic control mechanisms may be available, whereas for other packet-based technologies there are often no inherent traffic control mechanisms available for effective network design.

From the perspective of traffic management, any VCC in the ATM network can be viewed as a concatenation of queues of cells distributed in the NE buffers, which need to be managed in such a way that cell losses due to buffer overflows are minimized and cell delays due to the buffering process are also kept within acceptable bounds. This is depicted schematically in Fig. 5-1.

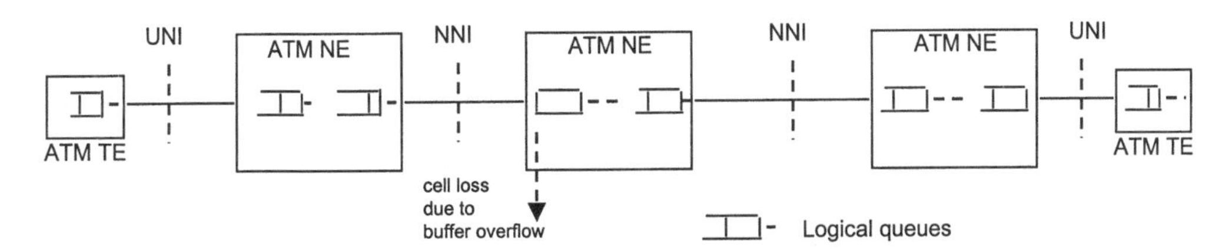

Figure 5-1. ATM traffic and congestion control.

From the traffic perspective, the individual ATM NEs (switches, cross connects, multiplexers, etc.) can be represented as a set of buffers that temporarily store the incoming cells, which are then switched to the appropriate output VCLs on the relevant transmission paths at some rate. The cell queues in the buffer(s) are said to be served in accordance with the algorithms designed in the NE to appropriately optimize the traffic performance of the NEs. It should be noted that however such algorithms are designed in actual NEs, the fundamental requirement of cell sequence integrity on a given VCC (or VPC) will need to be maintained in any ATM NE. This requirement generally implies that an overall First In First Out (FIFO) discipline is imposed on the queued cell flow on any connection processed by the NE.

The behavior of ATM NEs has been extensively modeled using the mathematical theory of queuing systems [5.1, 5.6]. The basic element in this approach is assumed to be a cell multiplexer function in which the cell buffers are modeled as a mathematical queue of cells. The cell arrival times into this queue are modeled according to a given probability distribution function, which depends essentially on the characteristics of the traffic source emitting the cells. Thus, typical ATM traffic sources can range from emitting cells at a constant rate (the so-called constant bit rate or CBR sources such as voice or video signals), to sources that emit cells in bursts of variable duration followed by intervals where no cells are emitted, also with variable duration. Such sources are also generically known as variable bit rate (VBR) sources, or on–off sources, and may result from typical data applications such as TCP/IP or Ethernet sources. Given the wide range of possible source traffic characteristics (or profiles, as they are sometimes called) and rates, it becomes evident that designing ATM NEs that can effectively handle all types of traffic behaviors requires extensive use of the queuing models as well as suitable assumptions of source traffic characteristics.

The primary objective of such simulation or empirical studies is to enable optimum dimensioning of the system buffers and selection of suitable resource management algorithms to obtain the traffic performance targets. In many cases, such design studies may utilize so-called "worst case traffic profiles" to ensure robustness of the design to potential extreme load situations that may arise in actual networks. If the traffic characteristics (e.g., cell arrival rates or distribution of arrival times) of the ATM VCCs are known or can be assumed, then the methodology of the queuing theory may be used to predict the probabilities of cell loss for buffers of various sizes, and the delay distributions due to the cell buffering delays in the cell multiplexer function, for any given rate of the egress transmission path. Where analytical solutions are not possible (as in the majority of cases of practical interest), simulation models may be employed to derive the cell loss and delay probabilities, buffer fill characteristics, and other parameters of interest in characterizing traffic performance of an ATM NE. The primary objective of such studies is to derive a design that minimizes cell losses due to buffer overflows, or to minimize cell buffering delay for a given level of cell loss. When cells are lost due to buffer overflow, the system may be considered to be in a congested state, and this should clearly be avoided as much as possible in order to maintain an acceptable QoS.

The main traffic control mechanisms adopted for ATM have as their primary objective the avoidance of congestion conditions in a network (or NE). Assuming that a combination of the traffic control mechanisms are in place and that the network has been adequately dimensioned for the anticipated traffic characteristics and loads, it may be understood that cell loss due to congestion conditions will be a relatively rare event. For example, a network may be designed to maintain (on average) a cell loss of better than 1 in 10^6 cells on an ATM connection. If, on the other hand, it is observed that congestion states occur with unacceptable frequency, therefore jeopardizing the QoS available to a network user, it will be necessary to deploy alternative strategies of traffic control or reengineer the network resources to minimize the onset of congestion conditions.

The precise definition of congestion state (or condition) is also somewhat arbitrary. An obvious definition may be the point at which any buffer in an ATM NE incurs cell loss due to an excess of cells arriving at the buffer over the cells actually able to be transmitted from the buffer. However, in many cases, it may be possible to incorporate a thresholding mechanism in the buffer design, so that a buffer fill exceeding such a threshold triggers a congestion state, with possibly some consequent action, before cells are actually lost due to physical buffer overflow. Attempts have also been made to specify several thresholds indicating, for example, conditions corresponding to 1) congestion onset, 2) mild congestion, 3) severe congestion, and so on. Although such approaches may be useful from an overall network resource management perspective, it should be noted that such distinctions may have little practical value in the design of high-speed networks such as ATM. When the arrival rates of ATM cells at a NE buffer is high, the actual time interval between the instantaneous buffer fill exceeding a congestion indication threshold and physical buffer overflow may be too short for any effective management action to avoid congestion.

As a consequence of such uncertainties, there has been an attempt in the literature to make a distinction between so-called "traffic control," which comprises the actions and mechanisms taken to avoid congestion, and so-called "congestion control," which is taken to mean those mechanisms designed to deal with recovery from a congestion state in the (typically rare) event that it occurs in a NE. For the purposes of this discussion, congestion control procedures will be treated as an important part of the overall traffic control mechanisms available to the ATM network designer. In all cases, the primary objectives of traffic control can be summarize as follows (Fig. 5-1):

1. To provide the negotiated QoS to all users of the network in an equitable manner
2. Maintain the network performance objectives
3. Optimize the use of the network resources, i.e., the bandwidth and buffer available in an ATM NE

Interestingly, it is the third objective of traffic control that has elicited the most contentious discussion with respect to ATM traffic control procedures, possibly because network resource optimization most directly impacts the relative cost/performance aspects in providing ATM-based services. In principle, the first two objectives can be relatively readily met by "overprovisioning" network resources (bandwidth and buffer capacity in ATM NEs) such that contention for bandwidth by the users is minimized. However, such a strategy can be relatively costly, particularly for broadband services. In practice, network operators will seek to reduce cost by reliance on the statistical nature of most ATM traffic sources by sharing network resources between more users than can strictly be accommodated from a simple summation of the bandwidth requirements of all users together.

In discussing the possibility of exploiting the statistical nature of some ATM traffic sources, it is useful to make a distinction between the so-called "deterministic multiplexing" and "statistical multiplexing" of ATM connections onto any given transmission path. Using the concept of the ATM multiplexer function from conventional queuing theory, a multiplexed (virtual path or channel) connection is said to be deterministic when the sum of the peak bandwidth of the input ATM VPCs (or VCCs) is less than or equal to the peak bandwidth of the output (multiplexed) cell stream. Conversely, it is said to be statistical when the sum of all the peak bandwidths of the input ATM VPCs (or VCCs) is greater than the peak bandwidth of the output cell stream of the multiplexer. The statistical "gain" achieved is the ratio of the sum of the input VCCs (or VPCs) bandwidths in the statistical multiplexing case to that in the deterministic multiplexing case.

Thus, if a multiplexer has n input VCCs (or VPCs), each with a peak bandwidth of p_i bits/sec (or cells/sec), and P_{out} bits/sec (or cells/sec) denotes the peak multiplexed output cell stream, then $n\Sigma_i p_I \leq P_{out}$ identifies the deterministic multiplex condition and $n\Sigma_i p_I > P_{out}$ signifies a statistical multiplexing condition. The ratio Σp_i (statistical)/Σp_i (deterministic) is a measure of the statistical "gain" being utilized. Alternatively, the statistical gain can also be viewed as the "additional" ratio of input connections that a statistical multiplexer can support over the deterministic multiplexer case, for a given level of QoS.

The concept of statistical multiplexing is not unique to ATM technology and may be applied to any virtual-channel-based packet mode technology such as frame relay, IP, etc. Statistical multiplexing (and the resulting statistical gain) simply utilizes the fact that for some applications not all the traffic sources into the multiplexer will be emitting cells continuously, so that cell arrival events into the multiplexer buffer may be approximated according to some (possibly empirically derived) probability distribution. Consequently, the instantaneously "unused" capacity (bandwidth and/or buffer space) in the multiplexer function may be shared by more additional sources than would be possible if all the input source cell streams had been continuous. Also, the more variable the input sources are, for example, in terms of on-off interval ratios, the more possibility there may be to utilize the unused capacity by multiplexing additional virtual channel connections, and hence to achieve more statistical "gain."

It needs to be kept in mind that use of statistical multiplexing generally involves an engineering trade-off with cell loss probability. Since the asynchronous cell arrival events on any of the input VCCs (or VPCs) of the multiplexer function are assumed to be in accordance with some probability distribution characteristic of the traffic source behavior, there is always the probability that multiple cells may arrive simultaneously, resulting in cell losses due to the finite buffer sizes of practical multiplexers. For typical bursty traffic sources that may emit bursts of cells, the probability of multiple bursts arriving coincidently could result in unacceptably high cell loss probability unless the buffer size is adequately dimensioned to accommodate multiple "worst case" burst lengths of cells. Note that even for the case of deterministic multiplexing, there remains a finite (but possibly small) probability of cell losses for finite buffer sizes, since simultaneous multiple cell (or burst) arrivals may also occur in this case, resulting in buffer overflows.

These considerations indicate that arriving at a workable trade-off between multiplexing gain and cell loss probability (more generally QoS) in practical networks requires a suitable level of understanding of the relevant source traffic characteristics, the overall network traffic flow patterns, and the behavior of NEs under load conditions. To achieve this, both simulation models and empirical studies have been undertaken for a variety of network applications. These studies are an important component of overall ATM network design, since the use of statistical multiplexing gain to significantly reduce the cost of providing ATM service, by effectively spreading the cost of bandwidth over a larger number of users for a given QoS level, may provide significant competitive advantage to a network operator or service provider.

5.2 TRAFFIC CONTROL FUNCTIONS OVERVIEW

The key traffic control functions developed to date for ATM may be grouped into the following main categories described briefly in subsections 5.2.1 through 5.2.5.

5.2.1 Network Resources Management (NRM)

This function includes the provisioning of sufficient relevant network resources (e.g., transmission path bandwidths as well as the internal buffer dimensioning within the ATM NEs) to be able to manage the anticipated network loads and segregate traffic flows according to service categories if required. It will be realized that, in general, network resources management is not particular just to ATM technology, but would be required in any case for the robust design of any telecommunications network. The particular challenge of ATM is that the precise mix of traffic and its distribution remains somewhat uncertain, making the actual dimensioning of potentially high speed, multimedia networks, difficult.

5.2.2 Connection Admission Control (CAC)

This function comprises the procedures that are taken by the network during the connection setup phase to establish whether the connection can be accommodated within the available resources without any degradation of the QoS of the connections already in service. The CAC function needs to be performed whether the (virtual channel or path) connection is being set up by on-demand signaling or by provisioning through a management interface. It should be noted that although connection admission control may sometimes be viewed as a network-wide function, typically it will need to be performed on a NE-by-NE basis, since, in general, each NE will know at any instant the available resources for additional connection requests. The detailed modeling of the CAC function within a NE is described later in relation to the other functional elements within an ATM NE.

Although the requirements for the CAC function as a basic traffic control function have been described most specifically for ATM, it will be clear that the concept is somewhat analogous to the more familiar "call blocking" actions in existing STM telephony during periods of high traffic load. However, it is important to bear in mind that the objective in both cases remains the same. This objective is to ensure that the network resources (e.g., bandwidth, buffer allocations) being utilized by existing VPCs or VCCs to maintain the (negotiated) QoS levels are not degraded by the requests for resources from the new connection.

The precise mechanisms whereby the CAC function is performed within ATM NEs is essentially implementation-specific, and may often be cited as a means to differentiate the performance of ATM NEs. Thus, although the CAC function is a basic traffic control (or resource management) requirement for ATM networks, its actual algorithmic implementation is left deliberately unspecified to allow for optimization and fine tuning of NE performance.

The connection admission control procedures may also be taken to include the means by which resources are allocated to a new VCC (or VPC) request by an ATM NE. This aspect of CAC introduces a significant difference between ATM resource management procedures and those used in STM based networks. Since the possibility of statistical multiplexing of VCCs (or VPCs) exists for ATM networks, network operators may choose to adopt a policy-based bandwidth allocation to ATM connections to obtain statistical gain advantages. For example, if a VCC requested a maximum bandwidth of 10 Mbit/sec, the network operator may actually arrange for the CAC function to allocate less bandwidth (say, 8 Mbits/sec) since it may know from past experience of the given traffic source behavior that such a VCC is only active for, say,

50% of the time on average. The "saved" bandwidth may then be used to support additional VCCs, thereby providing statistical gain, which translates to additional revenue.

With conventional STM-based technology, this flexibility in bandwidth allocation is not possible, since the admission of an additional connection generally implies a fixed allocation of 64 kbits/sec (or 56 kbits/sec) regardless of whether the STM connection is partially active or not. As noted earlier, this fundamental difference between STM and ATM is simply a consequence of the packet mode nature of ATM, and the related concept of "virtual" channels that allow for the possibility for statistical sharing of network resources.

By including the bandwidth allocation capability to VCCs (VPCs) as part of the CAC procedures, the network can exploit statistical multiplexing gain and optimize use of network resources based on known traffic-source behavior patterns. Simply put, a flexible CAC implementation can allow the inherent statistical multiplexing capabilities of ATM to be more effectively used for many network applications, while not unduly jeopardizing the QoS available to VCCs/VPCs.

It may also be of interest to mention here that although the concept of connection admission control is essentially a part of connection-oriented packet mode technology that enables support of QoS objectives, connectionless technologies such as IP have recently also incorporated similar concepts in order to support QoS capability. These protocols will be discussed in more detail later, but their adoption of analogous concepts serves to demonstrate the ubiquity of connection admission control procedures when it is necessary to ensure that acceptable levels of QoS are maintained.

5.2.3 Usage (and Network) Parameter Control (UPC/NPC)

This function comprises the set of actions taken by the network to monitor and control the offered ATM traffic at the UNI (UPC) and NNI (NPC), respectively. This includes the detection and enforcement of the offered traffic to protect the QoS of existing connections when the offered traffic on any given connection is found to violate the negotiated traffic parameters. The traffic-enforcing action performed by the UPC (or NPC) function is also sometimes referred to as "policing" of the ATM traffic flow. Inherent to the concept of traffic enforcement by the UPC (or NPC) function is the requirement that when a VCC (or VPC) is set up, the parameters by which the subsequent traffic flow will be enforced (or policed) are known by the network. This key concept, often termed a traffic contract, will be described in further detail below. However, it is clear that without any mutually agreed upon traffic parameters to enforce for a VCC or VPC, the UPC/NPC function is ineffective.

The relationship between the connection admission control function described above and the usage (or network) parameter control needs to be clearly kept in mind. The CAC function is used to ascertain whether the requested VCC or VPC with its associated traffic parameters, including bandwidth and QoS requirements, can be accepted by the network, essentially by checking if sufficient resources are available in the relevant NEs. Once the VCC/VPC is accepted, the CAC function may also assign the suitable traffic parameters to the new VCC/VPC, which may depend on the level of statistical multiplexing gain required as part of the network operator service. The UPC or NPC function may then use this traffic information to enforce the actual cell flow on the VCC or VPC to ensure that it does not violate the agreed upon parameters. For example, if the user had requested a maximum bandwidth of, say, 10 Mbit/sec on his VCC, but subsequently attempted, either by intent or by accident, to actually transmit at rates higher than 10 Mbits/sec, the UPC function monitoring the cell rate on this connection should detect the excess rate and enforce the violation, for example, by discarding cells in excess of the agreed upon maximum rate. By this means, the UPC or NPC protects the negotiated QoS of all the other connections by ensuring that some VCCs do not consume more than their legitimate share of resources.

It should be noted that the UPC function will need to be performed on every VCC or VPC in order for complete QoS protection to be ensured. Partially for this reason, and partially from a lack of understanding of the role of ATM, the UPC function has been somewhat controversial, with some early implementations dispensing with it altogether to simplify the implementation. For NNIs, the NPC function is optional. It is recognized that for NNIs demarcating administrative boundaries, it may be necessary to perform an NPC function to ensure that the traffic entering an administrative domain actually does fall within the negotiated traffic loads. However, for NNIs within an operator domain, it may not be necessary to enforce the traffic between NEs, since in general, the operator should be able to manage the available resources and traffic loads to maintain the target QoS levels. In such cases, the NPC function may be disabled.

As for the case of the CAC function, it is interesting to note that although the UPC function requirements have been standardized [5.7, 5.8], the algorithm whereby the UPC function is performed on the cell flow within a VCC or VPC is not specified. The method for performing the UPC or NPC function in an ATM NE is largely vendor-specific. The intent here was to allow for performance optimization of the UPC mechanism in any given implementation. However, many implementations currently have tended to be based on the so-called "leaky bucket" algorithm, which is described in more detail below.

It is also important to clearly distinguish between the UPC function and the concept of "traffic shaping." These terms have sometimes been used interchangeably in the literature, resulting in ambiguity. Traffic shaping results when traffic characteristics of the cell stream on a VCC or VPC is intentionally modified to "fit" to some predetermined flow pattern. For example, a bursty traffic source may intentionally "smooth" the cell flow by buffering cell bursts and reemitting them at a more regular intervals. Although traffic shaping is generally associated with an ATM traffic source, say, within the CPN, it may also be performed by any NE along the connection path if desired. In contrast, the UPC (or NPC) function monitors the traffic pattern on the VCC or VPC, which may or may not be "shaped," to determine whether the cell flow pattern falls within the allowed (and agreed upon) resources allocated to that connection. If it does, the UPC function takes no action. If on the other hand, the UPC or NPC detects a violation, it may take action, for example, by discarding cells if the maximum allowed bandwidth is exceeded. It should be noted that although both the traffic shaping and UPC functions may be combined in any NE, for example to minimize cell discard by smoothing the traffic flow prior to performing UPC test, the functions remain different in their intent.

The UPC (and NPC) function, combined with the CAC function, provide ATM the primary means to control traffic loads within the network, hence avoiding congestion and thereby providing the QoS objectives designed by the network operator. It is also interesting to note here that while UPC/NPC concepts were criticized initially for the resulting complexity, notably for data networking applications, analogous concepts have recently been proposed to provide QoS capabilities in IP-based data internetworking, as discussed in more detail later. Although this provides yet another example of the cross-fertilization of ideas between differing technology solutions, it also attests to the fundamental soundness of the conceptual framework for traffic control in ATM.

5.2.4 Priority Control

Priority control functions comprise those actions whereby the "selective" discard of low-priority traffic may be used by a NE to protect as far as possible the QoS for higher-priority traffic in the event of congestion. As described earlier, ATM provides two distinct mechanisms to indicate the relative priority of a cell flow. Explicit priority indication refers to the use of the cell loss priority (CLP) field in the ATM cell header to explicitly indicate the (relative) priority of any cell within a given VCC or VPC. Implicit priority indication refers to the use of the VPI/VCI values themselves to group sets of VCCs or VPCs into relatively low and high-priority connections. Since explicit priority based on use of the CLP bit indicates two levels of priority within any given VCC (or VPC), it may, in principle, be combined with implicit priority, which pertains to "groups" of VCCs/VPCs provisioned by management action, resulting in many possible levels of relative priority.

The use of the word "relative" priority is significant in both cases, however, since no specific cell loss probability values need be assigned to the priority levels at this stage. The quantitative values assigned to any priority level are engineerable parameters that may be determined by the network operator or service provider, either based on standardized QoS objectives or mutually agreed upon negotiated parameter values, depending on available network resources. In addition, as noted earlier, the congestion threshold for selective discard of cells subject to priority control is also essentially an engineerable parameter, which may or may not be under management system control in any given ATM NE.

Whereas the concept of explicit or implicit priority control may be viewed as a mechanism for congestion avoidance in packet-switched networks, it may also be argued that since priority control action (e.g., selective cell discard) does not occur until some congestion threshold is reached within the NE, the intent of priority control is essentially to mitigate the effects of congestion. In addition, it assists the network to recover from a congestion state. It should also be noted that priority control procedures are not unique to ATM, and have been used for other packet mode networks with some success. In this case, ATM technology essentially borrowed the concept of priority control from other packet mode technologies such as frame relay.

However, the use of priority for congestion control in ATM has been somewhat controversial, particularly as regards the use of the CLP field in relation to selective discard. Some experts have argued that the additional complexity introduced by supporting two priority levels is not warranted by any significant benefit of selective discard in practical networks. In particular, a number of European network operators have indicated they see no significant value in maintaining two levels of priority within any given VCC or VPC, and hence the ETSI ATM specifications do not support the use of the CLP field for congestion control. On the other hand, some North American operators have tended to support the use of priority control within their networks, based on their experience with other packet mode technologies such as frame relay, for example, which utilizes selective discard to mitigate the onset of congestion. It should be noted in this context that the ATM Forum specifications in general support the use of priority control including the CLP field for traffic control purposes.

Such differences of approach and opinion are not surprising when we consider that, in general, traffic control or resource management is very dependent on policy considerations. In effect, there is no "right" approach to traffic control, as noted earlier. Service providers and network operators may use a variety of traffic control strategies deriving from the network applications and architectures in question and the QoS requirements that are being sought. In this regard, ATM is no different to any other packet mode technology, but it may be recognized that ATM offers arguably much more choice in the mechanisms available for traffic control, thereby making it somewhat more difficult for the network operator to choose the appropriate resource management strategy in some cases.

5.2.5 Feedback Controls

Feedback control functions comprise the procedures that are used by the network and the users to regulate the traffic flow on VCCs or VPCs based on feedback information on the congestion state of the NEs. Feedback controls generally operate by introducing some form of flow control of the ATM cell stream on the VCC or VPC, and several different flow control mechanisms have been specified for use in ATM networks, as described below. More recently, other mechanisms for flow control have been proposed, so that it needs to be borne in mind that this area of study has been evolving rapidly in recent years, with the result that a number of techniques ranging from the very simple to very complex flow control mechanisms have been incorporated into this general category.

The simplest feedback control capability in ATM utilizes the codepoints for explicit forward congestion indication (EFCI) that have been defined in the payload type (PT) field of the ATM cell header, as noted earlier. If a NE in the path of the VCC or VPC experiences a congestion state (e.g., by crossing of a congestion threshold related to some level of buffer fill), the NE may set the EFCI codepoint to indicate "congestion experienced" for cells in the downstream direction. The downstream traffic sink is thereby alerted that congestion exists in the network, and may use this information to initiate a reduction in the cell flow from the source, for example, by a reduction in the "window size" of transmitted packets if window flow control is being used in the higher layers. This, for example, is the case when TCP/IP constitutes the "higher layers," since TCP incorporates window-based flow control. The EFCI procedures also require that the traffic sink sets the EFCI values for cells flowing upstream, along the return path of the bidirectional ATM connection. This action serves to alert the traffic source of the onset of a congestion state in the network, thereby enabling the source to take any appropriate action, such as reducing the rate of cells generated on the VCC or VPC until the congestion state is removed.

The simple EFCI function essentially acts as a possible "trigger" to initiate flow control actions at the higher layers (e.g., Layer 4 in the case of TCP), but does not introduce any flow control functions at the ATM layer itself. Since EFCI relies on any higher-layer flow control procedures if present, it is not applicable to many potential ATM applications. An additional limitation of EFCI use is that the procedures whereby the end systems can use EFCI information to initiate flow control actions such as window size reduction or reductions of cell generation rate at source have not been specified in detail to date. As a result, vendors are unclear as to how to interpret EFCI information or use it effectively. These considerations have resulted in some controversy as to the value of the EFCI function for traffic control, and as for the case of the use of priority control, differing approaches between some European operators who do not see the need for EFCI use and some North American operators who support the optional use of EFCI.

Readers familiar with frame relay (FR) technology will have noted that whereas the EFCI function is essentially equivalent to the forward explicit congestion notification (FECN) function in FR, there is no correspondence in ATM to the backward explicit congestion notification (BECN) function in FR. Although an equivalent function was originally proposed for ATM as well, it was not adopted for a variety of reasons. Apart from questions of its real value, it was recognized that for the major applications such as TCP/IP over ATM, the TCP window flow control mechanism was controlled by the receiver (traffic sink), and hence the EFCI information would be sufficient in such cases to initiate TCP flow control and thereby regulate the effective source rate. On the other hand, there is a body of opinion that maintains that it is more effective in some cases to quench the source traffic directly by utilizing a "backward" indication as in FR technology. However, this debate has to a large extent been overshadowed by the development of more powerful flow control mechanisms that are based on the use of the resource management (RM) cells and operate within the ATM Layer itself and are hence independent of any higher-layer flow control mechanisms that may or may not be present.

The definition of a dedicated resource management (RM) cell for traffic control purposes may be viewed as the means for providing a much more flexible and powerful mechanism for feedback controls. As noted earlier, the RM cell is identified by a preassigned codepoint (PTI = 110) of the payload type field in the ATM cell header. In addition, for VPCs, a preassigned value of VCI = 6 (binary 110) may be used to distinguish the RM cell from user cells in the cell stream. In either case, the RM cells may be used to carry traffic-control-related information in-band in any given VCCs or VPCs, and may be generated by the end terminals or intermediate NEs. As will be seen below, the RM cell is a special type of OAM cell dedicated for resource (i.e., traffic) management functions.

A more detailed description of the use of RM cells to provide feedback control mechanisms is given later, since these techniques are in general more complex and require an understanding of additional concepts, which are introduced below. Nonetheless, it will be intuitively obvious to the reader that the potential flexibility of using RM cells to transport feedback information on the congestion state of the network or other traffic-control-related parameters is substantial. When this is coupled to the fact that there are clearly many ways in which flow control can be performed using in-band feedback information, it is hardly surprising that there was extensive discussion and controversy, initially in the ATM Forum but also in the ITU-T, as to which mechanisms should be adopted for ATM. As noted earlier, there is no obvious "right" solution for every network situation, and the extensive simulation as well as experimental studies in support of the many contending proposals have demonstrated that many mechanisms can be devised to give equally good (or bad) results for typical network scenarios. To a large extent, the ongoing debate on the relative merits and drawbacks of the different feedback control techniques need not unduly concern us here. It is clear that continued research into improved flow control mechanisms will lead to new capabilities over those currently specified and available to the network designer. As always, there are likely to be engineering trade-offs in terms of complexity versus performance or flexibility versus complexity (and hence cost). Under these conditions, it becomes important to identify the point of diminishing returns relative to the QoS objectives required for the network applications. Thus, for example, if the desired application or services will work robustly with an average cell loss probability of, say, 1 in 10^6 cells, it would appear to be engineering overkill to demand a more complex flow control mechanism capable of ensuring cell losses better than 1 in 10^8.

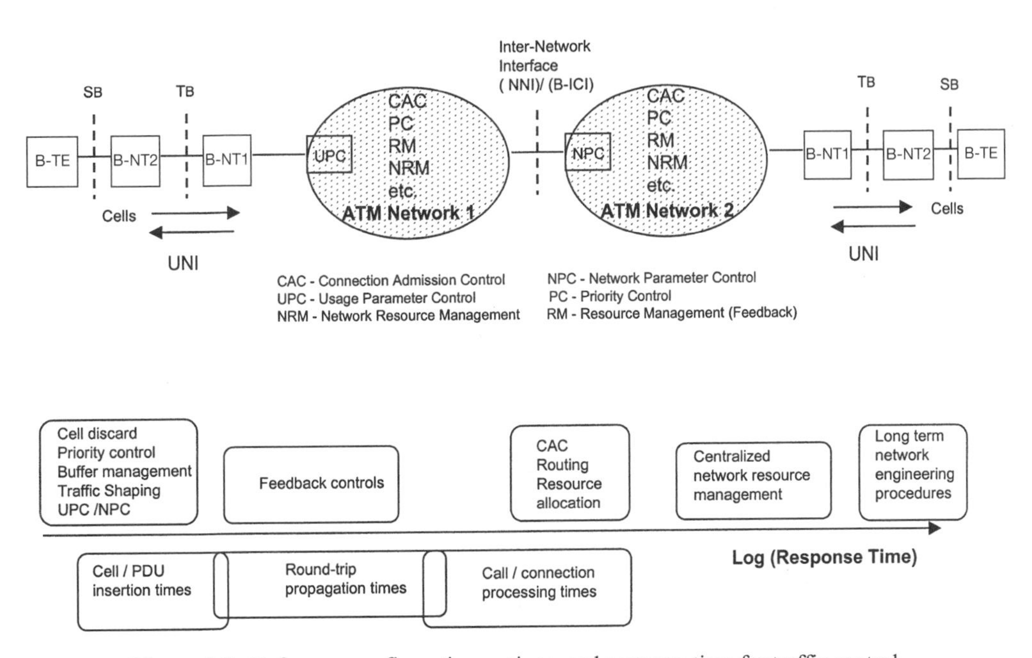

Figure 5-2. Reference configuration, actions, and response time for traffic control.

5.3 REFERENCE CONFIGURATION AND TIMESCALES FOR TRAFFIC CONTROL

The description of the five main categories of traffic control functions outlined above should not leave the impression that all functions are required all the time for every ATM network application or service. In practice, since there is no single traffic control mechanism that is optimized for all network applications, an appropriate combination of these functions may be used for any given scenario. To understand the relationships between the different functions and timescales over which they are effective, it is useful to have in mind a reference architecture for traffic control, as well as categorize the response times for their action. The reference configuration for traffic control may be based on the general B-ISDN reference model discussed earlier. From the traffic control perspective this is shown in Fig. 5-2.

In this reference configuration, traffic on any given VCC or VPC enters or leaves the ATM network from the customer premises network (CPN) across the UNI specified at the T_B reference point. The connection admission control (CAC) function within the network must determine whether the requested resources and QoS for the VCC or VPC are available and can be reserved without jeopardizing the QoS of the other existing virtual connections in the network. If accepted, the CAC may allocate the relevant bandwidth characteristics to the VCC or VPC based on the network operator/service provider policy. The UPC function at the entry point of the network may then monitor the cell flow in accordance with these traffic characteristics in order to ensure the user cell flow does not violate the agreed upon traffic characteristics and thereby endanger the QoS accorded to other existing VCCs or VPCs. If the UPC detects a traffic violation, it may enforce (police) the VCC or VPC by discarding excess cells to protect the network, otherwise it need take no action. Similarly, at the NNI between two administrative domains, for example, an NPC function may (optionally) be used to enforce the traffic entering an operator domain.

Depending on the types of network applications supported, other traffic control functions such as priority control (PC) or feedback controls based on RM cells may or may not be utilized by the network operator, as shown in Fig. 5-2. Overall network resource management (NRM) procedures may be used by each network operator to ensure that sufficient capacity may be called upon so that QoS is maintained and robustness to anticipated traffic loads has been suitably engineered.

The temporal relationships between the various traffic control functions may be represented schematically as shown in Fig. 5-2. The traffic control functions that need to have a response time of the order of cell insertion times include cell discard or priority control, UPC/NPC actions, buffer management within NEs, and traffic shaping. Since the cell interval (or duration) at a speed of 155 Mbit/sec is of the order of 2.7 μsec, it can be concluded that such functions generally need to be fast and are usually implemented in hardware. The feedback controls (either EFCI or RM cell based) will generally operate with response times of the order of the round trip propagation times for cell transport on the VCC or VPC. These will generally be in the order of 1–10 milliseconds, depending on network size. These functions are also generally implemented in hardware.

Progressing along the (logarithmic) timescale, the CAC function will need to operate in timescales of the order of connection (or call) processing times, which may be in the order of several tens to hundreds of milliseconds. For on-demand (switched) VCCs or VPCs, the CAC procedures are closely linked to the signaling applications within the NE, so CAC implementations are generally embedded in the core software of ATM switching systems. However, since CAC procedures need also to be invoked when provisioning semipermanent virtual channel (or path) connections (PVCs) via a management interface, it is clear that the management applications within the NE needs also to access CAC procedures. This relationship will be described in more detail in subsequent chapters.

For longer timescales, for example, of the order of hours or days, the centralized or distributed network resource management (NRM) procedures for overall traffic control of a network may be called into play. Such procedures are generally highly operator or service provider policy dependent, and may be part of the operating system function (OSF) software employed by the network operator for day-to-day operational management of the network. In the longer term, with timescales ranging from months to years, long-term network engineering procedures may be used to ensure that the network design and architecture are suitably dimensioned to meet anticipated growth in traffic loads while maintaining QoS objectives and robustness in periods of peak demand.

The reference configurations and response timescales for overall traffic control and resource management reviewed here essentially provide a broad conceptual framework within which more detailed study of traffic control can proceed. Nonetheless, it is interesting to reflect that the primary goals of ATM traffic control–that of providing robust QoS performance and optimization of network resource usage to minimize cost–can only be met by actions that range from timescales of the order of microseconds to years, and functions that may be embedded in virtually all parts of the network in both hardware and software. In this context, it should also be stressed that the specific traffic control functions adopted in ATM are derived from the results of a sound theoretical framework based on statistical queuing theory, and developed to model the behavior of packet mode networks in general. In addition, ATM has also benefited from pragmatic experience with other packet mode technologies by adopting "tried and tested" procedures, with the result that it provides a wealth of traffic control tools far exceeding those available to any other packet mode networking technology in use at present.

It should also be said that such an "embarrassment of riches" has its disadvantages as well. It has been stated by some experts that ATM traffic control functions may be overly complex and hence costly to implement in their entirety. The relative value of priority control and aspects of feedback controls have been questioned and interoperability issues between the different feedback-based mechanisms cited as a cause for concern. Moreover, the technical discussion has been public, often in open competition with connectionless packet technologies such as IP, resulting in the perception that ATM traffic control specifications are somewhat immature. That this is not the case will be clear in this review of the conceptual framework for ATM traffic control and in the detailed description of the functional elements below. However, it needs to be borne in mind that ATM, unlike other earlier packet mode technologies and particularly connectionless packet technologies such as IP, has been designed to provide suitable QoS to a wide range of traffic sources in a multimedia network embracing voice, video, and data applications. It is, therefore, hardly surprising that the resulting ATM traffic control procedures are likely to be more complex than those of networks that do not provide QoS capabilities for a range of applications.

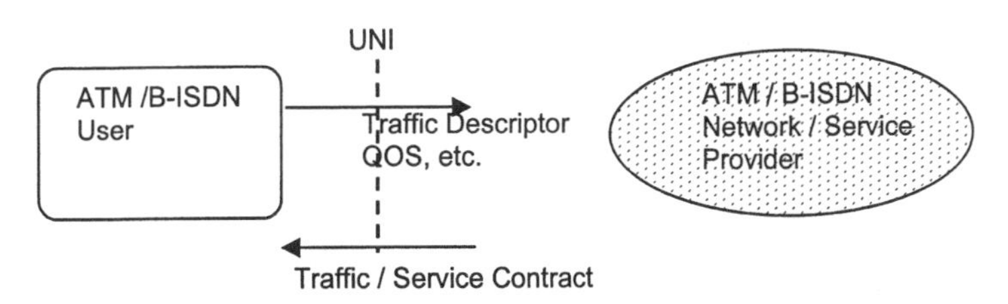

Figure 5-3. Traffic contracts, descriptors, and QoS.

5.4 TRAFFIC CONTRACTS, DESCRIPTORS, AND PARAMETERS

Having reviewed the broad framework for ATM traffic control, we now discuss in some detail the essential concepts underlying the framework. A key concept in ATM traffic control is the notion of a "Traffic Contract," as illustrated in Fig. 5-3. In general terms, the definition of the traffic contract between an ATM (or BISDN) user and the ATM (or BISDN) network (or service) provider is the negotiated set comprising all or some of the following items:

1. ATM transfer capability
2. Source traffic descriptor
3. QoS class or specific parameters
4. Maximum allocated cell delay variation tolerance (CDVT)
5. Any traffic control options required by the user (e.g., cell tagging)

In subsequent sections, we describe each of these elements in turn, but here we focus on the basic definitions of traffic descriptors and parameters as well as the relevant QoS parameters.

An ATM traffic parameter is defined as a particular qualitative or quantitative aspect of the traffic characteristics on a VCC or VPC. Examples of typical ATM traffic parameters are maximum cell burst size, peak cell rate, and so on. ATM traffic parameters provide a basic means to describe any traffic pattern to a level of accuracy dependent on the number of traffic parameters that can be specified, although, in practice, only a few traffic parameters have been generally specified for use in ATM. The most important traffic parameters for ATM include the peak cell rate, the maximum burst size, and the sustainable cell rate. These are defined below.

An ATM traffic descriptor is defined as the set of traffic parameters that can be used to describe the complete traffic characteristics of the ATM connection (VCC or VPC). Although, in theory, the ATM traffic descriptor is intended as a statistically complete description of the cell flow characteristics on a VCC or VPC, in practice, the traffic descriptor may only comprise the set of standardized traffic parameters as described above. The actual cell arrival probability distributions are not generally specified, given the uncertainties relating to practical ATM traffic source behavior. However, it should be clear that the traffic descriptor negotiated as part of the traffic contract should be sufficiently precise to enable the network to allocate sufficient resources to maintain the required QoS, even for worst-case situations. It is essentially for this reason that the "useful" traffic parameters comprising the traffic descriptor are defined in terms of "maximum" or worst-case traffic patterns anticipated for any given source (or application).

As noted earlier, ATM networks are intended to provide various QoS categories at the ATM Layer, in order to support a wide range of higher layer applications. The QoS categories may be defined in terms of specific ATM layer performance parameters including:

1. Cell loss ratio (CLR)
2. Cell transfer delay (CTD)
3. Cell delay variation (CDV)
4. Cell misinsertion ratio (CMR)

The precise definitions and the relationships between these QoS performance parameters will be dealt with in subsequent chapters and are not of primary concern to traffic control functionality, with the exception of cell delay variation (CDV), which, as described below, has a direct impact on the operation of UPC/NPC function. However, the QoS class or category comprising the set of QoS performance parameters is an important part of the traffic contract and, hence, needs to be mentioned in this section.

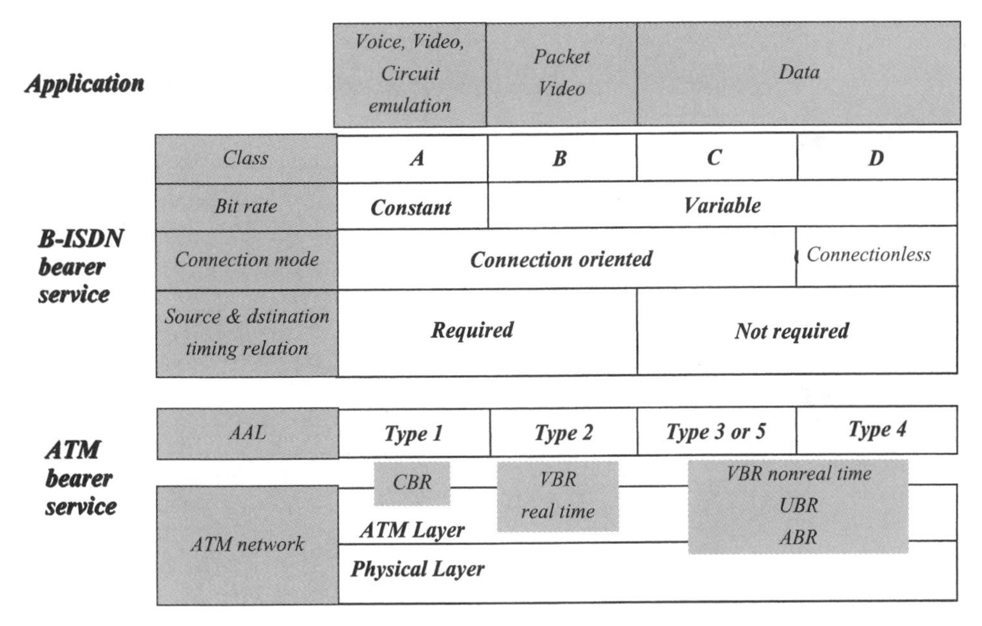

Figure 5-4. General classification of ATM services.

From the perspective of the traffic contract, the QoS may be specified in a number of ways. One way is to specify the individual QoS parameters values explicitly, e.g., CLR = 10^{-6}, etc. The other possibility is to specify QoS "classes" or categories, which are essentially made up of a set of QoS parameter ranges, generally corresponding to anticipated service requirements for typical ATM network applications, e.g., video, voice, etc. Unfortunately, both methodologies may be used from the perspective of existing standards, which has led to considerable confusion in the interpretation, as well as deployment, of QoS in ATM networks. The ATM Forum originally based the specification of QoS on a categorization of service classes such as CBR (e.g., voice and video applications), realtime VBR (e.g., packet video), and non-realtime VBR (e.g., data applications). This simple classification scheme, shown in Fig. 5-4, was originally only intended to be used to define the relationships between the AALs and the anticipated B-ISDN services by the ITU-T. Consequently, it was clearly not precise enough for the purposes of traffic control, and the ATM Forum specification was subsequently enhanced to allow the use of individual QoS parameters values as well as the original QoS service categories. In this scheme, the overall relationship between the individual QoS parameters such as CLR and the service categories may be represented as shown in Fig. 5-5. The acronyms in this representation will be defined below after we describe the traffic parameters more precisely.

However, more recently, the ITU-T has defined a set of ATM QoS classes based on target QoS performance parameter values in Recommendation I.356 [5.9] for a so-called "hypothetical reference model." This approach is different to the description adopted in the ATM Forum specifications currently, but it is important to note that there is no fundamental contradiction, since the individual QoS performance parameter definitions remain essentially the same. It is nonetheless unfortunate that different descriptions of the means to specify QoS in traffic contracts have been adopted in the ATM Forum specifications and ITU-T Recommendations, resulting in ambiguity and confusion on this important issue. This ambiguity is also somewhat compounded by the different terminology in use with respect to the service categories in relation to traffic control mechanisms.

<table>
<tr><th></th><th>CBR</th><th>VBR
real-time</th><th>VBR
non-real-time</th><th>ABR</th><th>UBR</th></tr>
<tr><td>Cell Loss Rate</td><td colspan="4">specified</td><td>unspecified</td></tr>
<tr><td>Cell Transfer Delay</td><td colspan="3">specified</td><td colspan="2">unspecified</td></tr>
<tr><td>Cell Delay Variation</td><td colspan="2">specified</td><td colspan="3">unspecified</td></tr>
<tr><td>Traffic Descriptors</td><td>PCR/CDVT</td><td colspan="2">PCR/CDVT
SCR/BT</td><td>PCR/CDVT
& others</td><td>PCR/CDVT</td></tr>
<tr><td>Flow Control</td><td colspan="3">no</td><td>yes</td><td>no</td></tr>
</table>

CBR : Constant Bit Rate
VBR : Variable Bit Rate
ABR : Available Bit Rate
UBR : Unspecified Bit Rate
PCR : Peak Cell Rate
CDVT : Cell Delay Variation Tolerance
SCR : Sustainable Cell Rate
BT : Burst Tolerance

Figure 5-5. ATM Forum service categories and traffic control. Note: The ATM Forum has adopted a different approach to the ITU-T in relating service categories to traffic control mechanisms.

In an attempt to categorize the various ATM statistical multiplexing mechanisms that have been adopted (or are being proposed), the ITU-T has introduced the concept of the ATM transfer capability (ATC) in its work on traffic control in Recommendation I.371 [5.7]. An ATM transfer capability may be defined as a combination of QoS commitments and ATM traffic parameters suitable for a set of (user or higher-layer) applications, and which are supported by a specific traffic multiplexing mechanism at the ATM Layer. It is also intended that an ATC that is to be used on a VCC or VPC be negotiated as part of the traffic contract during the connection set-up phase. It should also be noted that there need not be a one-to-one correspondence between the ATC and the B-ISDN service classes. Thus, any given ATC may be used to support several different services if suitably chosen. To date, four distinct ATCs have been defined in Recommendation I.371, and several other proposals are being actively studied for future enhancements. The four current ATCs are listed below and described in further detail in subsequent sections. They are:

1. *Deterministic Bit Rate (DBR) Transfer Capability.* The DBR ATC was formerly called Constant Bit Rate (CBR) in the ITU-T. The CBR terminology continues to be used in the ATM Forum specifications (e.g., UNI v.4.0).
2. *Statistical Bit Rate (SBR) Transfer Capability.* The SBR ATC was formerly called Variable Bit Rate (VBR) in ITU-T, and this terminology (VBR) continues to be used in the ATM Forum specifications. At present three versions of Statistical Bit Rate transfer capability, SBR 1, SBR 2, and SBR 3 are defined depending on the specific traffic parameter options that are selected.
3. *Available Bit Rate (ABR) Transfer Capability.* The ABR ATC is based upon a rate-dependent flow control mechanism using RM cells to carry feedback information, In this case, the same terminology and mechanisms are also used in the ATM Forum specifications.
4. *ATM Block Transfer (ABT) Capability.* The ABT mechanism also utilizes RM cells, but in

a different way from ABR. It was originally known as "fast reservation protocol," since the statistical multiplexing mechanism essentially depends on in-band reservation of bandwidth for cell bursts. At present, two versions of the ABT protocol have been defined. It should be noted that there is no equivalent of the ABT mechanism in the ATM Forum specification at present. The ABT ATC has been an European initiative essentially supported by ETSI specifications.

It should be noted that the ATM Forum specifications do not (at present) utilize the concept (or terminology) of an ATC to relate traffic parameters and multiplexing mechanisms. In fact, as noted earlier and described in Fig. 5-5, the relationship between the service categories (CBR, VBR, etc.) and the QoS parameters such as cell loss rate, etc., are either specified or not as shown. In addition, QoS parameter values may be individually assigned as part of the traffic contract to any arbitrary value suitable for the application or service in question. In this scheme, a user of the ATM network may request any combination of cell loss and cell transfer delay values for this particular application, for example. In practice, however, it should be recognized that for ease of management and control, a network operator or service provider may only offer a limited number of QoS classes. Consequently, it may be somewhat irrelevant for a user to request any QoS parameter value, as in practice this will be allocated to the nearest relevant QoS class offered by the network provider.

It should be borne in mind that although the terminology and descriptions of the QoS and traffic parameter categories are different in the ATM Forum specifications and the basic ITU-T Recommendations I.371 [5.7] and I.356 [5.9], there is inherently no contradiction in the basic elements comprising a traffic contract between the user and the ATM network. However, it must be admitted that the somewhat different terminology and descriptions used in these specifications has resulted in some confusion and ambiguity of interpretation. This is of particular relevance in the definition of ATM signaling protocols for on-demand switched VCCs or VPCs, since, as will be seen later, the traffic contract information will need to be transferred by the signaling messages, where an unambiguous interpretation is essential for interoperability of multivendor networks.

An alternative procedure for providing the QoS and traffic descriptor information required in the traffic contract is to specify the so-called "service type" for the VCC or VPC requested. For example, an end user may request a service type such as voice or a video-conference connection. In this case, the relevant QoS and traffic descriptor is assumed to be implicit in order to support the requested service type. Thus, a service type such as a video-conference will have a "prearranged" QoS objective and predefined traffic parameters, thereby freeing the user from having to explicitly specify these when setting up a call or service contract with the network provider. The concept of using service type as a part of the traffic contract will inevitably develop as the technology matures and applications become more commonplace. At present, however, no standardized "service type" definitions have been established to simplify the traffic contract information.

5.5 THE PEAK CELL RATE (PCR)

The concept of ATM traffic parameters was defined above, and its relation to the source traffic descriptor and traffic contract was discussed in general terms. We now define the important traffic parameters in more detail, starting with the fundamental parameter of key importance to ATM traffic control, the peak cell rate (PCR). As shown in Fig. 5-6, the peak cell rate of the ATM connection (VCC or VPC) is defined formally as the inverse of the *minimum* interarrival

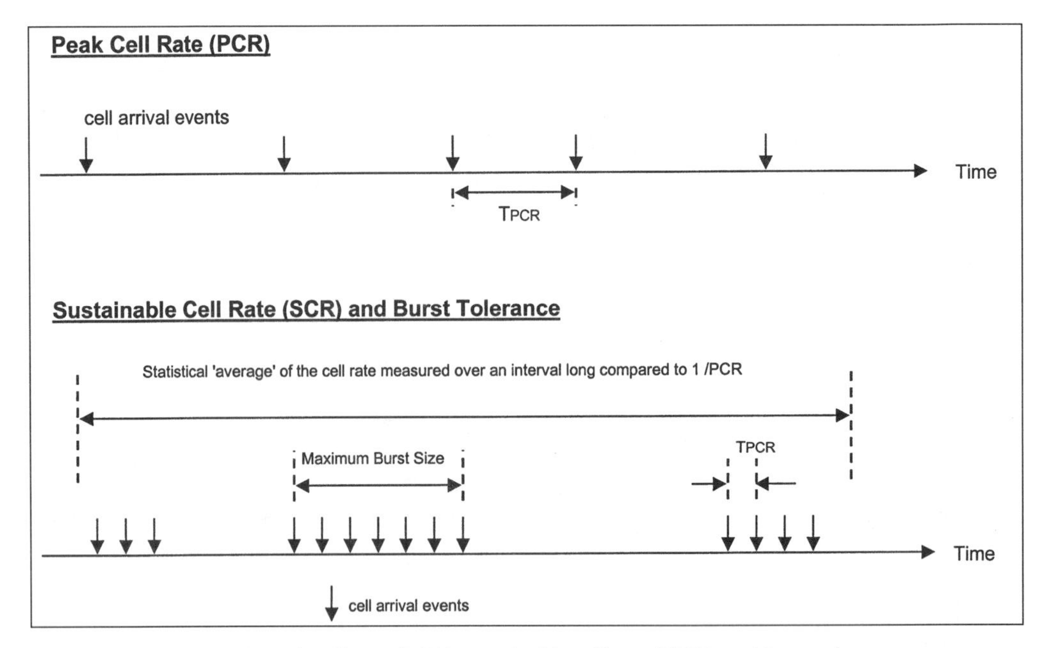

Figure 5-6. Peak cell rate (PCR), sustainable cell rate (SCR), and burst tolerance.

time T_{PCR} between the cell request primitive events at the SAP between the ATM layer and the physical layer:

$$\text{PCR} = 1/T_{\text{PCR}}$$

T_{PCR} is called the peak emission interval (PEI) of the ATM VCC or VPC.

The definition of the cell request primitive event is the request to send an ATM-PDU at the physical layer SAP for any given VCC or VPC. The interarrival time between such cell events physically corresponds to the inverse of the "instantaneous bandwidth" of the cell stream on the connection, resulting from the basic inverse relationship between time and bandwidth. Note that since the cell arrival events may be asynchronous (by definition of ATM), the instantaneous bandwidth corresponding to the cell stream may vary anywhere between some arbitrary minimum or zero, up to some "maximum" value represented by the PCR for the connection. If we are able by some means to monitor the time intervals at which the cells arrive at the physical layer SAP, the minimum interval would correspond to the maximum rate (or bandwidth) of cell flow on an instantaneous basis. Thus, the PCR in the source traffic descriptor specifies the "upper bound" on the rate at which traffic flows on the given ATM VCC or VPC.

The PCR is required to be (explicitly or implicitly) declared in any source traffic descriptor. It allows the CAC function in any NE to allocate sufficient resources (i.e., bandwidth) to meet the QoS objectives requested for the VCC or VPC, depending on the policy adopted for peak bandwidth allocation by the network operator or service provider. As a result of that policy, the CAC function will generally indicate to the UPC/NPC function the equivalent PCR parameter to be enforced by the UPC/NPC function for the VCC or VPC. The "equivalent" PCR parameter that the UPC/NPC will enforce may be equal to the actual PCR value declared in the source traffic descriptor or less, depending on the operator policy for obtaining statistical multiplexing gain in the network.

It should be noted that however it may be used in the network, the PCR constitutes the basic parameter for traffic control in the ATM network. It provides the essential information required by the network operator to allocate the appropriate "maximum bandwidth" to any VCC or VPC and thereby to provision the network resources adequately. In practice, a suitable "safety margin" may also be designed into the resource allocation process to ensure that even if the PCR is exceeded momentarily, the QoS requirements are not necessarily compromised.

5.6 CELL DELAY VARIATION (CDV)

In listing the important QoS performance parameters that may be specified as part of the traffic contract or service contract, it was mentioned that one particular parameter, the cell delay variation (CDV), has a direct and profound impact on traffic control, primarily in respect to the UPC or NPC action. To understand this, it should be noted that the definition of the PCR given above refers essentially to a somewhat idealized situation in which the effect of multiplexing of other connections at the ATM layer as well as the impact of various physical layer functions has not been taken into consideration in determining the actual cell arrival times on a given connection.

The relationship between the CDV and the PCR due to multiplexing of connections at the ATM layer is illustrated schematically in Fig. 5-7; the idealized ATM multiplexing function model described earlier is assumed, incorporating an arbitrary cell scheduling algorithm serving the buffering of cells from a number of ATM VCCs or VPCs with PCR, as shown. It is evident that due to the resulting variable cell fill of the buffers, a cell delay variation component will be added to the actual arrival time instants at the output of the multiplexed cell stream of the ATM multiplexer. Thus, the theoretical cell arrival events corresponding to the PCRs of the VCCs or VPCs shown by the dashed arrows in Fig. 5-7 will be modified by the superimposition of the variable delays caused by the instantaneously variable buffer fills in the multiplexer function due to all the VCCs or VPCs.

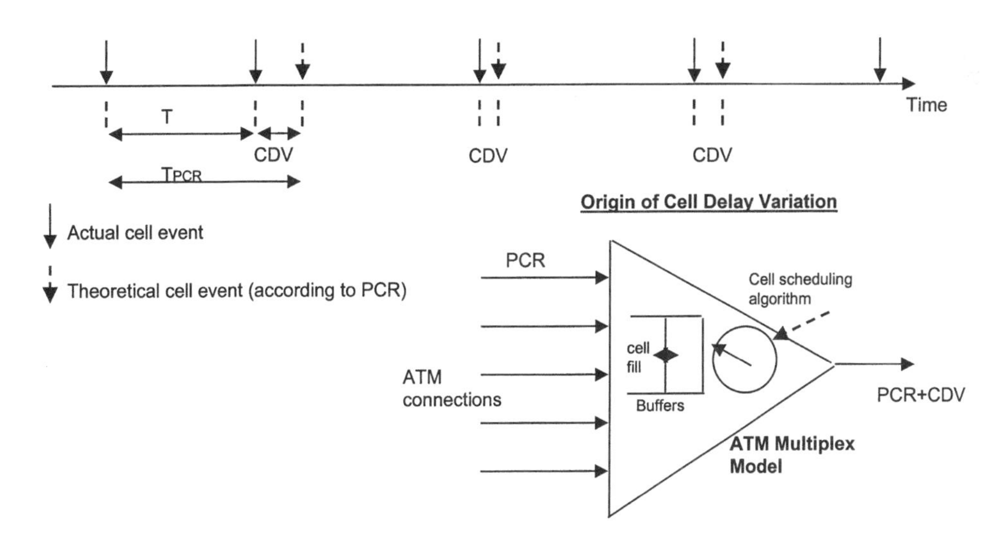

Figure 5-7. Cell delay variation (CDV). Cell multiplexing and other functions at the ATM layer alter the cell flow characteristics by introducing cell delay variation (CDV). CDV may be described by a CDV tolerance (CDVT) parameter specifying "maximum" or effective CDV at an interface. The CDV tolerance parameter, T_{CDV}, at the UNI is required as part of the traffic contract to enable the UPC to effectively enforce the conformance to the traffic contract.

The delay variation resulting from the asynchronous arrival times of cells on all the multiplexed ATM connections essentially manifests itself as a "jittering" of the cell arrival events on each individual connection, as illustrated by the solid arrows indicating the actual arrival times of the cell events on a VCC or VPC. It can be envisaged that the CDV superimposed on the theoretical cell events may result in a reduction of the actual "minimum" interarrival time corresponding to the PCR. In this case, the CDV has effectively resulted in an increase of the actual instantaneous PCR of the VCC or VPC, since, as shown:

$$T + \text{CDV} = T_{\text{PCR}}$$

where T and T_{PCR} are the actual and theoretical minimum interarrival times, respectively, as shown in Fig. 5-7.

This effect of the CDV, potentially resulting in an effective increase in the PCR (i.e., a decrease in the minimum interarrival time), must be taken into account in determining the parameters by which any UPC or NPC function should enforce the cell flow on the connection. It will be clear that unless the effects of CDV are taken into account, the UPC or NPC function responsible for measuring that the minimum cell interarrival events are consistent with the declared PCR in the traffic descriptor may discard cells in error if the minimum interarrival spacing is reduced by the CDV.

In order to take into account the effects of CDV due to cell multiplexing at the ATM layer, as well as other cell delay mechanisms due to the physical layer functions, such as insertion of OAM overhead, a corresponding traffic parameter called the cell delay variation tolerance (CDVT) was introduced. The CDVT parameter is intended to provide an upper bound to all combined CDV effects in "jittering" the cell arrival times at any interface. The notion of the CDVT parameter is analogous to the use of the "jitter mask" concept in transmission system technology in that it also provides a maximum effective upper bound to (bit) delay variation effects due, in that case, to bit stuffing mechanisms. For the case of ATM, the CDVT parameter T_{CDV}, is required to be specified as part of the traffic contract (e.g., at the UNI) to enable the UPC function to effectively enforce the conformance of the cell stream on the VCC or VPC to the traffic contract.

The relationship between the CDVT concept and the UPC function may be illustrated as shown in Fig. 5-8 by using the concept of an "equivalent terminal." It should be stressed that

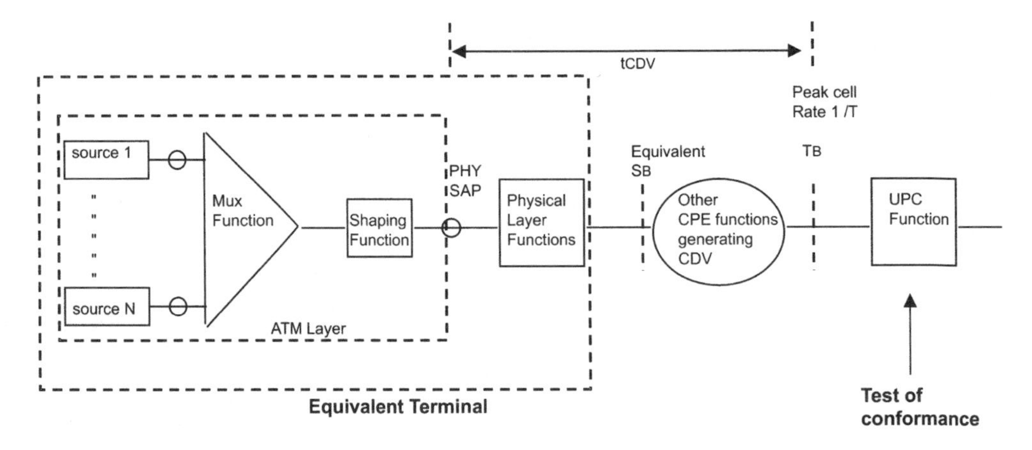

Figure 5-8. Equivalent terminal concept.

the modeling of the equivalent terminal functions in this context is intended only for the purposes of clarifying the relationship of the traffic parameters, and it is not intended to represent the other functions that may be associated with terminal equipment or customer premises equipment (CPE). With this in mind, it is seen that the equivalent terminal models the ATM layer multiplexing of a number of traffic sources, thereby generating cell delay variation superimposed on the individual theoretical cell arrival events in the multiplexer function. In addition, the equivalent terminal may also include a so-called "shaping function," whereby the multiplexed cell stream may be "smoothed" by means of a suitable buffering mechanism to restore the cell flow to the original "idealized" pattern. In effect, the shaping function may attempt to remove the effects of CDV on the multiplexed cell stream.

The use of such a traffic shaping function is considered an option that may or may not be used within an equivalent terminal (or NE) to effectively "smooth" the cell stream toward some desirable traffic pattern. Although attractive conceptually, it should be noted that the need for, and use of, additional traffic shaping as described above has lead to significant discussion as well as controversy, particularly with respect to its use within an ATM network (as opposed to in terminal equipment). In this context, it should also be pointed out that there is also ambiguity in the terminology between traffic shaping and the normal scheduling of the cell emission events to comply with the declared source traffic descriptor, as required by the concept of a traffic contract across any given interface (UNI or NNI).

To avoid this ambiguity and the resulting confusion, we will here define traffic shaping to refer to the (optional) function of smoothing of any cell stream over and above the normal scheduling of cell emission events to comply with any given traffic descriptor that is part of a traffic contract at an interface. In other words, the scheduling of the cell emission intervals is a requirement necessary to meet the constraints of the traffic contract, whereas traffic shaping refers to any additional (and optional) modification of the cell flow to further smooth the cell pattern for any desired purpose. Such a purpose may be, for example, to neutralize the effects of CDV, or to optimize throughput on a transmission path.

With this clarification in mind, it may be noted that some simulation studies have indicated that traffic shaping may result in enhanced utilization of the transmission path payload, as well as lower cell loss due to UPC or NPC action, although with the expense of overall transfer delay. These results essentially confirm the intuitive notion that for a given level of utilization and buffer size, there is a direct trade-off between cell loss probability and cell transfer delay due to buffering. As will be seen later, this fundamental relationship is the basis of using flow control mechanisms to reduce cell loss probability for congestion control. From this perspective, it will also be recognized that traffic shaping, in the sense described here, implies the need for additional buffering of the cell stream coupled with reemission to allow for the desired smoothing of the traffic flow. For highly "bursty" traffic flows with large burst sizes (in cells), this may imply the provision of large buffers at the points where traffic shaping is desired.

These considerations may indicate some of the reasons underlying the initial controversy over the relative merits and disadvantages of deploying traffic shaping in ATM networks. Fortunately, this discussion has to some extent abated due to the development of ATCs based on other flow control mechanisms, which essentially achieve the same ends as traffic shaping. Nonetheless, it is useful to bear in mind the distinction between the required traffic scheduling function necessary to comply with a traffic contract to be enforced at any given interface and the possible use of additional traffic shaping to modify cell flows to improve the overall network performance or throughput for any given application. In a sense, it will be recognized that since traffic shaping essentially implies a need for buffering, and this function is in any case an essential element of any ATM NE and therefore distributed throughout the network, the question of where and when to use traffic shaping has no clear answer. Thus, a network operator may choose to "combine" deliberate traffic shaping with the inherent buffering within any given NE, or may equivalently ensure that the traffic contracts at all specified interfaces are enforced by the UPC or NPC functions in a way suitable to optimize the network resources for the required QoS levels.

Returning to the description of the equivalent terminal concept shown in Fig. 5-8, it may be noted that the CDVT parameter is also intended to include the effects of variable cell delays caused by any physical layer functions within the customer premises equipment (or network) before the UNI defined at the T_B reference point. Apart from the inclusion of any OAM overhead, which may also lead to CDV, it must be stated that other mechanisms that may contribute to the overall CDV are not clearly defined, but may include the effects of generic flow control (GFC), for example, if this is used. In addition, the CDV generated within the equivalent terminal depends on the number of multiplexed VCCs or VPCs as well as the individual source traffic characteristics to a large extent, as may be recognized from the underlying cause of CDV described in Fig. 5-7. Consequently, it is often difficult to quantitatively characterize actual CDV performance parameters in pragmatic deployments. Hence, the CDVT parameter may be viewed as a "catchall" intended to serve as a CPE specific traffic characteristic that allows the UPC function to perform the "conformance test" to the specified traffic contract at the UNI, as shown in Fig. 5-3.

5.7 SUSTAINABLE CELL RATE (SCR) AND BURST TOLERANCE

It was noted above that the peak cell rate (PCR) traffic parameter constituted an essential element of information in the traffic descriptor used in the traffic contract for any ATM connection. By appropriate processing of this information, which may be transported either in the signaling messages or via the management interfaces, the CAC function allows a network to allocate sufficient resources to support a given QoS objective. It is entirely possible to engineer and implement a perfectly viable ATM network to support a range of applications based solely on the use of PCR for traffic characterization. However, it was recognized very early in the development of ATM that for traffic sources, which may emit cells at highly variable rates or as "bursts" of cells interspersed with idle periods, as shown in Fig. 5-6, for example, traffic contracts based solely on PCRs would result in somewhat inefficient use of the network resources, since the resources allocated to any given VCC or VPC would "overestimate" the bandwidth for highly bursty VBR sources. In other words, the network may not be able to share the bandwidth available during idle periods for statistical multiplexing gain, since it has no (quantitative) measure of the extent of the idle periods on the VCC or VPC.

Consequently, to utilize more fully the advantages of statistical multiplexing for VBR and bursty sources, it is clear that some measure of an "average" cell rate as well as a measure of the "burstiness" would need to be specified as part of the traffic contract to enable a network to gauge the extent of statistical multiplexing gain possible for a given level of cell loss. Although this concept may be intuitively apparent qualitatively by inspection of a typical "on–off" traffic pattern depicted in Fig. 5-6, the problem of specifying suitable precise traffic parameters to characterize such complex and varied traffic patterns resulted in extensive discussion and controversy in the standards bodies. It soon became clear that a simple average cell rate measure would be difficult to specify accurately, since such a parameter would vary considerably over time and hence be difficult to enforce. It would also be highly dependent on the type of traffic source (or application) as well. Thus, it seemed preferable to define some form of "worst case" or maximum measure of variability, and this measure is called the sustainable cell rate (SCR). Essentially, the sustainable cell rate is a measure of the maximum average cell rate for a variable bit rate (VBR) traffic source (or an equivalent terminal). The formal definition of the SCR traffic parameter has been given in terms of a mathematical algorithm for conformance testing, called the generic cell rate algorithm (GCRA), which is described in detail below. However, it is useful to have in mind a more physical understanding of the meaning of SCR and its relationship to typical data application traffic sources that may exhibit the so-called "on–off" behavior illustrated in Fig. 5-6.

From Fig. 5-6 it can be seen that the SCR parameter may be derived by a statistical "average" of the cell rate measured over an interval that is long compared to the inverse of the PCR, i.e., 1/PCR. If cells are emitted in bursts as shown, the "worst case" or a maximum value of this average would occur when bursts consisting of the maximum number of cells occurred "back-to-back" in two (or more) consecutive measurement intervals. It may be intuitively envisioned that such a traffic pattern, with maximum bursts of cells arriving on either side of a measurement interval boundary, would result in a measure approximating a maximum value of the average rate. This value is called the SCR traffic parameter.

It was noted earlier that a traffic parameter characterizing the "burstiness" of typical VBR traffic sources is also an important element for maintaining QoS when utilizing statistical multiplexing. It may be intuitively apparent that if the available buffer space in the NE is insufficient to handle, say, large "bursts" of cells arriving at statistically distributed intervals from numerous VCCs or VPCs, cells may be lost due to buffer overflow (congestion) and the QoS of the VCCs or VPCs may be degraded. Hence, a traffic parameter characterizing the "maximum" burst size that may be emitted by a traffic source constitutes important information in a traffic contract to enable adequate allocation of resources (e.g., in this case, buffer size) and the dimensioning of the NE buffer capability. As shown in Fig. 5-6, the traffic parameter characterizing the "maximum" burst size is called the intrinsic burst tolerance (IBT) by the ITU-T in Recommendation I.371 [5.7].

It should be noted that the same parameter is also termed the burst tolerance (BT) in the ATM Forum specifications for traffic management [5.8]. The difference is essentially terminological, since it will be noted that the same mathematical algorithm (based on the GCRA described below) is used in both the ITU-T Recommendations and the ATM Forum specifications to characterize the burstiness of variable bit rate traffic sources. It should also be noted that during a burst, the cell arrival intervals should not be less than the peak emission interval defined earlier in relation to the PCR. Thus, for the typical "on–off" traffic source depicted in Fig. 5-6, cells may be emitted in bursts of variable size up to the maximum value given by the IBT traffic parameter, with an interval between each cell arrival event within a burst less than or equal to the PEI. Between these cell bursts, there are periods with no cell arrival events. The statistical distribution of these periods may be characterized by the SCR traffic parameter, which serves as an upper bound to the average cell rate.

When the SCR and IBT parameters can be specified in the traffic contract for the VCC or VPC, the network operator or service provider may use the information to allocate resources corresponding to an average rate for a VBR service, thereby obtaining statistical gain over an allocation based solely on the PCR values. In practice, this implies that for a given available bandwidth that may be shared, a larger number of VBR VCCs or VPCs may be accommodated by an allocation based on the average (i.e., SCR) rather than the peak (i.e., PCR) cell rates. In calculating such an allocation based on SCR, it should also be noted that the SCR may be subject to CDV due to multiplexing as well as the other effects described above. Consequently, an associated CDVT traffic parameter will need to be factored in to take account of the maximum CDV experienced by the VBR VCCs or VPCs across any given interface (UNI or NNI). It is generally assumed that the CDVT values for a VBR VCC or VPC, characterized by a given SCR and IBT, are same as the CDVT parameter for a CBR connection for the same network scenario and level of multiplexing.

It may also be recognized that since the SCR and IBT traffic parameters included in a traffic contract correspond in a sense to "worst case" or maximum values, resource allocations based on these parameters also overestimate (or overprovision) the network resources to some extent. The network operator may choose to allocate even less resources than called for by the SCR and IBT parameters to attempt to obtain even more statistical gain. However, the trade-off in this case (as for PCR-based allocation) is an increased probability of cell loss or degraded QoS due to either buffer overflow, instantaneous contention for bandwidth, or both. As before, empirical experience of actual traffic behavior patterns may be essential in arriving at the optimum trade-off between an acceptable level of statistical gain and QoS degradation and network performance. To a large extent, and despite the availability of highly sophisticated traffic engineering simulation tools, this remains the key challenge for network traffic engineering.

5.8 THE GENERIC CELL RATE ALGORITHM (GCRA) AND THE "LEAKY BUCKET" ALGORITHM

In the somewhat descriptive definition above of the SCR and IBT traffic parameters, it was noted that their actual formal definition is given in terms of a mathematical algorithm termed the generic cell rate algorithm (GCRA). The term "generic" is intended in the sense that the GCRA may be used for any suitably defined traffic parameter. The intent of the GCRA is to provide an abstract and common methodology to allow for a test for conformance of a cell flow to any given traffic parameter. Clearly, such a formal test for conformance to a traffic parameter such as PCR or SCR needs to be independent of any specific implementation in a NE or terminal equipment. Thus, the GCRA mechanism is not intended to be implemented in order to ensure that any given traffic parameter described by it complies with the values declared in the traffic contract. Rather, it is that any implementation that monitors the cell flow for compliance should yield the same result as would the GCRA. In this sense, the GCRA should be viewed as a "reference" algorithm, against which any physical implementation in terms of counters and timers may be evaluated.

In searching for a suitable GCRA mechanism, it was recognized that the operation of a conformance test on the traffic flow can be visualized (somewhat picturesquely) as analogous to the action of a "leaky bucket," as depicted in Fig. 5-9. In this visualization, the cells are considered to be conforming to any given traffic parameter represented by the leaky bucket if the bucket does not overflow on cell arrival. It should be noted that the leaky bucket itself is in essence a simple counter of a certain size. The size is generically called the "limit parameter," which corresponds to the bucket size (capacity). The counter is decremented at a certain rate, corresponding to the "drain rate" of the leaky bucket. Typically, the drain rate may be 1 per unit of time.

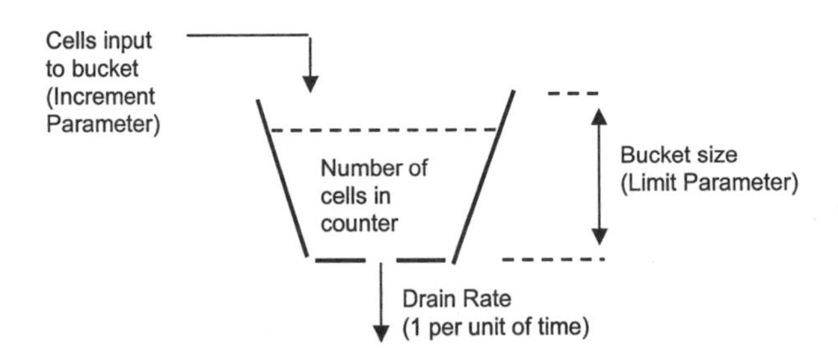

Figure 5-9. The "leaky bucket" algorithm.

The rate at which cells are added to the "bucket" (i.e., the counter) is generically called the "increment parameter." Thus, the operation of the leaky bucket counter may be characterized in terms of the increment and limit parameters for a given drain (decrement) rate. By selecting these leaky bucket parameters to correspond to characteristics of the cell flow described by a traffic parameter such as PCR or SCR, the leaky bucket operation may be used to "test" the conformance of the cell flow to the traffic parameter. The test criterion is taken as bucket overflow. It should be noted that each traffic parameter requires one leaky bucket representation, so that a traffic descriptor comprising more than one traffic parameter requires an equivalent number of "leaky bucket" counters for conformance testing.

The operation of the leaky bucket algorithm may also be equivalently but inversely described in terms of a credit-based mechanism. In this visualization, the cell arrivals consume credits at a certain rate (corresponding to the increment parameter of the leaky bucket). The credit (i.e., the counter values) are supplied at a given rate analogous to the drain rate of the leaky bucket. If a cell arrival event corresponds to all remaining credits being used up (corresponding to the bucket overflow situation), then the cell is considered to be nonconforming to the represented traffic parameter. Both the leaky bucket or the credit-based algorithms may be viewed as useful conceptual tools for visualizing the conformance of traffic parameters describing a cell stream, and it will be noted that in either case, the use of counters and timing mechanisms for cell arrival events are implied for the specific ATM VCC or VPC.

The generic cell rate algorithm (GCRA) is specified in Recommendation I.371 [5.7], as well as in the ATM Forum UNI v.3.1 [5.10] or v.4.0 [5.8], as a reference algorithm for the test of conformance of any traffic descriptor. It is essentially a mathematical (i.e., formal) representa-

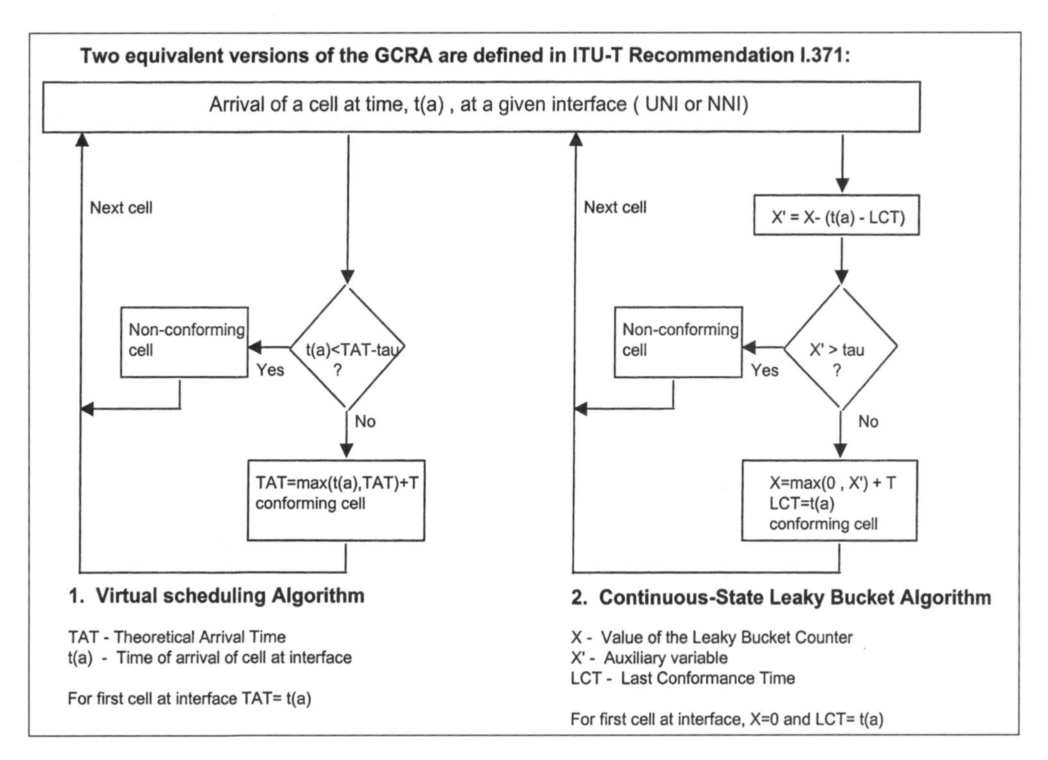

Figure 5-10. The generic cell rate algorithm (GCRA). The GCRA depends on only two parameters: an emission interval, T, and tolerance "tau" (in units of time). The upperbound of X (the capacity of the leaky bucket) is T + tau.

tion of such a leaky bucket mechanism. Since the conformance of any given traffic descriptor may be equivalently described either in terms of leaky bucket counter values or in terms of cell arrival events, two equivalent representations of the GCRA have been given in the ATM standards, as shown in Fig. 5-10. The two equivalent representations are termed

1. The virtual scheduling algorithm (VSA)
2. The continuous-state leaky bucket algorithm

The two algorithms are equivalent in the sense that for a given traffic stream on a VCC or VPC, both descriptions determine the same cells to be conforming to the corresponding traffic parameter. It will also be noted that the algorithms are described using the notation of the system description language (SDL) for representing a procedure or flow diagram in abstract terms.

Considering first the virtual scheduling algorithm (VSA), Fig. 5-10 indicates that it essentially compares the arrival event of a cell on any VCC or VPC at a given interface (UNI or NNI) to a theoretical arrival time (TAT), which is the nominal arrival time corresponding to the cell rate when the source is active. If the actual arrival time t_a of a cell on the connection is not earlier than the TAT and to a "tolerance" value denoted by τ (tau) then the cell is considered to be conforming. At the arrival time of the first cell $t_a(1)$, the VSA initializes the theoretical arrival time to TAT = $t_a(1)$. For every subsequent cell, the VSA compares the arrival time to TAT – τ, using the current value of TAT. If the arrival time of the nth cell, $t_a(n)$ is greater than or equal to TAT – τ, the cell is conforming and the value of TAT is increased by the increment T, where T is the inverse of the cell rate. If the arrival time of the nth cell is earlier than the current value of TAT – τ, the cell is nonconforming and TAT is left unchanged. The algorithm repeats for every consecutive cell arrival on the connection. As described earlier, the continuous-state leaky bucket algorithm is analogous to a finite capacity "bucket" whose capacity is denoted by $(T + \tau)$, constituting the upper bound of a counter. The contents of the bucket drain out at a continuous rate of 1 unit of content per unit of time (the drain rate). The bucket content is also increased by the increment T for each conforming cell, where T is the emission interval. The parameters X or X' denote the content of the bucket (i.e., the values of the leaky bucket counter), and LCT denotes the last conformance time. At initialization, corresponding to the arrival time of the first cell $t_a(1)$, the counter value $X = 0$, and LCT = $t_a(1)$. For subsequent cell arrival times $t_a(n)$, the bucket content is set to a provisional value X' which equals the content after the arrival of the last conforming cell (LCT) X minus the amount the bucket has drained:

$$X' = X - [t_a(n) - \text{LCT}]$$

If X' is less than or equal to the limit parameter τ, the cell is considered to be conforming. The bucket content is then set to $X' + T$, or to zero if X' is negative, and LCT = $t_a(n)$. However, if X' is greater than τ, the cell is considered nonconforming and the values of X and LCT are left unchanged. The algorithm then repeats for every consecutive cell arrival on the given VCC or VPC.

Either version of the GCRA may be used as a reference algorithm in designing a mechanism to test for conformance to a traffic contract and enforce it as part of the UPC or NPC function at the UNI or NNI, respectively. Although described here in terms of a cell stream, it should also be noted that the leaky bucket algorithm may be extended to apply to virtually any packet transfer mode technology. Thus, analogous conformance test algorithms have also been adopted for frame relay (FR) services and, more recently, for IP packet flows, as will be described later. In principle, either of the equivalent versions of the GCRA may be extended to provide a reference mechanism for conformance, provided the two parameters characterizing the algorithm can be determined for the traffic stream. These parameters are T, the emission interval, and τ, the tolerance (or limit) parameter.

5.9 ATM TRANSFER CAPABILITIES (ATC)

The concept of an ATM transfer capability (ATC) has already been described earlier in relation to the traffic contract, and the main currently defined ATCs were listed there to illustrate the concept. We now describe these ATCs in further detail, but is should be recognized that work in this area is still progressing rapidly, with new ATCs under active discussion. Moreover, as noted earlier, the ATC framework has yet to be incorporated into the ATM Forum's TM specifications, so that some differences in terminology as well as approach still bedevil the discussion on ATM traffic control in the general literature.

The notion of ATCs arose as a result of the recognition that there are many ways in which statistical multiplexing could be performed for VBR applications based on use of the traffic descriptors, resource management (RM) cells, and other functions of the ATM layer. To avoid confusion, it seemed desirable to attempt to categorize the various ATM statistical multiplexing protocols and combinations of traffic parameters and QoS commitments together with a traffic multiplexing mechanism at the ATM Layer. The clear advantage of such an approach is that a relatively small number of ATCs could be defined in a way that would allow the support of a much wider range of applications or services provided by the ATM layer. In other words, a number of B-ISDN services may use a given ATC. As will be seen later, this notion is analogous to the use of a limited number of AAL types to support a very large range of potential B-ISDN services or higher-layer applications by utilizing common requirements of a range of higher-layer services.

It is also clear that an ATC needs to be specified as part of the traffic (or service) contract during the connection setup phase or when provisioning. Typically, the same ATC will be used for both directions of a bidirectional ATM VCC or VPC. If a service-type methodology is used to implicitly communicate the traffic contract information, the ATC is assumed to be also implicit in the definition of that service type, although to date use of service type in connection setup has not been well defined, as noted earlier. It will also be recognized that not all ATM networks will provide all the possible ATCs defined here, and that some ATCs may be entirely unsuitable for some services. Consequently, in practice, a user may have a limited set of choices on which to call upon. In addition, it should be borne in mind that there are some limitations of the ATC approach. Thus, it is not yet clear how to multiplex VCCs with different ATCs into a single VPC, nor how to interwork between the various ATCs. The latter problem may be avoided by requiring that all VCCs and VPCs have the same ATC at all interfaces, so that different networks supporting a different set of ATCs do not need to interwork.

A key aspect of any ATC is the conformance definition applicable to all the parameters comprising the ATC. The underlying assumption is that if the user of the ATM layer service complies with the conformance definition in every respect, the QoS commitments negotiated as part of the traffic (or service) contract will be maintained by the network or service provider. Even in the event that some cells on the VCC or VPC are nonconforming to the conformance definition, and the network provider considers the connection to be thereby noncompliant to the traffic contract, the network provider may still choose to maintain the QoS commitments and not penalize the user by discarding cells, for example. Such a choice is essentially a matter of network operator policy and sound business sense rather than one of technology or traffic control.

5.10 DETERMINISTIC BIT RATE (DBR) TRANSFER CAPABILITY

The DBR ATC was based on the so-called constant bit rate (CBR) category defined earlier by the ITU-T, and this latter terminology continues to be used in the ATM Forum specifications for traffic management (TM) [5.8]. The DBR ATC constitutes the basic and essentially simplest ATC, and is considered as a "default" ATC in the sense that any ATM network should in principle be able to provide such a service. In addition, it should be noted that in principle any higher-layer application or service should be able to use the DBR ATC, even though this may be a somewhat inefficient use of network resources in practice if the cell flow is highly bursty.

The DBR ATC is intended primarily for services or applications that require a fixed amount of bandwidth that is available continuously during the connection lifetime. This bandwidth is determined by the peak cell rate (PCR), which must be specified together with its associated cell delay variation (CDV) tolerance value. Thus, with the DBR ATC, it is assumed that the traffic source may emit cells at the PCR for any length of time and the QoS commitments will be met. However, the source may also emit cells at rates below the PCR for periods of time, or may even stop cell emission for an interval. Examples of the type of services that may require the characteristics of DBR ATC are voice, circuit emulation service (CES), and CBR video signals.

The DBR ATC is applicable to either VCCs or VPCs, and the CLP function is not used, so that the VCC or VPC is associated with one QoS objective with respect to cell loss probability, irrespective of whether the CLP = 0 or 1. This implies that the tagging option on CLP is not used. Resource management (RM) cells are also not required in DBR, although RM cells may be present on a DBR VCC or VPC and are treated as part of the user data cell flow. For the DBR ATC, the traffic contract for connection setup (or by subscription, for the case of PVCs) must include one of the following three alternative traffic descriptors:

1. One PCR for the aggregate of the user data cells and the user OAM cells, with the associated peak emission interval.
2. Two separate PCRs. One PCR for the user data cells with the associated PEI, and a separate PCR for the end-to-end OAM cells generated by the user, with the associated PEI.
3. Service type, which implicitly refers to the relevant traffic descriptor required

In all cases, the DBR conformance definition also requires that the CDV tolerance at the interface (UNI or NNI) is also specified to allow for enforcement of the cell flow by the UPC or NPC function at that interface. The most common traffic descriptor for DBR is alternative 1 above, which is also the simplest, since no distinction is made from a traffic control perspective between any user-generated OAM cells and normal user data cells in the aggregate cell stream on any VCC or VPC. The use of the OAM cell functions will be discussed in detail later, but it should be noted that from the traffic management perspective, these cells are subject to the same QoS requirements and resource allocation as any other cells on the VCC or VPC. Consequently, bandwidth will need to be reserved to accommodate any potential OAM cell flow anticipated on the VCC or VPC in addition to the normal user cells.

For the case of alternative 2, a separate PCR for the OAM cell flow is specified to allow for a more precise indication to the network of how much bandwidth to allocate solely for the OAM cell flow purposes. Although this may seem desirable at first glance, the use of this alternative has been controversial, since it implies two separate tests for conformance for this additional information to be of any practical value. Thus, a dedicated GCRA for OAM cells would need to be provided, and since, in general, OAM cells are expected to constitute only a very small percentage of the total traffic under normal conditions, this additional complexity is generally viewed as somewhat wasteful. Nonetheless, some network operators have indicated a preference for basing

DBR ATC on alternative 2, since it provides more precise information on OAM requirements. As noted earlier, the use of service type to implicitly encapsulate the required traffic parameters and QoS requirements holds significant potential for the future as ATM based services and applications become more standardized. However, at this stage alternative 3 for DBR ATC is not sufficiently well defined to warrant widespread usage.

Considering the more commonly used case in which one PCR for the aggregate flow (user cell + OAM cells) and its associated CDV tolerance value are specified in the DBR ATC traffic descriptor, the conformance definition requires essentially that the maximum cell rate should not exceed the PCR adjusted by the CDVT value in accordance with the GCRA procedure described earlier. The conformance test is performed at the UNI by the UPC function, and optionally at the NNI by the NPC function on each DBR VCC or VPC involved. Nonconforming cells are subject to discard, although, as noted earlier, a network or service provider may choose to maintain QoS commitment even for an occasionally noncompliant connection, depending on policy or other factors. However, this aspect of traffic management policy may apply to any ATC, and is not restricted just to DBR-based services.

Although the DBR ATC strictly implies peak bandwidth allocation to any given VCC or VPC (plus the adjustment for CDVT and an additional "safety margin" to allow for measurement uncertainties, for example), it may be possible for the network or service provider to employ empirical or historical knowledge of the traffic characteristics to choose to allocate somewhat less than the PCR to any given VCC or VPC. This may result in a degree of statistical multiplexing gain for the service provider, with the inevitable trade-off for potentially degraded cell loss probability (i.e., higher CLR). This aspect is also largely dependent on network or service provider policy.

5.11 STATISTICAL BIT RATE (SBR) TRANSFER CAPABILITY

In order to engineer the ATM network to obtain the advantages of statistical multiplexing gain, it is clear that additional traffic characterization information to the PCR needs to be provided as part of the traffic (or service) contract. For this reason, the statistical bit rate (SBR) transfer capability has been defined, which specifies the traffic parameters sustainable cell rate (SCR) and intrinsic burst tolerance (IBT) in addition to the PCR to characterize the cell flow behavior on the connection in greater detail. The additional parameters may be used by the network operator to provide a more quantitative measure for the statistical multiplexing of VCCs or VPCs for a given set of QoS objectives (e.g., the CLR). The SBR ATC is intended but not restricted to applications generating "bursty" traffic for which a maximum average cell rate and a maximum burst size can somehow be defined. For example, the traffic pattern for a VBR application may be "shaped" (i.e., buffered and scheduled) to not exceed a maximum burst size of 100 cells and a worst case average of one-tenth of the PCR. Given such information as part of a traffic contract, the network provider may choose to allocate up to ten times the number of such VCCs or VPCs than would be possible based on PCR alone (i.e., a statistical gain of up to 10).

In addition, the cell loss ratio (CLR) may be engineered to a given level by dimensioning the network element buffer sizes to be able to accommodate many bursts of cells of length 100 arriving at any given instance with some assumed probability distribution. It will be intuitively obvious that the probability of cell loss will decrease as the buffer size increases to many times the burst size, although the trade-off is inevitably additional delay and delay variation (CDV). However, it is typically envisaged that for most applications likely to use the SBR ATC, such as VBR data applications, the cell delay performance is not critical. Nonetheless, the SBR ATC may or may not be used for applications that require bounded delay performance objectives. It is primarily for this reason that the ATM Forum's TM specifications have maintained a distinction between so-called "real-time VBR" and nonreal time VBR (n-r VBR) for the VBR traffic category, which corresponds to the SBR ATC, as noted earlier. The specific delay criteria for these two categories have not been specified to date in any quantitative way by the ATM Forum, but are clearly highly application-specific.

In order to negotiate the SBR ATC, the traffic source (or user) must describe (either in the signaling message or as part of a service contract) in the traffic contract one of the following traffic descriptors:

1. The PCR and the SCR with associated IBT
2. The PCR and the service type
3. The service type (with implicit traffic parameters)

In all cases, it is assumed in SBR that the users' OAM cell flows are included in the derivation of the declared traffic parameters. Consequently, both user OAM cells and data cells are aggregated for purposes of traffic enforcement in SBR ATC (or equivalently for the VBR category in the ATM Forum terminology). As noted earlier, the use of the "service type" as a traffic descriptor involves the implicit knowledge of the traffic parameters required (SBR and IBT) based on preassigned values negotiated at connection setup. To date, use of service type traffic descriptor is essentially a choice of individual service providers, and no generally agreed upon methodology has been defined. When the first option involving explicit negotiation of PCR, SCR, and IBT is selected, three separate types of the SBR ATC will have been identified, depending on the options and the configuration chosen for enforcement of the parameters. The three types of SBR, denoted SBR 1, SBR 2, and SBR 3, may be defined as follows:

1. SBR 1. For this case, the PCR is specified for the total user cell stream on the VCC or VPC for both CLP = 0 and CLP = 1, and the SCR and IBT are also specified for the total CLP = 0 + CLP = 1 cell stream. The cell stream for both CLP = 0 and CLP = 1 is usually denoted by the term "CLP = 0 + 1" cell stream, which implies that both the high-priority (CLP = 0) and the low-priority (CLP = 1) cells are considered in total.
2. SBR 2. For this case, the PCR is specified for the total user cell stream on the VCC or VPC for the total CLP = 0 + 1 cell flow, and the SCR and IBT traffic parameters are only specified for the CLP = 0 (i.e., high-priority) cell stream. For this case, the use of the tagging option, described in further detail below, is not used.
3. SBR 3. For this case, the PCR is specified for the total user cell stream on the VCC or VPC for the CLP = 0 + 1 cell flow, and the SCR and IBT traffic parameters are only specified for the CLP = 0 (i.e., high-priority) cell stream. Thus far, the SBR 3 configuration is identical to the SBR 2 capability described above. However, the difference is that the SBR 3 case allows the use of the tagging option for CLP = 0 cells that are in excess of the negotiated SCR/IBT traffic parameters in the traffic contract. The tagging option essentially allows the network to "tag" the CLP = 0 (high-priority) cells in excess of the contract by changing the CLP value to CLP = 1, i.e., marking them as low-priority cells that may be selectively discarded in the event of congestion at any NE.

The concept of selective cell discard and cell "tagging" has often been confused and misrepresented, and, consequently, has generated considerable controversy in the earlier literature on ATM traffic control. However, selective cell discard by a congested ATM NE is in essence no different from earlier use of low-priority packet discard used in packet mode technologies to minimize the impact of congestion. As noted earlier, for the case of ATM, use of the CLP func-

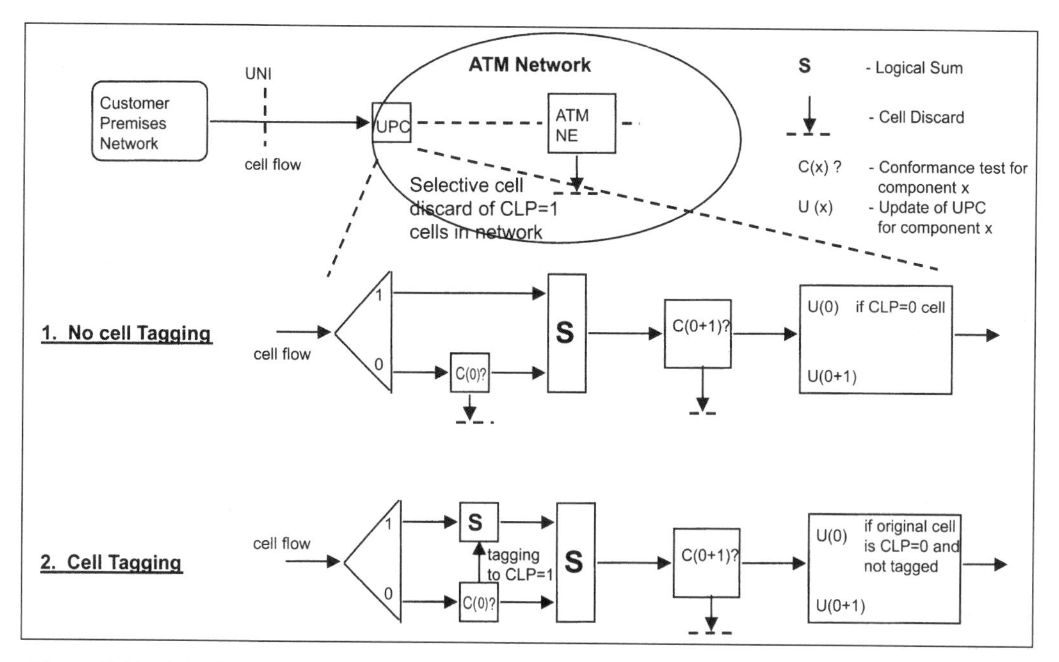

Figure 5-11. Selective cell discard and tagging. Use of CLP = 0 and CLP = 1 cells in a connection enables two levels of cell loss QoS (if part of traffic contract). Network may "tag" nonconforming CLP = 0 cells (e.g., excess cells) by converting them to CLP = 1 cells. The CLP = 1 cells may be selectively discarded before CLP = 0 cells (in event of congestion), i.e., CLP = 1 cells have higher cell loss ratio (CLR).

tion in the ATM cell header allows an ATM NE to distinguish high-priority (CLP = 0) cells from low-priority (CLP = 1) cells within any VCC or VPC cell stream. Consequently, if the selective cell discard function is provisioned for the NE, it may preferentially discard the CLP = 1 cells when congestion is experienced (i.e., a congestion threshold is crossed in any of the cell buffers of the NE, for example). In this way, it is anticipated that the effects of incipient congestion on the (relatively) high-priority CLP = 0 cells may be minimized.

The concept of cell tagging builds on selective cell discard by allowing a network element performing the UPC function to change the CLP value of a high-priority (CLP = 0) cell to a low-priority (CLP = 1) cell if and only if those cells have exceeded the negotiated traffic contract for CLP = 0 cells, as determined by the UPC function. This concept is illustrated in Fig. 5-11. At the UNI, the NE incorporating the UPC function enforces the traffic contract on any given VCC or VPC. For the case where no tagging of CLP = 0 cells is allowed, those CLP = 0 cells which are in excess of the bandwidth allowed by the negotiated traffic contract will be discarded by the UPC function. However, for the case where tagging is enabled, the NE may tag the nonconforming CLP = 0 cells (i.e., only the CLP = 0 cells in excess of the allowed bandwidth) by converting them into low-priority CLP = 1 cells. These tagged cells may then be transported through the ATM network just as any other CLP = 1 cells, provided the requisite congestion threshold is not exceeded in any given node along the connection path.

Tagging of the CLP = 0 cells from a user implies that the ATM network has changed the priority of the user cells from high to low priority. Note that the network is not allowed to change CLP = 1 cells to CLP = 0 cells under any circumstances. Thus cells, once tagged, cannot be changed back to high-priority CLP = 0 cells under any circumstances, even if the subsequent cells do conform with the traffic contract. The tagging action is considered irreversible. This was one of the factors that contributed to the controversy over tagging, since it was considered that modifying the users intended high-priority cells may adversely impact the application in question. However, since tagging is only applicable to the CLP = 0 cells that exceed the allowed rate in the traffic contract (which would be discarded if the tagging function were to be disabled), it would appear preferable for the network to be able to deliver these tagged cells if sufficient capacity were available.

In essence, cell tagging is useful only when there is available capacity in the network for the low-priority (CLP = 1) cells to be transported, due either to provisioning, the statistical nature of most ATM applications, or both. Thus, tagging relies on the reasonable assumption that there is, in general, sufficient residual capacity in the ATM network to allow the occasional "excess" traffic to get through on average, since the network will typically be operating below its peak load. In practice, most packet mode networks, including ATM, will need to be somewhat over-provisioned in terms of capacity to support adequate cell loss and allow for the wide variations in traffic loading over time resulting from the very bursty VBR nature of most traffic sources. Under such circumstances, and assuming the number of tagged cells is small compared to the normally conforming CLP = 0 and CLP = 1 cells (for which sufficient bandwidth has anyway to be allocated, based on the traffic contracts), the tagging option is considered useful. It enables the tagged CLP = 0 cells in excess to be carried with a probability equivalent to the CLP = 1 low-priority cells, whereas in the absence of tagging they would have been discarded by the UPC function.

However, opponents of tagging have pointed out that whereas this may be the case for very small numbers of excess cells, if users were to exceed significantly the traffic contract for the high-priority CLP = 0 cells, the network would rapidly experience severe congestion, resulting in impaired QoS for other connections. This viewpoint is therefore based on a concern that allowing the possibility of tagging would tempt user applications to violate the traffic contract for the CLP = 0 cells in the hope that the network would attempt to deliver these cells as tagged (lower priority) traffic in any case. While this remains a valid concern, and often gives rise to questions regarding the value of not only tagging but also the use of high- and low-priority cells within a virtual connection, the issue is essentially one of individual network operator policy for providing ATM service. The decision whether or not to use tagging may also involve the tariffing mechanisms underlying any given ATM service. For example, if the service is priced on a usage basis, use of the tagging option may be desirable despite the (relatively) small additional cost and complexity involved. However, for tariffing based on "flat-rate" or usage-independent schemes, tagging may not prove to be cost-effective. Such regulatory and tariffing policy considerations are often complex and outside the scope of this book, but need to be factored in when considering the use of traffic control functions such as tagging and selective cell discard.

It should also be noted that there appears to be a regional dimension to the debate on use of tagging and cell priority (CLP). In the ETSI specifications, which are considered mandatory for the participating organizations in the European Union, the use of the tagging and selective cell discard functions is precluded. This essentially reflects the earlier strong objections in the ITU-T of a number of European network operators regarding the value of the tagging option for ATM services. However, most North American network operators and vendors generally support the use of selective cell discard and tagging options, and this is reflected in the ATM Forum specifications as well as the ANSI T1S1 standards for ATM as a choice for the ATM service provider. Since both the ETSI and ATM-F specifications are ostensibly based on the ITU-T ATM standards, this dichotomy provides an interesting example of regional differences of approach in the use of ATM technology. Fortunately for this case, interoperability between ATM connections originating in the different domains need not necessarily be compromised, since it is generally possible to select the requisite option by signaling or provisioning, as will be described further in subsequent sections.

Irrespective of whether tagging is used or not in utilizing statistical bit rate (SBR) transfer capability (or the equivalent VBR traffic categories using the ATM Forum specification terminology), it is important to recognize that this ATC allows for a great deal of flexibility in engineering ATM networks to provide statistical gain at a minimum of additional complexity. From the user side, this additional complexity results from the need to specify and comply with the additional traffic parameters of SCR and IBT in characterizing the traffic source. However, this may not always be possible, since existing applications that may benefit from ATM may not always be readily characterized by means of a maximum average rate such as the SCR. From the perspective of the ATM network, the additional complexity implied by the use of SBR ATC (or VBR) results from the need to enforce the additional traffic parameters SCR and IBT in the UPC/NPC functions to verify compliance with the negotiated traffic contracts. This involves separate GCRA-based enforcers (e.g., leaky bucket counters) for both SCR and PCR parameters (with associated CDVT parameters) to test for conformance to the traffic contract. In practice, this requires two separate enforcers per SBR connection (as opposed to one for the DBR case).

The criteria for conformance to the SBR ATC may be tabulated as follows:

	PCR Enforcement	SCR Enforcement
SBR 1 (VBR 1)	GCRA with $T_{PCR\ 0+1}$ CDVT	GCRA with T_{SCR} IBT + CDVT
SBR 2 (VBR 2)	GCRA with T_{PCR0+1} CDVT	GCRA with $T_{SCR=0}$ IBT + CDVT tagging not allowed
SBR 3 (VBR 3)	GCRA with $T_{PCR=0+1}$ CDVT	GCRA with $T_{SCR=0}$ IBT + CDVT tagging allowed

Definition of the SBR ATCs (or the equivalent VBR1, VBR 2, and VBR 3 traffic categories specified by the ATM Forum TM 4.0 specification [5.8]) was essentially the first and simplest attempt to provide a quantitative means for statistical multiplexing gain in ATM networking. By allocating bandwidth based on the SCR values rather than PCR values, and ensuring sufficient buffering capability in NE to accommodate multiple simultaneous arrivals of many cell bursts, the network could be engineered to support many more VCCs (or VPCs) than if based on simply PCR allocation. This possibility is independent of whether or not the network performs the conformance test on the negotiated traffic parameters at either UPC or NPC, since the conformance enforcement is primarily intended to protect the QoS provided to all the relevant VCCs or VPCs. Thus, a service provider may simply use the SCR parameter as a means to assess "average" utilization of a VCC (or VPC) by a user, and provision connections accordingly. If the service provider can trust the traffic source not to exceed the declared traffic descriptor, then enforcement may not even be necessary at UPC or NPC functions. However, this constitutes an idealized situation and, in practice, prudent ATM service providers may need to ensure that some users do not violate their traffic contracts and thereby jeopardize the QoS provided to compliant traffic sources.

5.12 AVAILABLE BIT RATE (ABR) TRANSFER CAPABILITY

Although the SBR (or VBR) ATM transfer capability described above allows for the engineering of a flexible degree of statistical multiplexing gain in a network, the uncertainties associated with characterizing actual traffic sources by means of traffic parameters such as SCR and IBT soon led to questions regarding its usefulness in the design of typical data networking applications. It was recognized that more efficient utilization of the residual capacity available in the network could be achieved if the ATM VCC (or VPC) could be flow controlled in some way. Consequently a number of mechanisms to provide for flow control at the ATM layer, based on the use of the dedicated resource management (RM) cells defined for that purpose, were proposed and extensively discussed. Initially, the various proposals for flow control using RM cells to provide feedback were studied in the ATM Forum Traffic Management Working Group, where a set of basic requirements were set up to evaluate the proposals.

Broadly, the flow control mechanisms studied could be grouped into two main categories. One category was based on flow controlling the VCC or VPC using cell rate information fed back from NEs along the connection path. This was termed "rate-based" flow control. The other category, termed "credit-based" flow control, relied on the use of so-called credit information fed back to a traffic source that corresponded to the actual buffer space available for the cells to be sent to the NEs along the connection path. In this scheme, cells would only be sent if a corresponding number of "credits", signifying the requisite number of buffer spaces, were available. Each node would feed back the available buffer space for any given VCC (or VPC) as a number of credits encoded, for example, in an RM cell. If no buffer capacity were available, no credits would be returned and no cells would be sent, thereby eliminating potential cell loss due to buffer overflow.

Both the rate-based and credit-based flow control mechanisms were extensively studied (and often hotly debated) by the ATM Forum, and both categories have advantages and disadvantages (in various network scenarios) in terms of traffic performance and relative complexity. However, the rate-based approach eventually gained favor and was adopted, initially by the ATM Forum in the TM specification [5.8], and then also by the ITU-T in a somewhat simplified description in Recommendation I.371 [5.7]. The rate-based flow control approach using RM cells to carry in-band feedback was termed the "available bit rate" (ABR) transfer capability (or service category in the ATM Forum terminology), since it essentially operates by modifying the rate at which cells are generated on a given VCC or VPC in accordance with the "available" resources along the connection path (at any given instance). It is anticipated that if the traffic source can adjust the rate at which it generates cells in accordance with the available resources (bandwidth and buffer capacity) along the connection path, very low cell losses may be maintained even at relatively high utilizations of the transmission path bandwidth (i.e., the physical link bandwidth).

Since the ABR transfer capability implies that the cell transfer characteristics provided by the network may change in the duration of the connection in accordance with available resources, it is clear that the higher-layer applications using it must be able to adapt to the changes in cell rate and delay. Thus, the ABR ATC is not suitable for applications such as voice or constant bit rate video transport, since the cell delay variation (CDV) may be essentially unbounded. ABR is essentially intended for data networking applications, where delay and changes in cell transfer rate are not of critical importance to service quality, but low cell loss is. It may also be recognized that the ABR ATC applies on a per-connection basis, since the RM cells are generated in-band with the user data cells and have the same VPI and VCI values along the connection path. In addition, for the rate-based flow control mechanism to work robustly, the ABR procedures at the traffic source, destination, and intermediate NEs need to be defined and implemented. These procedures are called the source, destination, and switch behaviors in the ATM Forum description of the ABR service category. In the ITU-T description of the ABR ATC, the source, destination, and switch behaviors are not currently specified, but the semantics of the RM cell parameters and conformance definitions at standardized network interfaces are provided, leaving the

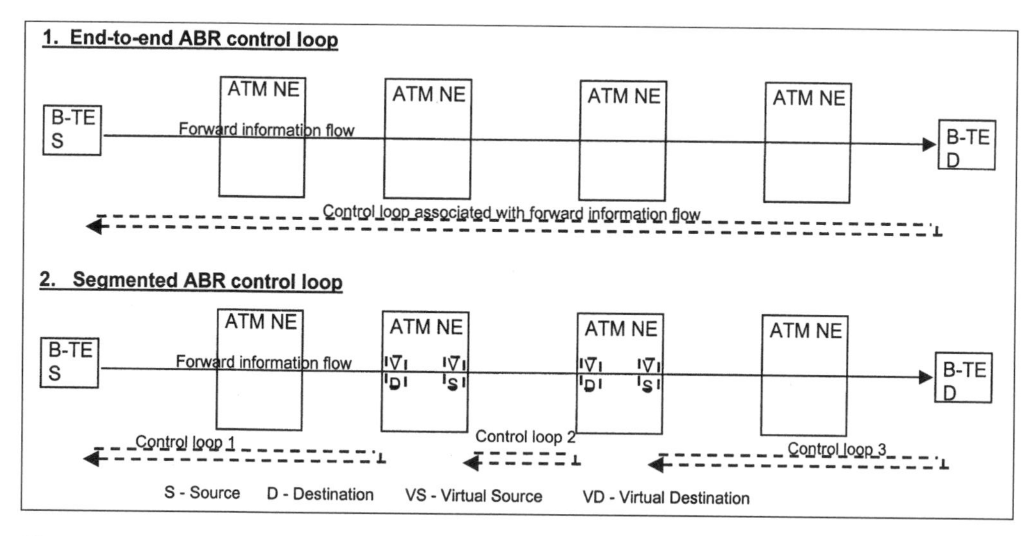

Figure 5-12. ABR flow control model and operation. ABR flow control occurs between source TE and destination TE based on information in RM cells. Any intermediate NE may be configured as a "virtual source" and "virtual destination" to segment the control loops. Source periodically generates RM cells, which indicate current cell rate (CCR) and other parameters. Depending on the available capacity, NE sends back an "allowed cell rate" (ACR) or "explicit rate" (ER), which may vary between minimum cell rate (MCR) and PCR.

behaviors as essentially implementation-specific actions to be performed by the NEs in question.

It is important to distinguish between the generic flow control (GFC) mechanisms described earlier and the RM cell-based flow control mechanism underlying the ABR ATC. It was noted that the GFC was not carried out on a per-connection basis, whereas ABR is intrinsically a per VCC or VPC flow control, since the RM cells carrying the cell rate information are carried along with the user data cells in the individual VCCs or VPCs. The GFC protocol in the ATM cell header uses a relatively simple on–off instruction set to regulate the flow of cells on a given transmission path, and is only applicable at the interfaces at the S_B and T_B reference points. In contrast, ABR uses rate control information in dedicated RM cells across all interfaces to vary the cell rates over a range of values and is consequently substantially more complex procedurally than GFC.

The basic ABR flow control model and operation is shown in Fig. 5-12. To simplify the illustration, only one direction of a (normally) bidirectional ATM VCC (or VPC) is shown and it should be understood that identical considerations apply for the reverse direction of the ATM VCC or VPC. In fact, ABR requires bidirectional connectivity to operate, since the control loop uses the return direction for the feedback information in the RM cells. The source terminal B-TE (S) sends user data cells and forward RM cells to a destination terminal B-TE (D). The destination terminal returns the RM cells as backward RM cells after performing the destination behavior to form the control loop associated with the forward information flow. Similarly, there will be a control loop (not shown) associated with the other information flow direction, so that, in practice, each terminal must incorporate both source and destination behaviors.

Two possible configurations have been identified for ABR, as indicated in Fig. 5-12. For the so-called end-to-end configuration, the control loop extends between the ATM VCC (or VPC) end points as one entity. For the so-called segmented ABR configuration, the control loop may be broken up into an arbitrary number of subloops between selected intermediate Nes, as shown. For the segmented case, it is immediately evident that intermediate ATM NEs will need to act as virtual destination (VD) and virtual sources (VS) in order to provide the control loops between the NEs. This functionality is referred to as "virtual source/virtual destination" (VS/VD), simply in order to distinguish it from the actual traffic source and destination at the connection end points. However, the actual procedures the VS/VD functions need to perform remain essentially the same as those required by the source and destination end points of the VCC or VPC. In addition, there will need to be appropriate "coupling" between the VD and VS functions within the NE along the information flow direction to ensure that the different segments of the ABR control loops operate consistently. This follows from the requirement that if the cell rates in consecutive segments are not matched, cell loss may result in the NEs between the two segmented control loops. This aspect of the segmented ABR configuration has not been generally resolved to date, and is viewed as an implementation-specific requirement within the buffer management system of the NE.

We consider here the description of the single ABR control loop, with the understanding that extension to the VS/VD case implies suitable coupling of the resulting segmented ABR control loops. For the ABR connection establishment, the user will negotiate (either by service contract or on-demand signaling) the maximum required bandwidth for the ABR VCC (or VPC). This parameter may correspond to the PCR for the ABR connection. In addition, a minimum cell rate (MCR) parameter may also be negotiated at connection establishment. The primary intent of the MCR parameter is to enable the network to allocate a minimum usable bandwidth for the ABR connection, and does not preclude that the actual rate at which cells are sent falls below the MCR value negotiated. If a user does not specify a MCR value at connection establishment, the default MCR is taken to be zero. If a nonzero MCR is negotiated, it implies that the network must reserve at least sufficient bandwidth to support that value, which in turn implies that connection admission control (CAC) function will need to be performed for the connection request. For MCR = 0, the CAC criteria need not apply. It should be noted that even if a user negotiates a nonzero MCR value, the instantaneous cell rate may fall below the MCR value or even be zero, but the network will have reserved sufficient resources to support the negotiated MCR for the duration of the connection. The values of the PCR and MCR may also be different for the two directions of the bidirectional ABR connection in general.

For the duration of the ABR connection, the instantaneous cell rate sent by a source may vary between the PCR and the MCR values (or below) according to the rate control information fed back in the RM cells from the network or the destination. The ABR procedures allow the intermediate NEs (or the destination NE) to either:

1. Insert or modify the explicit cell rate information in the forward or backward RM cells as they traverse the NE. This explicit cell rate (ECR) value represents the NEs indication of the cell rate it can support at any given instant on the connection, and may vary depending on the congestion state and traffic load on the NE at that instant. ATM NEs that are able to calculate and indicate the ECR values in RM cells are called explicit rate (ER) ATM switches.
2. Generate a backward RM cell called a backward explicit congestion notification (BECN) cell directly in the reverse direction containing the ECR value calculated by the NE based on its traffic load or congestion state at that instant. It is important to distinguish between a backward RM cell that has been "returned" by an ABR destination (or VD) on receiving a forward RM cell, and a BECN cell that is sent by an intermediate NE independently of the forward RM cells to directly communicate the ECR information to the source at any required instant.
3. Set the explicit forward congestion indication (EFCI) bit in the ATM cell header of any

user data cell, as described earlier when describing the payload type (PT) codepoints, to indirectly inform the destination of the presence of congestion in the NE in question. Note that if the EFCI bit is set to indicate congestion in the forward direction, downstream NEs are not allowed to change the values and the destination terminal will simply retain the EFCI setting in the user data cells it sends in the reverse direction to allow the source to obtain the congestion information as well. It will be recognized that the EFCI procedure provides a simple binary indication of the presence (or absence) of congestion and is independent of the RM cell information and hence of the ABR mechanism. ATM ABR NEs providing EFCI capability as well as ER capability have been called "binary" ATM switches in ATM Forum terminology, although the term "binary" here refers to the ability to use either ER or EFCI procedures or both, and not (somewhat confusingly) to the binary nature of EFCI function.

For the ABR service, it is assumed that NEs allocate and share the available bandwidth between the ABR connections in accordance with an allocation policy that depends not only on user bandwidth requests but also on network operator policy for determining fair sharing of the available bandwidth left after the MCR requirements are met. The criteria for "fairness" have been extensively discussed, and although some general guidelines have been provided for information purposes by the ATM Forum's TM specification v.4.0 [5.8], there is no clear mechanism defined in the ABR procedures to ensure fair sharing of the available bandwidth in all circumstances. In practical implementations, fair sharing of bandwidth seems possible only if some form of "per-VCC" resource allocation (e.g., queuing) is performed within the ABR NEs. If this is not done, some ABR connections may monopolize a larger proportion of the available bandwidth at the expense of other ABR VCCs sharing common resources, thereby negating the desired fairness criteria. This phenomenon has been referred to as "connection beat down" or "bandwidth hogging," since a so-called "greedy" source may end up using a greater proportion of the shared resources than fair allocation policy demands. The provision of fair bandwidth sharing constitutes a significant technical challenge for ABR-capable NEs and remains a weak point in the scalability of flow-control-based ATCs for very large numbers of connections.

Although an ABR connection may specify an MCR on connection establishment as noted earlier, which enables the network to allocate sufficient resources to the VCC to support at least MCR for the duration of the connection, the ABR procedures also indicate that the network can instruct a source to reduce the cell rate below its MCR value if required. This action may be required under conditions where insufficient capacity is available due, for example, to congestion conditions. In order to maintain the control loop for ABR operation, a source (or VS) is required to send RM cells in-band with the user data cells at sufficient intervals to allow the control loop to stabilize. For this purpose, the parameter N_{RM} is used to specify the number of user data cells between each RM cell inserted in the cell stream of a given VCC or VPC. N_{RM} is a settable parameter currently between the range 2–256 to enable a reasonable trade-off between the overhead due to RM cells (e.g., for small values of N_{RM}) and looser control resulting from high values of N_{RM}. Values of N_{RM} between 32 and 256 (in factors of two) are commonly used in typical ABR implementations.

For ABR the user data cells are denoted by CLP = 0 (high priority) and tagging of user cells is not permitted according to current procedures. The in-band RM cells are also denoted by CLP = 0 and these are referred to as in-rate RM cells. However, the ABR procedures also allow the use of RM cells with CLP = 1 when it is not possible to send an in-band (CLP = 0) RM cell. The RM cells with CLP = 1 are called out-of-rate RM cells. The out-of-rate CLP = 1 RM cells may only be inserted up to a fixed rate of 10 cells/sec, and are primarily intended to maintain the control loop at a minimal level if normal in-band CLP = 0 RM cells cannot be generated on the ABR VCC or VPC. Since there is always the possibility that out-of-rate RM cells may be used by a source (or VS), it is clear that the requisite bandwidth should be available for them even in the case where the MCR is specified as zero. However, the negotiated QoS for the ABR VCC or VPC is only committed for the conforming CLP = 0 RM cells and user data cells, so that selective discard of the out-of-rate CLP = 1 RM cells is always permissible.

Considering the parameters described so far, the typical ABR VCC or VPC may be viewed in terms of a minimum cell rate when an MCR > 0 is specified (but which may go to zero). together with a variable cell rate up to the maximum specified by the PCR parameter. The fluctuation of the variable cell rate component will depend on the (fair) sharing of the available capacity allocated by the network to the ABR service (or ATCs), in accordance with some allocation policy and algorithm. Each ABR-capable explicit rate NE along the connection path incorporates an algorithm to enable it to calculate the appropriate explicit cell rate (ECR) that can be supported for the required QoS (i.e., CLR) to be maintained for the traffic loading at that instant. Each NE (or VS/VD) may insert the calculated ECR value into the traversing forward or backward in-rate RM cells, which appear every N_{RM} user cells on the ABR VCC or VPC. The NEs may only overwrite the ECR value if it is larger than its own current ECR information. Thus, the final value of the explicit cell rate parameter (ECR) that is returned to the source (or VS) by the ABR control loop represents the minimum value that can be supported along the ABR connection path for the period of the round-trip delay from source to destination and return.

It is important to recognize that the finite round-trip delay involved in returning the explicit cell rate information back to the source implies that any consequent reaction by the source in reducing (or increasing) its cell rate to respond to the instantaneous available bandwidth may be out of step with the congestion state in the NEs. As for all feedback control loops, this aspect may strongly influence the stability and robustness of the ABR mechanism. If the rate of change of the effective bandwidth available to ABR VCCs is very large compared to the round-trip times involved, the ABR control mechanism may not be effective in maintaining very low cell losses at high utilization of the transmission path rate, and stability may be adversely affected. On the other hand, if the round-trip delays are relatively small compared to instantaneous changes in the available bandwidth, stable control may be readily achieved, but at the expense of relatively high processing overhead of the RM cells.

In addition to the round-trip time for the ABR control loop (termed the fixed round-trip time or FRTT in the ATM Forum definition of ABR service), a number of other parameters may affect the performance of the ABR control loop. For example, these may include the rate at which a source (or VS) reduces (or increases) its cell generation rate in response to the explicit rate information in the backward RM cell, or the behavior of the destination (or VD) in processing the RM cells. It may therefore be apparent that a full description of the ABR control loop operation may require a significant number of parameters and their interactions in order to achieve stable, low cell loss performance under all realistic network traffic scenarios. Nonetheless, a number of detailed simulation and empirical studies have demonstrated that with judicious selection of ABR control loop parameters and algorithms for source and destination behavior, the ABR objectives of very low cell loss (e.g., $< 10^{-6}$) at high utilization may be robustly achieved for a range of network applications.

However, in this respect there is a marked difference of approach in the description of the ABR service category developed by the ATM Forum and that adopted by the ITU-T for the ABR ATC, even though the same RM cell mechanism is used in both cases. The ATM Forum ABR service category specification defines a set of parameters that enable a relatively complete algorithmic description, as well as reference source and destination (or VS and VD) behaviors to

General RM Cell Format

ATM Header (5 octets)	RM Protocol Identifier (1 octet)	Function Specific Fields (45 octets)	Res. (6 bits)	CRC 10 (10 bits)

RM Cell Format for ABR

Field	Octet(s)	Bit(s)
Header	1-5	all
Protocol ID (=1 for ABR)	6	all
Message Type: Direction (=0 forward, =1 backwards)	7	8
Message Type: BECN cell	7	7
Message Type: Congestion Indication (CI)	7	6
Message Type: No Increase (NI) bit	7	5
Message Type: (Not used for ABR)	7	4
Message Type: (Not used for ABR)	7	3
Message Type: Reserved	7	1-2
Explicit Cell Rate (ECR)	8-9	all
Current Cell Rate (CCR)	10-11	all
Minimum Cell Rate (MCR)	12-13	all
Queue Length (QL)	14-17	all
Sequence Number (SN)	18-21	all
Reserved	22-51	all
Reserved	52	3-8
CRC-10	52	1-2
CRC-10	53	all

Figure 5-13. Resource management cell format for ABR.

be observed. In addition, it specifies the parameters to be signaled for on-demand switched ABR VCCs or VPCs, as well as those that may remain fixed or provisioned in any given implementation. The intent of such a detailed approach is to provide a range of key ABR parameters and associated procedures to enable implementation for various applications. These parameters and procedures are listed in the next section.

On the other hand, the current approach adopted for the ABR ATC by the ITU-T essentially provides a more general description of the ABR control loop operation with more detailed description of the semantics of the RM cell handling. Only a limited subset of the parameters specified by the ATM Forum are considered and the source and destination (or VS and VD) behaviors are not mandated, since these are considered to be implementation-specific. Despite these differences in approach, not to mention terminology, it is important to recognize that there is no basic inconsistency between the ATM Forum ABR Service Category description and the ITU-T's ABR ATC approach, since the latter may be viewed as a subset of the former. In essential respects, particularly as regards the semantics of RM cell processing, both approaches are the same. It is not the intent here to provide the reader with an exhaustive description of all aspects of ABR operation, but it may be useful to briefly describe the significant parameters involved in order to gain an understanding of the relative complexity involved. This description will be based on the approach defined by the ATM Forum specification.

5.13 ABR RM CELL STRUCTURE AND PARAMETERS

The flow control mechanism to support ABR operation uses explicit cell rate information written by the intermediate ATM NEs in the payload of the RM cell. As noted earlier, the RM cell was generically defined for traffic control purposes, and ABR is simply a specific instance of its use. Other uses of RM cells are also under consideration as described later. The general RM cell format is shown in Fig. 5-13. The RM cell is essentially a special case of ATM layer management cells and has the same general format as the OAM cells used for other maintenance purposes and described in subsequent chapters. It will be recalled that the RM cells in a cell stream are distinguished from the user data cells by a preassigned codepoint of the payload type field (PTI = 110) in the ATM cell header for a VCC. In addition, the preassigned VCI value VCI = 6 has been reserved for distinguishing RM cells from user data cells for the case of any given VPC. These codepoints are necessary, since the RM cells in any given VCC or VPC cell stream have the same VPI and VCI values as the user data cells for every VCL and VPL along the connection path.

As illustrated in Fig. 5-13, the first octet (i.e., octet 6) in the RM cell payload is denoted as the RM protocol identifier and serves to identify the specific resource management mechanism (or protocol) in question. For example, for the ABR ATC (or service category) the RM protocol ID is equal to 1 (PID = 0000 0001). This value indicates to the NE monitoring the RM cells in the cell stream of a connection that the ABR ATC is being used, and the NE may process the subsequent information in the function-specific fields following the protocol ID accordingly. The defined function-specific fields (octet 7 onward) are also listed in Fig. 5-13. The message type field signifies directionality of the RM cell with respect to the originating source. Thus message type bit 8 = 0 is the forward RM cell and bit 8 = 1 denotes the backward RM cell. This bit is set by the destination (or VD) only and not by intermediate ABR NE's (which may, however, process other fields described below). On the other hand, the backward explicit congestion notification (BECN) cell bit (= 1) denotes that the RM cell has been generated in the backward direction by any intermediate NE or destination (or VD) unrelated to the "paired" forward and backward RM cells. This distinction between the normal forward (direction bit = 0) and paired backward (direction bit = 1) RM cell and the independently generated BECN cell (BECN bit = 1) from any NE along the connection path was described earlier. Clearly, the intent of the BECN cell is to allow NEs to return instantaneous cell rate information to the source if desired when no backward RM cell is available for this purpose, and hence "tighten up" the cell rate control loop in the event of rapid changes in the resource availability of any given NE.

Two other functions are also defined in the message type field. The congestion indication (CI) bit is used to signal the presence of congestion in any NE along the connection path. Its function is analogous to the explicit forward congestion indication (EFCI) function in the ATM cell header PT field, as described earlier. Thus CI = 1 signifies that a congestion threshold has been exceeded somewhere in the network, and requires the ABR source to decrease its current cell generation rate accordingly. When the destination (or VD) receives a forward RM cell with the CI bit set to 1, it must also set CI = 1 in the backward RM cell, in exact analogy with the EFCI function. Finally, the message type field includes the no increase (NI) bit (bit 5), whose function is to indicate to the ABR source (or VS) not to increase its current cell rate. Thus, when the NI bit is set to 1, a source (or VS) must maintain its current cell rate value at the indicated level. The NI bit may be set by any intermediate NE (e.g., to signal an impending congestion state) or by the destination (or VD). When the destination (or VD) receives a forward RM cell with NI = 1, this value is copied to the backward RM cell. In addition, a source may also set the NI = 1 to indicate that it does not need to increase its current cell rate. The remaining bits of the message type field are not used in the ABR operation and bits 1–2 are reserved for future use.

The subsequent fields of the ABR RM cell are used to code the cell rate information required for the feedback control loop. Octet numbers 8–9 encode the explicit cell rate (ECR) value. The ECR value may range between the minimum cell rate (MCR) or zero and the PCR negotiated for the ABR VCC or VPC. The source may set the ECR value in this range, whereas any intermediate NE may reduce the ECR value to the allowed cell rate (ACR), which the NE may be able to support at that instant. If the ACR calculated by any intermediate NE is larger than the value in the ECR field, the NE does not modify the existing ECR value. Consequently, the ECR value returned to the ABR source reflects the maximum rate at which the source should transmit cells on the VCC or VPC in order to maintain its QoS. It should be noted that although the terminology "allowed cell rate" (ACR) and "explicit cell rate" (ECR) is sometimes used interchangeably in describing ABR, there is a distinction in that ECR is effectively the minimum of ACRs along the connection path. ABR operation will not work effectively if a source (or VS) ignores the ECR restriction returned to it in the backward RM (or BECN) cells.

Octets 10–11 encode the value of the current cell rate (CCR), which is set by the source at its present cell rate. As an option, it has been proposed that this value may be used by NEs to compute a suitable ECR value. For a BECN cell, the CCR is set to zero. In a similar vein, the minimum cell rate (MCR) value encoded in octets 12 and 13 may also be used in estimating an ECR value by NE or allocating bandwidth between ABR VCCs or VPCs. However, it should be noted that the MCR value is negotiated at connection set-up time, either via on-demand signaling or by subscription. Consequently, the value of its inclusion in the RM cells may appear superfluous and not central to ABR operation. As for CCR, the MCR value is set to zero for BECN cells, and its use for computing ECR is optional.

The coding format for the cell rate information in all the above fields (ECR, CCR, and MCR) is the same and is based on a 14 bit binary floating point representation, which is also used to represent PCR values. The coding employs a 5-bit exponent and a 9-bit mantissa to represent the cell rates, together with a 1-bit zero–nonzero flag (nz). The MSB of the 16-bit (2-octet) word is reserved, followed by the nz bit and the 5-bit exponent. The final 9 bits encode the mantissa. In this representation, the cell rate is defined as:

$$R = 2^e \,(1 + \mathrm{m}/512) \times \mathrm{nz} \text{ cells/sec}$$

where $0 \leq \mathrm{e} \leq 31$, $0 \leq \mathrm{m} \leq 511$, and $\mathrm{nz} = 0$ or 1.

Two other optional fields have been defined in the ABR RM cell structure. Octets 14 to 17 (four octets) may optionally encode a queue length (QL) parameter. The QL denotes the maximum number of cells queued for the VCC or VPC at NEs along the connection path. If the NE supports the optional QL capability, it may enter this value in the QL field or leave it unchanged if not. A source (or VS) simply enters zero. The next four octets (octet numbers 18 to 21) encode a sequence number (SN), which may optionally be used by a source (or VS). If supported, the value in the SN field may be increased by one (modulo 2^{32}) for each forward RM cell generated by the source, whereas the destination (or VD) returns the same SN value unchanged in the paired backward RM cell. The intermediate NEs are not allowed to change the SN value, and BECN cells set the SN value to zero. Although use of the SN function may allow for the detection of lost (i.e., discarded) RM cells in the cell stream, it may be apparent even from this brief description that use of the QL function and the SN fields is not directly pertinent to efficient ABR control loop operation. Consequently, there has been controversy over inclusion of these options and it is important to recognize that both the queue length and sequence number functions are not used for the ABR service category as defined by the ATM Forum at present. In fact, these functions may be more pertinent to other flow control mechanism, as will be seen later, and hence may be viewed by many as competitive to ABR and therefore not required in its operation. However, in the more general description of ABR ATC provided by the ITU-T in Recommendation I.371 [5.7], the options of the QL and SN functions are described, although the relationship to their use in ABR is not clearly defined.

In addition to the somewhat controversial issue of the QL and SN options, the specification of several ABR parameters in the ATM Forum's ABR specification finds no counterpart in the ITU-T description of ABR ATC at present, as pointed out earlier. The main parameters required for robust ABR operation are listed below, but the description is not intended to be comprehensive. Further parameters may be found in the ATM Forum specification [5.8], if required.

The key variable parameters include:

1. Rate Increase Factor (RIF). This parameter describes the rate at which a source (or VS) may increase cell generation after receiving (and processing) the RM cell. The RIF varies by power of two within the range 1/32,768 up to unity.
2. Rate Decrease Factor (RDF). This parameter describes the decrease in the rate of the user data cells by the source (or VS) after receiving (and processing) the RM cells. The RDF varies by a power of two within the range 1/32,768 to unity, as for the RIF.
3. N_{RM}. This settable parameter signifies the maximum number of user data cells between each forward RM cell, as described earlier.
4. The Allowed Cell Rate (ACR). This rate in units of cells/sec defines the current rate at which a source (or VS) is allowed to send cells. This term is often used to signify the computed explicit rate (ER) derived by a NE for insertion into the ECR field.
5. Fixed Round-Trip Time (FRTT). This parameter is a measure of the total round-trip delay for an RM cell from a source (or VS) to a destination (or VD) and back. It is obtained by summing both propagation and processing delays along the connection path, in units of 1 μsec within the range 0 to 16.7 sec. It is interesting to note that in the ATM Forum specification, the FRTT value may be derived from the accumulated time for a signaling set-up message during the connection set-up phase of an ABR VCC or VPC, even though the RM cell round-trip delay should correspond to the in-band propagation of the RM cell. Consequently, the FRTT derived from the signaling procedure may represent a significant overestimation of the in-band round-trip delay. Since the FRTT is considered an important parameter for optimizing the performance of the ABR control loop, a more accurate estimate of its value may be desirable.
6. Transient Buffer Exposure (TBE). This parameter signifies the maximum number of cells a source (or VS) should send at start-up before the first backward RM cell is returned. Its units are in cells, within the range 0 to 16,777,215. The need for the TBE parameter has been questioned, since it has been argued that sources (or VS) may initially send cells at low [e.g., MCR or initial cell rate (ICR) $\ll$ PCR] rates, thereby avoiding cell discards due to transient buffer overflows.

In addition to the above settable parameters and the PCR and MCR defined earlier, a number of other parameters such as initial cell rate (ICR) and tagged cell rate (TCR), which is constant at 10 cells/sec, have been specified in order to "fine tune" ABR implementations. The need for some of these additional parameters has been questioned, and it may be noted that in practice it is quite possible to implement robust ABR operation even without recourse to using all the specified parameters. Consequently, some of these may be viewed as optional, having somewhat less impact on the basic ABR performance. The question also arises as to which of the parameters (in addition to the PCR, which is mandatory) are required to be included in the signaling messages and which are considered optional. The required parameters include the PCR, which is mandatory in any ATC, the MCR, the ICR, TBE, FRTT, RDF, and RIF. The optional parameters include

N_{RM} and three others (not defined here). If these optional parameters are not signaled, the specified default values may be assumed.

Together with the ABR parameters, it is also required to specify the behaviors of the ABR source (or VS) and destination (or VD) in order to ensure interoperability and stable operation between sources and destinations from different vendors. The primary intent of these procedures is to specify how the ABR sources and destinations process the cell rate information in the RM cells, as well as the handling of the RM cells, to ensure stability and robustness of the ABR control loop. These procedures will not be listed in detail here, since many of these pertain to the actual implementation of ABR, and their essence may in any case be inferred from the brief description of the ABR parameters given above.

From the perspective of the intermediate NE, the ABR procedures (termed the "switch behavior") allow for a number of options, as noted earlier. It is anticipated that an ABR-capable NE will implement at least one of the options given below:

1. EFCI Marking. This procedure involves setting of the EFCI bit in the ATM cell header of only user data cells, independently of any RM cells.
2. Relative Rate Marking. This procedure involves setting of the CI and NI (= 1) for forward and backward RM cells only in the event of congestion.
3. Explicit Rate Marking. This procedure requires the setting of the ECR field in the forward or backward RM cells with the appropriate cell rate value computed in accordance with any implementation-specific algorithm in the NE.
4. Virtual Source and Virtual Destination Control. For the case where it is required to segment the ABR control loop at any given NE, it is required that the NE provide the complete VS/VD behaviors. This essentially involves implementation of the source and destination procedures for all ABR VCCs or VPCs with appropriate coupling between the ingress and egress queues.

It will be apparent that the above NE options are listed in order of increasing complexity, and allow a network operator to select the level of ABR control offered in any given network service from the simplest EFCI-based control to complete segmented ABR control loops between any given set of NEs. Although this wide range of service flexibility available with ABR may appear highly desirable from the perspective of a service provider, it is also problematical in the sense that it may be difficult to select which option to use in any given application. This is because at present there is little empirical network experience available with ABR services for wide area networks (WANs) on which to base commercial service offerings. Nonetheless, as more fully featured ABR-capable ATM NEs become available to network operators, it is apparent that the potential flexibility of ABR services will be more widely utilized to enable fine tuning of performance to suit specific ATM network applications.

5.14 ATM BLOCK TRANSFER (ABT) CAPABILITY

It was pointed out earlier that ABR is not the only standardized ATC to make use of RM cells to obtain statistical multiplexing gain. A class of traffic control mechanisms generically termed "fast reservation protocols" had been proposed long before ABR was defined, notably by France Telecom, as a means for obtaining statistical multiplexing capability for highly bursty traffic sources such as data. The essential concept underlying fast reservation protocols is to provide an in-band mechanism to reserve bandwidth and buffer capacity just prior to sending a burst of user

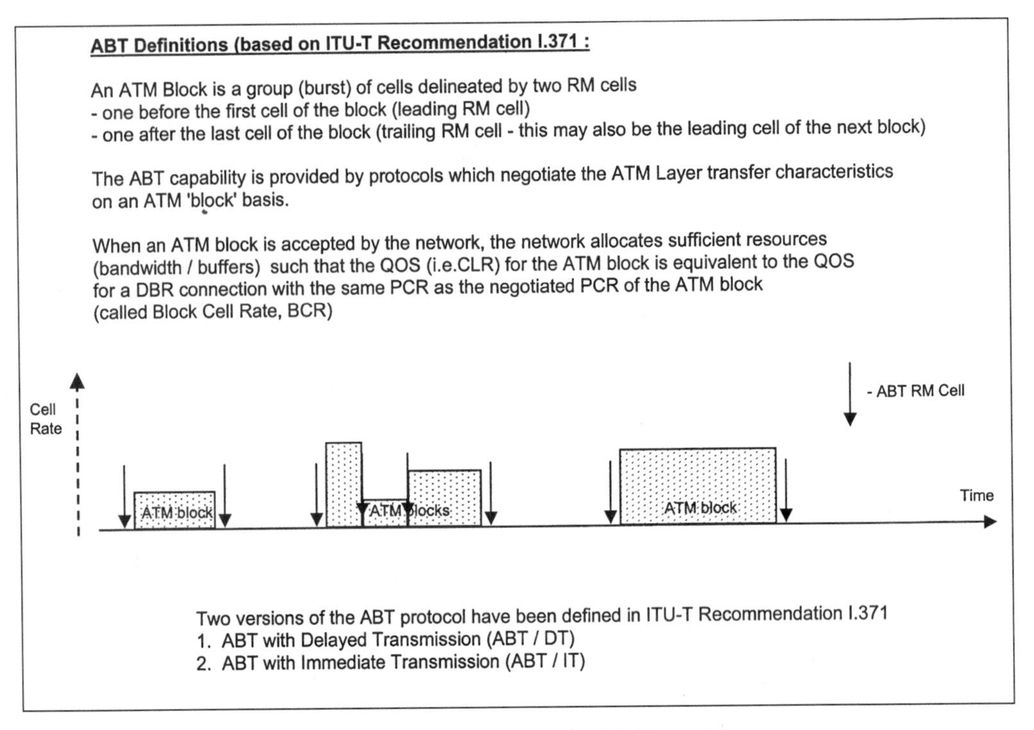

Figure 5-14. ATM block transfer (ABT) capability.

data cells on the VCC or VPC. If each NE along the connection path can accommodate the in-band request for resources, the cell burst will traverse the VCC or VPC with low probability of cell loss. Immediately following the passage of the burst of cells through the NE, the reserved resource is released to allow it to be shared by other connections transversing the NE (e.g., other subsequent bursts). Thus, the fast reservation mechanism is based on the concept of an in-band "signaling" message that temporarily reserves resources for a burst of cells and then releases those resources for statistical sharing among other connections. This mechanism is quite different from the modification of the cell rate along the connection path by means of flow control messages as for ABR.

The ATM block transfer (ABT) capability uses RM cells to provide the in-band fast reservation of resources. Essentially, the ATM block is defined as a group of user data cells delineated by two RM cells: a leading RM cell and a trailing RM cell. The blocks themselves may be comprised of one or more bursts of cells with a cell rate called the block cell rate (BCR), which may vary from block to block as shown schematically in Fig. 5-14. When resources are allocated (if available) to accommodate a given ATM block, it is anticipated that the QoS (i.e., the cell loss probability) is equivalent to that available to a deterministic bit rate (DBR or CBR) VCC or VPC that has the same PCR as the block cell rate. In this sense, the ABT transfer capability may be viewed as a "piece-wise DBR" mechanism, where the PCR (i.e., the BCR) is individually allocated for each ATM cell block transversing the NE. It should also be noted that the trailing RM cell, which is responsible for releasing the reserved resource for the preceding block, may also act as the leading RM cell for a subsequent concatenated ATM block.

Two distinct version of the ABT protocol have been defined by the ITU-T [5.7]. The first, termed ABT with delayed transmission (ABT/DT), incorporates the use of an acknowledgement message from the network that sufficient resources have been reserved before sending the ATM cell block on the VCC or VPC. The second version, termed ABT with immediate transmission

(ABT/IT), essentially transmits the block directly following the leading RM cell, without waiting for an acknowledgement of resource reservation from the network. For this case, it is assumed that if a given NE along the connection path is unable to provide sufficient resources to support the cell block, it will be discarded. However, for both ABT/DT and ABT/IT, it has been proposed that the use of a negotiated sustainable cell rate (SCR) traffic with associated IBT parameter during connection set-up may ensure that an "average" bandwidth (and intrinsic burst tolerance) is available to allow a high proportion of the block to be delivered, given reasonable bandwidth allocation in accordance with traffic engineering guidelines. In practice, the use of the SCR/IBT traffic descriptor for supporting ABT/IT has not been clearly demonstrated for WAN applications but, as noted earlier, it may be fair to assume that an SCR + IBT traffic descriptor, if available, can be used to allocate an estimate of minimum bandwidth to the VCC or VPC in dimensioning bandwidth requirements for the network.

Before considering details of the ABT protocols and associated RM cell structure and semantics, it is useful to consider its scope and relationship with the ABR ATC in the ATM standards. It was noted that the basic concepts underlying fast reservation protocols had been extensively discussed in the ITU-T well before the introduction of the notion of ATM layer flow control was introduced in the ATM Forum, leading to the eventual specification of the ABR service category. However, the fast reservation proposals, although elegant in concept as a means for handling cell bursts, were considered by many as too complex for widespread acceptance. Consequently, this approach had a somewhat limited constituency. However, when the rate-based ABR mechanism was introduced into the ITU-T following its general acceptance by the large community of interest represented by the ATM Forum, it was recognized that the level of complexity and scope was essentially of the same order as that implied by fast reservation. Consequently, the impetus for an ATC utilizing the fast reservation concept was revived roughly in parallel with the specification of ABR in the ITU-T standard, where it was agreed that both the rate-based ABR mechanism and the fast-reservation-based ABT may constitute different but equivalent ATCs to provide statistical multiplexing capability in ATM networks. However, attempts to introduce the ABT mechanism into the ATM Forum TM specification by its supporters have been unsuccessful. Thus, it is important to note that although the ABT mechanism has been standardized by the ITU-T, there is currently no counterpart to it in the ATM Forum's service categories, constituting a significant difference between the two bodies.

Given the currently limited appeal of the ABT ATC, it is not the intent here to give a detailed account of its operation, and interested readers may find further details in [5.7]. However, a broad description of the ABT parameters and associated RM cell semantics is useful, as the techniques have relevance to some network applications. Thus, some simulation studies have shown that ABT provides relatively good performance characteristics for applications such as LAN hubs or when the number of NEs involved is low, such as in campus networks. The essential difference between the ABT/DT and ABT/IT protocols is the absence of the acknowledgement phase in the latter, which primarily impacts the QoS characteristics (both cell loss and cell delay) available to the ABT/IT version. However, both versions may be considered together in this overview, while pointing out the procedural differences where appropriate.

The traffic parameters for ABT that must be negotiated at VCC or VPC set-up (either through signaling or by means of management interface) include the following:

1. The maximum PCR and associated CDV tolerances in both directions (of a bidirectional ABT connection) of the CLP = 0 + 1 cell components, including any end-to-end OAM cells that may be present. As an option, the OAM cell component may be specified separately, as for the DBR ATC.
2. The SCR and associated IBT parameters together with the CDV tolerance for the total (CLP = 0 + 1) cell flow. The SCR may be zero.

General RM Cell Format

ATM Header (5 octets)	RM Protocol Identifier (1 octet)	Function Specific Fields (45 octets)	Res. (6 bits)	CRC 10 (10 bits)

RM Cell Format for ABT

Field	Octet(s)	Bit(s)
Header	1-5	all
Protocol ID (=2 for ABT/DT; =3 for ABT/IT)	6	all
Message Type: Direction (=0 forward, =1 backward)	7	8
Message Type: Traffic Management cell	7	7
Message Type: Congestion Indication (CI)	7	6
Message Type: Maintenance bit	7	5
Message Type: Req / Ack	7	4
Message Type: Elastic / Rigid bit	7	3
Message Type: Reserved	7	1-2
BCR for CLP=0+1 flow	8-9	all
User OAM BCR	10-11	all
Reserved	12-13	all
Block Size	14-17	all
Sequence Number (SN)	18-21	all
Reserved	22-51	all
Reserved	52	3-8
CRC-10	52	1-2
CRC-10	53	all

Figure 5-15. RM cell format for ABT.

3. The peak renegotiation rate, as inferred from the specification of the PCR for the ABT RM cell flow and its associated CDVT.

It should also be noted that in ABT, the sequence integrity of the (leading and trailing) RM cells with respect to the user cells in the ATM blocks is required to be maintained. This is in contrast to the ABR case, where there is no requirement to maintain sequence integrity between the ABR RM cells and user cells, although in both cases the relative sequence integrity between successive RM cells needs to be maintained.

At connection set-up, the block cell rate (BCR) may be set to zero, and the above parameters are considered static during the ABT connection duration, although the actual BCR may vary within the range $0 \leq \text{BCR} \leq \text{PCR}$ as it is dynamically negotiated (only for ABT/DT) for successive ATM cell blocks during the connection lifetime. For ABT/IT, there is clearly no negotiation of BCR, and blocks may be discarded if sufficient resources are not available at any NE along the connection path. Apart from this possibility, the traffic descriptors are the same for both ABT/DT and ABT/IT. The network may commit bandwidth for the total (CLP $= 0 + 1$) cell flow in either direction when a nonzero SCR value is specified at the ABT connection establishment. This committed bandwidth is related to the reserved resources (i.e., bandwidth and buffer space) at least equal to the specified SCR to enable the requisite cell loss probability to be maintained by ABT. However, if the SCR is not specified or is zero, the network need not commit resources for the ABT VCC or VPC, which implies that a QoS is not guaranteed for this case.

The RM cell structure and semantics are shown in Fig. 5-15, where the similarities and differences with the ABR RM cells should be carefully noted. ABT uses the same general RM cell format as ABR, but the protocol identifier field coding of 2 for ABT/DT and 3 for ABT/IT distinguishes it from ABR. The message type field (octet 7) includes the codepoints for directionality (bit 8 = 0 for forward and = 1 for backward) and other functions specific to ABT operation, as described below. The traffic management function of bit 7 (octet 7) is intended to distinguish between the RM cell generated by a user (source) for renegotiation of the BCR value from the RM cell generated by the network for traffic control purposes. Such an ABT RM cell is called a "traffic management" cell and is identified by the codepoint bit 7 (octet 7) = 1. The detailed procedures for the Traffic Management ABT cell are currently under consideration but, in general, may be used by the network to provide ABT-specific information for the user.

The congestion indication (CI) bit (bit 6, octet 7) of the "message type field" is used to indicate whether or not the BCR renegotiation has been successful or not. Thus, the coding CI = 0 indicates that the BCR modification has been successful, whereas the coding CI = 1 indicates that the BCR modification has not been accepted (i.e., there is congestion in the network). Note that the semantics of the ABT CI field is somewhat different from that described earlier for the ABR CI field. The maintenance bit (bit 5, octet 7) in the message type field is used to distinguish between ABT RM cells used for BCR negotiation by the network or user from those used for ABT maintenance procedures, which at this stage are still under consideration. Thus, the ABT RM cells that are used for normal BCR modification are coded maintenance bit = 0, whereas those intended for ABT maintenance are coded maintenance bit = 1.

The request/acknowledge bit (bit 4) in the message type field (octet 7) is used to indicate whether the ABT/DT RM cell is a request or an acknowledgement message. Specifically:

1. If a user sends a BCR modification request, the Req/Ack bit = 0
2. If a user sends an acknowledgement to a request or a BCR modification by the network, the Req/Ack bit = 1
3. If used by the network as an acknowledgement of a BCR modification, the Req/Ack bit = 1
4. If sent by the network as a request for a BCR modification, the Req/Ack bit = 0

The Req/Ack field is not used for ABT/IT operation. The elastic/rigid bit (bit 3 of octet 7) is used to indicate whether or not the BCR request by a user may be modified (i.e., decreased) by the network if sufficient capacity is not available for the block. Thus, when the elastic/rigid bit = 0, the source indicates that the network may modify the BCR request (i.e., the request is "elastic"), and when elastic/rigid bit = 1, the network should not modify the requested BCR (i.e., the BCR request is "rigid").

The block cell rate (BCR) field (octets 8–9) for the total cell flow (CLP = 0 + 1) is used to encode the BCR requested by the source or the BCR allocated by the network on negotiation. As noted above, this value depends on the setting of the rigid or elastic mode. In the elastic mode, the BCR value may be less than or equal to the requested BCR, whereas in the rigid mode, the BCR value will be equal to the requested value. For the case when the SCR parameter is specified, the allocated BCR should not be less than the SCR. The coding of the cell rate in the BCR field follows the same coding algorithm as that described earlier for the ABR RM cell fields (5-bit exponent and 9-bit mantissa). The user OAM BCR field (octets 10–11) is used to encode the BCR for the total user OAM cell flow, for the case where this flow is specified separately from the user data cell flow. The coding scheme for this rate is identical to that used for the BCR field.

The block size (octets 14–17, four octets) and sequence number (octets 18–21, four octets) fields are only used for the ABT/IT protocol and are not used for ABT/DT version. The block size field is used to encode the length of the ATM cell block (in number of cells) and may be used by the ABT/IT NE to allocate resources if required. The sequence number field may optionally be used by an ABT/IT source by incrementing by one each subsequent ABT/IT RM cell. The ABT/IT destination simply copies the sequence number value into every RM cell sent in response to the source RM cell, and no other intermediate NEs are permitted to alter the sequence number. The sequence number may be used to detect lost RM cells but, in addition, other uses have been proposed for this field, which may not be related to the ABT ATC, as indicated below, and these have generated discussion concerning its usefulness in ABT. If the sequence number field is not used, its value is set to 0.

For both ABT/DT and ABT/IT, a distinction is generally made between the QoS commitments at the cell level and at the block level. The cell level QoS (cell loss and delay) apply to the case where an equivalent DBR QoS is provided to the block, assuming that the BCR = PCR of the DBR VCC, with sufficient bandwidth allocated to enable this. The block level QoS commitment is only applicable if the SCR value is negotiated at ABT connection set-up. For this case, it is anticipated that a network may allocate a minimum bandwidth equivalent to the SCR value, so that BCR values of at least this amount may be negotiated with an engineerable block-discard probability. However, the actual values may be different for ABT/DT and ABT/IT, in general.

5.15 UNSPECIFIED BIT RATE (UBR) AND GUARANTEED FRAME RATE (GFR)

The ATCs (or ATM layer service categories, using the ATM Forum specification terminology) described so far all assume that their use provides a contracted QoS in terms of cell loss probability or delay, etc., from the service provider. However, a number of early implementations of ATM services simply offered a so-called "best efforts" type of service, in which essentially no maximum cell loss or delay bounds were specified as part of the service contract. User data (e.g., the higher-layer data frames) were delivered to a destination, provided sufficient capacity was available when the data frame was transmitted. In a number of these primitive ATM service offerings, the end user's terminal simply generated cells at the physical line rates (transmission path rate), so that not even a PCR was required and no traffic enforcement (i.e., UPC or NPC) was invoked, other than basic dimensioning of the network capacity to accommodate the anticipated number of user traffic streams. These "best efforts" ATM services were categorized as the unspecified bit rate (UBR) service category by the ATM Forum specification and, in effect, operated (and many still do) much as legacy data networks based on internetworking protocols such as TCP/IP.

For small networks with limited numbers of traffic sources the "best efforts" or UBR-based ATM networks can work surprisingly well, primarily due to the highly bursty nature of the data traffic and the substantial buffering and relatively high cell throughputs available even on early ATM NEs. Thus, buffers of up to several hundreds of cells and throughputs in excess of several Gbits/sec were not uncommon in initial ATM switch implementations. Consequently, such networks were essentially highly "overengineered" and cell losses due to contention for resources relatively rare. However, as user expectations and network application requirements for more stringent QoS evolved, it was recognized that the basic UBR service category may become insufficient to meet these needs. Although the SBR, ABR, and ABT service categories described earlier clearly provide the means to obtain guaranteed QoS, their relative complexity compared to the simple UBR approach elicited a number of proposals for simpler alternatives that could build on the UBR concept. These proposals, initially originated in the ATM Forum and more recently in the ITU-T, sought to enhance the basic UBR service category by introducing the concept of a minimum cell rate (MCR), which could be used to commit resources to the UBR VCC or VPC. The ensuing service category was initially called "UBR plus" (UBR+) and, more recently, "guaranteed frame rate" (GFR), to convey the notion of a minimum committed rate of data (frame) transfer.

The guaranteed frame rate (GFR), as well as the more basic UBR+ service category, specify a PCR and a MCR for the VCC or VPC as part of the service (or traffic) contract. The network may optionally enforce these traffic parameters at the UPC or NPC function exactly as it would do for the case of the SBR ATC described earlier in order to protect the QoS of other users. More importantly for the UBR+ concept, these traffic parameters provide the network operator with a means to dimension the necessary resources within the network to enable a committed QoS. Thus, as noted earlier, the connection admission control (CAC) function may be used with the MCR parameter to limit the number of GFR (or UBR+) VCCs or VPCs that can be accepted by the network without jeorpardizing the cell loss probabilities. In this respect, the GFR service category may be viewed as a special case of the SBR (or VBR) ATC described earlier, with the MCR traffic parameter as essentially equivalent to the sustainable cell rate (SCR) traffic parameter. However, the notion of ensuring a minimum "guaranteed frame rate" for this service category has prompted proposals for including the option of enforcing the traffic at the (data) frame level (in addition to performing normal cell-based UPC/NPC), so as to provide for a "frame level" service assurance.

It should be recognized that the option of frame-level traffic enforcement for GFR is not essential to the GFR service objective of at least ensuring a committed QoS for the minimum cell rate. Besides introducing additional complexity into the UPC or NPC functionality, which is not strictly necessary, this concept has been questioned in that it requires processing at the frame level in ATM NEs that essentially only process user data at the cell level. That such frame-level processing is readily possible is due to the use of the AAL Type 5 protocol for the encapsulation of typical data applications, as will be described in detail later. However, it may be recalled here that the payload type (PT) field in the ATM cell header contains the ATM user-to-ATM user indication codepoint, which is used by the AAL Type 5 protocol to delineate the end of the AAL Type 5 protocol data unit (PDU). Thus, the cell with the PTI = 001 identifies the last cell of the AAL Type 5 PDU, thereby delineating the encapsulated data frame even at the ATM layer. Consequently, applications using AAL Type 5 may readily perform frame-level traffic processing using this essential property of AAL Type 5 protocol, by enhancing the traffic enforcer to detect and then enforce the AAL Type 5 PDUs in the cell stream of a given GFR VCC or VPC.

The potential to provide for frame-level traffic enforcement for applications utilizing the AAL Type 5 PDU delineation mechanism has also been somewhat confused with another traffic control capability utilizing the same property of AAL Type 5 for congestion control. This mechanism has been variously termed early packet discard (EPD) or partial packet discard (PPD), and is based on the capability of applications utilizing AAL Type 5 to delineate the data frame at the ATM layer using the PTI ATM user-to-ATM user codepoint for end of frame indication. For early packet discard (EPD), the NE detects and then discards the entire packet (frame) if the congestion threshold has been exceeded rather than just some cells of the VCC or VPC at random. Discard of the entire data frame on congestion makes sense, since, in any case, if even one cell of the packet is discarded due to congestion, the entire packet will be corrupted and need to be retransmitted by the higher layers to ensure data integrity. Discard of the entire packet alleviates the congestion condition much more rapidly, assuming the packet is discarded prior to the ATM NE cell buffer congestion points.

In the partial packet discard (PPD) variant of this congestion control device, a part of the data frame may have been queued in the buffer before the congestion threshold was exceeded, but the remainder of the AAL Type 5 encapsulated packet may be discarded to alleviate the congestion condition more rapidly. Simulation studies have shown that the EPD and/or PPD congestion control mechanisms may be useful for more rapid recovery from congestion states, but these devices should not be confused with the GRF option of frame level traffic enforcement. However, it is clear that since both options are based on the same mechanism of frame delineation in AAL Type 5, they may be combined in any given GFR implementation if required.

5.16 CREDIT-BASED FLOW CONTROL CAPABILITIES

In introducing the ABR capabilities earlier, it was mentioned that so-called credit-based mechanisms to incorporate the advantages of cell flow control at the ATM Layer were also extensively studied as alternatives to the cell rate-based flow control protocols, which came to be known as ABR. The basic credit-based flow control approach uses the RM cells to feed back available buffer capacity information in the form of "credits" to the traffic source, which then sends only the equivalent number of cells to the NE in question. By this means, any cell loss due to buffer overflow can be eliminated, since, in the ideal case, buffers should never overflow, assuming that the credits can be made to accurately represent available buffer capacity. In effect, the credit-based flow control mechanism is similar to earlier so-called "window-based" flow control mechanisms, such as TCP window flow control, for example. The "window" represents the number of octets of data that can be accommodated without loss by the receiver's buffer capacity at any given time. A number of variants on the basic credit concept have been considered, ranging from credits representing per-VCC or VPC buffer space availability (so-called per-VCC queuing) to credit information representing shared buffer capacity over several VCCs or VPCs. In addition to their conceptual simplicity (which may not necessarily translate to actual implementation simplicity, unfortunately!), the primary strength of credit or window-based flow control systems is the potential to eliminate cell loss due to congestion, and thereby simplify network traffic engineering, particularly for the per-VCC approach. Nonetheless, it was recognized that the proposed credit-based protocols would require somewhat different buffer management architectures than those typically implemented in most ATM NEs, and that for WAN applications at very high cell rates, the per-VCC buffering requirements would need to be prohibitively large to accommodate the large numbers of cells in flight. Moreover, for optimum effectiveness, the credit-based flow control loop needs to operate on a per-hop basis from NE to NE, in contrast to the ABR approach, which could be used end-to-end or segmented as convenient without undue loss in performance.

For these and possibly other reasons, the ABR service category was selected by the ATM Forum for inclusion in its TM v. 4.0 specification after a long (and often emotional) discussion period. Despite this setback, supporters of the credit-based flow control school have continued to refine, test, and promote the protocols in independent trials under various consortia. The credit-based mechanism was further developed under the term "quantum flow control" (QFC) and, more recently, variants of this methodology have been proposed to the relevant ITU-T study group where it has been termed the "controlled transfer" (CT) capability and is being intensively studied for specific network applications by its proponents. These proposals are still in a somewhat preliminary state, and their scope of application clearly overlaps those addressed by the other established ATCs such as ABR and ABT. This inevitably prompts the question as to why it is necessary to specify even more mechanisms to cater to the same network applications as existing and possibly simpler ATCs. At the time of writing, this still remains an open question.

5.17 GENERIC UPC/NPC AND CAC REQUIREMENTS

The basic concepts relating to usage parameter control (UPC) and network parameter control (NPC) have been discussed in general earlier in terms of their use for traffic and resource management. The connection admission control (CAC) function was also described earlier in these terms. It is useful to summarize here the main requirements for these functions, taking into account the ATM transfer capabilities or service categories described in the previous sections. The requirements for the UPC/NPC and CAC are essentially common to all of these ATCs in general, with the possible exception of the very basic UBR service category, where neither CAC nor UPC may exist for some initial network deployments. It is also interesting to note that whereas the ATM Forum specifications regard the UPC function as somewhat of an option, even for the CBR service category, the prevailing view in the ITU-T appears to be that UPC is essential for robust operation of the ATM network when QoS needs to be maintained. To some extent, this divergence of perspectives reflects the differences in network engineering approach between those primarily interested in customer premises networks (CPNs or private networks) and those concerned more with public network operations.

The UPC and NPC are described as the set of functions in an ATM NE used to monitor and control (i.e., enforce) traffic entering a network to protect network resources from noncompliant users (who may be unintentional or malicious), whose behaviors may affect the QoS of other established (and compliant) VCCs and VPCs. The UPC or NPC function should be capable of detecting violations of the negotiated traffic parameters in the traffic contract of any of the supported ATCs and take appropriate action. These actions could include:

1. Cell passing (e.g., for conforming cells)
2. Cell discard (e.g., for nonconforming cells in excess of the traffic contract)
3. Cell tagging, wherein excess CLP = 0 cells are overwritten or "tagged"to CLP = 1 cells and then passed. This option is available for the SBR 3 capability
4. Cell rescheduling, wherein a "traffic shaping" function is combined with the UPC as an option
5. Releasing the connection. This action may be initiated by the UPC or NPC at the control plane connection level in certain circumstances

The selected ATC and associated traffic contract determine which of the above UPC or NPC actions are applicable. In general, the NPC function is considered a network option, although it may be more important at NNIs associated with administrative boundaries. These are also known as internetwork interfaces (INIs) or as broadband intercarrier interfaces (B-ICI). Although the actual algorithm or mechanisms for the UPC or NPC are implementation-specific and not subject to standardization, it is interesting to note that the leaky bucket type of algorithm described earlier is often used for traffic enforcement. One instance of the leaky bucket algorithm is required for each enforced traffic parameter. The settable (internal) parameters for the UPC or NPC operation are derived from the negotiated traffic parameters by the CAC function and may depend on specific network operator policy considerations. The CAC function feeds these internally derived parameters to the UPC/NPC function, as modeled in detail later. To assess the relative performance of the UPC or NPC functions, two performance attributes have been specified, although no quantitative values have been assigned thus far to act as benchmark objectives. The performance of the UPC or NPC depends on:

1. Response time—defined as the total time to detect the traffic violation on the given VCC or VPC and take the appropriate action described above.
2. Transparency—a measure of the accuracy with which a given UPC or NPC implementa-

tion takes control actions on nonconforming cells in a cell stream, and avoids inappropriate control actions on the conforming cells in the cell steam of a given VCC or VPC.

Since any impairments to the conforming traffic characteristics will be considered part of the overall network performance degradation, the UPC or NPC implementation should attempt to minimize their affect on the traffic flow. This implies minimizing the response time and maximizing the transparency.

The CAC function specifies the set of actions at the call set-up phase or during the call renegotiation phase in order to establish whether the requested VCC or VPC should be accepted or rejected. It should be noted that the CAC procedures need to be invoked regardless of whether the connection set-up is initiated by means of the on-demand signaling procedures or by means of management action as for a semipermanent virtual connection (PVC). The connection request is accepted by the CAC function only when sufficient resources are available to establish the call through the whole network with the requested QoS while maintaining the QoS of the existing VCCs and VPCs. In this respect, the CAC function is an essential element in each NE for achieving network-wide traffic control. As for the case of the UPC and NPC functions, the actual algorithms utilized by any given CAC implementation are not standardized and may differ from NE to NE as well as depend on the network operator policy considerations for bandwidth allocation. As a consequence, for multivendor or multioperator configurations, the call-blocking probability for the VCCs or VPCs may be limited by the NE or operator policy that uses the most stringent call acceptance criteria.

In general, the CAC function may utilize various combinations of the

1. Traffic descriptors
2. ATC required for the connection request
3. Required QoS Class or parameters

in order to derive

1. Whether the connection request (or renegotiation) can be accepted or not
2. The internal traffic control parameters to be used by the UPC or NPC to monitor the traffic contract
3. The allocation of the network resource (e.g., bandwidth) and possibly routing of the connection, depending on the network operator policy

These considerations imply that, in general, the CAC functionality may be quite complex and require substantial software processing in the NE.

5.18 GENERAL COMMENTS ON ATM TRAFFIC CONTROL

In considering the previous sections on ATM traffic control, the reader may justly be forgiven for being somewhat perplexed and possibly even confused at the bewildering range of mechanisms, options, ATCs, and the seemingly excessive complexity involved in ATM traffic control. The question often arises as to why it is necessary to support several mechanisms that all purport to provide statistical gain while maintaining an acceptable level of cell loss. It is certainly true that aside from the case of basic UBR capability, which arguably involves virtually no traffic control, the available ATCs described:

- Statistical bit rate (SBR) in three versions
- Available bit rate (ABR) with several modes of operation
- ATM block transfer (ABT), with both ABT/DT and ABT/IT versions
- Guaranteed frame rate (GFR) or UBR+
- Potential variants of credit-based flow control such as the (currently) loosely defined controlled transfer (CT)
- Other potential schemes being currently researched

are all intended for essentially the same types of potential network applications. The objectives are to provide statistical multiplexing gain with acceptably low cell losses and high bandwidth utilization for highly bursty, variable bit rate (VBR) traffic sources such as data communications, image transfers, etc.

In principle, all these traffic control protocols are intended to provide varying degrees of statistical multiplexing gain to improve the overall network efficiency of bursty VBR traffic sources. The actual amount of statistical gain or network efficiency improvement that may be achieved in practice will greatly depend on the interplay of a number of complex factors. These include:

1. How effectively the ATC mechanisms are implemented in a given NE
2. The network operator or service provider policy considerations and engineering or dimensioning of the network resources
3. The application requirements fit to the selected ATC
4. The source and destination behavior implementation
5. How accurately the actual traffic characteristics of the applications may be determined or predicted

It is usually not possible to lay down definitive guidelines as to which ATC to select for any given set of applications and how best to engineer the resulting network. Coupled with the myriad choices available, this has resulted in the perception that ATM traffic control is excessively complex and hence not desirable, or even workable. However, that this is clearly not the case is evident by the numerous ATM networks operating worldwide, offering ATM-based services for a range of applications. In practice, many of these deployments are designed from the traffic control perspective in a phased approach, using empirical data of the traffic characteristics to improve network efficiency while maintaining QoS using the simplest of traffic control tools. Thus, initial service offerings may have been based on a "best efforts" or UBR service category with overprovisioning to maintain low cell loss, or on the basic DBR (or CBR) service category with peak bandwidth allocation per VCC or VPC. It will be recognized that the DBR ATC is capable of supporting any type of network application, although with somewhat low efficiency for very bursty traffic sources. Nonetheless, experience gained from such initial phases of deployment may allow the service provider to further fine tune the network to increase overall efficiency and hence revenue.

Armed with such empirically gathered data on traffic characteristics, the next phase may incorporate some form of SBR ATC or the GFR service category, which may be considered as a specific instance of SBR ATC with SCR equivalent to the negotiated MCR traffic parameter. This phase of the traffic engineering design may be based on more quantitative estimates of the number of actual VCCs or VPCs possible based on the MCR values. At this stage, the traffic control complexity is minimal, involving basic UPC and CAC functionality, as described previously. If significantly higher network utilizations are desired, it may then be necessary to deploy either the ABR or ABT ATC, selecting the simplest option (e.g., EFCI mode for ABR) of the selected ATC. However, this phase does involve a significant step-up in the traffic control functionality required in the network elements, but with correspondingly better performance. It should also be recognized that although the ABR (and ABT) mechanisms are relatively complex, continuous improvements in the technology to implement these protocols will result in substantial cost reductions, particularly with more widespread deployments in the future. Thus, what may appear somewhat complex and somewhat unnecessary now will eventually become commonplace as the underlying technology evolves.

When compared to any other packet mode networking technology, such as, for example, TCP/IP, frame relay, etc., ATM arguably has the most comprehensive set of resource management capabilities available for providing flexible and scalable QoS. Moreover, the extensive analytical and empirical studies on the traffic performance aspects of ATM underlying these capabilities should provide a degree of confidence in their use not possible with earlier packet mode technologies. It therefore seems ironic, but not surprising, that these very aspects are often cited as evidence that the technology is "immature" or unnecessarily complex, notably by proponents of competing networking technologies. This view may result from a misunderstanding that all the ATM traffic control mechanisms defined to date are required in every case. However, as noted earlier, this is not the case for practical networks in which a phased approach can be very effective.

The comprehensiveness of the traffic control mechanisms incorporated into ATM, ranging from the basic cell discard priority and EFCI congestion control to the notion of UPC/NPC, CAC, and ABR ATCs, is not surprising when it is considered that ATM was intended and designed to integrate all foreseeable traffic sources and characteristics, unlike earlier packet mode technologies. It is also important to recognize that many of the mechanisms used by ATM have been also used in other packet mode technologies. For example, cell discard priority and explicit forward congestion notification concepts have been utilized by frame relay technology, whereas flow control mechanisms have been utilized with success in X.25 and TCP/IP technologies. In turn, traffic control concepts such as UPC/NPC and CAC introduced by ATM have been borrowed by other packet mode technologies such as IP, as described in subsequent chapters. This interaction between the basic traffic control mechanisms of the various packet mode technologies used for wide area networking provides an interesting field of study in its own right, while lending credence to their general utility in traffic control and resource management.

CHAPTER 6

ATM Signaling

6.1 B-ISDN SIGNALING PRINCIPLES

A fundamental aspect of ATM is that it is a connection-oriented packet mode technology. As noted earlier, "connection oriented" implies that a VCC or VPC has to be set up between a source and destination prior to any data transfer, either by using on-demand signaling protocols for so-called switched virtual channel (SVC) connections or network management action through the management interface for the case of permanent virtual channel (PVC) connections. For on-demand signaling, the ATM VCC or VPC may be released after the user data transfer phase has been completed, exactly analogous to switched voice calls in existing telephony networks. Since ATM is the transfer mode chosen for the B-ISDN concept, and B-ISDNs are essentially extensions of the existing 64 kbits/sec ISDN or public switched telephony networks (PSTNs), it is natural to use essentially the same principles and protocol structures for ATM signaling as for existing ISDN signaling. In essence, ATM signaling protocols can be considered to be extensions of the ISDN signaling protocols, but may be somewhat more complex in detail since ATM is anticipated to support many more types of services and applications than was the case for narrowband ISDN.

The fundamental B-ISDN signaling principles can be summarized based on the simple network scenario depicted in Fig. 6-1. A customer network or terminal, usually termed the "calling party" in signaling terminology, initiates a call using signaling messages transferred across the UNI to an ATM network. The signaling messages are exchanged between signaling entities (SE) in the end terminal and the ATM NE that are capable of generating, terminating, and processing the information contained in the signaling messages. The ATM network interprets the destination or so-called "called party" address and routes the call to its destination by exchanging signaling messages across the NNIs, exactly analogous to the way existing narrowband telephony calls may be routed. Finally, at the destination NE, the called party is signaled across the UNI to set up (or reject) the call. Once the calling party is notified that the VCC or VPC has been set up as requested, it may begin the data transfer phase, after which the call may be released using another set of messages across the UNIs and NNIs to enable the network resources to become available to subsequent calls.

The basic principles used for ATM signaling, as for narrowband ISDN signaling, may be summarized as follows:

1. It is based on the concept of cmmon channel signaling (CCS), which uses a separate dedicated VCC to transport the signaling messages across all interfaces. The B-ISDN signaling messages flowing on the common (i.e., dedicated) signaling channel may be used to establish, maintain, and clear one or more ATM VCCs or VPCs, either associated with the common channel or not associated with it. This is identical in concept to the common channel (e.g., D-channel) signaling mechanism used in narrowband (64 kbits/sec) ISDN

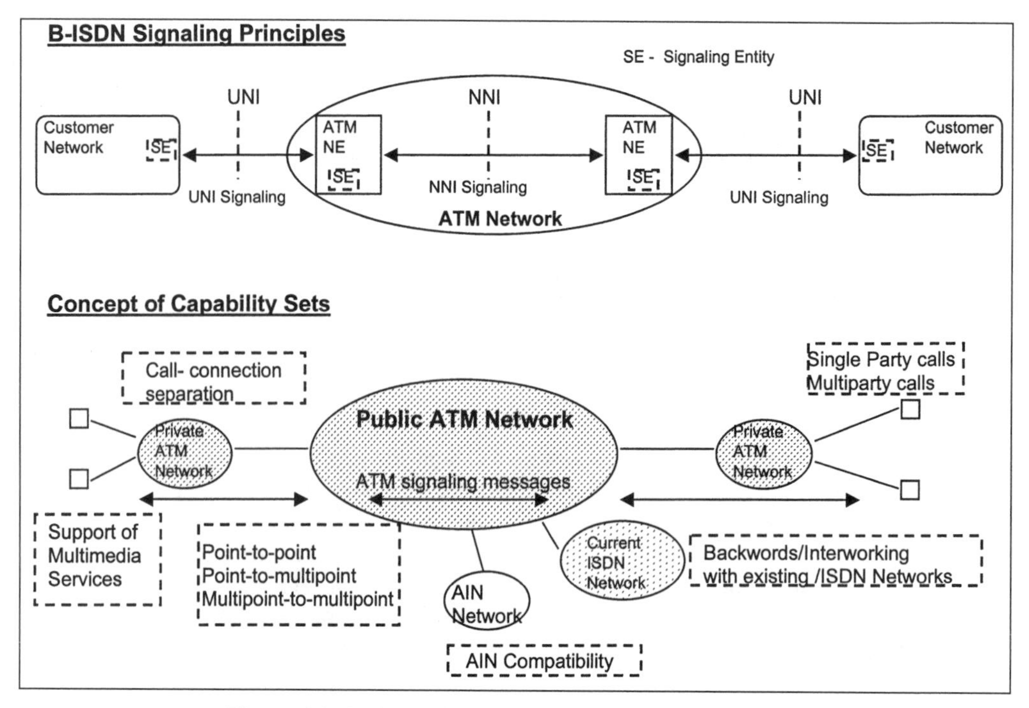

Figure 6-1. B-ISDN Signaling principles and capability sets.

for both the UNI (Digital Subscriber Signaling 1: DSS1) and the network Signaling System 7 (SS7) cases.

2. As may be already recognized, ATM signaling is "message- (i.e., packet-) based" in the sense that a dedicated set of signaling messages are used to transfer the signaling (also called control) information between the signaling entities (or applications) for either side of any given interface. The coding, semantics, and detailed procedures for processing of the signaling or control information need to be precisely defined by standards to allow for universal interoperability in the setting up, maintaining, and releasing of the VCCs or VPCs across all interfaces in question. However, it should be noted that the actual signaling message sets and procedures may be different for the UNI and NNI cases. In addition, as will be described in more detail later, for the case of NNI signaling, there are different signaling (and routing) protocols defined, ostensibly, for public carrier ATM networks and private (or enterprise) networks. This further complicates the detailed description of signaling between ATM NEs, but the basic principles of message-based signaling are common across all protocols, even though detailed semantics and procedures may differ.

The extraordinary flexibility and power that results from the adoption of message-based common channel signaling principles will come as no surprise to readers who have some general acquaintance with the existing narrowband (64 kbit/sec) ISDN-based signaling widely deployed in modern telephony networks. This system, which forms the basis of the so-called Signaling System Number 7 (SS7 for short, the number 7 deriving from an evolutionary naming terminology that originated in the earlier CCITT standards body) allows for the continuing development of a myriad of telecommunications services and functions that may be provided to network users. The same essential signaling principles may be extended to the evolving broadband ISDNs based on ATM, with the possibility of enabling service providers or network operators to provide not only the existing types of telecommunications services but also additional capabilities such as those indicated in Fig. 6-1. Thus, as noted earlier, ATM enables the support of multimedia services that integrate voice, video, and data into a single transport technology, multicast calls, and the development of the so-called advanced intelligent networks (AIN) based on distributed intelligence and databases throughout a network, as in current intelligent network (IN) architectures.

The flexibility and potential power of message-based common channel signaling technology also poses a challenge in managing the resulting complexity if all the complex and varied requirements for ATM need to be supported by the signaling protocols. This would require the processing of large amounts of signaled information by the NEs, which may make the provision of basic connectivity service costly. To avoid this, it was agreed that the ATM signaling standards should be developed in so-called capability sets (CS), as indicated in Fig. 6-1. The primary intent of a capability set was to allow for support of a given set of requirements or functions by means of the signaling procedures employed across any given interface. Thus, capability set 1 (CS1) was intended to support simple point-to-point (pt–pt) calls between single parties with a basic set of bearer services. CS2 allowed for the enhancement to point-to-multipoint (pt–mpt) calls with multimedia capability and additional service category (or ATC) support and so on. In this way a "phased" approach to introducing additional complexity into the signaling protocols may be managed in the network. In practice however, despite the attempt to introduce many additional capabilities in such a controlled way, the large number of ATM service categories and other (e.g., QoS) requirements considered important has resulted in significant burden on signaling protocols as well as charges that the ATM signaling protocols are overly complex. Whether or not this is the case, it will be recognized below that signaling message structures and semantics are designed in such a way that not all available capabilities need be supported in any given configuration. Thus, only those signaling functions thought necessary for any given set of network applications need to be implemented at any given stage, with additional capability added by means of software upgrades as the network evolves.

Before entering into the detailed description of signaling protocols, it is useful to consider briefly why they can be thought of as being both complex as well as essentially simple in principle, and the underlying reasons for their flexibility. Since the signaling information required to set up, clear, and maintain VCCs or VPCs are carried in messages over a separate dedicated "common channel," the control network can be considered as a separate network with its own architecture and dedicated resources. In this model, the control (i.e., signaling) network may not only be used to set up and clear VCCs or VPCs on demand, but also exchange control information while the VCC or VPC is being used to exchange the user data, without disturbing the user data flow. In this way, the control network may be used to access distributed databases, carry maintenance information, or alter the characteristics of the VCC or VPC while the call is in progress. As can be envisaged, the availability of an independent network dedicated to control processes resulting from the common channel signaling concept enables a range of applications to be devised in addition to the basic on-demand connection control. This independence between the control architecture for the VCCs or VPCs and the so-called "bearer channels" carrying the user data cell streams has also led to the concept of separation of the "call" from the "connection" (in this case, a virtual connection). In this so-called "call–connection separation" concept, a "call" is viewed as an association between two arbitrary signaling entities, set up via the control network. Subsequently, connections (virtual) may be set up, added to, or even modified as required by the given call, thereby enabling even more flexibility in the use of the control network. This concept of call–connection separation is particularly useful in the extension of AIN capabilities to ATM or B-ISDNs, which is currently under consideration in the ITU-T.

Not only does the availability of a dedicated control network provide B-ISDN with unprecedented flexibility, but the fact that variable-length signaling messages are used allows the exchange of potentially vast amounts of signaling-related information between ATM NEs and users as well. Thus, information relating to required QoS, traffic characteristics and related ATCs, higher-layer applications, routing, and many other aspects may be coded into the signaling messages to be duly processed by the NEs in setting up, clearing, or modifying an on-demand VCC or VPC. To a large extent, the common perception of complexity in ATM signaling arises from this wealth of required information, its coding, semantics, and associated procedures, which need to be specified in minute detail to ensure interoperability and robustness in processing the signaling information in NEs. However, even though the amount of such signaling information or parameters is large and keeps growing as new requirements and capabilities are added in response to market needs, the fundamental mechanisms by which this information is coded and handled by the signaling message sets remain the same. Consequently, by clearly understanding the structure and semantics of the signaling message sets and the basic procedures involved, it is possible to obtain a working knowledge of the signaling framework, which may be used as a basis for further detailed study of the relevant specifications, if required.

An additional complication that needs to be clarified at this stage concerns the use of different signaling message sets (and even terminology) at the UNI and NNI, and differences in the protocols that have arisen as a result of standardization initiatives undertaken somewhat independently in the ITU-T and the ATM Forum based on differing general requirements. Here, apart from the purely technical considerations that are relatively easy to understand, it must be admitted that historical factors and differing perspectives on ATM network applications come into play, and these are somewhat more difficult to rationalize. These differences in perspective primarily result from the quite different mandate and make-up of these bodies but need not concern us in this discussion. It may be noted that an extensive effort is under way to narrow differences in the protocols where they may exist, and to provide clear guidelines for interworking between the protocols in various networking scenarios. These issues will become clearer later, after the basic protocol differences are described in more detail below.

Setting aside any differences of detail between the ITU-T Signaling Recommendations and the ATM Forum specifications at this stage, it is not difficult to first understand why the signaling message sets and associated procedures at the UNI may need to be different from those required between ATM NEs, i.e., at the NNI. At a typical UNI, the signaling procedures are essentially used to request the setting up (or clearing) of an ATM VCC or VPC with specified QoS, traffic parameters, and service category, among other parameters. A relatively simple set of signaling messages and associated set of procedures may be used to support the user signaling requirements for most applications between a "user" (which may be an end-terminal or an enterprise network) and a wide area network (WAN). The protocols may be inherently asymmetric. However, for the case of NNI signaling, this is clearly not the case, since the signaling protocols may incorporate many additional functions required for maintenance of the control network, as well as for reliability, database access, charging, and so on. As a consequence, the signaling protocols across NNIs tend to be far more complex than those required for the UNI case.

These differences between UNI and NNI signaling requirements are clearly exhibited for the case of narrowband (64 kbit/sec-based) ISDN in the ubiquitous telephony network. Here UNI signaling is typified by the so-called Digital Subscriber Signaling 1 (DSS1) system, comprising a relatively simple message set with the associated procedures defined in exhaustive detail by the ITU-T. However, for NNI signaling between the ISDN switches in carrier networks, the much more complex and highly reliable Signaling System Number 7 (SS7) protocols are used in most advanced telephony networks. The dedicated SS7 control networks employ a different protocol architecture and procedures designed to provide extremely high performance and reliability in the handling of the signaling messages. To enable these stringent requirements to be fulfilled, a much larger message set and more complex procedures need to be defined, which include functionality for the maintenance and testing of the SS7 control network.

The same concepts have been extended to the case of ATM-based B-ISDNs by the ITU-T. Thus, for the UNI signaling, the extension of the DSS1 for ATM is called Digital Subscriber Signaling Number 2 (DSS2). DSS2 simply extends the basic DSS1 message set by incorporating the additional signaling information necessary for ATM, such as traffic descriptors, QoS classes, etc., while preserving essentially the same generic message structures and coding rules. It should also be noted that the UNI signaling specifications developed by the ATM Forum [6.2] are essentially based on the DSS2 standards, although there are a number of differences of detail, which may create interoperability issues if not accounted for, but need not concern us at this stage.

Similarly, for NNI signaling for the case of ATM, the B-ISDN concept is to extend the ISDN SS7 control architectures and associated signaling protocols by incorporating those parameters and protocols necessary to support robust ATM signaling for WANs. Thus, the narrowband ISDN User Part (ISUP) call control protocols are extended to B-ISDN User Part (B-ISUP) call control protocols by adding the requisite parameters and procedures for ATM. Moreover, since the signaling messages may be themselves transported over dedicated (i.e., common channel) ATM VCCs, only a part of the Message Transfer Part (MTP) protocols are required for the B-ISDN case, as will be described in further detail below. From this broad perspective, the control architecture and signaling protocols for ATM B-ISDNs may be viewed essentially as a logical extension of the existing narrowband ISDN SS7 based architectures to incorporate ATM-specific attributes.

However, for the case of NNI signaling, it must be noted here that the ATM Forum has developed two distinct approaches depending on whether the network in question may be considered a "public carrier" network or a "private ATM" network. Disregarding for the moment the thorny question as to whether such a distinction exists in the increasingly deregulated telecommunications environment evident today, it is still useful to understand the initial reasons for this divergence. The ATM Forum recognized the importance of providing enterprise networks ATM solutions using SVC capability. However, for such scenarios it was clear that the public-carrier-oriented SS7 control architectures were unnecessarily complex and hence costly. Consequently, a so-called "private NNI" (PNNI) signaling and routing specification was developed to address enterprise and campus signaling requirements. As described in detail later, the basic approach adopted for PNNI signaling was to extend the UNI message set and signaling procedures to be applicable between ATM NEs. In addition, it included source routing methodology similar to that used in IP internetworking to provide the potential for so-called "dynamic"(or automatic) routing through the network. The initial so-called Inter Switch Signaling Protocol (ISSP), also known as PNNI version 0 (see [7.2]), simply extended the DSS2-based UNI v.3.1 [5.10] signaling procedures across the NNI, while providing for static (i.e., preconfigured) routing. Subsequently, a dynamic routing protocol was added and the signaling functionality further enhanced to provide the PNNI v.1 specification (see [7.3]). As will be seen in more detail subsequently, the resulting control architecture is quite different to SS7-based protocol architecture developed by the ITU-T for B-ISDN signaling.

In parallel with the above "simplified" signaling specifications for private NNI use, the ATM Forum also developed specifications derived from the ITU-T B-ISDN control architecture for the so-called Broadband Inter Carrier Interface (B-ICI). The B-ICI signaling architecture is intended as a "subset" of the B-ISDN (SS7-derived) control architecture intended for public networks [6.1]. The significant differences between this B-ISUP-based approach and that adopted for private NNIs inevitably requires the necessity of interworking between PNNI-based and B-

ICI- (or B-ISUP) based networks. Recently, efforts have been initiated to specify these interworking requirements in the ATM Forum and ITU-T. It is also important to recognize that although private NNI (PNNI) signaling was originally intended primarily for use in enterprise ATM networks, its widespread availability and useful feature set has resulted in many public network operators utilizing it to provide SVC services, particularly in North America. This increasing use of PNNI signaling and routing in public as well as private networks, together with the use of B-ICI-based control architectures in some public carrier networks, has resulted in considerable uncertainty and confusion among operators as regards to which signaling protocol architecture should be utilized, as well as some urgency in specifying the interworking between the different protocols being deployed.

6.2 ATM SIGNALING PROTOCOL ARCHITECTURES

In this section, we review the basic signaling protocol architectures that have evolved for ATM signaling for both UNI and NNI. As noted above, ATM signaling may be viewed as an extension of the currently used 64 kbit/sec (narrowband) ISDN signaling procedures and uses the same essential message structure. As shown in Fig. 6-2, the ATM signaling messages contain so-called information elements (IEs) that carry the signaling specific information such as:

1. Called party and calling party addresses
2. Traffic descriptors (bandwidth requirements)
3. QoS classes or parameters
4. VPI/VCI values (connection identifier)

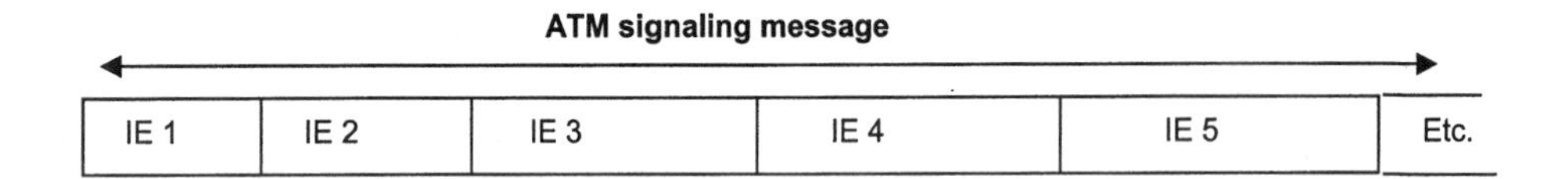

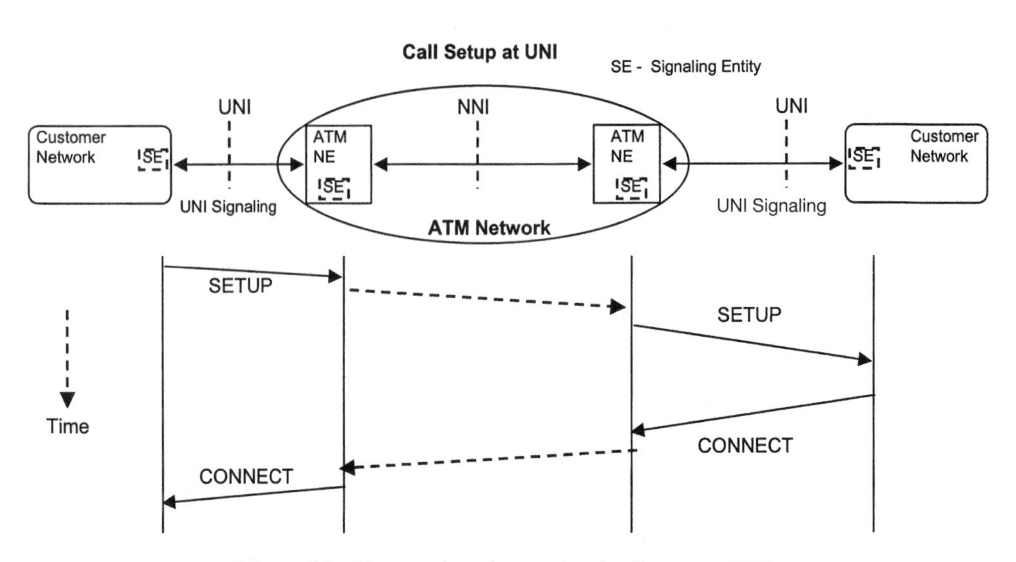

Figure 6-2. Message-based control and call setup at UNI.

and so on. In the ITU-T B-ISUP signaling message set, these information elements are termed "parameters," but it should be noted that this is essentially a terminological difference, and the same type of signaling information may be carried in B-ISUP parameters. In either case, the structure, coding rules, and semantics of each IE or parameter in each signaling message must be defined in precise detail in the signaling standards for such message-based signaling protocols to interoperate.

The different information elements (or parameters) are concatenated together to form ATM signaling messages of varying lengths. In principle, a large number of consecutive IEs may be added to a given signaling message, potentially resulting in considerable complexity since these IEs will need to be processed by the call processing entity (software) in the NEs. These processing requirements may often determine the call (or connection) performance of the NEs, so that it is important to use only those IEs that are necessary to perform the specific signaling functions of a given message. Thus, not all the signaling messages contain all the information elements, and each IE may be specified as either mandatory (M) or optional (O) in any given signaling message. The signaling protocols use separate messages for:

1. Setting up a connection—the so-called SETUP message for UNI and PNNI protocols and Initial Address Message (IAM) in B-ISUP-based protocols
2. Clearing a VCC or VPC. so-called RELEASE message in both UNI and NNI protocols

and so on. A more complete list of the signaling messages is described below. It should be noted that there are differences in terminology between the UNI and NNI messages, even though the basic function may be the same.

An example of a message-based call set-up at the UNI is shown schematically in Fig. 6-2. It should be noted that the signaling standards and specifications often use such "time–space" diagrams to indicate the time relationships of the flow of the signaling messages through the network. In this example, the SE of a user requests a VCC or VPC by sending a SETUP message across the UNI to the ATM NE (its local "exchange"). This message must contain the destination address and other required IEs indicating the characteristics of the requested VCC or VPC. Assuming that sufficient resources are available, as determined by the CAC functions along the connection path, the connection set-up information will be routed through the ATM network towards the destination ATM NE in much the same way as existing voice calls are routed. The far end destination ATM NE will then send a SETUP message to the destination terminal SE, which, if able to accept the call, will return a so-called CONNECT message back to the calling party to establish the call. These signaling messages propagate through "dedicated," or "common channel" ATM VCCs which are preestablished for signaling in the network. It should be noted that the signaling messages propagating across the NNIs inside the network are deliberately not specified at this stage. These will be defined later, depending on which NNI signaling protocol is being used.

The simple connection set-up procedures summarized above may be considered as the basis for extension to the other signaling functions required across UNIs, such as releasing VCCs or VPCs, or notifying users that calls are proceeding, or that the called party has been alerted, and so on. The functions of the various UNI signaling messages are described below, but it is useful here to consider the overall protocol stacks that are used at the UNI and NNI, and their relationship to the existing (narrowband) ISDN protocols. In Fig. 6-3 the UNI signaling protocol architecture for both ISDN and B-ISDN are compared. For the 64 kbit/sec (narrowband) ISDN case, the physical layer (i.e., the transmission path) may be a conventional 64 kbit/sec TDM channel or the D-channel of the 2B + D ISDN configuration for the basic rate interface (BRI). The signaling messages defined for the ISDN call control layer in ITU-T Recom-

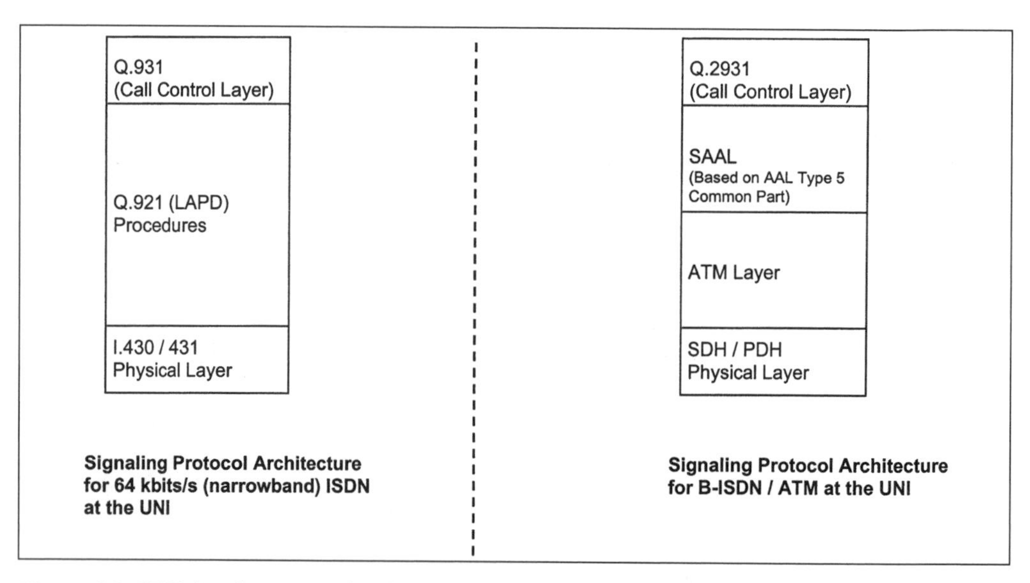

Figure 6-3. UNI signaling protocol architecture: ISDN and B-ISDN. In B-ISDN the ATM layer and AAL are used to transport the call control messages between ATM NEs.

mendation Q.931 [6.9] are encapsulated in the Q.921 link access protocol over D-channel (LAPD) frames (strictly the protocol data units at layer 2) and transmitted over the physical transmission media as defined in ITU-T Recommendations I.430 [6.10] and I.431 [6.11], for example. The call control layer functionality in this model is responsible for the generation as well as the interpretation (sometimes called "parsing") of the various information elements (IEs) in the signaling messages. The means by which this processing is done by the software within the SEs is implementation-dependent, and may be quite complex for fully featured ISDN interfaces.

For the B-ISDN (or ATM) UNI, it may be noted that the protocol architecture is essentially similar, except that here the dedicated ATM VCCs are used to carry the signaling messages generated and processed by the functions associated with the call control layer, as specified by ITU-T Recommendation Q.2931 [6.8]. Note that in the ITU-T nomenclature, the prefix "2" in the Q-series recommendations indicate that the associated protocols relate to broadband (i.e., ATM) aspects of ISDN. For the ATM case, the signaling messages that are generated (as well as interpreted on receipt) by the call control (Q.2931) layer functionality are encapsulated in a signaling-specific ATM adaptation layer (AAL), often termed the signaling AAL (SAAL). The various functions of the SAAL protocols are described in more detail in subsequent chapters, but it needs to be noted here that the main purpose of the SAAL is to ensure the reliable and assured transfer of the Q.2931-based signaling messages. The SAAL utilizes the so-called "common part" (i.e., service-independent) of the AAL Type 5 protocol, but adds additional capability above this to achieve reliable transfer of messages. The SAAL encapsulated Q.2931 messages are then segmented into ATM cells with the preassigned VCI value (i.e., VCI = 5 in any VPL) for signaling, or in a dedicated VCC set up only for signaling. As noted earlier, the physical layer (i.e., transmission path) functionality for ATM may be based either on the conventional PDH or SDH transmission hierarchies. Thus, for B-ISDNs the SAAL + ATM layer protocols perform the same functions as the Q.921 (LAPD) protocol used to carry signaling messages in the ISDN case, implying that in B-ISDN, the ATM Layer and SAAL functions are used to transport the call control messages between the NEs across any given interface.

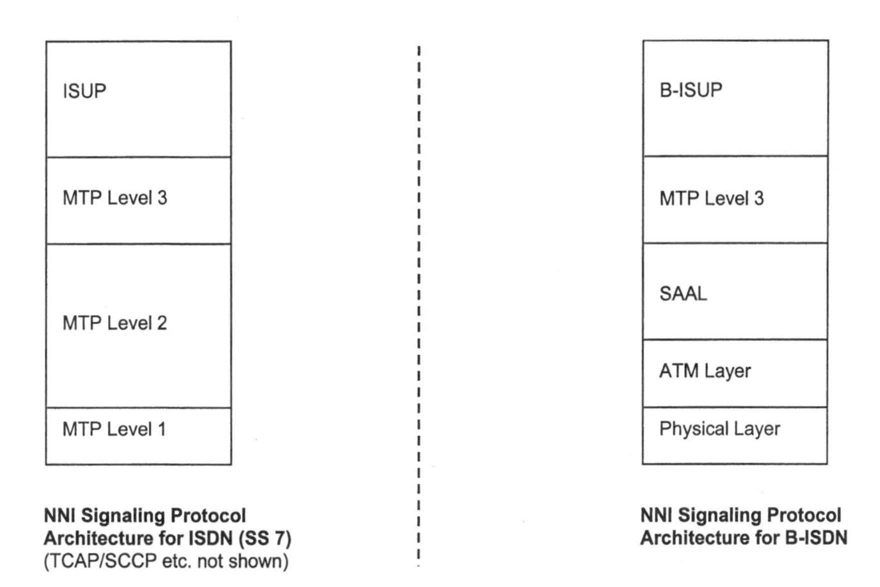

Figure 6-4. NNI signaling protocol architecture: ISDN and B-ISDN. For B-ISDN NNI signaling, the lower layers of the message transfer part (MTP) are replaced by ATM/SAAL. The call control protocol (ISUP) is extended to accommodate ATM/B-ISDN (B-ISUP).

This same basic concept of using ATM to carry the signaling messages for B-ISDN is also applicable for the case of NNI signaling. However, here the NNI signaling protocols architecture is somewhat more complex, as shown in Fig. 6-4, corresponding to the more complex SS7-based NNI signaling protocol architecture of 64 kbit/sec narrowband ISDN. In this architecture, the call control functions of the ISDN user part (ISUP) layer use the services provided by the so-called message transfer part (MTP) levels 1–3 to carry the signaling messages between the NEs. In SS7 nomenclature, the MTP level 1 protocol corresponds to the physical layer (transmission) functions used also in ISDN UNI signaling, as noted earlier. Similarly, MTP level 2 is essentially the equivalent of the LAPD protocol indicated above in ISDN UNI signaling, with some additional functionality to enhance the reliability and maintenance capabilities necessary for robust public carrier network applications. For the B-ISDN protocol stack, the MTP Level 2 functions are replaced by the ATM layer + SAAL functions, as was the case for UNI signaling. However, it should be noted here that the SAAL specified for ATM NNI signaling is slightly different from the SAAL used for the simpler UNI signaling protocol stack. This is primarily because the NNI SAAL protocol is required to provide layer services to the MTP level 3 protocols, rather than directly to the Call Control layer, which in the NNI case is B-ISUP.

It is important to recognize that the B-ISDN signaling protocol architecture retains the use of the message transfer part level 3 (MTP 3) protocols, which constitute a significant element of the ISDN SS7 architecture. The functions provided by the MTP level 3 protocols are a major factor contributing to the high reliability and flexibility of the SS7-based signaling approach. Although it is outside the scope of this text to provide a detailed account of the MTP 3 protocols, it should be noted that the MTP level 3 protocol provides a highly reliable connectionless layer service to the B-ISUP Call Control layer and other higher layers utilized for the full SS7 capability. The fact that MTP 3 is essentially a connectionless packet transfer mode allows for the signaling messages to be routed through the SS7 network in either "associated" or "nonassociated" mode with respect to the actual connection path of the resulting VCC or VPC. In practice, the degree of flexibility implied by nonassociated signaling may not be required for most applications involving simple VCC establishment and release, so that the "associated signaling" capability of MTP 3 is generally adopted as, for example, specified in the ATM Forum's B-ICI specification [6.1].

The capabilities provided by the MTP level 3 protocol clearly add complexity to NNI signaling in comparison to the direct UNI case described earlier, and constitutes one of the main reasons for the development of the alternative "private" NNI signaling approach adopted by the ATM Forum. In this context, it should also be stated that the capabilities provided by the call control (B-ISUP) protocols layer are also significantly more than those available for UNI signaling, for the reasons noted earlier. As will be described in more detail later, the B-ISUP protocols incorporate additional messages for maintenance and test of the signaling network that were considered unnecessary for supposedly simpler private network requirements. Nonetheless, to meet the stringent performance and reliability requirements expected of conventional public carrier networks, the maintenance and test capabilities built into the MTP 3 plus B-ISUP protocol architecture clearly warrant the additional complexity involved in its implementation. It is interesting to reflect that with the increasing use of the private NNI signaling (and routing) protocols in what may normally be considered as carrier networks, there is a trend to include additional capability into the PNNI protocols to provide some of the reliability attributes expected of the traditional SS7-based signaling. As will become more evident after the more detailed description of PNNI signaling and routing protocols, it is becoming less clear as to which of these protocol architectures is actually simpler to implement or preferable in any given network signaling application.

Aside from these more general considerations, we return to the description of the overall B-ISDN signaling protocol architecture. In Fig. 6-5, the combined UNI and NNI protocol architectures (or "protocol stacks" as they are often referred to) are shown as they would be colocated within the call processing signaling entities of the ATM NEs. The signaling protocol architecture across the UNI also depicts another function that will be described below. This is the so-called meta-signaling entity (m), which actually is defined as a function in the layer manage-

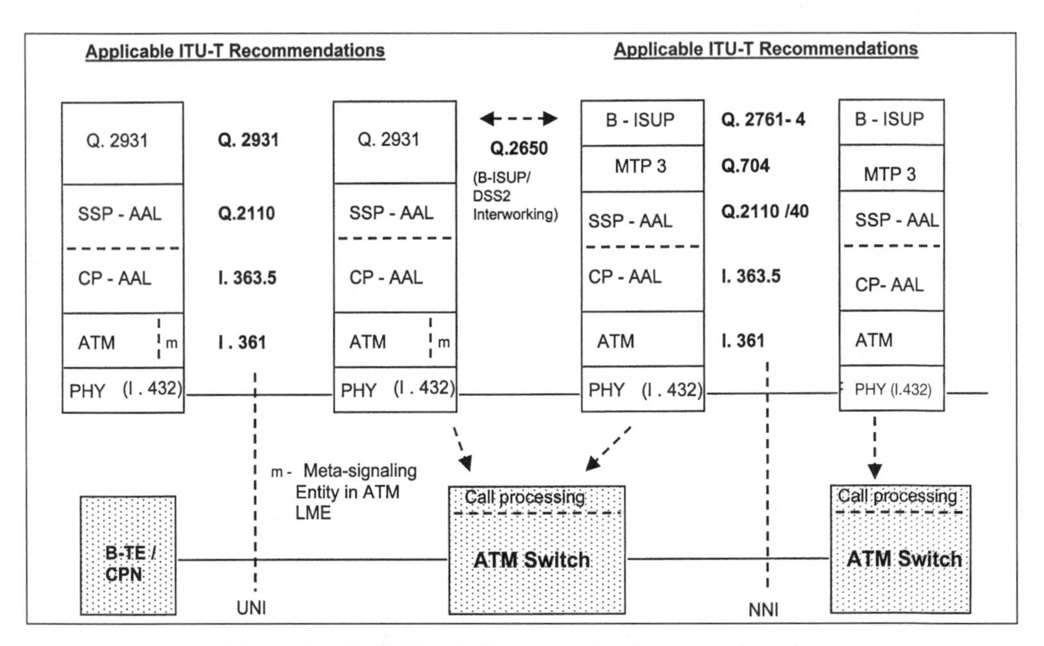

Figure 6-5. B-ISDN signaling protocol architecture overview.

ment entity (LME) of the ATM NE, although it is related to signaling (or control plane functions). It should also be noted that the signaling AAL (SAAL) is further sublayered into the service-specific part (SSP) and the common part of AAL (AAL Type 5) in this diagram. A more detailed description of this sublayering of the SAAL is given later when we discuss specific AAL protocols for both user and control plane applications. Fig. 6-5 also includes the relevant ITU-T Recommendations which describe the protocols relating to any given layer, as well as the ITU-T Recommendation Q.2650 [6.12] which addresses the requirements for "interworking" between the DSS2 (Q.2931 [6.8]) and B-ISUP call processing within the NE. This arrangement is directly analogous to the ISDN case, where a "local exchange" must "translate" the user-side DSS1 signaling information into the network- (or trunk-) side SS7 messages in order to route the call. For "tandem" ATM NEs (i.e., those with only NNI or "trunk" interfaces), there is clearly no need to perform DSS2 to B-ISUP interworking.

Based on the relationships between the signaling protocol architectures at the UNI and the NNI and the translation between them required, as shown in Fig. 6-5, the general modeling of the signaling functions within the ATM NEs can be represented as shown in Fig. 6-6. Here, the on-demand switched VCCs between two terminal equipments (TE) in two remote customer premises networks (CPNs) are connected via two ATM NEs employing B-ISUP-based NNI signaling in accordance with the protocol architectures described above. Ignoring for the moment the metasignaling virtual channel (MVC) also shown in Fig. 6-6, we note, as mentioned earlier, that in ATM the preassigned VCI value VCI = 5 in each virtual path link (VPL) has been defined as the (common) virtual channel for carrying the point-to-point signaling messages between the SEs in each NE (and TE). The call processing SE associated with the CPN (or TE) uses the UNI message set (Q.2931) to request VCC or VPC set up to the far end terminal equipment (or CPN). Assuming for now that sufficient resources are available in order to support the QoS and traffic parameters, etc. of the VCC or VPC requested, the signaling entity in the "local" ATM NE processes the call and assigns the appropriate VPI/VCI values by means of the switch connection control functions to the connection memory (CM) of the NE. The connection memory serves as the "look-up" or context tables that enable the cross-connection (or translation) be-

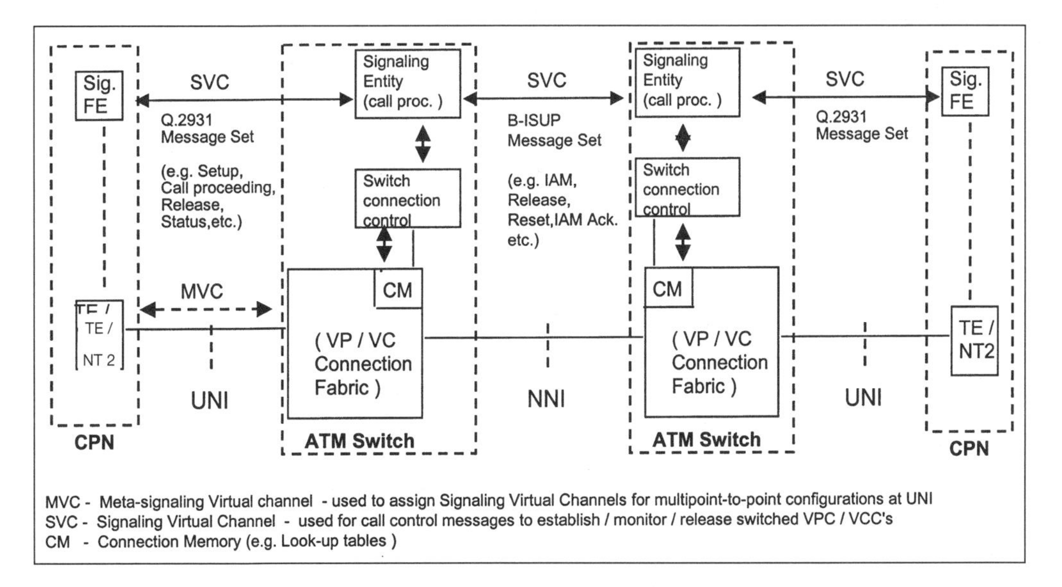

Figure 6-6. Signaling procedures overview. In ATM, the preassigned VCI value VCI = 5 (in each VPL) has been defined as the virtual channel for carrying point-to-point signaling messages.

tween the incoming and outgoing VPI and VCI values corresponding to the incoming and outgoing VPL and VCL, respectively.

The call processing in the signaling entity (or application) also routes the call onto the next ATM NE using the B-ISUP NNI message set via the B-ISDN SS7 signaling network. It should also be noted that each ATM NE assigns the virtual connection links (VCL) on a bidirectional basis, using the same VPI and VCI values for both directions of the VPC or VCC. Once the switch connection control entity has configured the connection memory (or context tables) in this way and the end-to-end connection has been set up, the data transfer phase may commence. During the data transfer phase, the VP or VC connection (switching) fabric essentially switches the user ATM cells by "translating" the VPI and VCI values in accordance with the connection memory from the incoming VPL or VCL to the outgoing VPL or VCL on any given transmission path. The selection of the relevant ingress and egress physical transmission paths and the contained VPLs or VCLs depend on the routing policy or protocols in use for call routing through the SS7-based network, by analogy with the existing telephony-derived routing methodology. The routing aspects relating to ATM connectivity are discussed later, and may be considered somewhat separately from the signaling protocols, even though the two are clearly related for the purposes of connection set-up.

The general functional model for call and connection processing described above should not be considered as a detailed or rigorous functional breakdown of the ATM NE, but rather as an aid to understanding how the call processing entities are used to implement the standardized message-based signaling protocols defined for the UNI and NNI. In addition, it will be noted that the NNI signaling protocols considered above are based on the ITU-T SS7 B-ISUP architecture. The PNNI protocol architecture, which is based on an extension of the UNI protocol stack, will be considered later.

6.3 METASIGNALING

It was mentioned above (see Figs. 6-5 and 6-6) that a so-called "metasignaling" function has been defined for use across the UNI. Metasignaling refers to the protocol by which an ATM terminal equipment (TE) indicates to the network that it requires an ATM VCC to be assigned to it for purposes of signaling. It is important to stress that this VCC is intended *only* for transport of signaling messages (across the UNI) between the signaling entities (or applications), and not for user data. In principle, the metasignaling concept is analogous to the "off-hook" signal of conventional telephony. The essential concept of metasignaling is that when a user (or TE) requires a dedicated virtual channel connection for signaling, it sends a standardized "metasignaling cell" across the UNI to the "local"ATM NE. The metasignaling cell is distinguished by a preassigned VCI value, as noted earlier (VCI = 1 for any VPI), and is intended to be processed by the layer management entity (LME) in conjunction with the signaling functions in the NE. The metasignaling cell contains the information regarding the terminal (i.e., the address and service profile, etc.) to enable the NE to assign a dedicated VCC to it for signaling. Note that this VCC is different from the preassigned VCI = 5 which is always available for point-to-point signaling.

The question arises as to why it is at all necessary to use metasignaling when a preassigned virtual channel (VCI = 5 in any VPL) is always available to a terminal for signaling. As depicted in Fig. 6-7, the intent of metasignaling is to provide a mechanism for configurations such as multipoint-to-point signaling, or when a user intends to use a different dedicated VCC for signaling from the default (VCI = 5) VCLs. For the simple multipoint-to-point network configuration at the UNI shown in Fig. 6-7, the intent of metasignaling is to provide the information to the ATM NE to distinguish the TE requesting the VCC for signaling by incorporating the relevant

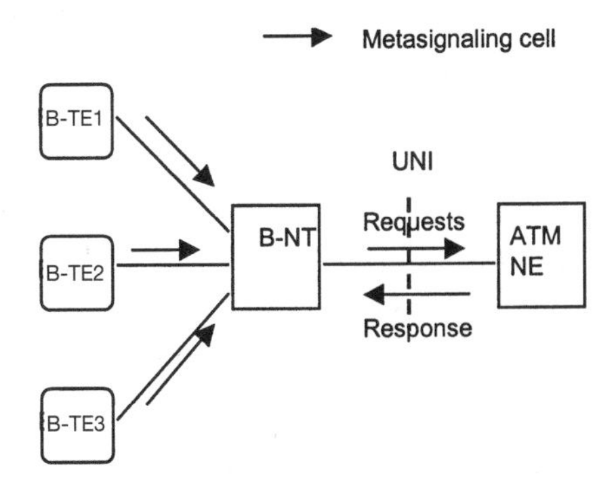

Figure 6-7. Metasignaling.

identifier (or address) within the metasignaling cell requesting the VCC for signaling purposes. The response from the ATM NE is a metasignaling cell that includes a connection identifier (i.e., VCI value) to enable the broadband network termination (B-NT) functional element to set up the dedicated VCC for signaling by the terminal.

The metasignaling capability was introduced to provide an additional degree of flexibility for ATM UNI signaling, particularly for multipoint-to-point configurations. However, it does introduce some additional complexity into both the end terminals and the NE generally thought to be unnecessary for most applications. For this reason, the metasignaling function has not been typically implemented. It is significant to note that metasignaling is not required in either the existing ATM Forum specifications or the Capability Set 1 (CS1) ITU-T Recommendations, despite the fact that the metasignaling procedures have been defined for some time now (in ITU-T Recommendation Q.2120 [6.13]). Thus, even though metasignaling capability is not currently used in the predominantly point-to-point UNI signaling configurations, it is useful to bear in mind its potential for future use in more complex ATM signaling network configurations.

6.4 ATM CONNECTION TYPES

In describing ATM VCCs or VPCs up to now, we have generally not made any formal distinctions between the possible connection types that may be required in any specific network application. One fundamental distinction that has been almost tacitly assumed in the discussion on signaling is clearly between the so-called permanent virtual circuits (PVC, also sometimes referred to as semipermanent virtual circuits) and the on-demand switched virtual circuits (SVCs) that are the subjects of signaling procedures. The basic differences between these two connection types is illustrated in Fig. 6-8. The primary characteristic of PVCs is that they are configured (or provisioned) by the network management system (or TMN) through the management interface (or local craft interface) of the NEs in the connection path, in response to a service request from a user. Consequently, PVCs may require relatively long set-up times, and may be left active for relatively long periods (e.g., on the order of days or months) without change. Initially, PVCs may have been manually configured, with the appropriate VPI/VCI values statistically set up in the ATM NE (cross-connects). More recently this process may be automated by means of sophisticated network management applications software, enabling the PVC provisioning process to be made easier and less prone to error.

In contrast, it may be seen that on-demand switched virtual connections (SVCs) are inherently dynamically "provisioned" by means of the information carried in the signaling messages ex-

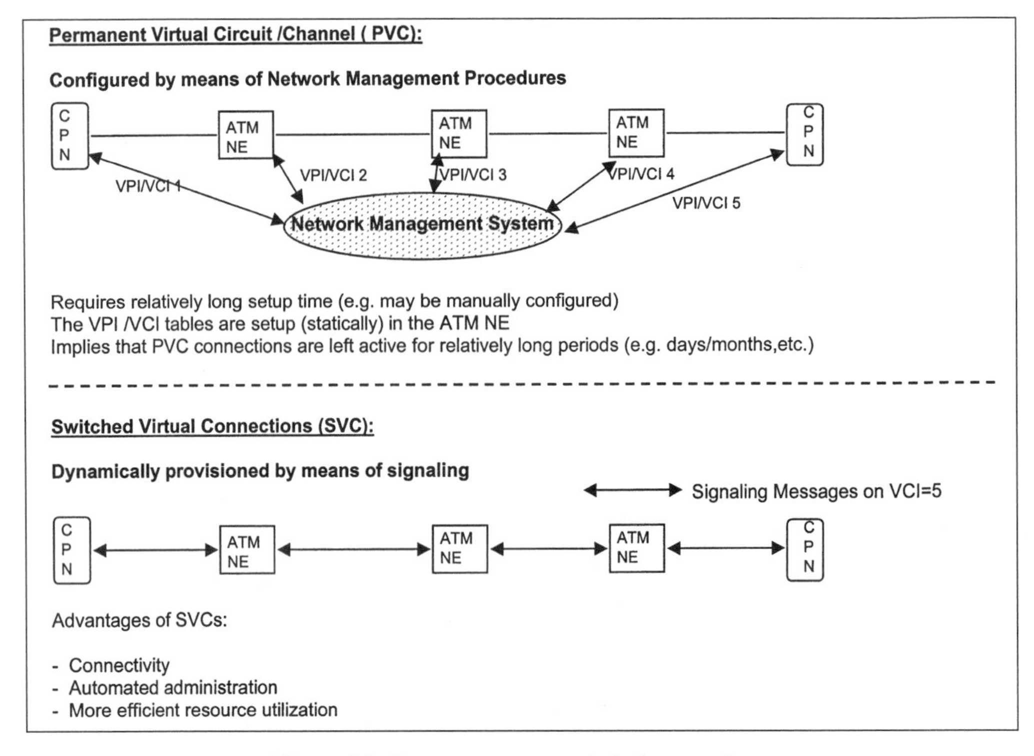

Figure 6-8. Permanent versus switched connections.

changed between the ATM NEs (switches). The fact that the VCCs or VPCs may be "automatically" set up and released clearly has advantages in comparison with PVCs. The major advantages of SVCs are:

1. More efficient resource utilization, since bandwidth may be made available for other connections when the data transfer phase is over
2. The connectivity administration is automated with associated simplification and hence costs less
3. Ubiquitous connectivity between all end terminals reachable by the network
4. Enhanced service offerings (and hence revenue opportunities) by utilizing advanced intelligent network (AIN) capabilities such as credit card calls, etc.

In these and other respects, the SVCs are no different from existing voice or ISDN connections in the global telephony networks.

There is a third type of connection, not shown in Fig. 6-8, that has some of the attributes of PVCs as well as of SVCs, and may therefore be viewed as somewhere between conventional PVCs and SVCs, as described above. This type of connection has been termed the so-called soft permanent virtual circuit (SPVC), and is increasingly important for a growing number of network applications. Soft PVCs are essentially PVCs that use the ATM signaling procedures between two points along a connection path (typically between the network side of two UNIs) to set up the VCC (or VPC). The mechanism underlying SPVCs will be described in more detail when discussing PNNI signaling later, but it is interesting to note here that in using the inherently automated signaling procedures instead of the management interfaces to provision a PVC, the

SPVC capability essentially bridges the control plane and management/TMN plane functionality while retaining the advantages of both.

In addition to the basic distinctions between permanently provisioned and on-demand switched VCCs or VPCs, ATM connections may also be categorized in terms of their connectivity configuration, as shown in Fig. 6-9. The two most common connection types in this sense are:

1. Point-to-point (pt–pt) connections
2. Point-to-multipoint (pt–mpt) connections

Two other connection types (not shown in Fig. 6-9) that have also been defined but are currently less used are:

3. Multipoint-to-point (mpt–pt) connections
4. Multipoint-to-multipoint (mpt–mpt) connections

The essential attributes of simple pt–pt connections, whether they are unidirectional or bidirectional, should be evident by now. However, in order to support the pt–mpt connections, which may be essential for some ATM network applications such as video (e.g., TV) distribution, it should be noted that the "branch"-point NEs should be able to copy the user data cells in order to send them to the "leaves" of the pt–mpt VCC or VPC. In such configurations, the VCL (or VPL) prior to the branch point is termed the "root" of the pt–mpt connection. The "copy" function at the branch nodes requires the identical replication of the incoming cells, which are then transmitted on the leaf VCLs (or VPLs) along the relevant connection paths. It should be

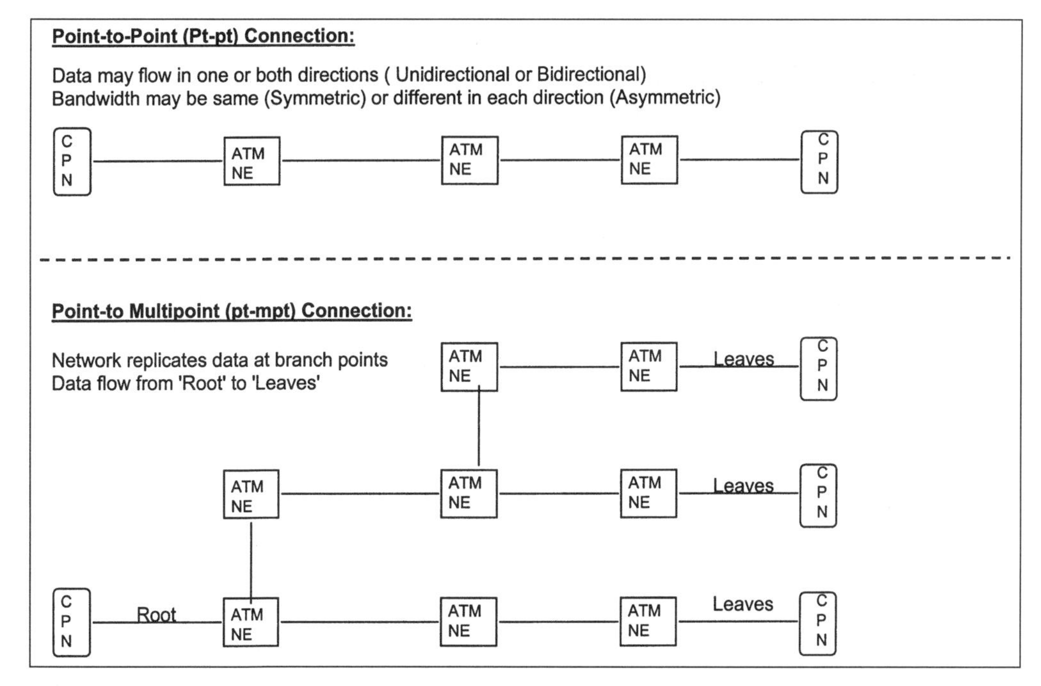

Figure 6-9. Point-to-point and point-to-multipoint connections. Point-to-multipoint capability is important for some ATM applications (e.g., video distribution).

pointed out that even though the pt–mpt connections may be established as bidirectional (that is, with bidirectional assignment of the VCI and VPI values at each NE), typically, the user data flow is unidirectional from the root toward the leaves. The reverse path (in the direction from the leaves toward the root) may be used as a return path for the associated ATM layer OAM cell flows (as described in detail subsequently), but most current applications of the pt–mpt connection type consider only a unidirectional user data cell flow from the root toward the leaves.

The reasons for this can be understood when we consider the implications of the multipoint-to-point configuration, which essentially corresponds to the "reverse" direction of the pt–mpt connection type described above. For the unidirectional mpt–pt VCC or VPC (not shown in Fig. 6-9), it is evident that the cell flows from the "leaves" toward the "root" will need to be "merged" at the branch NEs. The merging function implies that the user data cells from the leaves should be interleaved onto the same outgoing VCL (or VPL) in the direction of the root. In effect, the merging of the various cell streams into one outgoing cell stream may be viewed as the logical "opposite" of the copy function required for the (unidirectional) pt–mpt connection type considered earlier. However, the difference is that in order to separate (or deinterleave) the user data cells from the end-point leaves in the merged cell flow, the VCI (or VPI) values cannot be used, and it is hence necessary to use additional end-point identifiers in the individual cells at a layer above the ATM layer functions. Consequently, the handling of the merged cell flow may require additional functionality at the root end of the connection to deinterleave (so to speak) the cells interleaved as a result of the "merging" operation.

It is important to distinguish the normal multiplexing/demultiplexing at the ATM layer from the cell merging (or interleaving) operation required to realize a unidirectional mpt–pt connection or, for that matter, a bidirectional pt–mpt connection type. As described earlier, normal ATM multiplexing and demultiplexing utilizes the VPI and/or VCI values associated with each logical virtual path (or channel) link (VPL or VCL) to distinguish cells belonging to any given VPC or VCC that comprise the concatenation of the VPLs or VCLs. However, since a merged cell flow has the same VPI and VCI values by definition, it is clear that an additional multiplexing "identifier", typically at the AAL or higher layers, is needed in each cell in order to distinguish which leaf that particular cell originated from. In this sense, merging may require the use of demultiplexing above the ATM layer, or some other mechanism to separate the cell flows. It should be noted that although the cell merging function is clearly more complex than the simple cell replication required for pt–mpt connections, the use of bidirectional pt–mpt VCCs to more efficiently utilize some ATM network applications such as emulated LANs (LANE) and IP over ATM has resulted in recent interest in merging capability in ATM NEs.

Although also not shown in Fig. 6-9, the case of the multipoint-to-multipoint connection type constitutes the most general (and complex) configuration, enabling bidirectional connectivity between multiple end-points. The mpt–mpt connection type may be viewed as a superimposition of bidirectional pt–mpt connections to realize the mpt–mpt configuration, and it approximates the connectivity available to terminals on a "shared" media environment such as a LAN, where any terminal may communicate with any other on the shared media, subject, of course, to the constraints imposed by the MAC protocols. It is interesting to note that although such a capability is relatively simple to realize in a "shared media" paradigm, the mpt–mpt connectivity still poses technical challenges in the inherently pt–pt ATM configurations.

6.5 THE UNI SIGNALING MESSAGE SET

The brief survey above of the signaling protocol architectures for UNI and NNI, as well as the description of the connection types, has indicated how ATM mechanisms are related to and have evolved from existing narrowband ISDN signaling. It should therefore come as no surprise to note that the basic UNI signaling message set summarized in Fig. 6-10 is essentially the same as that defined for the ISDN case. The UNI signaling messages may be subdivided into three basic categories:

1. Call establishment messages
2. Call clearing messages
3. Miscellaneous messages essentially for maintenance purposes

The primary function of each message type is summarized in Fig. 6-10 and may in fact be readily inferred from their names. Thus, the SETUP message is used to initiate the B-ISDN (ATM) call (or connection) and clearly constitutes an essential message that must carry all the information required to enable the network to set up and route the virtual connection. The contents and coding of this and other selected UNI messages are described further below. It should be noted that the CONNECT and CONNECT ACKNOWLEDGE messages may be used in pairs to essentially "symmetrize" the essentially asymmetrical UNI signaling procedures. These messages perform related functions in indicating call acceptance. In addition to the function of each of the message types, it should be noted that Fig. 6-10 also specifies the direction of the messages across the UNI.

The functions of the two call clearing messages RELEASE and RELEASE COMPLETE are also self-evident if it is pointed out that the "connection identifier" is the local VPI and VCI values that identify the VCL across the UNI. The "call reference" value is an arbitrary identifier associated with the call, which allows the messaging related to any given call (or connection) to be correlated. The two maintenance messages STATUS and STATUS ENQUIRY are essentially

Call estalishment messages:	
ALERTING	Sent by called user to network and network to calling user to indicate that called user alerting has been initiated
CALL PROCEEDING	Sent by called user to network or network to calling user to indicate requested call establishment initiated and no more call information will be accepted
CONNECT	Sent by called user to network and network to calling user to indicate call acceptance
CONNECT ACKNOWLEDGE	Sent by network to called user to indicate call awarded. Also sent by calling user to network to allow symmetrical call control procedures
SETUP	Sent by calling user to network and network to called user to initiate B-ISDN call
Call Clearing messages:	
RELEASE	Sent by user to request network to clear connection, or sent by network to indicate that connection is cleared
RELEASE COMPLETE	Sent by user or network to indicate the equipment has released its call reference value and connection identifier
Miscellaneous messages:	
NOTIFY	Sent by user or network to indicate information relating to call/connection
STATUS	Sent by user or network in response to STATUS ENQUIRY message to report certain error conditions/maintenance information
STATUS ENQUIRY	Sent by user or network to solicit a STATUS message for monitoring Error conditions on signaling links for maintenance purposes

Figure 6-10. UNI signaling message set. *Note:* Additional messages are defined in ITU-T Recommencdation Q.2931 for B-ISDN/ISDN interworking. The ATM Forum UNI v3.1 and v4 do not specify all Q.2931 messages.

designed to enable the monitoring of the (common-channel) signaling virtual channel links as indicated. Thus, on receipt of the STATUS ENQUIRY message the NE returns a STATUS message indicating whether or not an error (or fault) condition has been detected on the signaling links. This basic monitoring mechanism has drawbacks since it relies on "polling" the NE with the STATUS ENQUIRY message and is not "event"-driven by the fault or error condition itself. Despite this processing overhead of periodic polling to monitor the signaling VCLs, this mechanism is widely used and, as described later, has even been extended for use in the monitoring of frame relay PVC links in providing ISDN-based frame relay services. It may also be recognized that the miscellaneous messages category may be used to exchange connection-related information while the call is "in progress," as it were, as in the case for the NOTIFY message, thereby enabling their use in providing additional services to simplify connection management.

The ten basic UNI messages described in Fig. 6-10 are not the only signaling messages that have been standardized for B-ISDN use. There are some additional messages that have been defined by the ITU-T in Recommendation Q.2931 [6.8] for purposes of interworking between the narrowband (64 kbit/sec based) ISDN and the ATM-based B-ISDN, but these are outside the scope of the present discussion. For our purposes, it is more important to recognize that in deriving UNI signaling specifications versions 3.1 [5.10] and 4.0 [6.2] from Recommendation Q.2931, the ATM Forum does not specify the use of all the Q.2931 message set. In addition, there still exist some differences between the ATM Forum UNI signaling specification version 4.0 [6.2] and the ITU-T Recommendation Q.2931 in terms of detailed procedures, which may require that consideration is given to interworking between these protocols to avoid interoperability problems. These differences have been documented in [6.2] to assist implementors and network providers, and need not unduly concern us in understanding the essential ATM signaling mechanisms.

It is not the intent here to describe each of the UNI signaling messages and associated procedures in painstaking detail, since it is possible to obtain a clear understanding of ATM signaling methodology by considering the structure of a few essential messages. However, for the network element designer interested in implementing the complex software necessary for call processing, the detailed semantics, coding, and procedures related to each message and its contents have to be taken into consideration to ensure interoperability. These details may be obtained from the ITU-T [6.8] and ATM Forum signaling specifications [6.2], as required. Here we consider as examples two specific signaling messages: the SETUP and the CONNECT messages used for call establishment. The possible contents of the SETUP message are shown in Table 6-1. As noted earlier, the signaling messages are made up of a concatenation of so-called information elements (IEs), which may be either mandatory (M) or optional (O), depending on the network applications and services that are being supported. However, the first four IEs are always present in every UNI signaling message. These are:

1. Protocol Discriminator (length = 1 octet). An IE that specifies which "version" of the signaling protocol is being used to ensure backward compatibility in the event of protocol enhancements.
2. Call Reference Value (length = 4 octets). As mentioned earlier, the call reference value is an arbitrary number assigned to the call having local significance to enable correlation of all messages belonging to a given call. All messages relating to a call (or connection) will have the same call reference value, and this value is released when the call is torn down.
3. Message Type (length = 2 octets). This IE is uniquely coded to identify the signaling message type (e.g., SETUP or RELEASE, etc.), so that the call processing software may process the message accordingly. The message type code points are described later.
4. Message Length (length = 2 octets). Since signaling messages may be of different lengths

TABLE 6-1. SETUP message content

Message type: SETUP
Significance: global
Direction: both

Information element	Direction	Type	Length
Protocol discriminator	both	M	1
Call reference	both	M	4
Message type	both	M	2
Message length	both	M	2
AAL parameters	both	O	4–21
ATM user traffic descriptor	both	M	12–20
Broadband bearer capability	both	M	6–7
Broadband high-layer information	both	O	4–13
Broadband repeat indicator	both	O	4–5
Broadband low-layer information	both	O	4–17
Called party number	both	O	4–*
Called party subaddress	both	O	4–25
Calling party number	both	O	4–*
Calling party subaddress	both	O	4–25
Connection identifier	both	O	4–9
End-to-end transit delay	both	O	4–10
Notification indicator	both	O	4–*
OAM traffic descriptor	both	O	4–6
QoS parameter	both	M	6
Broadband sending complete	both	O	4–5
Transit network selection	u→n	O	4–*

depending on the number and sizes of the IEs included, an indication of the message length is necessary for processing the messages. The Message Length IE codes the length (in octets) of the message, not including the above octets and itself.

It will be noted from Table 6-1 that following the above four IEs there may be a number of subsequent IEs dedicated to specific aspects of the information required for setting up suitable connections. These include necessary items such as "called party number" (i.e., the destination address), the ATM Traffic Descriptor, QoS parameters, Connection Identifier (the VPI and VCI values), and so on. In addition, other IEs such as AAL parameters, Called Party Sub-Address, and Transit Network Selection, to mention but a few more self-evident IEs, may be included, and, if so, may need to be processed by the NE in establishing the connection. It is not necessary to describe all the listed IEs here in any detail since these descriptions can be found in the signaling specifications, if required. However, it may already be apparent to the reader that the more IEs included in the message, the more processing overhead incurred in the NE. Consequently, it may be useful to include only those essential IEs necessary to support a given ATM service in order to optimize call performance. Typically, however, the SETUP message is likely to be the most lengthy and complex, since it generally includes all the information necessary to route the call as well as enable QoS and traffic control functions to be invoked.

As another example of typical UNI message content, Table 6-2 shows the makeup of the CONNECT message. It may be noted that this message may include fewer IEs than the SETUP message list in Table 6-1, as may be expected from the function of the CONNECT mes-

TABLE 6-2. CONNECT message content

Message type: CONNECT
Significance: global
Direction: both

Information element	Direction	Type	Length
Protocol discriminator	both	M	1
Call reference	both	M	4
Message type	both	M	2
Message length	both	M	2
AAL parameters	both	O	4–21
Broadband low-layer information	both	O	4–17
Connection identifier	both	O	4–9
End-to-end transit delay	both	O	4–10
Notification indicator	both	O	4–*
OAM traffic descriptor	both	O	4–6

sage, which is essentially to indicate call acceptance by the called party (or user). As for the SETUP message (and, for that matter, all of the UNI message set) the CONNECT message also includes the Protocol Discriminator, Call Reference, Message Type, and Message Length IEs described above. The use of the Message Type, its length, followed by a (variable) number of IEs provide a generic structure for all signaling messages based on the well-known type–length–value (TLV) coding format.

It is important to recognize that the use of a TLV coding format is fundamental to all signaling message structures, as well as to the IE format described later. Consequently, the general UNI signaling message format is as shown in Fig. 6-11, where the conventional bit and octet numbering scheme is used to locate the individual fields. The Protocol Discriminator field (octet 1) is coded as shown (for the current UNI signaling version) and is followed by a field that indicates the length (in octets) of the call reference value IE. The Call Reference flag is used to identify which signaling entity end originated a call reference value. The originating side sets the call reference flag to 0, whereas the destination end sets it to 1. The Message Type IE identifies the signaling messages, as mentioned earlier, and is followed by the Message Length IE, in accordance with TLV format. For signaling messages, the "value" in the TLV scheme consists of the variable length information elements that follow the Message Length IE. As noted earlier, the message length codes the length in octets of the subsequent IEs, excluding itself and the protocol discriminator, call reference, and message type IEs prior to itself. If there are no IEs following the message length fields, it is coded as zero. It should be noted that the information in octets 1 to 9 appear in the order specified in Fig. 6-11 in all Q.2931 signaling messages.

The codepoints used for the UNI message types are listed in Table 6-3 for octet 1 for the Message Type IE and in Table 6-4 for octet 2 of this IE. It may be noted that the all zeros codepoint allows for the possibility to "escape" to a national-specific message type if required, further increasing the flexibility of the coding scheme. Also, by comparing with the earlier basic UNI message set (Fig. 6-10), it will be noticed that there are additional message types such as SETUP ACKNOWLEDGE, RESTART/ACK, and INFORMATION also included here.

These message types, although not presently included in the ATM Forum specifications, may be utilized for some signaling applications. In addition to these pt–pt signaling message types, it should be noted here that there are other message types that have been specified for pt–mpt signaling that will be described later. The coding of octet 2 of the Message Type IE is used to indicate how the message should be processed in the event of errors, as well as specific message ac-

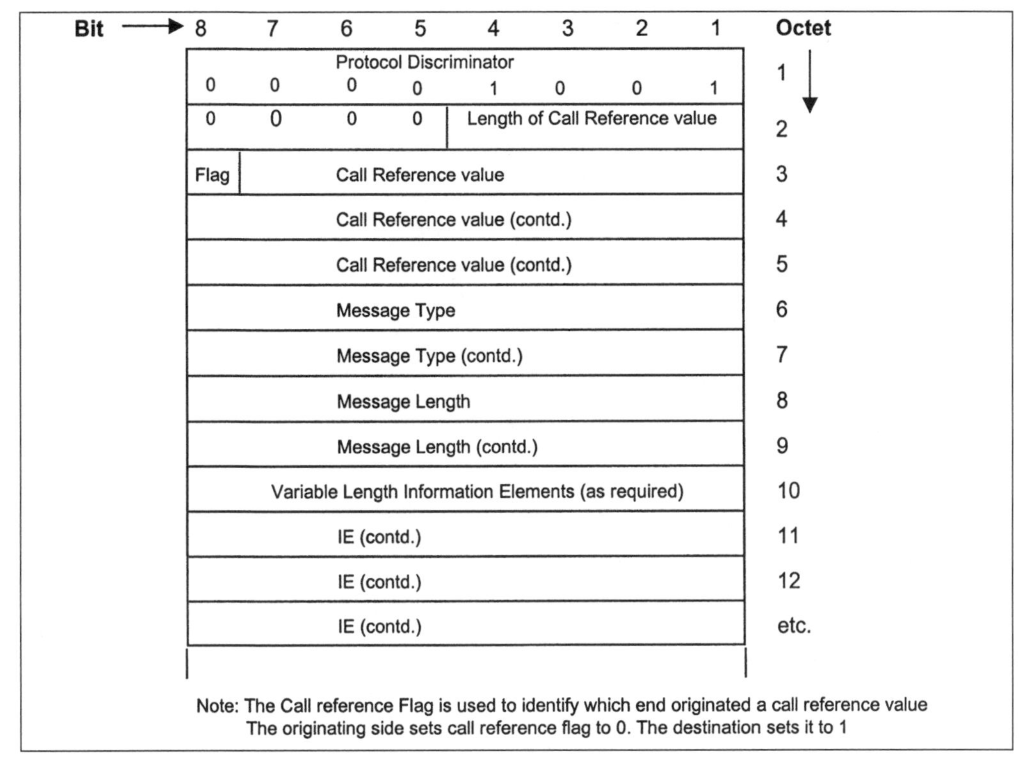

Figure 6-11. General UNI signaling message format. The information in octets 1–9 appears in all Q.2931 signaling messages in the order specified. The variable length information elements (IEs) following these depend on the specific message type.

tions that may be taken such as "call clearing" or "discard." Thus, by interpreting both octets of the message type IE, the signaling entity is able to identify which message should be processed and any actions pertaining to it if error conditions are encountered.

6.6 UNI INFORMATION ELEMENTS (IEs)

The UNI IEs that are concatenated together to constitute the various UNI signaling messages are also based on the general "type–length–value" coding format that forms the basis of the message structures themselves. The generic UNI IE format is shown in Fig. 6-12, wherein it may be recognized that the first two octets essentially correspond to a "message type" coding that identifies the specific IE (in octet 1) and the actions to be taken under various conditions (e.g., if the IE is not recognized). The IE identifier and instruction fields are followed by the IE length fields (2 octets), which are used to code the length in octets of the subsequent fields that carry the "content" of the IE. Each of the numerous IEs specified to date are structured in exactly the same way, using the basic TLV format. It may be evident that since both the signaling message set and the set of IEs use TLV coding formats in general, it may be considered that ATM signaling is based on a "nested" TLV format, which permits a phenomenal level of flexibility as well as efficiency in the exchange of signaling-related information between the call processing entities.

The coding of the IE identifiers for most commonly used IEs in ITU-T Recommendation Q.2931 [6.8] is given in Table 6-5. It should be noted that as new requirements and functions are

TABLE 6-3. Codepoints for octet 1 for the message type IE (including message compatibility instruction indicator)

Message type (octet 1) Bits 8	7	6	5	4	3	2	1	
0	0	0	0	0	0	0	0	Escape to nationally-specific message type; see Note 1.
0	0	0	-	-	-	-	-	Call establishment message:
			0	0	0	0	1	- ALERTING
			0	0	0	1	0	- CALL PROCEEDING
			0	0	1	1	1	- CONNECT
			0	1	1	1	1	- CONNECT ACKNOWLEDGE
			0	0	0	1	1	- PROGRESS
			0	0	1	0	1	- SETUP
			0	1	1	0	1	- SETUP ACKNOWLEDGE
0	1	0	-	-	-	-	-	Call clearing messages:
			0	1	1	0	1	- RELEASE
			1	1	0	1	0	- RELEASE COMPLETE
			0	0	1	1	0	- RESTART
			0	1	1	1	0	- RESTART ACKNOWLEDGE
0	1	1	-	-	-	-	-	Miscellaneous messages:
			1	1	0	1	1	- INFORMATION
			0	1	1	1	0	- NOTIFY
			1	1	1	0	1	- STATUS
			1	0	1	0	1	- STATUS ENQUIRY
1	1	1	1	1	1	1	1	Reserved for extension mechanism when all other message type values are exhausted; see Note 2.

1. When used, the message type (excluding the message compatibility instruction indicator) is defined in octet 10 of the message, and the content follows in the subsequent octets, both according to the national specification.

2. In this case, the message type (excluding the message compatibility instruction indicator) is defined in octet 10 of the message, and the content follows in the subsequent octets.

introduced, additional IEs will be specified in order to support these capabilities, requiring the definition of new IE identifiers. Thus, Table 6-5 is intended to provide the reader with a "snapshot" of commonly used existing IEs for UNI signaling and convey a general idea of the available range of capabilities. It is also outside the scope of this text to describe each and every IE function and associated procedures in detail, since the existing signaling specifications may be referred to for that purpose when required. However, it is important to understand that each IE will be uniquely identified on receipt of the relevant message and processed by the call processing/signaling entity according to the instructions encoded in the IE instruction field and the contents of the IE in accordance with the TLV procedure. The IE instruction field is only interpreted in case of unrecognized IE or contents, in which case (as for the case of the message set) the IE action indicator codepoints specify whether to discard, ignore or pass on the IE, and so on. The "coding standard" field is used to specify to the NE which (standards) authority has been used for the coding of the content fields. Thus, coding standards identify ATM-Forum-specific or ISO-specific variations in the coding of the IE, in addition to the normal ITU-T-based coding.

Although the functions of many of the IEs listed may be self-evident from their titles listed in Table 6-5, some are not immediately obvious and may require a brief description here. Thus, the Narrowband Lower- and Higher-Layer Compatibility as well as Narrowband Bearer Capability

TABLE 6-4. Codepoints for octet 2 for the message type IE

Flag (octet 2) Bits		
5		
0		Message instruction field not significant (= regular error handling procedures apply)
1		Follow explicit instructions (these supersede the regular error handling procedures)
Message action indicator (octet 2) Bits		
2	1	
0	0	Clear call
0	1	Discard and ignore
1	0	Discard and report status
1	1	Reserved

IEs are intended for ISDN-to-B-ISDN interworking scenarios to enable compatibility checking of the lower- and higher-layer narrowband protocols as well as bearer channels. Similarly, the Broadband Low-Layer Information (B-LLI) and Broadband High-Layer Information (B-HLI) IEs are intended primarily for compatibility verification of terminals for the broadband case. Other IEs such as "Call State," "Progress Indicator," "Cause," and "Restart Indicator" are primarily intended for maintenance purposes in conjunction with the STATUS ENQUIRY and STATUS messages described earlier. The Transit Network Selection IE enables the user to indicate choice of network operator (e.g., toll carrier) if available, whereas the End-to-End Transit Delay IE may be used to accumulate call set-up delay values for performance assessment or, as indicated in the description of the ABR ATC earlier, the round-trip delay estimates for feedback control purposes. The Broadband Locking Shift and Broadband Nonlocking Shift IEs are intended to provide "escape" mechanisms for the use of alternative (e.g., national-specific) coding schemes and functions in the event these are required. Such mechanisms enable greater flexibil-

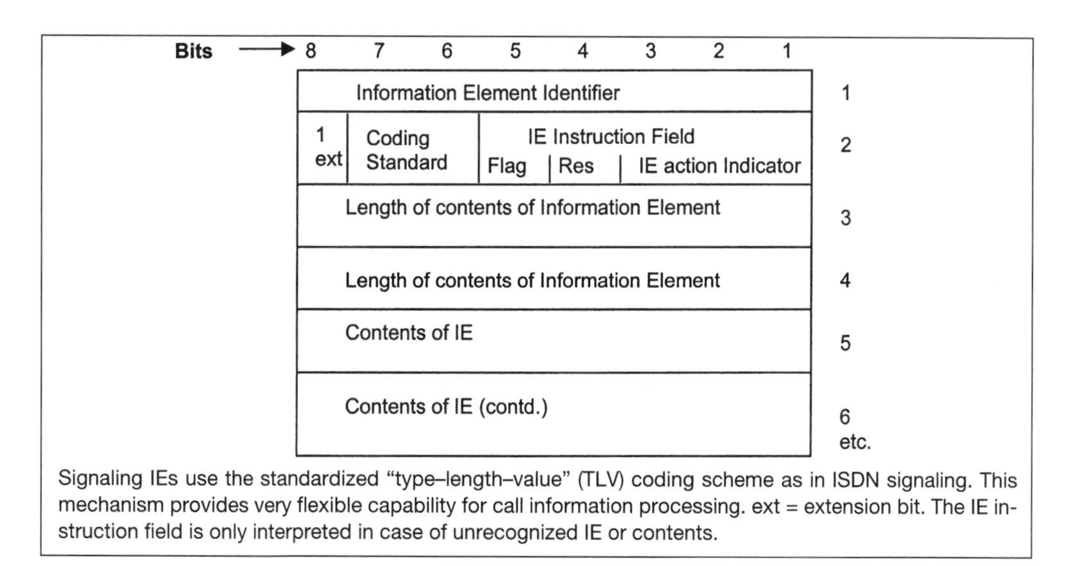

Signaling IEs use the standardized "type–length–value" (TLV) coding scheme as in ISDN signaling. This mechanism provides very flexible capability for call information processing. ext = extension bit. The IE instruction field is only interpreted in case of unrecognized IE or contents.

Figure 6-12. General UNI information element format.

TABLE 6-5. General information element format—information element identifiers

Bits 8	7	6	5	4	3	2	1	
0	1	1	1	0	0	0	0	Called party number
0	1	1	1	0	0	0	1	Called party subaddress
0	1	1	1	1	0	0	0	Transit network selection
0	1	1	1	1	0	0	1	Restart indicator
0	1	1	1	1	1	0	0	Narrowband low-layer compatibility
0	1	1	1	1	1	0	1	Narrowband high-layer compatibility
0	1	1	0	0	0	0	0	Broadband locking shift
0	1	1	0	0	0	0	1	Broadband nonlocking shift
0	1	1	0	0	0	1	0	Broadband sending complete
0	1	1	0	0	0	1	1	Broadband repeat indicator
0	1	1	0	1	1	0	0	Calling party number
0	1	1	0	1	1	0	1	Calling party subaddress
0	1	0	1	1	0	0	0	ATM adaptation layer parameter
0	1	0	1	1	0	0	1	ATM traffic descriptor
0	1	0	1	1	0	1	0	Connection identifier
0	1	0	1	1	0	1	1	OAM traffic descriptor
0	1	0	1	1	1	0	0	Quality of service parameter
0	1	0	1	1	1	1	0	Broadband bearer capability
0	1	0	1	1	1	1	1	Broadband low-layer information (B-LLI)
0	1	0	1	1	1	0	1	Broadband high-layer information (B-HLI)
0	1	0	0	0	0	1	0	End-to-end transit delay
0	0	1	0	0	1	1	1	Notification indicator
0	0	0	1	0	1	0	0	Call state
0	0	0	1	1	1	1	0	Progress indicator
0	0	0	0	1	1	0	0	Narrowband bearer capability
0	0	0	0	1	0	0	0	Cause

ity to support additional features or services that may not be necessary for the basic standardized signaling interfaces, or may be regional in nature. It is important to note here that since each IE may be identified individually by means of the IE identifier, the IEs may be concatenated in any order in a given signaling message. Only the "type" and "length" functions need to be in the order specified.

Although a detailed knowledge of the contents and semantics of each IE is primarily of interest to ATM signaling specialists and designers implementing call processing software, it is useful for the general reader to examine some examples of typical UNI IEs to gain an understanding of the structure that ensues from use of the TLV coding format. The three examples of IEs selected here from ITU-T Recommendation Q.2931 to illustrate the principles described earlier are:

1. ATM Traffic Descriptor IE
2. Called Party Number IE
3. Connection Identifier IE

Fig. 6-13 shows part of the ATM Traffic Descriptor IE that is incorporated in the SETUP message. It will be noted that the part shown relates to the deterministic bit rate (DBR) ATC described earlier, since it includes only the peak cell rates (PCR) for the forward and backward di-

Bits 8	7	6	5	4	3	2	1	Octet
ATM traffic descriptor								
0	1	0	1	1	0	0	1	1
Information element identifier								
1 ext	Coding Standard		IE Instruction Field: Flag	Res.	IE Action Ind.			2
Length of ATM traffic descriptor contents								3
								4
Forward Peak Cell Rate Id. (CLP=0)								5
1	0	0	0	0	0	1	0	
Forward Peak Cell Rate (for CLP=0)								5.1
								5.2
								5.3
Backward Peak Cell Rate Id. (CLP=0)								6
1	0	0	0	0	0	1	1	
Backward Peak Cell Rate (for CLP=0)								6.1
								6.2
								6.3
Forward Peak Cell Rate Id. (CLP=0 + 1)								7
1	0	0	0	0	1	0	0	
Forward Peak Cell Rate (for CLP=0 + 1)								7.1
								7.2
								7.3
Backward Peak Cell Rate Id. (CLP=0 + 1)								8
1	0	0	0	0	1	0	1	
Backward Peak Cell Rate (for CLP=0 + 1)								8.1
								8.2
								8.3

Figure 6-13. ATM traffic descriptor information element.

rections for both the CLP = 0 (i.e., high-priority) and CLP = 0 + 1 (i.e., total) cell flows. Clearly, the other traffic parameters required for the various ATCs described earlier may be simply added to the traffic descriptor IE as required for support of the chosen ATC. In this respect, it should be noted from Fig. 6-13 that each traffic parameter in the IE has its own unique identifier value coded in the leading octet of the field in question. For example, the forward PCR for CLP = 0 is identified by the value 1000 0010 in the leading octet (octet 5 in Fig. 6-13). This is an example of an additional level of "nesting" of the TLV format approach, although in this case the "length" is not required, since it is fixed at 4 octets (including the identifier field). In principle, since each traffic parameter has its own individual identifier code, they may be concatenated in any order in the ATM Traffic Descriptor IE, just as the different IEs may be concatenated in any order in the signaling message.

Figure 6-14 depicts the Called Party Number IE, which provides the addressing information necessary to route the call to its destination. It will be noted that, as before, the first octet (octet 5) defines the "type" of addressing information and coding format used in the following octets to enable the ATM NE to perform the routing analysis needed to establish connections. In ATM, as in conventional telephony, addressing (i.e., the type of numbering plan used) and call routing are intimately related, and constitute an important as well as complex aspect of the technology. Although a detailed study of routing mechanisms is outside the scope of this text, the relationships between ATM addressing and routing will be discussed further after we describe the routing protocols developed for PNNI signaling. However, it is useful to note here the difference between routing and signaling protocols in relation to the called party IE. In general, the signaling messages simply carry the destination address (or routing) information in the IE. The signaling entity software in the NE is responsible for interpreting the message and extracting the routing information, which may then be processed by the routing software to determine the selected (or optimized) path for routing the call. The routing methodology (or algorithms) used may range from statically configured (or preassigned) paths to complex, dynamic, "automatic" routing mechanisms such as those used in connectionless (e.g., IP-based) networks. Whatever the actual mechanism, and even though signaling and routing protocols are often lumped together, it is important to clearly distinguish the functions of each.

As a third example of a typical IE, Fig. 6-15 shows the connection identifier IE, which is used to exchange the VPI and VCI values used to identify the virtual path (VPL) and virtual channel (VCL) links between the NEs that make up the connection. For the UNI case, this is the logical link between the ATM connection end-point on the terminal and the first NE on the network side of the UNI. It may be noted that the virtual path level field is actually denoted as a virtual path connection identifier (VPCI) value to distinguish it from the normal VPI value. The VPCI is used to allow the possibility of virtual path level cross-connect equipment between the UNI and the "local" ATM NE containing the signaling entity. Since an ATM VP cross-connect NE may translate the VPI values from ingress to egress VPLs without terminating (or interpreting) the signaling messages, the VPCI value that identifies the VPL between the call processing NE and the UNI is used for this case. However, if a VP level cross-connect is not present between the UNI and the first SE, the VPCI value equals the VPI value. The VPCI field is also not used for the case of VP-associated signaling, if indicated by the appropriate coding in octet 5 of the IE. The VCI value may be selected by the user initiating the call, or by the network side of the UNI. Codepoints in the "preferred/exclusive" field in octet 5 of the IE specify the selection procedure for the VCI values. Since the VPI (and VCI) values are the same in both directions of the VPC or VCC (for bidirectional assignment of VPI/VCI values), the selection procedure should be the same in either direction for each VPL or VCL along the connection path.

The underlying "nested" type–length–value (TLV) coding principles are the same for all the defined IEs that make up any given UNI signaling message, as illustrated by the above exam-

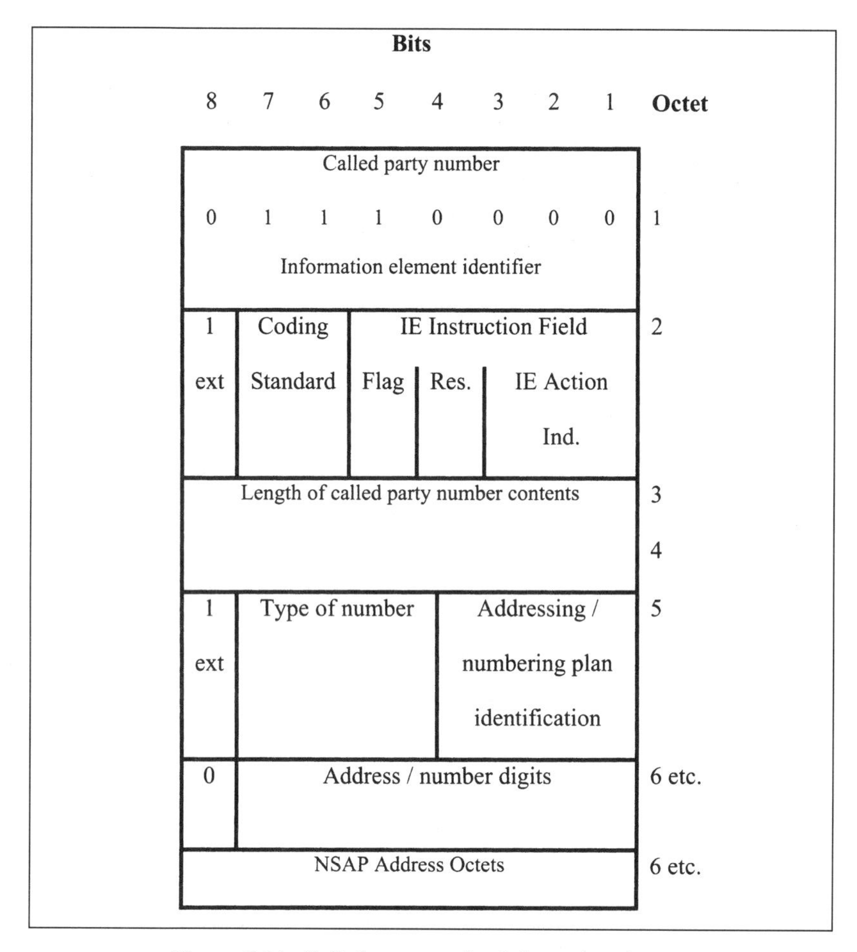

Figure 6-14. Called party number information element.

ples. Even with the somewhat brief descriptions of the structure of the IEs given above, it may be apparent that the methodology is capable of enormous flexibility, as well as economy, for the exchange of signaling-related information between the NEs. In fact, as new signaling functionality and service requirements are continuously being proposed and added to ATM technology, new IEs are in the process of being added to the signaling standards. As long as the underlying TLV coding mechanisms are adhered to and understood, such enhancements may be made backward compatible and considered as straightforward "add-ons" to the basic signaling "capability set" required for on demand switched VPCs and VCCs.

6.7 EXAMPLES OF MESSAGE FLOWS AT THE UNI

Having described the basic set of "nested" TLV structured signaling messages and the constituent IEs, it is useful to consider selected examples of their use for call setup and clear message flows across the UNI. Fig. 6-16 illustrates a typical message flow diagram for the basic "call accept" situation at the UNIs. It should be noted that the signaling messages required across the various NNIs are deliberately not defined here, and will be considered later in de-

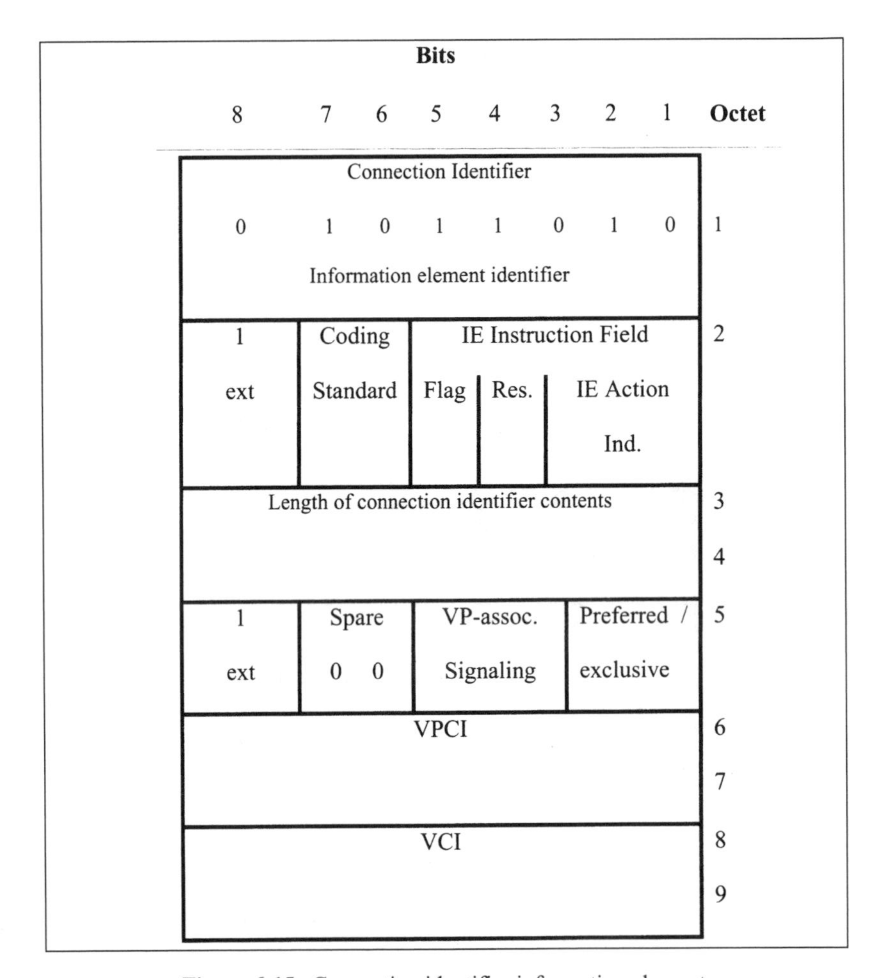

Figure 6-15. Connection identifier information element.

scribing NNI signaling protocols. In this simple scenario, the terminal (or CPN) on the left is assumed to initiate the call by transmitting the SETUP message to its local ATM switch. This NE processes the call request and routes the necessary information to the destination ATM NE. It also transmits a CALL PROCEEDING message back to the initiating terminal with the appropriate VPI (or VCI) values to identify the VPL or VCL for user data transfer.

The destination ATM Switch sends a SETUP message to the destination terminal (or CPN), which responds with the CALL PROCEEDING message as above. If the called terminal can accept the connection, it responds with the CONNECT message, which is "translated" across the NNI signaling messages to the originating terminal or NE. The edge ATM NEs may transmit the CONNECT ACKNOWLEDGE messages to inform the destination terminal it has received the CONNECT information, as may the originating terminal, to complete the Call Accept message sequence. It will be recognized that the total time for the call set-up process will greatly depend on the time required to process the signaling information in the NEs along the connection path, as well as the propagation delays incurred, which are generally only significant for very long paths. Typically, the connection set-up times are likely to be dominated by processing of the IEs in the SETUP message, which is generally the most complex message (particularly for the more "flexible" ATCs). Processing of the NNI signaling information may also be the rate-limiting

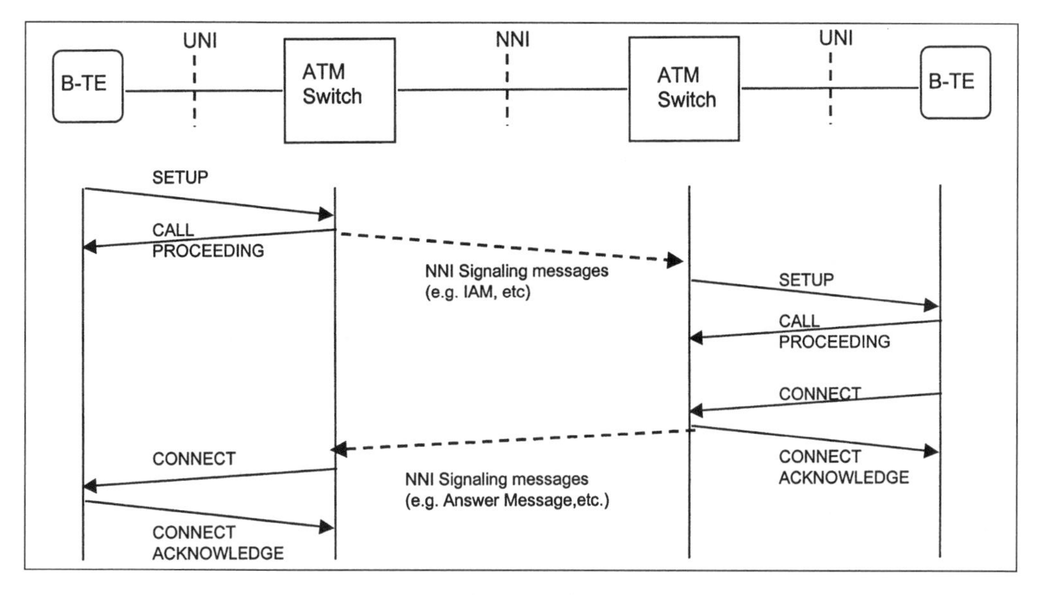

Figure 6-16. Example of message flows—call accept (UNI).

step, particularly for the case where PNNI protocols are in use, as will be described in more detail subsequently. Consequently, for a given level of call processing capability in the NE, the call performance may be maximized by limiting the signaling information to only that required for any given service and avoiding complex processing requirements for that information.

For the case where the call may be rejected or cleared by the destination end point, the typical message flows across the UNI are as depicted in Fig. 6-17. Assuming that the destination terminal cannot accept (or clears) the call, it sends a RELEASE message to the near end edge NE, which returns a RELEASE COMPLETE message, while translating the information across the ATM network to the calling terminal. The RELEASE COMPLETE message from the calling B-TE essentially acknowledges that the process is completed. Just as the SETUP message IEs are used to allocate network resources to the VPCs or VCCs by the CAC, so the information in the RELEASE messages are used to free up the scarce resources so that they may be available for subsequent connection requests. Although only the UNI case is under consideration here, it may be noted that a RELEASE message may be initiated by any NE along the path in the event that defect conditions may be encountered.

6.8 POINT-TO-MULTIPOINT CONNECTIONS

The signaling message set described so far has been used for the case of simple point-to-point (pt–pt) VPCs or VCCs. However, in the earlier description of ATM connection types, the importance of point-to-multi-point (pt–mpt) connectivity was noted for the support of applications requiring a multicast capability. In order to provide pt–mpt on-demand connectivity, additional signaling messages have been defined, initially in the ATM Forum and subsequently by the ITU-T, for both UNI and NNI signaling. For the UNI case, the typical message flow diagram for a basic pt–mpt connection is shown in Fig. 6-18. In general, the pt–mpt connection is set up sequentially after the basic pt–pt connection is established by sending the so-called ADD PARTY message, which contains the destination address of the additional terminal to be added to make

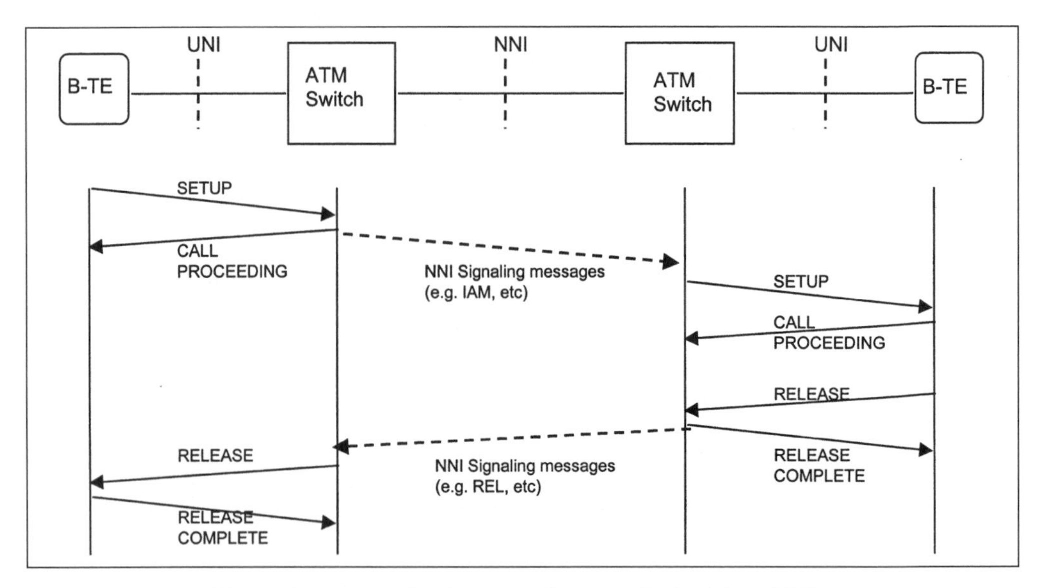

Figure 6-17. Example of message flows—call reject/clear (UNI).

up the pt–mpt VPC (or VCC). The far end "branching" ATM NE will then send a normal SETUP message to the new terminal in order to add it to the pt–mpt connection. This process may be repeated any number of times by means of the ADD PARTY messages to include additional destination TEs to the multicast tree configuration.

The ADD PARTY message used sequentially to set up the pt–mpt VPC or VCC is essentially equivalent to the SETUP message except that it may be sent repeatedly (with different end-point addresses) to build the pt–mpt multicast tree structure. Similarly, the ADD PARTY ACKNOWLEDGE message is analogous to the CONNECT ACKNOWLEDGE message as shown, but is used only at the "root" side of the pt–mpt tree to indicate that the new terminal has been added to the pt–mpt tree. It will be noted that at the "leaf" end of the pt–mpt connection, the familiar pt–pt message set is used.

The pt–mpt signaling procedures using the ADD PARTY and ADD PARTY ACKNOWLEDGE messages are essentially sequential in nature, although it is possible in principle to send a number of ADD PARTY messages from the initiating (root) terminal without waiting for the ADD PARTY ACK messages for each individual leaf in turn in order to reduce the overall set-up time, particularly for a large number of leaf terminals. However, processing of the numerous ADD PARTY messages received by the NE may essentially prove to be the rate-limiting step in determining the pt–mpt connection set-up performance. For large multicast trees, this processing load may prove significant. In order to alleviate some of these problems, as well as enhance the capability of pt–mpt connection types, the signaling procedures for pt–mpt have been enhanced to include the so-called leaf-initiated join (LIJ), whereby a leaf TE may elect to "join" into an existing pt–mpt connection. The LIJ enhancements may enable larger multicast trees to be constructed more rapidly, as well as provide additional flexibility to the leaf terminals to enter or leave the multicast tree without involving the root terminal element. A more detailed description of the LIJ procedures is outside the scope of this overview, but may be found in the LIJ specifications developed by the ATM Forum in the UNI signaling specifications (v.4.0) [6.2]

In addition to the ADD PARTY and ADD PARTY ACKNOWLEDGE messages used during the call set-up phase, other messages specific to pt–mpt connections have also been specified. These include:

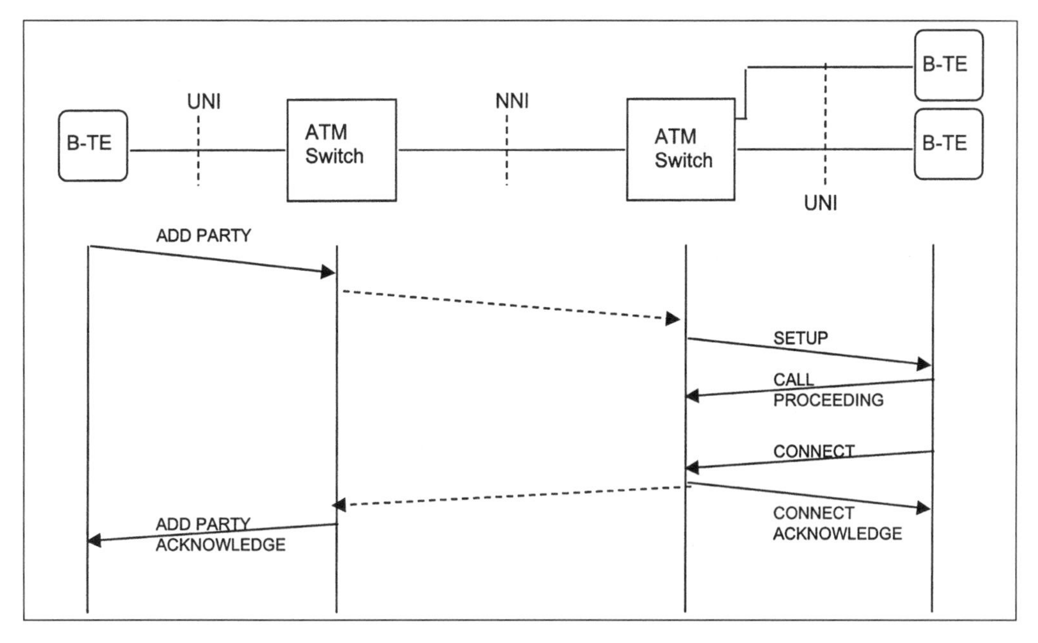

Figure 6-18. Point-to-multipoint connection—add party message.

1. ADD PARTY REJECT message: used by the called entity when the pt–mpt call cannot be accepted
2. DROP PARTY message: used by the calling entity to tear down the pt–mpt call
3. DROP PARTY ACKNOWLEDGE message: used in conjunction with the above messages
4. PARTY ALERTING message: used for pt–mpt calls in the same way as the ALERTING message for pt–pt calls, to indicate that called user alerting has been initiated

It should be stressed that the pt–mpt signaling message set listed above performs essentially the same functions as those for the basic pt–pt signaling case and are structured and coded in the same way (using TLV principles) as the basic message set. The IEs listed earlier are used as appropriate for the most part, although pt–mpt-specific IEs may be included for specific functions such as the leaf-initiated join (LIJ) procedure, which clearly has no counterpart in the simple pt–pt case. Consequently, it is not necessary to describe here the detailed structure of the pt–mpt message set and associated IEs. By recognizing that the pt–mpt connection type is constructed by the stepwise addition of the "leaf" end points, or by the enhancement of LIJ procedures, the function of each pt–mpt message and the required IEs may be inferred.

In principle, the pt–mpt signaling mechanism described so far makes no distinction between unidirectional and bidirectional connection types. For most cases, the ATM NEs will set up the VPLs (or VCLs) using bidirectional assignment of VPI and VCI values as required for pt–pt VPCs and VCCs. Thus, the transfer path in the "leaf-to-root" direction is automatically set up (in any case) and needs to be present for the transfer of in-band OAM information, as will be described later. However, the presence of a bidirectional transfer path (e.g., for OAM purposes) in this case need not imply that "user" cell flow occurs in the reverse direction, since the user cell bandwidth in the leaf-to-root direction may be set to essentially zero in the ADD PARTY message, thereby suppressing the reverse user data flow. This situation constitutes the more usual case for pt–mpt connections, corresponding to an essentially "unidirectional" multicast application in which user data flow occurs from the root to the leaves of the multicast tree.

The unidirectional (multicast) user data transfer requires the capability to replicate (copy) cells at the branching NEs, which is generally relatively straightforward to implement. However, it will be recognized that for the truly "bidirectional" pt–mpt connection case in which user data transfer occurs in both root-to-leaf and leaf-to-root directions, a "merge" capability is required at the branching NEs. As noted earlier, the merge function essentially interleaves the cells from the separate VPLs (or VCLs) into a *single* outgoing VPL or VCL. While this functionality may be readily incorporated in the branching NEs, the deinterleaving of the cell streams from the different leaf end points requires additional logical identifiers above the ATM layer. As will be described later, such a capability may be incorporated into certain AALs or higher-layer protocols, but this may not always be available. Hence, at present, use of bidirectional pt–mpt connections is not generally specified. However, recently there has been increased interest in the use of bidirectional pt–mpt connectivity with respect to some network applications such as LAN emulation (LANE) and multiprotocols (e.g., IP) over ATM (MPOA), as described in more detail later. It has been pointed out that such applications may benefit from the merge capability necessary for the mpt–pt connectivity, resulting in proposals to specify bidirectional pt–mpt signaling procedure enhancements in order to support such networking applications.

6.9 THE NNI SIGNALING MESSAGE SET—B-ISUP

Just as the UNI signaling procedures, message sets and IEs are based on an extension of the narrowband (or 64 kbit/sec) ISDN signaling protocols (generally referred to as DSS1 signaling), the ATM signaling across NNI is an extension of the SS7 protocols based on the ISDN user part (ISUP) architecture. As described earlier relating to the general B-ISDN signaling protocol architecture, the NNI call processing functionality lies in the broadband–ISUP (or B-ISUP) layer, which is responsible for processing and interpreting the information contained in the NNI (SS7) signaling messages. In common with the UNI case, the messages themselves are transported over preassigned ATM VCCs (using for example VCI = 5 in every VPI for the case of associated signaling) using AAL Type 5. However, unlike the UNI case, NNI signaling uses the connectionless transport capabilities of the message transfer part level 3 (MTP 3) to route the NNI signaling messages between the relevant signaling entities (termed signaling transfer points, or STPs, in conventional SS7 terminology). The essential functionality of the MTP 3 and lower layers will be dealt with later, and here we focus on the message set and associated procedures relating to the call processing B-ISUP Level of the NNI signaling protocol stack.

The basic B-ISUP signaling message set is listed in Fig. 6-19 with a brief description of the essential function of each of the messages. There are two aspects of the B-ISUP message set that are immediately apparent from even a cursory inspection of this list. The first point to note is that there are many more message types than for the UNI case (28 basic types are listed here compared to the 12 or so for the UNI DSS2 case). However, this aspect is only to be expected, since it was already pointed out earlier that the requirements for signaling and control within the network are clearly much more stringent than may be envisioned for the UNI. Thus, as will be noted from Fig. 6-19, a significant number of the B-ISUP messages are associated with operational and maintenance or test procedures for connections, necessary for the reliable operation of public carrier networks. B-ISUP assumes that such functions are likely to be of as much value for ATM-based networks as they are for existing SS7 networks designed primarily for voice telephony. Whether this is the case still remains an open question.

The second point that emerges from Fig. 6-19 is that a terminology different from that used for DSS2 messages is used for the B-ISUP messages. For example, in B-ISUP, the term initial address message (IAM) is the equivalent to the SETUP message term used in DSS2, and the answer message (ANM) corresponds to the CONNECT message encountered for the UNI case. (Note also the use of lowercase letters for the B-ISUP messages names, in contrast to the capital letters for UNI message types). However, the "release" and "release complete" messages retain the same names as for the UNI case. These differences in terminology, as well as the detailed descriptions of the associated procedures and underlying models between DSS2 and SS7 protocols, may at first glance appear somewhat unnecessary and even trivial. The reasons are rooted in the different historical evolution of signaling concepts for the UNI and NNI in the ITU-T, based not only on the different technical requirements as indicated earlier, but also on historical factors that precede the ISDN model for network control architecture. Although these evolutionary aspects need not concern us unduly in describing the underlying signaling mechanisms for ATM NNIs, it must be admitted that the terminological, as well as architectural differences have resulted in some confusion, and even a perception that B-ISUP protocols are excessively complex.

Despite the differences in terminology and approach, it is important to recognize that the fundamental underlying mechanisms for B-ISUP-based call processing and DSS2 call processing remain the same. Both use common channel signaling procedures, in which the message structure, coding and formats are based on the nested "type–length–value" approach, which allows flexible and economical coding of large amounts of signaling-related information. As a consequence, use of the common TLV coding formats enables direct translation between the DSS2

Message	Meaning
1. Address Complete (ACM)	Indicate all address signals required for routing the call to called party have been received
2. Answer Message (ANM)	Indicates above as well as call has been answered
3. Blocking Acknowledge(BLA)	Response to Blocking message indication resource has been blocked
4. Blocking (BLO)	Maintenance message sent to block a resource
5. Consistency Check End (CCE)	Indicates end of a consistence check procedure (maintenance purposes)
6. Consistency Check End Ack.(CCEA)	
7. Consistency Check Request (CSR)	
8. Consistency Check Request Ack. (CSRA)	
9. Confusion (CFN)	Sent in response to any message if exchange does not recognize message
10. Call Progress (CPG)	Sent in either direction during setup or active phase of call indication an event
11. Forward Transfer (FOT)	Sent to request operator assistance at incoming international exchange
12. IAM Acknowledge (IAA)	Sent in backwards direction to indicate IAM message has been accepted
13. Initial Address Message (IAM)	Sent in forward direction to initiate connection setup
14. IAM Reject (IAR)	Sent in backwards direction to indicate IAM not accepted/resource unavailable
15. Network Resource Management (NRM)	Sent to modify network resources associated with call
16. Reset Acknowledge (RAM)	Sent in response to Reset message indication resources released
17. Release (REL)	Sent in either direction to indicate call/connection release
18. Resume (RES)	Sent in either direction to indicate calling or called party is reconnected
19. Release Complete (RLC)	Sent in response to Release message to indicate resources made available
20. Reset (RSM)	Sent to release a resource for some cause
21. Subsequent Address (SAM)	May be sent after IAM to convey additional address information
22. Segmentation (SGM)	Sent to convey additional segment of overlength message
23. Suspend (SUS)	Sent in either direction to indicate temporary disconnection
24. Unblocking Acknowledge (UBA)	Sent in response to Unblocking message indication resource unblocked
25. Unblocking (UBL)	Sent to unblock engaged resource
26. User Part Available (UPA)	Sent in either direction in response to test message to indicate UP available
27. User Part Test (UPT)	Sent in either direction to test status of UP marked as unavailable
28. User-to-User Information (USR)	Used to transport user-to-user signaling independent of call control

Figure 6-19. NNI signaling message set: B-ISUP.

and B-ISUP message information within the call processing software of the NE. It may also be noted from Fig. 6-19 that the additional perceived complexity of the B-ISUP message set is largely due to the inclusion of a number of messages for test, operational, and maintenance features deemed necessary for reliable control of the network. Apart form these maintenance aspects, which may or may not be deployed in any given network depending on operator's choice, the basic message set used for call establishment and clearing operate in essentially the same way as discussed above for UNI signaling. However, since B-ISUP call procedures are designed for operation between peer ATM NEs, the protocols are inherently symmetric.

The general B-ISUP message structure is shown in Fig. 6-20, which also depicts the general structure of the constituent information elements. For the case of B-ISUP these are called "parameters" (another basic terminology difference between DSS2 and B-ISUP protocols). Ignoring for the moment the routing label fields which actually relate to the MTP 3 (connectionless) routing address and other control information, the basic TLV format may be clearly discerned in the B-ISUP message structure. The "message compatibility information" field in the first octet of the "value" part is included primarily for protocol version control and compatibility verification purposes and hence provides additional flexibility for functional enhancements. The B-ISUP message contents are made up of a concatenation of "parameters" which are the equivalent of the IEs used in DSS2 UNI signaling. It will be noted that each B-ISUP parameter has the TLV format comprising parameter name, length, and parameter compatibility field, followed by the content of the signaling information. Similarly to the DSS2 IEs, some B-ISUP parameters may contain additional level(s) of TLV nesting to distinguish different information types which may be coded in the parameter. An example is the traffic descriptor parameter, which may comprise several separate traffic parameters for any given ATC (or service category) as for the DSS2 IE. It is useful to compare Fig. 6-20 for B-ISUP with the corresponding Figs. 6-11 and 6-12 for DSS2, while making the appropriate allowances for the difference in the terminology used. In this way, the essential similarity in the TLV coding format may be readily noticed, despite some differences in detailed format or the use of different code sets to describe the signaling information.

We now turn to the B-ISUP messages listed in Fig. 6-19 for a general description of their purpose, although as for the UNI case, a detailed description of all the associated call processing procedures is outside the scope of this treatment, but may be of interest to designers of B-ISUP call processing software. These details may be obtained from the relevant ITU-T documents [6.3–6.6], if required, but the intent of the present description is to assist the reader in finding the underlying similarities in all the messages. For a typical call establishment phase in B-ISUP the messages used include:

1. Initial Address Message (IAM)—to initiate the call
2. IAM Acknowledge (IAA)—to indicate receipt of IAM
3. IAM Reject (IAR)—if call blocking occurs
4. Subsequent Address Message (SAM)—only used if additional address information to that included in the IAM needs to be sent
5. Answer Message (ANM)—to notify call acceptance
6. Call Progress (CPG)—to notify call processing is being performed

For clearing a call, B-ISUP utilizes the following messages:

1. Release (REL)—to notify release of the connection
2. Release Complete (RLC)—to signal that resources are freed

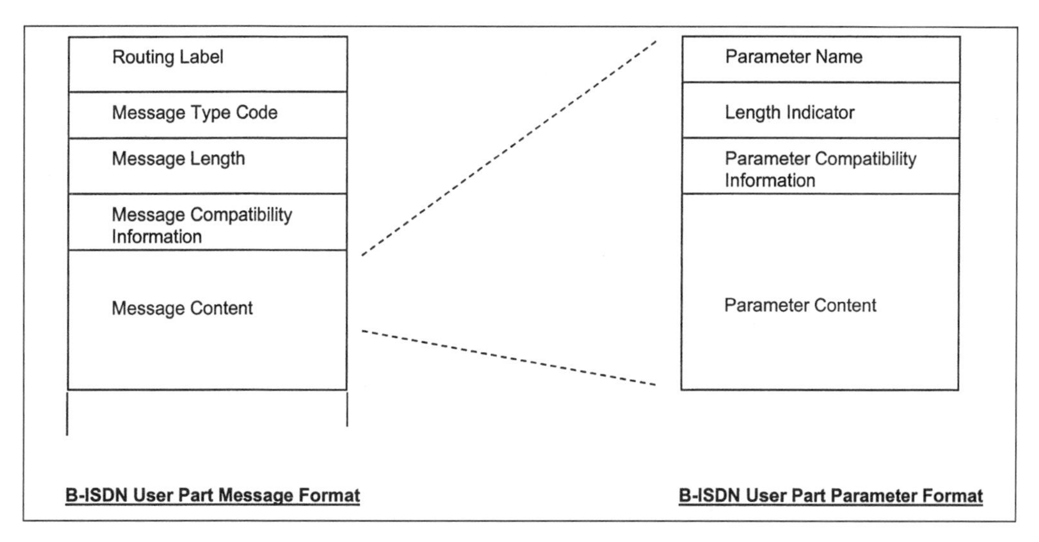

Figure 6-20. General B-ISUP message and parameter structure.

3. Reset (RSM)—may be used to release resource for causes other than normal call disconnection or under defect conditions generally executed under management action
4. Reset Acknowledge Message (RAM)—sent in response to the Reset (RSM)

In addition to the normal call establishment and call clearing procedures, B-ISUP also includes functionality for the temporary suspension and resumption of a given connection (for test or other reasons) using the following messages:

1. Suspend (SUS)
2. Resume (RES)

Additional test and operational capability is also made possible by using the message set related to the so-called blocking functions. These include:

1. Blocking (BLO)—use to block a resource (i.e., a VPC or VCC)
2. Blocking Acknowledge (BLA)—in response to the BLO message
3. Unblocking (UBL)—used to unblock the resource
4. Unblocking Acknowledge (UBA)—in response to the UBL message

The blocking/unblocking functions may be useful for various testing and administrative purposes, particularly when used in conjunction with the in-band OAM capabilities described later, for example for verifying connection continuity using simple OAM cell loopback procedures.

Other operational/test capabilities provided in B-ISUP include a "consistency check" procedure to verify the consistency of the VPI/VCI values assigned along the connection paths together with other parameters of interest, for example. This capability uses the message set:

1. Consistency Check Request (CSR)
2. Consistency Check Request Acknowledge (CSRA)
3. Consistency Check End (CCE)
4. Consistency Check End Acknowledge (CCEA)

As noted earlier, B-ISUP procedures in general utilize symmetric protocols based on "command-positive acknowledgment" procedure for added reliability, but it may be noted not all the messages are paired with the "acknowledge" counterpart (e.g., the Resume message). Other useful operational and maintenance-oriented capabilities provided by B-ISUP include the following messages:

1. Confusion (CFN)
2. Forward Transfer (FOT)
3. Network Resource Management (NRM)
4. Segmentation (SGM)
5. User Part Test (UPT)
6. User Part Available (UPA)

The brief description of these messages in Fig. 6-19 summarizes their primary function in the B-ISUP capability set for the most part. Use of these and other enhancements being incorporated in the SS7-based networks depends to a large extent on the sophistication required by the network provider and the SS7 management capabilities available. However, it will be evident that the B-ISUP procedures provide substantial capability for operational and maintenance purposes if required, and these capabilities continue to be developed as additional requirements are established.

The User-to-User Information (USR) message provides a capability to transparently transport "user" or CPN signaling information across a network without incurring any processing in the intermediate NEs. An example of such information may be signaling information required by the remote ATM PABX in an enterprise network. It should be noted that the current B-ISUP standard limits such user-to-user signaling information to a maximum of 136 octets, so as not to burden the signaling virtual channel with excessive amounts of user signaling information. However, with the increasing need for interworking between private NNI (PNNI) signaling and B-ISUP, recent studies have indicated a need to enhance USR message capability, since, as will be seen later, some PNNI signaling messages may exceed this length significantly.

The B-ISUP message type codeset is given in Table 6-6, as defined in ITU-T Recommendation Q.2763 [6.5]. These codepoints enable the B-ISUP call-processing software in the NE to identify the type of NNI signaling message and thereby process the contained information (i.e., the individual parameters in the B-ISUP message) accordingly. Also, for reference here, the individual B-ISUP parameter name codeset for the basic set of parameters is shown in Table 6-7.

In keeping with the TLV coding mechanisms, the NE call processing software identifies each individual B-ISUP parameter (if present) using these code points and processes them accordingly. As for the B-ISUP message types, most of the parameter names indicate their essential function, and may be compared to the DSS2 IEs listed earlier, bearing in mind that differences in terminology exist in the naming of these entities. It may also be noted that B-ISUP defines significantly more parameters (57 are listed here) than the IEs specified in DSS2 for the UNI case. This may be expected, given the requirements inherent in the SS7 capability, although not all networks may support all of the defined functionalities in B-ISUP.

As for the UNI DSS2 signaling procedures discussed earlier, a detailed description of all the B-ISUP messages and the associated parameters is outside the scope of this treatment, and in fact, not necessary to a general understanding of the underlying signaling protocols. However, as for the UNI, it is useful to consider some examples of typical B-ISUP message contents, bearing in mind that all the defined messages will be structured similarly based on the generic nested TLV format shown in Fig. 6-20. The Initial Address Message (IAM) contents, the primary function of which is to set up the VPC or VCC across the network, are shown in Table 6-8, together with the possible length in octets. It should be recognized that not all the listed parameters need to be included in every IAM, and since each parameter is identified by a unique parameter name code point as in Table 6-7, they may occur in any order in the IAM format. However, certain parameters such as Called Party Number and ATM Cell Rate (i.e., Traffic Descriptor) are clearly essential for proper routing of the call and their absence would result in call blocking.

Some other examples of B-ISUP messages contents are shown in Table 6-9. These include the Release (REL) and Release Complete (RLC) messages, as well as the Subsequent Address (SAM) and User-to-User Information (USR) messages. The procedures for interpreting each of the parameters in every B-ISUP message have been defined in detail by the ITU-T in [6.3–6.6], as noted earlier. As new functionality is added to the technology, the parameters and associated procedures are enhanced and updated in new releases of these documents by the ITU-T. It need hardly be pointed out that with a technology evolving as rapidly as ATM, the enhancements to signaling functionality need to be tracked regularly to ensure that the services and applications required in any given deployment are provided for by the signaling capability set at that stage.

6.10 BROADBAND INTER CARRIER INTERFACE (B-ICI)

It was noted earlier that the ATM Forum has also developed specifications for signaling across the NNI public carrier interfaces. These are contained in [6.1]. It is important to note that the signaling messages and procedures, as well as the basic protocol architecture adopted in the B-ICI

TABLE 6-6. B-ISUP message type codeset

Message type	Code	Message type	Code
Address complete	0000 0110	Subsequent address	0000 0010
Answer	0000 1001	Suspend	0000 1101
Blocking	0001 0011	Unblocking	0001 0100
Blocking acknowledgement	0001 0101	Unblocking acknowledgement	0001 0110
Call progress	0010 1100	User part available	0011 0101
Confusion	0010 1111	User part test	0011 0100
Consistency check end	0001 0111	User-to-user information	0010 1101
Consistency check end acknowledgement	0001 1000	Reserved for ECT	0011 1001
		Reserved, used in N-ISUP	0000 0011
Consistency check request	0000 0101	Reserved, used in N-ISUP	0000 0100
Consistency check request acknowledgment	0001 0001	Reserved, used in N-ISUP	0000 0111
		Reserved, used in N-ISUP	0001 1001 to 0010 1011
Forward transfer	0000 1000		
IAM acknowledgement	0000 1010	Reserved, used in N-ISUP	0010 1110
IAM reject	0000 1011	Reserved, used in N-ISUP	0011 0000
Initial address	0000 0001	Reserved, used in N-ISUP	0011 0001
Network resource management	0011 0010	Reserved, used in N-ISUP	0011 0011
Release	0000 1100	Reserved, used in N-ISUP	0011 0110
Release complete	0001 0000	Reserved, used in N-ISUP	0011 0111
Reset	0001 0010	Reserved for extension of name code	1111 1111
Reset acknowledgement	0000 1111		
Resume	0000 1110		
Segmentation (national use)	0011 1000		

TABLE 6-7. B-ISUP parameter name codeset for the basic set of parameters

Parameter name	Code	Parameter name	Code
AAL parameters	0100 0111	Original called number	0010 1000
Access delivery information	0010 1110	Origination ISC point code	0010 1011
Additional calling party number	0001 0000	Origination signaling identifier	0000 0010
Additional connected number	0001 0001	Progress indicator	0011 0101
ATM cell rate	0000 1000	Propagation delay counter	0011 0001
Automatic congestion level	0010 0111	Redirecting number	0000 1011
Backward narrowband interworking indicator	0001 0100	Redirection information	0001 0011
		Redirection number	0000 1100
Broadband bearer capability	0101 0000	Redirection number restriction	0100 0000
Broadband high-layer information	0100 0110	Resource identifier	0011 1001
Broadband low-layer information	0100 1111	Segmentation indicator (national use)	0011 1110
Call diversion information	0011 0110		
Call diversion may occur	0010 0110	Subsequent number	0000 0101
Call history information	0010 1101	Suspend/resume indicators	0010 0010
Called party number	0000 0100	Transit network selection (national use)	0010 0011
Called party subaddress	0001 0101		
Called party indicators	0001 0111	User-to-user indicators	0010 1010
Calling party number	0000 1010	User-to-user information	0010 0000
Calling party subaddress	0001 0110	Reserved for FPH	0101 0001
Calling party category	0000 1001	Reserved, used in N-ISUP	0000 0000
Cause indicators	0001 0010	Reserved, used in N-ISUP	0000 0001
Charge indicator	0001 1001	Reserved, used in N-ISUP	0000 1101
Closed user group information	0001 1010	Reserved, used in N-ISUP	to
Connected line ID request	0001 1011	Reserved, used in N-ISUP	0000 1111
Connected number	0010 0001	Reserved, used in N-ISUP	0001 1000
Connected subaddress	0010 0100	Reserved, used in N-ISUP	0001 1110
Connection element identifier	0000 0110	Reserved, used in N-ISUP	0010 1111
Consistency check result information	0100 1010	Reserved, used in N-ISUP	0011 0000
		Reserved, used in N-ISUP	0011 0010
Destination signaling identifier	0000 0011	Reserved, used in N-ISUP	0011 0011
Echo control information	0011 0111	Reserved, used in N-ISUP	0011 1000
Forward narrowband interworking indicator	0001 1100	Reserved, used in N-ISUP	0011 1011
		Reserved, used in N-ISUP	to
In-band information indicator	0001 1111	Reserved, used in N-ISUP	0011 1101
Location number	0011 1111	Reserved, used in N-ISUP	0100 0001
Maximum end-to-end transit delay	0000 0111	Reserved, used in N-ISUP	0100 0010
MLPP precedence	0011 1010	Reserved, used in N-ISUP	0100 0011
MLPP user information	0100 1001	Reserved, used in N-ISUP	0100 0100
Narrowband bearer capability	0001 1101	Reserved, used in N-ISUP	0100 0101
Narrowband high layer compatibility	0011 0100	Reserved, used in N-ISUP	0100 1011
Narrowband low layer compatibility	0010 0101	Reserved, used in N-ISUP	to
National/international call indicator	0010 1001	Reserved, used in N-ISUP	0100 1110
		Reserved, used in N-ISUP	1100 0000
Notification	0010 1100	Reserved for extension of name code	1111 1111
OAM traffic descriptor	0100 1000		

specification, are based on the SS7 model using B-ISUP protocols for call processing, as discussed above. A key objective of the B-ICI version 2 specification was to simplify the SS7 B-ISUP protocols by selecting a suitable subset of the functions incorporated in the ITU-T standards, not only for B-ISUP call processing, but also for the MTP 3 functionality necessary for the SS7 architecture. Thus, B-ICI simplified MTP 3 protocols by specifying only the so-called "associated signaling" mode of SS7, whereby the signaling messages follow essentially the same transmission path as the VCCs or VPCs they control, and also simplify MTP 3 fault and traffic management functions. In addition, B-ICI v2.0 [6.1] specifies only an essential subset of the B-ISUP messages and parameters listed earlier. However, for the selected messages and parameters B-ICI uses essentially the same coding, semantics, and procedures as defined by the ITU-T for B-ISUP [6.3–6.6]. In principle, this approach allows for interoperability between ITU-T B-ISUP and ATM Forum B-ICI-based implementations, since, as already noted, it is not necessary for all the parameters and messages to be used in any given deployment. Nonetheless, when considering network deployments that involve interoperation between B-ICI v.2-based and ITU-T B-ISUP-based implementations, it would be prudent to verify in some detail which functions and procedures have been selected and which have been eliminated. This is because despite the best intentions, sometimes relatively minor differences of semantical interpretation in implementing protocols result in interoperability problems.

Although the ATM Forum B-ICI v.2 specification is based on, and considered a "subset" of, the B-ISUP/SS7 signaling procedures, some specific requirements for support of functions in related ATM Forum specifications, such as traffic management and PNNI, are also incorporated where appropriate. Particular areas where differences are likely to arise include ATM addressing (with the associated routing) and support of ABR and UBR service categories. In addition, it was noted earlier that the ATM Forum's PNNI signaling and routing specification has adopted a different approach to the B-ISUP/SS7 procedures, which may result in interoperability problems

TABLE 6-8. Message type: initial address

Parameter	Length (octets)	Parameter	Length (octets)
AAL parameters	?–22	MLPP precedence	10–11
Additional calling party number	6–15	Narrowband bearer capability	11–?
ATM cell rate	8–21	Narrowband high layer compatibility	11–?
Broadband bearer capability	7–11	Narrowband low layer compatibility	11–?
Broadband high-layer information	?–17	National/international call indicator	5–6
Broadband low-layer information	10–?	Notification	5–6
Called party number	7–15	OAM traffic descriptor	6–7
Called party subaddress	7–27	Original called number	6–15
Calling party number	6–15	Origination ISC point code	6–7
Calling party subaddress	7–27	Origination signaling identifier	8–9
Calling party category	5–6	Progress indicator	11–?
Closed user group information	9–10	Propagation delay counter	6–7
Connected line ID request	5–6	Redirecting number	6–15
Connection element identifier	8–9	Redirection information	5–7
Echo control information	5–6	Segmentation indicator (national use)	5–6
Forward narrowband interworking indicator	5–6	Transit network selection (national use)	6–?
		User-to-user indicators	5–6
Location number	7–15	User-to-user information	7–136
Maximum end-to-end transit delay	6–7		

TABLE 6-9. Other B-ISUP message types

Message Type	Parameter	Length (octets)
Release	Access Delivery	5–6
	Automatic congestion level	5–6
	Cause indicators	6–?
	Destination signaling identifier	8–9
	Notification	5–6
	Progress indicator	11–?
	Redirection information (national use)	5–7
	Redirection number (national use)	7–15
	Redirection number restriction (national use)	5–6
	Segmentation indicator (national use)	5–6
	User-to-user indicators	5–6
	User-to-user information	7–136
Release complete	Cause indicators	7–?
	Destination signaling identifier	8–9
Subsequent address	Destination signaling identifier	8–9
	Subsequent number	6–14
User-to-user information	Destination signaling identifier	8–9
	User-to-user indicators	5–6
	User-to-user information	7–136

between B-ICI and PNNI signaling. Recently, recognition of these interworking issues have resulted in work on a so-called ATM Internetwork Interface (AINI), which seeks to specify the interworking between PNNI and B-ICI based signaling and routing protocols, as well as ATM addressing interworking aspects. Since PNNI-based signaling is already widely deployed and network operators are considering enhancements to utilize the SS7-based capabilities for ATM, the need for interworking is likely to provide a significant incentive for the AINI specifications [7.1].

6.11 MESSAGE TRANSFER PART LEVEL 3 (MTP 3) FUNCTIONALITY

In describing the basic SS7 based signaling protocol architecture, it was pointed out that the Message Transfer Part Level 3 (MTP 3) [6.17] protocols were used to transport the B-ISUP messages, whereas the MPT Level 1 and 2 protocols used in narrowband ISDN were replaced by a dedicated (or preassigned) ATM VCC to carry the signaling messages encapsulated in the signaling AAL (SAAL). Since the MTP 3 protocols for B-ISDN are the same as for the narrowband ISDN case, it is not the intention to describe the protocol here. However, it is useful to summarize the main functions of MTP 3, since its operation is central to B-ISUP-based call processing. The primary purpose of the Message Transfer Part (MTP) protocols are to provide reliable (i.e., assured) transfer and routing of the B-ISUP signaling messages between the call processing entities [sometimes called Service control points (SCPs), in SS7 terminology] located in various NEs throughout the network. The MTP 3 protocols provide reliable connectionless transfer for the encapsulated B-ISUP messages between the signaling transfer points (STPs) in the network using a globally unique addressing scheme called "pointcodes" to identify each STP in the network. The MTP 3 pointcodes constitute the "routing label" indicated earlier in Fig. 6-

20 and are generally administered by the national telecommunications standards agencies, as well as the ITU-T to maintain global uniqueness.

It is interesting to note that MTP 3 provides a connectionless service to the B-ISUP protocol layer over a connection-oriented (although dedicated or preassigned) ATM plus SAAL protocol layer. Such a protocol architecture is analogous to the case of using the connectionless Internetworking Protocol (IP) over ATM for transporting applications such as e-mail or FTP. In fact, this parallel has led to some proposals to use IP instead of MTP 3 for transport of signaling information as a potential "simplification" of the SS7 architecture. Although the B-ISUP plus MTP 3 architecture may be viewed as a dedicated connectionless network for signaling and control applications only, it is important to recognize that a major consideration is that such a network be extremely robust and reliable. Since failure of the signaling network in any way would have catastrophic consequences, a large part of the cost and complexity of SS7 networks results from providing a high degree of reliability and robustness under load conditions. This is achieved by engineering a high level of redundancy in each element of the SS7 network, while providing mechanisms for operations and management (OAM) and testing, as well as traffic management and rerouting in the event of failures.

The MTP 3 [6.17] procedures may be grouped into four main categories:

1. Signaling Message Handling. These are the procedures for message discrimination, distribution, and routing of B-ISUP or other higher-layer applications such as TCAP, etc.
2. Signaling Links Management. This group of procedures relate to the testing and restoration of the (dedicated) virtual channel links, activation and deactivation, and the blocking and unblocking of the signaling resources. Note that these blocking and unblocking procedures relate to the signaling virtual channels and should not be confused with the B-ISUP blocking/unblocking messages used for the maintenance of resources such as VCCs or VPCs.
3. Signaling Route Management. These procedures relate to the connectionless routing capability based on the destination and origination pointcodes, as indicated above, to enable connectionless transfer of B-ISUP messages to the appropriate call processing entity.
4. Signaling Traffic Management. This group of procedures relate to providing flow control, rerouting in the event of failures, and changeovers between links for load balancing purposes to avoid congestion and resultant packet loss in the signaling network.

It may be evident that the MTP 3 functions summarized above result in a somewhat complex protocol, which has sometimes led to charges that this functionality may not be required for many ATM network applications such as, for example, IP or frame relay over ATM. While this remains a network operator choice to a large part, it is important to recognize that the MTP 3 procedures provide a powerful and highly reliable "tried and trusted" capability for B-ISUP message transfer, as well as providing inherent backward compatibility with the existing narrowband ISDN based SS7 networks designed for public carrier services.

6.12 ATM ADDRESSING AND B-ISUP ROUTING

It was noted earlier that addressing (or numbering) and routing of user information are intimately related, as may be evident by considering the ubiquitous telephony network. This is the case for both connection-oriented networks such as B-ISDN/ATM and connectionless networks such as the IP-based Internet. For connection-oriented transfer modes such as the PSTN and ATM, the destination address (or "called party number") is processed by the network elements by analyzing the digits dialed, for example, during the call establishment phase to set up a route for the ensuing VPC or VCC comprising a concatenation of VP or VC links between the NEs along the chosen route. After receipt and processing of the CONNECT and Answer (ANM) messages, the data transfer phase may commence. For connectionless transfer, each packet contains a (globally unique) address which is processed by the individual routing NEs, which compute a path to the destination and forward the packet accordingly. It need hardly be stated that both routing methodologies have been in use for many decades, resulting in a substantial body of knowledge to optimize the routing algorithms.

The addressing structure developed for narrowband ISDN for general use in public switched telecommunications networks (PSTNs) was defined by the ITU-T in Recommendation E.164 [6.14] and has been widely adopted and extended for national-specific purposes. The ITU-T has also specified the use of the E.164 addressing procedures for B-ISDNs based on ATM in Recommendation E.191 [6.15], so that the E.164 numbering scheme is assumed to be included in the signaling messages [e.g., SET UP and Initial Address Message (IAM)] in order to route the call. The E.164-based numbering scheme is essentially a generalization of the well-known "country code–area code–exchange code–local subscriber number" type of hierarchical addressing mechanism that has proved its effectiveness in conventional telecommunications. It should be noted that, apart from the definition of the "country code" designations by the ITU-T, administration of the addressing or numbering plans is generally the responsibility of the individual national telecommunications administrations and network operators. For example, in North America, this process has resulted in the North American Numbering Plan (NANP), which is in general used by all operators on the continent and has resulted in prompt and ubiquitous connectivity, despite extensive deregulation and competition in the industry. In general, the telecommunications authority in each country will structure the E.164-based number in accordance with its own (hierarchical) network topology.

In the initial work on ATM addressing and routing in the ITU-T, it was assumed that the recommended E.164- and E.191-based numbering methodology would be generally used for all ATM networks operated by regulated service providers, based on extensions of their existing numbering plans. In effect, B-ISUP signaling and routing methodologies generally assume this to be the case, with call routing based on digit analysis of the E.164/E.191 number with (generally statically) configured hierarchical routes. Call routing using the hierarchical digit analysis may be highly efficient, since the numbering plans are generally geographically constricted within any given administrative domain.

However, the ATM Forum subsequently selected an addressing format based on the ISO Network Service Access Point (NSAP) methodology. This was felt to allow much greater flexibility for encoding other types of addressing schemes of interest to the data communications industry, which viewed the telecommunications-oriented E.164 scheme as unnecessarily restrictive. It was noted before that ATM, despite its origins in traditional telecommunications practice and evolution from the existing narrowband ISDN/PSTN architectures, has also been adopted by the traditional data communications industry to a significant extent. In performing such a bridging role between networking paradigms that often display different architectural requirements, ATM needs to reconcile sometimes conflicting requirements. Nowhere is this more evident than in the area of addressing and routing, where both the ISDN/PSTN-based E.164/E.191 numbering scheme and the data-communications-based ISO NSAP addressing format have been adopted in the ATM Forum for signaling and routing specifications. Moreover, in order to promote interoperability between networks adopting the different addressing formats, the ITU-T has included the use and transport of NSAP ATM addressing in the UNI signaling message structure, although, at present, call routing in B-ISUP/SS7-based networks assumes the use of E.164/E.191 numbering. Other NSAP-based addresses are simply transported transparently across the B-ISUP- (or B-ICI-) based networks, without being used for call routing. In this respect, they may considered as the so-called "private numbering plan" (PNP) information, which may be transparently transported across the public ATM service provider networks in the signaling messages.

The generic ISO NSAP addressing format consists of a concatenation of an Initial Domain Part (IDP) field and a Domain-Specific Part (DSP) field [6.7]. For ATM addressing, the specific NSAP-based address formats selected by the ATM Forum are shown in Fig. 6-21 and are called ATM End System Addresses (AESA). There are four basic AESA types specified by the ATM Forum, as shown in Fig. 6-21. Each AESA is 20 octets long. The first (leftmost) field of the Initial Domain Part (IDP) of any NSAP formatted address is known as the Authority and Format Identifier (AFI) and is 1 octet long. The function of the AFI field is to specify which address registration authority is responsible for the addresses encoded in the IDP and DSP, as well as the coding format used in subsequent fields of the NSAP (AESA) address. Thus, the IDP field of a generic NSAP address consists of two parts: the AFI, which specifies the addressing administration authority and subsequent format, and the Initial Domain Identifier (IDI), which identifies the specific addressing type or domain in question. The Domain-Specific Part is similarly divided into three subfields as shown: these are called, the Higher-Order-DSP (HO-DSP), the End System Identifier (ESI) and the Selector (SEL) fields. The functions of these fields are described below for specified AESA types.

1. **Authority and Format Identifier (AFI).** The 1 octet AFI coding is administered by the International Standards Organization (ISO) for all NSAP IDP designations, and is used to specify the coding and format of the subsequent fields known as the IDI. The AFIs selected for the four AESA are coded as follows:

AFI coding		
Bits	Hexadecimal	IDI Format
0011 1001	39	DCC ATM Format
0100 0101	45	E.164 ATM Format
0100 0111	47	ICD ATM Format
0100 1001	49	Local ATM Format

Data Country Code (DCC) ATM Format. The AFI value 39 specifies the IDI as a Data Country Code, which is a two-octet field coded to specify the country in which the address is registered. The DCC values are administered by ISO (ISO 3166 [6.16]) and are encoded using the binary coded decimal (BCD) syntax, in which each group of four bits in an octet can code a decimal number. The coding is left-justified and unused bits are coded as hexadecimal "F" to fill the two octets. For example, DCC = 840 is assigned to the United States, where the subsequent DSP codings are administered by the American National Standards Institute (ANSI).

2. **International Code Designator (ICD) ATM Format.** The AFI value 47 in the IDP of the NSAP address format identifies the Initial Domain Identifier (IDI) to be the two-octet International Code Designator (ICD), which is also coded using the binary coded decimal (BCD) convention. Since the ICD was originally intended to be used as a coding identification (and not specifically an organization addressing) scheme, there has been debate regarding its usefulness as an addressing mechanism for ATM networks. More recently, the British Standards Institute (BSI),which administers the ICD values on behalf of ISO, has specifically requested that ICD codes should not be generally used for ATM addressing purposes and has defined a specific ICD value (ICD = 0124) to be used for this purpose. However, some organizations that have already

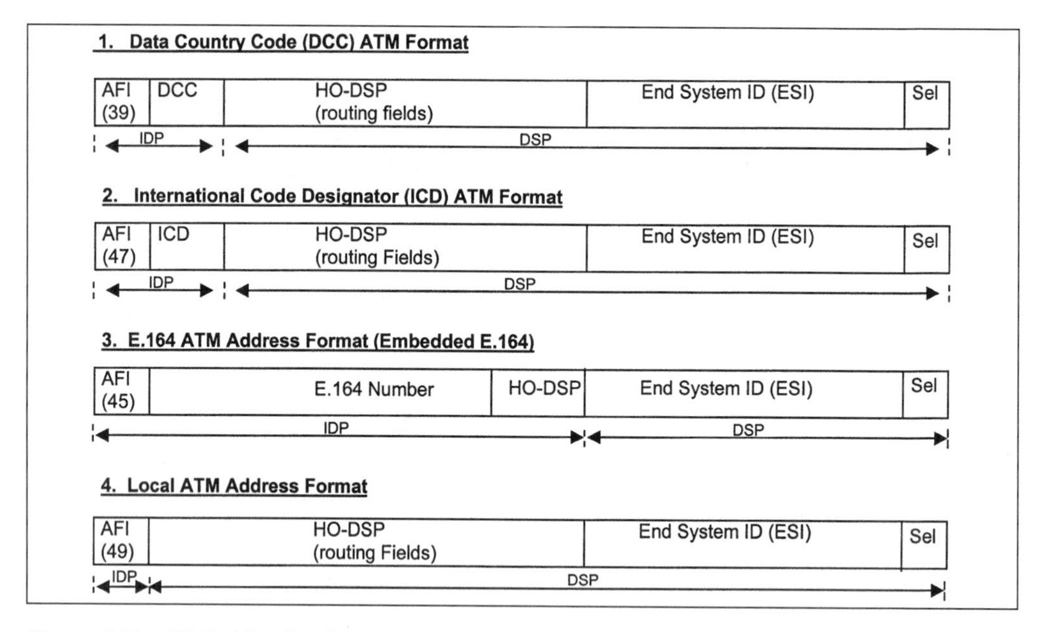

Figure 6-21. ATM addressing formats. AFI = authority and format identifier; IDP = initial domain part; DSP = domain-specific part; HO-DSP = higher-order domain-specific part; Sel = selector.

been allocated other ICD values may choose to use them for ATM addressing, but it remains questionable whether the ICD AESA will be generally used for ubiquitous connectivity.

3. **E.164 ATM Format** (also called the Embedded E.164). The AFI value = 45 in the IDP identifies the IDI to be the E.164/E.191 international B-ISDN number, as administered by the ITU-T. The E.164 address may be up to 15 (decimal) digits long and is coded using binary coded decimal (BCD) encoding into eight octets (each group of four bits is used to specify a decimal digit, with unused digits signified by leading zeros and a single semioctet of all "ones" added to complete the eight octets). As noted earlier, the E.164/E.191 number has the general form Country Code (CC) + National Destination Code (NDC) + Subscriber Number (SN), where the NDC may be further broken down into Area Codes, Service Access Codes (e.g., an 800 or 900 number) and Destination Network (e.g., transit or regional operator) Codes, such as is done in normal telephone numbering schemes. Although each country uses its own national numbering plan, the ATM Forum specifications only recognize the international E.164 format.

It is important to note that since the E.164 address constitutes a complete address format in its own right, a distinction is made between the so-called NSAP "embedded E.164" (E.164e) and the so-called "native" E.164 (E.164 N) number in ATM addressing. Typically, both types of addresses may be used in different networks, and are distinguished in signaling messages by codepoints in the Numbering Plan Identifier (NPI) field of the appropriate IE or B-ISUP Parameter. It may be noted that if the other DSP fields in the embedded E.164 AESA are all coded zero, then there is essentially no difference between the native E.164 and embedded E.164 from the perspective of call routing. However, it must be admitted that the use of both the native E.164 and the NSAP embedded E.164 in the ATM signaling specifications has resulted in some confusion as to the interoperability of ATM addressing schemes in general. In order to alleviate some of the confusion surrounding ATM addressing (arguably of its own making), the ATM Forum has recently issued documentation explaining the structure and use of the various AESA types and the interworking requirements that may be needed in various networking scenarios. However, it may be envisioned that both the NSAP embedded E.164 AESA and the native E.164 numbers will be the most widely used formats by ATM service providers.

4. **Domain-Specific Part (DSP).** The DSP of any AESA is subdivided into the Higher-Order Domain-Specific Part (HO-DSP), the End System Identifier (ESI), and the Selector (SEL) subfields as shown.

5. **Higher-Order DSP (HO-DSP).** In general, the coding of the HO-DSP field is determined by the authority specified in the Initial Domain Identifier (IDI) or its delegates. Thus, for the case of the DCC AESA, the authority identified in the DCC administration allocates addresses and coding of its HO-DSP. Typically, address space will be allocated to organizations and ATM service providers. For the DCC and ICD AESAs, the length of the HO-DSP field is 10 octets. For the E.164e AESA, the HO-DSP is four octets, and for the "local" ATM AESA it is 12 octets long. Clearly, the allocation of addresses in the HO-DSP should be structured to permit efficient routing in accordance with topologically hierarchical significance. However, it may be noted that for the E.164 AESA, the use of the HO-DSP is not defined, since the ITU-T does not at present specify any additional routing capability beyond the E.164/E.191 international number.

6. **End System Identifier (ESI).** The ESI field in all the AESAs is six octets long. It may be used to identify an end system or terminal in the sense used in data communications. Typically, it may encode an IEEE MAC address or serial number including identification of an individual manufacturer, which may be globally unique. The ESI may be used by an end system (or terminal) to autoconfigure its address by means of management interface protocols.

7. **Selector (SEL).** The Selector is fixed at one octet for all AESA types, and may be used by an end system or terminal for internal use. However, it is not used for routing purposes in the ATM network.

8. **Local ATM Format.** The AFI value = 49 identifies the so-called local ATM AESA, which is intended for private (i.e., proprietary) addressing use. The IDI is therefore null and the DSP is 19 octets long, with the HO-DSP field of 12 octets. The local or private AESA is intended for use in private voice telephony over ATM (VTOA) applications, but since such addresses are unregistered and essentially proprietary, interworking with other ATM addressing domains may require special configuration. The local AESA HO-DSP coding is not specified at present. The essentially proprietary intent of the local ATM address format implies that such an address will not be used for routing in any public ATM service provider networks, although it may be transported transparently through such a network—for example, in a subaddress field or in the user-to-user IE.

In principle, an ATM service provider may select one (or more) of the AESA types described above for routing within its network, although routing complexity considerations would generally militate against use of more than one addressing scheme in any given network. However, use of the different AESA types in different interconnected networks will result in the need for address interworking, for example by means of an interworking "gateway" NE at the boundaries of the different addressing (and hence routing) domains. In general, this process may be facilitated by using the so-called Numbering Plan Identifier (NPI) or Numbering/Addressing Plan Identifier (NAPI) code set in the SETUP or IAM messages. The codepoints in this field of the IE indicates to the NE that the address is destined for another routing domain so that (if provisioned) the message will be directly routed through the relevant gateway NE to that particular domain. However, other AESA interworking scenarios are also possible; for example, by address resolution servers providing for some form of address translation function.

As noted earlier, the key motivation behind the specification of the NSAP-based AESA formats in addition to the established native E.164 international address originally intended for ATM was to allow more flexibility, particularly for private ATM networks or new independent service providers. It is clear that the flexibility inherent in the NSAP format enables this objective to be met in the AESA types described above. However, it is also important to note that the close relationship between call routing efficiency and the (hierarchical) structure of an address imposes constraints on the ways in which the AESAs may pragmatically be used. In this respect, the E.164 numbers (either native or embedded) may be considered optimum, since their simple geographically structured hierarchy enables efficient localization of an end point in large WANs. For these reasons, it is likely that the E.164/E.191 addressing scheme will continue to be widely used for the majority of public carrier ATM service providers. The main contender is likely to be the DCC AESA, intended primarily for data communications networks or private (enterprise) campus networks. In this respect, it should be noted that the NSAP AESA hierarchy is essentially structured as a hierarchy of registration authorities, apart from the End System ID and Selector fields, which are only used for routing within the LAN or end system. Thus, the DCC field identifies a country (or administrative region) that administrates the structure of the HO-DSP field, which is then essentially the main WAN routing field that needs to be defined. This is also the case for the ICD AESA format, but since this structure is actually intended for code designations rather than actual numbering plans, it seems unlikely that the ICD AESA will find widespread use for ATM addressing and routing, and may find use primarily in niche applications.

For networks deploying B-ISUP-based signaling, it seems likely that the native E.164 format will continue to be used for efficient routing, based on current call routing practices in the SS7 network. The routing algorithms tend to be proprietary or administratively configured, depending on the network operator policy and individual manufacturers' implementations in the switch. In such network applications, the embedded E.164 AESA seems not to provide much added value for call routing purposes. However, for private network routing, the NSAP-based AESAs, such as DCC or local ATM format, may offer advantages for routing schemes based on different hierarchies.

CHAPTER 7

Private NNI (PNNI) Signaling and Routing

7.1 INTRODUCTION

In the previous chapter, it was noted that in developing specifications for signaling in "private" (or enterprise) ATM networks, the ATM Forum adopted a different signaling protocol architecture from the SS7-based ITU-T architecture intended primarily for conventional "public" carrier networks. It may be recalled that the primary rationale underlying this divergence was the notion that typical enterprise networks would not require the complex maintenance and testing capabilities as well as the AIN flexibility designed into the SS7 architecture. Such robustness and flexibility, considered desirable by regulated, service-oriented network operators, would likely prove too costly for enterprise applications. In addition, it was felt that although public network operators could support the maintenance and administrative overhead implied by the inherently statically routed SS7 protocols, the corresponding private network routing should be automated as much as possible to minimize administrative overhead. As a consequence, the ATM Forum also developed a complex "automated" or so-called dynamic routing protocol to be used in conjunction with the PNNI signaling protocols between the private ATM switches. In this chapter, we describe both the PNNI signaling and routing protocols in some detail, while highlighting the key differences (and similarities) with the DSS2 plus SS7 protocol architecture described earlier.

7.2 BASIC PNNI SIGNALING AND ROUTING CONCEPTS

In describing the PNNI control architecture, it is useful to clearly distinguish between the signaling protocols and the routing protocols. The separation between the signaling functionality and the routing functionality is depicted in the general PNNI protocol architecture reference model shown in Fig. 7-1. The PNNI signaling protocols comprise the messages and associated procedures that convey signaling-related information between the NEs, together with the associated call processing functions similar to those described previously. In fact, the PNNI signaling protocols (including both message set and IEs) are an extension of the basic UNI protocols with additional IEs for carrying routing information as well as some other NNI-specific functions described in more detail below. On the other hand, the PNNI routing protocols which include (1) topology/link parameter exchange, (2) topology database, and (3) route determination algorithm functional blocks are based on a hierarchical dynamic source routing mechanism using so-called "link state update" messages exchanged between the NEs. In essence, the PNNI routing functionality is used to "automate" routing in ATM networks, derived from the philosophy of connectionless routing used in Internetworking Protocol (IP) networks. Specifically, the PNNI routing protocols are analogous to the so-called Open Shortest Path First (OSPF) based routing protocols widely

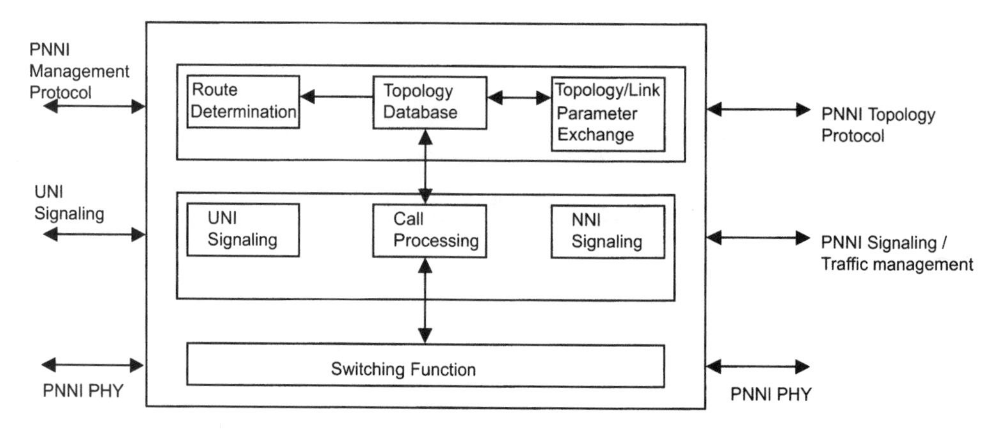

Figure 7-1. Private NNI (PNNI) signaling and routing, general PNNI architecture reference model. *PNNI signaling protocol:* based on UNI signaling (ATMF v4.0) with additional IEs carrying routing information. *PNNI routing protocol:* Based on hierarchical dynamic source routing using link state update messaging.

used in IP routing. The functional separation between the PNNI signaling and routing protocols shown in Fig. 7-1 is also highlighted by the fact that separate preassigned VCI values are used to identify the VCLs carrying the signaling and routing messages. Thus, as noted earlier, VCLs with value VCI = 5 are used to transport the PNNI signaling messages between NEs, whereas a VCL with a preassigned value VCI = 18 (in any VPI) are used to carry the dedicated routing messages between the NEs. Of course, within the NEs, the relationship between the signaling and routing functions remains the same as described previously: the routing function computes an optimum path through the network based on the destination address and other relevant (e.g., topology and link status) information, whereas the primary function of signaling is to set up (and clear) the ATM VCC or VPC from source to destination with the appropriate network resources.

Before undertaking a detailed description of the PNNI signaling and routing procedures, it is useful to bear in mind a general overview of the basic underlying routing concepts. Fig. 7-2 depicts a generic PNNI network and summarizes the concepts of link state routing and source routing as used in PNNI protocols. In link state routing protocols, each node periodically exchanges routing messages (called "hello" packets) with all neighboring nodes. The primary function of the hello packets is to identify the sending nodes as well as the status of the links or transmission paths in question. By processing the information in the hello packets, each NE is able to construct a list of links to its direct neighbors called a link state update (LSU). The LSU information from each NE is then sent to all the peer NEs (in the routing hierarchy), using a separate routing message, in a process called "flooding." This "flooding" of the nodal, as well as link state information to all the NEs at a given hierarchical level allows each node in the PNNI hierarchy to develop a topological "map" of the network in its routing database. The routing database can be automatically updated in the event of any changes that may affect subsequent connection set-up procedures (or resource allocation). In this way, the topology database may be used by the route determination algorithms in the NE to compute a suitable path through the network to route the call.

It may be recognized that the processes briefly described above essentially "automates" the call routing functionality in PNNI networks. For the case of the SS7 network, this is typically carried out administratively by preconfiguring optimum paths. However different the PNNI and SS7 routing procedures may be, it is important to recognize an essential similarity: in order for the call routing protocols to be efficient (and scaleable), the addressing hierarchy must corre-

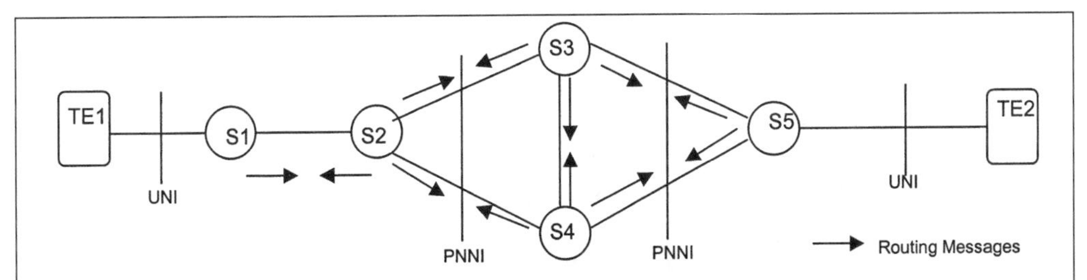

Notes. 1. Concept of "link state routing." Each node periodically exchanges "hello" packets with neighboring nodes. It constructs a "link state update" (LSU) list of links to direct neighbors, "floods" LSUs to all other neighbors, develops a hierarchical topological map of the network, and requires address hierarchy to correspond to network hierarchy for routing. 2. Concept of "source routes." Ingress node chooses a complete path to the destination, then adds full path to the message itself. Transit nodes follow given path unless blocked (then reroute message by alternate procedures, e.g., "crankback"). Source route IE added to Q.2931 SETUP and ADD PARTY messages as "designated transit list" (DTL).

Figure 7-2. Basic PNNI routing concepts.

spond to the hierarchy in the network topology. As pointed out earlier in our discussion of addressing for ATM networks, this requirement is common to any routing process, whether based on exchanging (standardized) routing messages between NEs and processing the resulting information to compute a suitable path, or statically preconfiguring paths based on administrative procedures. Nonetheless, it is useful to note that this aspect of PNNI routing is often obscured by the abstract nature of the hierarchical routing procedures described in the ATM Forum PNNI routing specifications, which were intended to apply to any general routing hierarchy, irrespective of the address type.

A significant point of difference between the SS7 routing and PNNI routing is the use of the "source routing" concept. Referring again to Fig. 7-2, in PNNI source routing, the ingress NE (e.g., S1) computes a complete path to the destination NE (e.g., S5), and then includes the full path information within the connection set-up message itself. The source route information is known as a designated transit list (DTL), and is included as an IE in the PNNI SET UP (or ADD PARTY) messages, as described in further detail below. The transit NEs then route the SETUP message according to the DTL path to establish the connection. If the path of the signaling message is blocked for any reason, the PNNI signaling protocols also have an additional capability known as "crankback," which enables the "retry" of the connection set-up via an alternative route.

The use of automated link state routing messages and source routing, together with a signaling message set based on extension of UNI signaling, implies that PNNI routing and signaling procedures are substantially different from those used in the SS7/B-ISUP architecture. As indicated earlier, the potentially low administrative overhead promised by PNNI routing together with its widespread availability has resulted in its use by a number of public network operators as well as the intended enterprise network applications. This immediately raises the question of interworking between the PNNI and B-ISUP (or equivalently, B-ICI) procedures. Although it may clearly be undesirable to mix both types of signaling approach within one network to avoid undue complexity, interworking between PNNI- and B-ICI-based administrative domains is possible by means of "gateway" NEs, which perform the necessary protocol translation at the domain boundaries. Nonetheless, general interworking between PNNI and B-ICI domains remains an open question at this stage. A number of interworking scenarios have been catalogued and analyzed in the ATM Forum's ATM Inter Network Interface (AINI) specifications in an attempt to assist deployment planning [7.1]

The dynamic link state routing procedures outlined above were developed by the ATM Forum from an earlier and much simpler version of PNNI signaling and routing known variously as the Interim Inter Switch Protocol (IISP) or PNNI version 0 [7.2], which have also been widely implemented by a number of vendors. The PNNI version 0 or IISP specification uses statically configured routing, and hence does not rely on the complex automated link state and topology database update mechanisms included in the later PNNI version 1 [7.3] routing protocols outlined above. In that sense also, the PNNI version 1 protocols are not backward compatible with the earlier version 0 statically routed protocols. Interworking between networks operating the earlier IISP (or PNNI version 0) protocols and those based on PNNI version 1 protocols may again be obtained straightforwardly by configuring these networks as lying within separate routing domains.

As noted earlier, a number of the fundamental concepts underlying the PNNI routing protocols are derived from the routing procedures commonly used in the earlier connectionless IP-based networks, often referred to as the Internet. Interestingly, these concepts have been applied to an essentially connection-oriented technology such as ATM signaling via the mechanism of source routing. The routing procedures automatically use the topology and link state information to compute routing tables and derive the designated transit list of nodes along a connection path, and then insert this information into the signaling messages to set up the VCC or VPC accordingly. The "borrowing" of concepts from connectionless routing provides another example of the way in which data internetworking technologies such as ATM and TCP/IP interact by adapting concepts and mechanisms from each other where appropriate. We will encounter many such examples as we consider other aspects of these, and other, internetworking technologies.

7.3 PNNI SIGNALING PROTOCOLS

As noted earlier, PNNI signaling procedures are based on the same message set used in the case of UNI signaling, with extensions of the relevant information elements (IEs) for the PNNI specific functions. The basic PNNI message set and the related signaling protocol stack are summarized in Fig. 7-3. The similarity of the PNNI signaling protocol stack to the UNI signaling protocol architecture described earlier should be noted and contrasted with the SS7/B-ISUP protocol architecture. As for the UNI case, PNNI signaling messages are encapsulated in the Service-Specific Coordination Function (SSCF) and the Service-Specific Connection-Oriented Protocol (SSCOP) sublayers. The primary function of the SSCF + SSCOP sublayer is to provide reliable, assured data transfer based on flow control and retransmission of protocol data units (PDUs) in the event of loss or corruption of information. The signaling AAL uses AAL Type 5 Common Part Convergence Sublayer (CPCS) for error detection and segmentation and reassembly. These common AAL functions are described in more detail subsequently. As before, the signaling messages are carried on a dedicated common-channel link designated by VCI = 5 between the ATM switches.

The PNNI protocol architecture sublayers the call control functions into the so-called PNNI Protocol Control and Call Control sublayers as shown. The PNNI Protocol Control sublayer is responsible for the control of the signaling state machines whereas the Call Control sublayer is responsible for processing and generation of the individual signaling IEs in the messages. This distinction, although only highlighted explicitly in the PNNI signaling architecture, may also be made for the UNI and B-ISUP protocols as well, since it serves essentially to distinguish various aspects of the overall call control processes. The signaling message set listed for PNNI is consistent with the ATM Forum defined UNI Signaling Specification [6.2], which includes the additional messages used for point-to-multipoint (pt–mpt) call control. Although in general the sig-

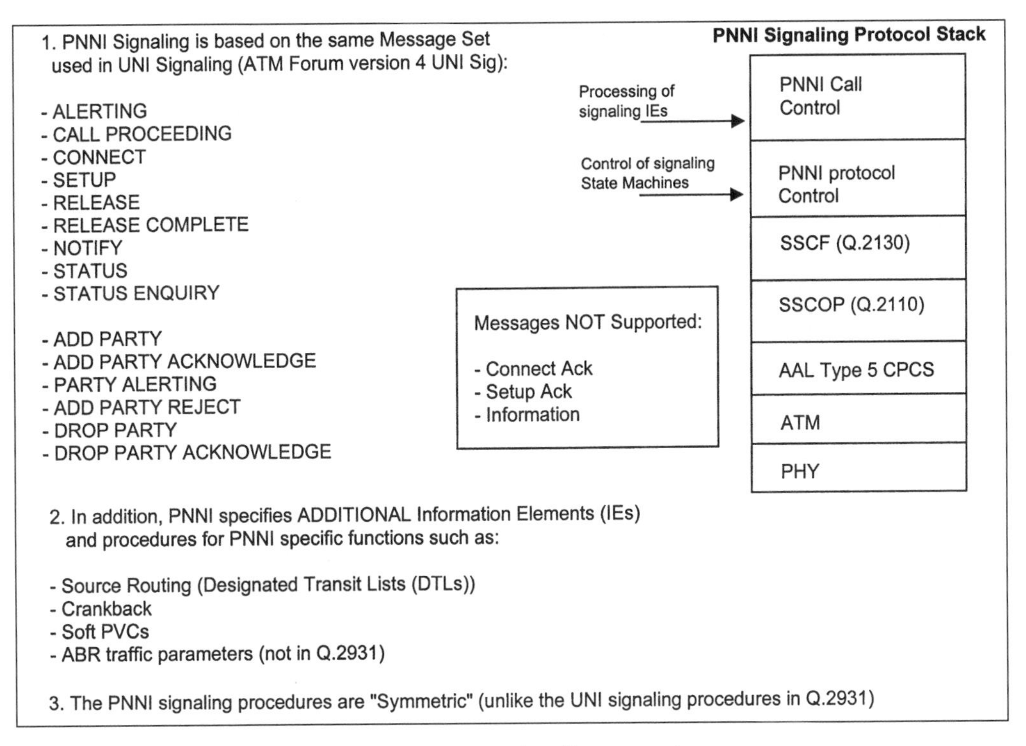

Figure 7-3. PNNI signaling protocols.

naling procedures for PNNI are the same as those defined for the UNI case, the PNNI procedures are symmetrical, reflecting the fact that PNNI signaling occurs between peer entities (ATM switches), as opposed to between terminal and switch for the (unsymmetrical) UNI case. However, PNNI signaling does not include certain messages such as:

1. CONNECT ACKNOWLEDGE
2. SETUP ACKNOWLEDGE
3. INFORMATION

In addition to the above deviations from the UNI signaling procedures, PNNI specifies some key additional IEs to support functions specific to the PNNI. These include:

1. Source routing using Designated Transit List (DTL) IE
2. Crankback capability
3. Soft PVC (SPVC) capability
4. Additional traffic parameters for ABR that are not included in the basic DSS2 (Q.2931 [6–8]) signaling procedures. This was required to obtain consistency with the ATM Forum versions of the ABR service category in their traffic management specifications [5.8].

Since PNNI signaling procedures are generally similar to those for the UNI case described earlier, it is not the intent here to discuss each message type in detail, since these can be readily obtained from the specifications. As before, the scope here is to obtain a broad understanding of how the signaling protocol works by selected examples of the structure and content of some messages and IEs, notably those specific to PNNI. As for all signaling messages, it is important to bear in mind that PNNI also uses the nested "type–length–value" (TLV) general coding format for the message set and included IEs.

As an example, the possible IEs in the PNNI SETUP message are shown in Fig. 7-4. The individual IEs may be of the mandatory (M) or optional (O) type as noted previously and, in general, the SETUP message tends to be the most complex to process, since it generally needs to include all the necessary information for proper call set-up and resource allocation. Moreover, the PNNI SETUP message may be relatively long since it needs to include the DTL IE necessary for source routing procedure, which may be between 33 to 546 octets in length, depending on the number of intermediate NEs to be traversed. Also, as noted previously, the first four IEs listed:

1. Protocol Discriminator
2. Call Reference
3. Message Type
4. Message Length

are mandatory and always appear in that order in all messages in accordance with the TLV coding format. The Protocol Discriminator signifying PNNI signaling messages is defined as the binary value 11110000. The other IEs listed may subsequently be concatenated in any order in the message, since each IE is uniquely identified by its own IE type codepoint as listed below for some selected IEs.

Information Element	Length (octets)	Type
Protocol discriminator	1	M
Call Reference	4	M
Message Type	2	M
Message Length	2	M
AAL Parameters	4 - 21	O
ABR Additional Parameters	4 - 14	O
ABR Setup Parameters	4 - 36	O
Alternative ATM Traffic Descriptor	4 - 30	O
ATM Traffic Descriptor	12 - 30	M
Broadband Bearer Capability	6 - 7	M
Broadband Higher Layer Information	4 -13	O
Broadband Repeat Indicator	4 - 5	O
Broadband Lower Layer Information	4 -17	O
Called Party Number	min - 25	M
Called Party Soft PVPC or PVCC	4 - 11	O
Called Party Subaddress	4 - 25	O
Calling Party Number	4 - 26	O
Calling Party Soft PVPC or PVCC	4 - 10	O
Calling Party Subaddress	4 - 25	O
Connection Identifier	4 - 9	O
Connection Scope Selection	4 - 6	O
Designated Transit List	33 - 546	M
Endpoint Reference	4 - 7	O
End-to-end Transit Delay	4 -13	O
Extended QOS Parameters	4 - 25	O
Generic Identifier Transport	4 - 33	O
Minimum Acceptable ATM Traffic descriptor	4 - 20	O
Notification Indicator	4	O
QOS Parameters	4 - 6	O
Transit Network Selection	4 - 9	O

Figure 7-4. PNNI signaling message example—SETUP.

Although a large number of possible IEs may be included in the PNNI SETUP message as listed in Fig. 7-4, the only mandatory IEs listed are

1. ATM Traffic Descriptor
2. Broadband Bearer Capability
3. Called Party Number (i.e., destination address)
4. Designated Transit List

Apart from the DTL necessary for source routing, the function of each of these IEs is the same as described previously and hence need not be elaborated on here. Depending on the application and functionality required, the other IEs may also be present in the SETUP message. Clearly, the more IEs that are included, the more processing (and hence call set-up latency) will be required.

Another example of a typical PNNI signaling message is the RELEASE message shown in Fig. 7-5. In addition to the mandatory cause IE, this message may also include the crankback IE if the network supports this function. The crankback function is described in more detail below. As noted earlier, the RELEASE message is sent either when a connection is cleared (by either node) or when a call is blocked due to a variety of causes (e.g., insufficient bandwidth resources). It should be noted that IEs such as the Generic Identifier Transport (GIT) and Notification Indicator are only used to carry particular types of information relating to specific applications and hence may not be generally used. The use of these and other special types of signaling IEs are outside the scope of this discussion. The codepoints identifying selected IEs, their maximum length in octets, and the maximum number of times the IE may appear in any given PNNI signaling message are shown in Table 7-1, which enables the reader to obtain an overview of the large number of IEs that have been defined to date. In this context, it should also be pointed out that as additional functionality is being added to PNNI signaling and routing by the ATM Forum

Information Element	Length (octets)	Type
Protocol Discriminator	1	M
Call Reference	4	M
Message Type	2	M
Message Length	2	M
Cause	6 - 34	M
Crankback	4 - 72	O
Notification Indicator	4	O
Generic Identifier Transport	4 - 33	O

1. The PNNI signaling procedures allow for transport of 64 kb/s ISDN signaling messages "transparently" across the PNNI network.

2. The Protocol Discriminator value for P-NNI signaling messages is defined as binary 11110000

3. In general the coding rules, message formats, IE formats and codepoints are the same as those for UNI signaling standards (ATM Forum UNI Signaling Specification version 4 based on ITU-T Recommendations Q.293, Q.2961,Q.2971 except for P-NNI specific functions)

4. The P-NNI specific IEs such as Designated Transit Lists (DTLs) and Crankback add additional complexity to processing of P-NNI messages.

Figure 7-5. PNNI signaling message example—RELEASE.

TABLE 7-1. Information elements used in PNNI

Bits 8	7	6	5	4	3	2	1	Information element	Maximum length	Maximum number of occurrences
0	0	0	0	0	1	0	0	Narrowband bearer capability[a,b]	14	3
0	0	0	0	1	0	0	0	Cause[a]	34	2
0	0	0	1	0	1	0	0	Call state	5	1
0	0	0	1	1	1	1	0	Progress indicator[a]	6	2
0	0	1	0	0	1	1	1	Notification indicator	[c]	[c]
0	1	0	0	0	0	1	0	End-to-end transit delay	13	1
0	1	0	0	1	1	0	0	Connected number	26	1
0	1	0	0	1	1	0	1	Connected subaddress	25	1
0	1	0	1	0	1	0	0	Endpoint reference	7	1
0	1	0	1	0	1	0	1	Endpoint state	5	1
0	1	0	1	1	0	0	0	ATM adaptation layer parameters	21	1
0	1	0	1	1	0	0	1	ATM traffic descriptor	30	1
0	1	0	1	1	0	1	0	Connection identifier	9	1
0	1	0	1	1	1	0	0	Quality of service parameter	6	1
0	1	0	1	1	1	0	1	Broadband high-layer information	13	1
0	1	0	1	1	1	1	0	Broadband bearer capability	7	1
0	1	0	1	1	1	1	1	Broadband low-layer information[b]	17	3
0	1	1	0	0	0	0	0	Broadband locking shift	5	
0	1	1	0	0	0	0	1	Broadband nonlocking shift	5	
0	1	1	0	0	0	1	1	Broadband repeat indicator[a]	5	3
0	1	1	0	1	1	0	0	Calling party number	26	1
0	1	1	0	1	1	0	1	Calling party subaddress[a]	25	2
0	1	1	1	0	0	0	0	Called party number	25	2
0	1	1	1	0	0	0	1	Called party subaddress[a]	25	2
0	1	1	1	1	0	0	0	Transit network selection	9	1
0	1	1	1	1	0	0	1	Restart indicator	5	1
0	1	1	1	1	1	0	0	Narrowband low-layer compatibility[b]	20	2
0	1	1	1	1	1	0	1	Narrowband high-layer compatibility[a]	7	2
0	1	1	1	1	1	1	1	Generic identifier transport[a]	33	3
1	0	0	0	0	0	0	1	Minimum acceptable ATM traffic descriptor	20	1
1	0	0	0	0	0	1	0	Alternate ATM traffic descriptor	30	1
1	0	0	0	0	1	0	0	ABR setup parameters	36	1
1	1	1	0	0	0	0	0	Called party soft PVPC or PVCC	11	1
1	1	1	0	0	0	0	1	Crankback	72	1
1	1	1	0	0	0	1	0	Designated transit list[b]	546	10
1	1	1	0	0	0	1	1	Calling party soft PVPC or PVCC	10	1
1	1	1	0	0	1	0	0	ABR additional parameters	14	1
1	1	1	0	1	0	1	1	Connection scope selection	6	1
1	1	1	0	1	1	0	0	Extended QoS parameters	25	1

[a]This information element may be repeated without the Broadband repeat indicator information element.

[b]This information element may be repeated in conjunction with the Broadband repeat indicator information element.

[c]The maximum length and the number of repetitions of this information element are network-dependent.

version 2 of the specification, enhancements to the IE set as well as their contents may be anticipated.

7.4 SOURCE ROUTING IN PNNI SIGNALING

As noted earlier, a significant point of difference between signaling protocols designed for public carrier networks based on SS7 and private NNI signaling is the use of "source routing" methodology for the latter. As shown in Fig. 7-6, in the source routing technique the originating ATM NE computes an "optimum" path for the VCC (or VPC) using the PNNI routing database built up by the topology and link state information exchanged between nodes in the network, as well as the destination AESA in the UNI SETUP message. Each NE is uniquely identified by a logical node identifier and hierarchy level, as described in more detail below. Once the originating node determines a suitable path to the destination, it encodes the list of selected NEs in the so-called Designated Transit List (DTL) as an information element in the PNNI SETUP message. In effect, the selected route is coded in the DTL IE as an ordered stack of node (and, optionally, port) identifiers. Each intermediate NE then processes the DTL IE to forward the SETUP (or ADD PARTY in the case of pt–mpt connection types) message to the next identified NE in the list of logical node identifiers. The intermediate nodes identified in the list confirm the logical node identifier and set up the connection on a VCL-by-VCL basis before forwarding the SETUP message to the next node in the DTL stack. The DTL IE may be up to 546 octets in length and may be repeated up to 10 times in a SETUP message. Thus, for large PNNI-based networks, substantial processing of the SETUP message may be involved, which may result in significant call set-up latency.

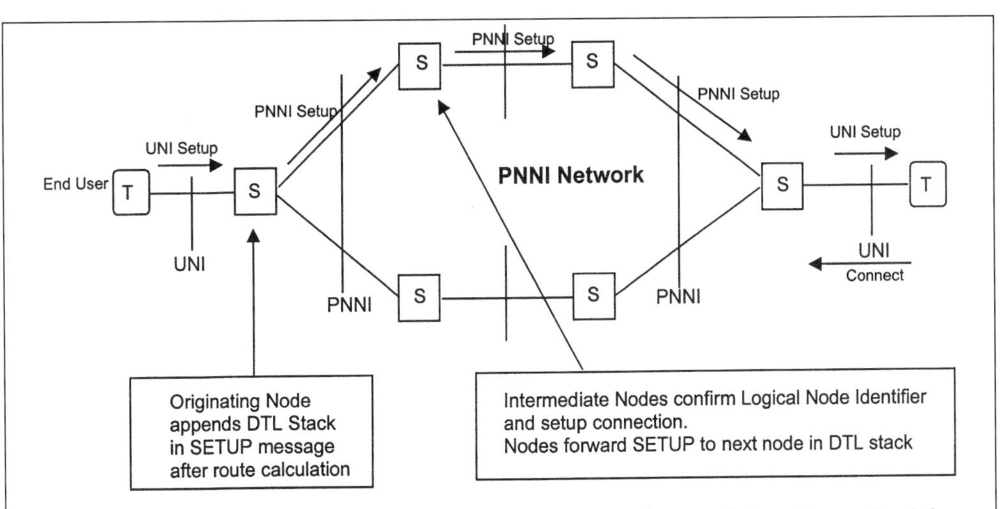

Notes. 1. In source routing, the originating ATM node computes optimum path based on routing information (topology, link states) provided by PNNI routing protocols and ATM end system address (AESA). 2. Each node is identified by logical node identifier and hierarchy level in the source route. 3. The selected route is coded in the Designated Transit List (DTL) IE as a ordered stack of node (and optionally port) identifiers. 4. Each intermediate node uses the DTL to forward the SETUP (or ADD PARTY) message to the next identified node in the list. DTL is based on "last in–first out" rule. 5. The DTL IE can be up to 546 octets long and be repeated up to 10 times in a SETUP message.

Figure 7-6. Source routing in PNNI.

The Designated Transit List IE structure is shown in Fig. 7-7, based on the ubiquitous TLV format used for all IEs and signaling messages. The "coding standard" field identifies the IE as one specified by the ATM Forum specifications (coding standard = 11), whereas the "IE Instruction Field" is used to encode the actions to be performed by a NE in the event that the IE cannot be processed or recognized, a similar mechanism to that encoded in other signaling procedures. It is important to recognize that the DTL IE includes an "ordered" list in which the processing of the octet groups is position-dependent. In effect, the DTLs are structured according to the last in–first out rule for processing of the stack of PNNI node identifiers within a hierarchical level. The last in–first out procedure for processing of the DTL IEs or individual node ID within a given DTL IE implies that the last IE or node ID in the stack is processed first, assuming, of course, that it corresponds to the node or hierarchical level in question. Thus, each DTL IE contains a list of node identifiers at a single level of the routing hierarchy in the order in which they are traversed, with the "last" or border node at the bottom of the list (i.e., the first ID in the list). Similarly, the stack of DTL IEs (up to 10) representing each level in the routing hierarchy are stacked such that the first (i.e., bottom of the stack) DTL IE corresponds to the last (hierarchical) peer group to be traversed by the SETUP message. The stack of DTL IEs ordered in this way, which permits a complete path from source to destination encompassing all levels of the routing hierarchy, is termed a "hierarchically complete source route".

Within each DTL IE corresponding to a given routing peer group or hierarchical level, the

bit #	8	7	6	5	4	3	2	1	Octet #
	1	1	1	0	0	0	1	0	1
	Designated Transit List Information Element Identifier								
	1 ext	Coding Standard (=11 for ATMF)		IE Instruction Field					2
	Length of DTL Contents								3
	Length of DTL Contents (continued)								4
	Current Transit Pointer								5
	Current Transit Pointer (continued)								6
	Logical Node / Logical Port Indicator (up to 20 times)								7
	Logical Node Identifier (22 octets)								7.1 to 7.22
	Logical Port Identifier (4 octets)								7.23 to 7.26

Notes. 1. The Designated Transit List (DTL) is an "ordered" list in which processing of the octet groups is position-dependent. The DTL is structured as a "last in–first out" stack (e.g., of node identifiers). 2. The Current Transit Pointer field is encoded to "point" to ID of current node (or, optionally, port). Consequently, each node on path must advance the Current Transit Pointer value to next node in the path. 3. The Logical Node Identifier (22 octets) is binary coded to uniquely identify each ATM node along the path which the call/connection is to transit. 4. The (optional) Logical Port Identifier (four octets) is binary encoded to uniquely identify a (physical) port on the node. The combination of Logical Node and Port Identifier should uniquely identify the PNNI route. 5. The DTL stack represents the "source routing""capability in PNNI. The originating node inserts the source route DTLs. Intermediate nodes may add (or substract) DTLs as appropriate. If the DTL stack represents all required levels of the PNNI hierarchy to reach the destination, it is known as the "hierarchically complete source route."

Figure 7-7. PNNI IE example—Designated Transit List.

Current Transit Pointer field is encoded to indicate or "point to" the node (and optionally to the port) identifier of the NE that is being traversed by the SETUP message. Consequently, each NE on the path must advance the Current Transit Pointer field value to the next NE (or port) before forwarding the SETUP (or ADD PARTY) message to that NE (or port). As may be recognized, if on receipt of the DTL IE a NE finds that the current transit pointer value is indicating a node ID that does not correspond to its own, there is likely a routing error, and the call may be blocked (or cranked back, as described below) with the relevant cause indicated in the RELEASE message. Thus, the Current Transit Pointer provides a mechanism to confirm that the SETUP message is being routed along the intended path listed in the DTL. The logical node identifier is a 22 octet binary value that uniquely identifies each ATM NE and the optional four octet logical port identifier is binary encoded to uniquely identify a physical port on the node. Consequently, the combination of logical node and port identifiers serves to identify the route through the PNNI network.

The PNNI NEs may also add or remove the DTL IEs in addition to processing their contents, depending on the function of the node in the PNNI routing hierarchy. Thus, the DTL procedures enable the exit border node of a given hierarchical level to remove the DTL IE corresponding to that level from the stack of DTL IEs after the SETUP message has traversed the peer group in question. Conversely, the NE at the entry to a given peer group (or routing hierarchical level) may add a DTL IE to the stack of DTL IEs in the SETUP (or ADD PARTY) message to enable the message to traverse the selected path through the hierarchical routing level. The DTL IEs are added or removed in accordance with the basic last in–first out processing rule. It will be recognized that the same procedures may be extended for source routing across several routing domains or peer groups.

The DTL IE represents the point of contact between the PNNI routing protocols and algorithms and the PNNI signaling functions. The routing protocols are intended for automated topology discovery and subsequent construction of the routing database. On receipt of the destination address in the UNI SETUP message, the routing algorithm computes a suitable path based on the network link state and topology database. The result of this computation is the DTL or stack of DTL IEs, which is then inserted in the PNNI SETUP message. Unless the stack of DTL IEs constitutes a hierarchically complete source route, nodes located at the entry of different hierarchical levels, or nodes at the entry of different routing domains or peer groups may add a DTL IE to the stack, whereas nodes at the exit may remove DTL IEs as appropriate. Each node must advance the Current Transit Pointer value to the next node ID before forwarding the SETUP message onto the next designated NE along the path. The SETUP message thus threads its way through the designated path, establishing the (bidirectional) VCLs in each traversed NE.

7.5 PNNI CRANKBACK FUNCTION

The source-routed connection set-up procedures described above may be unable to progress a call through the network for a variety of reasons. In such cases, signaling procedures have been defined that enable a PNNI NE to attempt to automatically reroute the call through an alternative path. This capability is called crankback and the intent is to reduce call blocking probability by incorporating mechanisms to automatically reroute a blocked call within the network. Note that the crankback function is not applicable when the call is rejected by an end user or destination for whatever reason, since this is not viewed as an intranetwork call blocking situation. Typically, the crankback function may be invoked for the following categories of routing faults:

1. Reachability errors or DTL processing errors resulting from changes in connectivity or link failures
2. Resource errors resulting from insufficient bandwidth or other network resources to meet the requested QoS and bandwidth for the connection

In such cases, and assuming the crankback capability is supported, the crankback procedures may be initiated by including a crankback IE in the first call clearing message, such as RELEASE, RELEASE COMPLETE, or ADD PARTY REJECT.

The crankback process is shown in Fig. 7-8, where it is assumed that the original SETUP message progresses as shown until it encounters a "failed" link or node. As noted above, the failure may not necessarily be a physical failure, but may be due to a number of routing or resourcing errors. In either case, the blocked NE in question responds to the SETUP message with a RELEASE message in the return direction, which must include a crankback IE (described in more detail below) as well as a cause IE, which encodes a cause for the call blocking. The RELEASE (or other suitable call clearing) message propagates back to the originating PNNI NE. The originating node then attempts to reroute the call via an alternative path by sending a "reroute" SETUP message, as depicted in Fig. 7-8. As will be seen below, the crankback IE in the RELEASE message contains the identifier of the "blocked" node (or link), so that the rerouting NE may construct a suitable DTL that does not include the failed node or link in the selected alternative path. The basic crankback mechanism described above may also be used iteratively in the event that successive rerouting attempts are also unsuccessful. In addition, extension of the crankback process to multiple levels of the routing hierarchy, pt–mpt calls, or to different routing domains are also possible and have been specified.

The essential structure of the crankback IE is shown in Fig. 7-9, including a summary of the associated procedures. It should be noted that even though call blocking normally results in a separate cause IE in the call blocking message, when crankback is enabled, the "crankback cause" specified in the crankback IE takes precedence over the cause IE value in determining actions at the rerouting NE. The "crankback level" field indicates the routing hierarchy level of the NE to which the call is returned for rerouting, and the crankback procedures allow for reroute attempts at the next higher level of the hierarchy, although this is more complex since it will involve additional manipulation of the DTL stack to determine the alternative path. The detailed

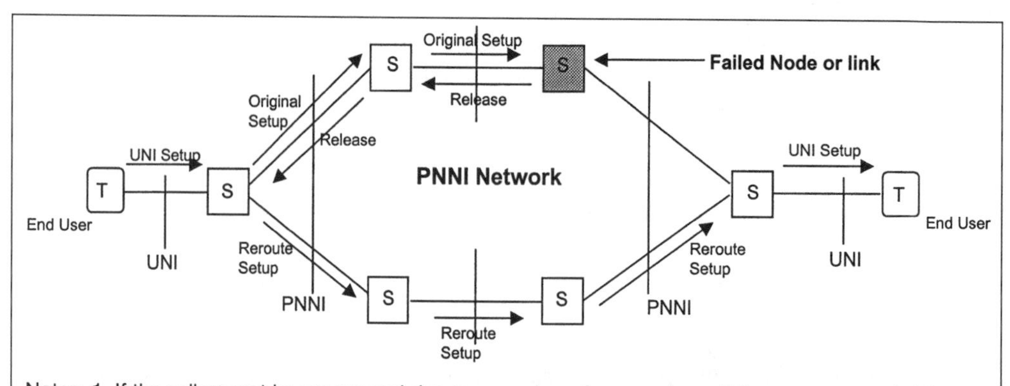

Notes. 1. If the call cannot be progressed due to a number of causes (e.g., DTL errors, unreachable, failure), crankback may be initiated to a preceding node for a call rerouting attempt. Crankback function in PNNI essentially "automates" alternative routing capability. 2. The Crankback IE may be included in the RELEASE, RELEASE COMPLETE, or ADD PARTY REJECT messages. 3. Crankback procedures may be iterative.

Figure 7-8. PNNI crankback function.

Bit #	8	7	6	5	4	3	2	1	Octet#
	1	1	1	0	0	0	0	1	1
	Crankback Information Element Identifier								
	1 ext	Coding Standard		IE Instruction Field					2
	Length of Crankback Contents								3
	Length of Crankback Contents (continued)								4
	Crankback Level								5
	Blocked Transit Type								6
	Blocked Transit Identifier (Format, length depends on type)								6.1 etc.
	Crankback Cause								7
	Crankback Cause Diagnostics								7.1 etc.

Notes. 1. The crankback function allows for a "blocked" call to be rerouted from a preceding PNNI node if the Crankback IE is included in the first call clearing message (RELEASE, RELEASE COMPLETE, or ADD PARTY REJECT). Call rejection by the destination is not subject to crankback procedures. 2. The Crankback Level indicates the PNNI hierarchy level to which the call is returned for rerouting. 3. The Blocked Transit Type identifies whether a node or link, etc. is blocked. 4. Causes for crankback may result from a) reachability or DTL errors, b) resource errors, or c) failures. 5. The node to which the call is "cranked back" for rerouting must insert the rerouting DTL stack. 6. Crankback Cause Diagnostics values should be included where possible.

Figure 7-9. PNNI IE example—Crankback.

procedures for this extension are outside the scope of this brief account of the basic crankback function, but may be found in the ATM Forum's PNNI specification [7.3]. The Blocked Transit Type field is used to indicate whether the node or the link (either ingress or egress) is blocked, whereas the Blocked Transit Identifier field is used to identify the specific NE or logical link and physical port that is causing the call blocking. The Crankback Cause field encodes the nature of the blocking, whether due to various resource unavailability problems, routing or reachability errors, link failure, or processing errors. In addition, the intent of the Crankback Cause Diagnostic field is to provide further information on the cause of the call blocking, which may assist the rerouting NE to update its routing database and thereby make more appropriate choices in selecting alternative paths to reroute the call. This optional function may be viewed as an adjunct to the normal routing protocol database updates, or as a management tool that may be used by the NE management system to convey status information if required. However, use of such information may not always warrant the additional processing complexity involved.

7.6 PNNI "SOFT" PERMANENT VIRTUAL CIRCUIT (SPVC) FUNCTION

Although signaling protocols are generally designed for "on-demand" connection setup or release, whereas so-called permanent (also called semipermanent) virtual circuits are usually associated with network management action, perhaps resulting from a customer service order, there is no essential reason why PVCs may not be set up using a signaling protocol. Use of signaling procedure to "configure" a PVC across the network considerably simplifies the operational burden on the management system, since the procedure may be viewed as effectively automated by the call processing software. This advantage of using signaling procedures to establish a PVC under administrative control was initially recognized in the operation of frame relay service and were incorporated in FR signaling standards. It was also recognized there that the signaling procedures could be enhanced to automatically reconfigure the PVC in the event of a failure at a node or link along the connection path. These procedures were collectively called Soft PVC (SPVC), though this should not be confused with semipermanent virtual circuits, and the underlying concepts were subsequently incorporated into other signaling protocols such as PNNI and more recently in B-ISUP. The SPVC capability may be used for either VCC or VPC configuration, and the PNNI SPVC procedures described here have been extended to the case of pt–mpt connections as well.

The basic network scenario for (pt–pt) SPVC is depicted in Fig. 7-10, where it is assumed that the network Operations System Function (OSF) initiates the SPVC connection (VPC or VCC) between the originating PNNI NE at the left side and the terminating NE on the right, on receipt of a UNI service order or SETUP message. The mechanism by which this may be achieved and a more detailed description of the management interface protocols between the NE and the OSF (labeled Q in Fig. 7-10, the so-called OS–NE interface) will be dealt with later, since this interaction does not significantly affect the PNNI SPVC signaling procedures being considered here. For the SPVC capability, the procedures for connection setup or release are essentially the same as for normal on-demand signaling except that since the configuration is assumed to be under administrative control, no negotiation of end-to-end parameters is possible. In addition, the traffic descriptors and QoS parameters are not negotiated but provisioned administratively. Thus, the SPVC SETUP (and other) messages may be somewhat simplified by eliminating the IEs relating to AAL parameters, ATM subaddresses, broadband lower- and higher-layer information and any other end-to-end information.

In the SPVC procedures, one of the two "endpoints" at the network side of the UNI is considered to own the SPVC in the sense that it (i.e., its network management system) is responsible

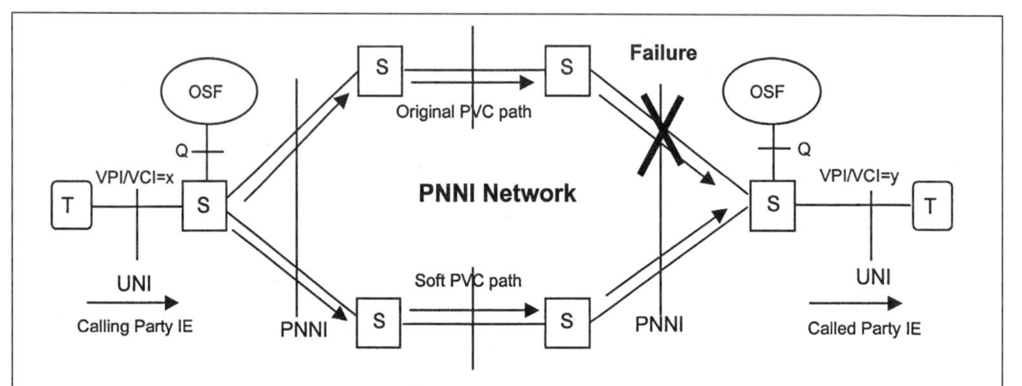

Notes. 1. The "soft" Permanent Virtual Circuit (PVC) function enables the "automatic" reconfiguration of a PVC between two UNIs in the event of a failure in the PNNI network. 2. The VPI/VCI values of the terminating VPL/VCL are conveyed transparently to the end node in the soft VPVC or VPCC information element to enable the reroute to terminate at the same VPL/VCLs at each end. 3. The calling endpoint is responsible for establishing (and releasing) the soft PVC. 4. The soft PVC procedures are extendable to point-to-multipoint connections. 5. The PNNI soft PVC mechanism is the same as used in frame relay soft PVC procedures. 6. The mechanisms by which the remote failure are made known to the origninating node are not specified in PNNI. Routing protocols or other mechanisms (e.g., OAM) may be used.

Figure 7-10. Soft PVC function.

for establishing or releasing the SPVC connection. This endpoint (or NE) is termed the calling endpoint, and it is also responsible for reestablishing the connection in the event that there is a failure of a link or NE, as shown in Fig. 7-10. The main difference between the SPVC and a normal on-demand call setup is that the identity of the two endpoints needs to be communicated to the other end, so that if the SPVC needs to be reestablished in the event of failure, the rerouted path terminates at the same VPL or VCL across the UNI at the called end of the connection. This may be achieved by conveying the VPI and VCI values of the terminating VPL and VCL transparently to the terminating NEs in the respective SPVC information elements during the SPVC establishment phase. Since SPVCs are administratively initiated, the management system (or OSF) is assumed to provide the VPI and VCI values, as well as the ATM addresses (AESAs) that will be used for the logical links to be used between the terminating PNNI NEs and the CPN or terminal equipment across the UNI. It should be noted that the SPVC procedures assume that the VPL/VCL across the UNI at either end (signified by the VPI/VCI values x and y in Fig. 7-10) remain constant for the duration of the SPVC connection. Thus, in case of a failure on the original PVC path, the rerouted SPVC connection must terminate on the original end node and be able to cross-connect to the same VPL or VCL across the UNI (i.e., VPI/VCI = y in the example shown). In this way, the SPVC function may be viewed as enabling the automatic reconfiguration of a PVC between two UNIs in the event of a failure in the PNNI network.

Although the SPVC protocols do not specify the detailed mechanisms by which the remote failure is made known to the originating node to initiate the SPVC connection reestablishment, the most common mechanism is based on receipt of the RELEASE message. Typically, the vast majority of failures result from physical layer faults such as cable breaks, hardware failures or defective physical connections. In such cases, all VPL or VCLs on the physical transmission path will fail, including the dedicated signaling and routing channels (VCI = 5 and VCI = 18 on all VPLs). Generally, failure of these control virtual channels will result in the release of all connections on that transmission path, with consequent sending of the RELEASE message to the calling NE. The calling NE may then initiate a SPVC reestablishment through an alternative path as described above. It may be noted that while SPVC reroute initiation by RELEASE message works well for physical layer failures, it may not be applicable for an ATM layer failure affecting an individual (or group of) VPCs or VCCs that does not include the preassigned signaling (or routing) VPL or VCL. However, for such cases, other mechanisms, such as the use of the ATM layer in-band OAM flows, may be used, as will be discussed in more detail later. In this context, it may also be recognized that the SPVC function enabling reestablishment of a VPC or VCC via an alternative route is closely analogous to the so-called "automatic protection switching" (APS) capability, successfully used in various transmission systems such as SDH. As will also be described later, an analogous automatic protection switching capability is also available at the ATM layer for individual VPC or VCC rerouting. In common with the SPVC mechanism described here, the intent of all these protection capabilities is to enhance the reliability (or availability) of the connection, although at different layers of the B-ISDN protocol stack.

The Soft PVC Information Element structure is shown in Fig. 7-11, where in fact it will be noted that there are two separate SPVC IEs—one for the Calling Party SPVC IE and one for the Called Party SPVC IE—distinguished by different values of the IE identifier codepoints. The calling party SPVC IE is used to transfer the VPI or VPI/VCI values used at the calling connecting point (i.e., the NE that "owns" the SPVC) to the called NE transparently through the PNNI network. Similarly, the called party IE is used to transparently transfer the VPI and VPI/VCI values used at the called NE UNI to the calling NE in the CONNECT message. It may be noted that the VPI and VCI value fields identifying the individual VPL and VCL across the respective UNIs are preceded by VPI and VCI Identifier fields, with different codepoints indicating that the subsequent values represent the VPI or VCI value. In addition, the Called Party SPVC IE in-

bit #	8	7	6	5	4	3	2	1	Octet #
	1	1	1	0	0	0	1	1	1
	Calling party soft PVPC or PVCC								
	1 ext	Coding Standard		IE Instruction Field					2
	Length of Calling (or Called) Party Soft PVPC or PVCC Contents								3
	Length of Calling (or Called) Party Soft PVPC or PVCC Contents (contd.)								4
	Selection Type (only in Called Party Soft PVC IE)								5
	1	0	0	0	0	0	0	1	6
	VPI Identifier								
	VPI Value								6.1
	VPI Value (contd.)								6.2
	1	0	0	0	0	0	1	0	7
	VCI Identifier								
	VCI Value								7.1
	VCI Value (contd.)								7.2

Notes. 1. Two IEs are defined for soft PVPC (or PVCC) functions: IE for Calling Party Soft PVPC (or PVCC). The message type coding for this is 11100011, as shown above and IE for Called Party Soft PVPC (or PVCC). The message type coding for this is 11100000. 2. Selection Type Field (present only in Called Party IE) indicates whether VPI/VCI has a) any value, b) required value, and c) assigned value. 3. Soft PVC procedures require the VPI/VCI values of the terminating VPL/VCLs be indicated by the NE management system. 4. For point-to-multipoint PVCs, the soft PVC procedures are similar to point-to-point, except that the IEs are included in the ADD PARTY messages.

Figure 7-11. PNNI IE example—soft PVC.

cludes an additional field called the Selection Type. The purpose of the Selection Type field is to indicate whether the following VPI and VCI values have:

	Codepoint
a) any value	0000 0000
b) required value	0000 0010
c) assigned value	0000 0100

In the first case, the NE may select any available VPI or VCI value, and the selected values are encoded in the called party SPVC IE in the CONNECT message, as indicated earlier. In the second case, the required VPI and VCI values should be selected if available, and indicated in the CONNECT message with the Selection Type field coded as "assigned value." If the required values are not available, the call is cleared with a RELEASE COMPLETE message indicating cause. Clearly, the selected VPI/VCI values assigned across either UNI are maintained by the management system to be used in the event of an SPVC reroute.

7.7 PNNI ROUTING FUNCTIONS

As outlined earlier, the routing protocols for private network node interfaces have been designed to enable the ATM NEs to automatically exchange links state and topology information in order to construct suitable routing databases and thereby select suitable connection paths between source and destination endpoints. The desired outcome of such a route computation is a (hierarchically complete) source route which is encoded as an ordered list of NE identifiers (and optionally, logical link identifiers) in the DTL stack IE of the PNNI SETUP message. The SETUP message then traverses this designated path to establish the end-to-end VPC or VCC, as described earlier. The capability of such a routing mechanism to dynamically adjust to changes in topology and link status among other benefits is primarily intended to simplify the administrative burden of manually configuring routes within the network, a significant advantage for enterprise networking. In order to achieve this, PNNI routing protocols have essentially adapted the underlying concepts of link-state-based connectionless routing protocols such as the so-called Open Shortest Path First (OSPF) approach used for IP routing developed by the IETF [7.4]. The essential concept underlying such an approach is that the network nodes exchange routing packets (called hello messages) between neighbors containing the relevant nodal and link status information on a periodic as well as event-driven basis. The nodes are thus able to "discover" and update the network topology and construct routing databases. Using additional routing messages, the nodes may share the topology information with peers to ensure consistent routing in the network. Subsequently, any one of a number of suitable routing algorithms, such as open shortest path first, may then be used within a node to forward the data packet along the appropriate logical (and physical) link.

Although originally intended for the connectionless Internetworking Protocol (IP) arena, these basic notions have been adapted to the case of (connection-oriented) ATM networks in the PNNI routing protocol. Although the underlying principles remain similar, the ATM Forum essentially generalized the link state routing protocol substantially to allow for additional functionality, diverse PNNI configurations, and hierarchical network topologies. In doing this, the PNNI routing protocol is described in a somewhat abstracted and generalized representation in the specifications, which has led to some confusion, as well as charges of excessive complexity in implementations. A detailed description of every facet and function of the PNNI routing protocols is clearly outside the scope of this text, as the primary intention here is to focus on the essential functions and procedures involved in path selection and construction of the DTL stacks, rather than the abstract intricacies of routing theory. In this context, it is useful to bear in mind that whereas the concept of routing hierarchy in PNNI has been generalized to apply to any topological hierarchy or interconnection of different peer groups, the most useful hierarchy from a pragmatic perspective is one in which the network hierarchy corresponds directly to the AESA hierarchical structure. As noted earlier in our discussion of ATM addressing, in this case, routing may be performed highly efficiently, and the routing procedures are relatively easy to understand. We will focus on this case here.

The main PNNI routing functions and the associated routing messages (or packets) defined for the exchange of routing information between NEs are summarized in Fig. 7-12. These functions include:

1. The discovery of neighboring NEs and the status of the (logical and physical) links by the periodic exchange (with neighbors) of the so-called "hello" packets and associated procedures.
2. The synchronization of the topology and link state databases between all the NEs within a given peer group (or hierarchical level) once this has been constructed and in the event of changes that affect route computation. The topology/link state database synchronization is performed by exchanging the "database summary" packets with their associated procedures.
3. The "flooding" of the PNNI topology state elements (PTSE) between the NEs within a given peer group to update the routing databases, using the PNNI topology state packets

1. The main PNNI routing functions are:
   1.1 Discovery of neighbors and links status (using "hello" packets and protocols)
   1.2 Synchronization of the topology database (using the "database summary" packets and protocol)
   1.3 Flooding of PNNI topology state elements (PTSEs) using the "PNNI topology state packets")
   1.4 Peer group leader (PGL) election
   1.5 Summarization of topology state information
   1.6 Construction of routing hierarchy
2. PNNI Routing Control Channel (RCC):
   - PNNI uses VCI = 18 (in any VPI) as the preassigned VCC to carry PNNI routing messages (for lowest level or neighbor nodes)
   - For higher or different levels, nodes may set up a separate (semipermanent or switched) VCC to carry the routing messages
3. PNNI Routing Messages (Packets):

Packet type code	Packet name
1	Hello
2	PNNI Topology State Packet (PTSP)
3	PTSE Acknowledgment
4	Database Summary
5	PTSE Request

Figure 7-12. PNNI routing functions.

(PTSPs). The PTSEs contain both "link state" parameters and "nodal state" parameters , pertaining to the characteristics of the PNNI nodes and links. The link state parameters can be grouped into "attributes" (e.g., administrative priority or weight) or into "metrics" (e.g., delay). The PTSEs that make up the topology database in the NE are subject to aging, and may therefore be discarded after a given interval if not refreshed by new PTSEs. Consequently, the flooding mechanism in PNNI routing is a continuing process between all nodes in any given peer group. The generation of the PTSPs containing the PTSEs may be both periodic as well as event-driven.

4. The election of a peer group leader NE. As will be described below each peer group selects a "leader" NE to represent the group at the next level in the hierarchy as the so-called "logical group node." The PNNI protocols include a mechanism that enables the transfer of the peer group leader role between nodes based on leadership priority criteria. This mechanism does not directly affect the routing procedure and hence is not further discussed here.
5. The NEs contain functionality to summarize the topology state information so that it may be represented at the next higher level in the routing hierarchy. The summarized information may include address summarization (e.g., in the case of E.164 addresses, this may be the local exchange area code) and topology summarization.
6. The construction of the routing hierarchy, as a result of the address/topology summarization process and the representation of a peer group by a selected (peer group leader) NE. These procedures imply a functionality in the NE that enables the routing hierarchy to be created and maintained. For the most common case where the (E.164) address hierarchy corresponds to the network topology, such a process would result in the creation of NEs fulfilling the roles of "local exchanges," "toll exchanges," "international gateways," and so on. However, for private or enterprise network scenarios, the above descriptions would not be valid, and a flatter hierarchy may be sufficient.

The PNNI routing protocols utilize five distinct packet types, each identified by its own Packet Type Code, as shown in Fig. 7-12. These are:

1. Hello Packet
2. PNNI Topology State Packets (PTSP), which contain the individual PNNI Topology State Elements (PTSEs)
3. PTSE Acknowledgment Packet
4. Database Summary Packet
5. PTSE Request Packet

A more detailed description of the routing packet structures and associated procedures is given later. Here it should be recalled that all the PNNI routing packets are encapsulated within the AAL Type 5 (Common Part Convergence Sublayer, CPCS) and transported between NEs over dedicated or preassigned ATM VCCs, in essentially the same way as are the signaling messages described earlier. However, for the lowest level (in the routing hierarchy), or for neighbor NEs, the preassigned VCI = 18 is used to carry routing messages. For the transfer of routing messages between the higher levels of the routing hierarchy, the PNNI specification requires that separate dedicated switched VCCs (for each hierarchical level) should be established between the representative nodes, in order to maintain the independence between the levels. The procedures for establishing these dedicated routing VCCs between the nodes representing the various hierarchical levels are essentially the same as for the normal connection setup, and hence will not be described in any further detail here. The transfer of the routing messages between the nodes may also be performed over permanent virtual channel connections (PVC) configured by the network management system.

7.8 PNNI ROUTING HIERARCHY

As pointed out earlier, the concept of hierarchical routing in PNNI has been described in general and somewhat abstract terms in the PNNI routing protocol specifications in order to be applicable to any network configuration. However, this generality may result in confusion unless it is recalled that typically the "hierarchy" corresponds to the well-known addressing hierarchy commonly used in the ubiquitous telephony network. The use of a network topology that corresponds closely to the familiar (Recommendation E.164-based [6.15]) international telephone number comprising country code + operator code + city code + area code + local exchange subscriber number results in very efficient routing of connections. Essentially the same model should be kept in mind in considering the use of the PNNI routing hierarchy concept in any realistic network deployment. In addition, for the typical enterprise (private) network applications intended, the ATM address structure and hence corresponding network topology may generally be assumed to be somewhat simpler, with fewer hierarchical levels. With this in mind, the concepts underlying the general PNNI routing hierarchy description may be more readily understood.

The basic routing definitions and terminology used in the description are summarized in Fig. 7-13. The concept of a "logical node or link" is introduced, signifying a PNNI node or link (between nodes) at a given hierarchical level that can be identified by a unique logical address called the Logical Node ID. In PNNI routing, the Logical Node IDs are based on the ATM End System Address (AESA), as further detailed below. The notion of a peer group (PG) has already been used intuitively earlier, and may be defined as a collection of logical nodes in which all

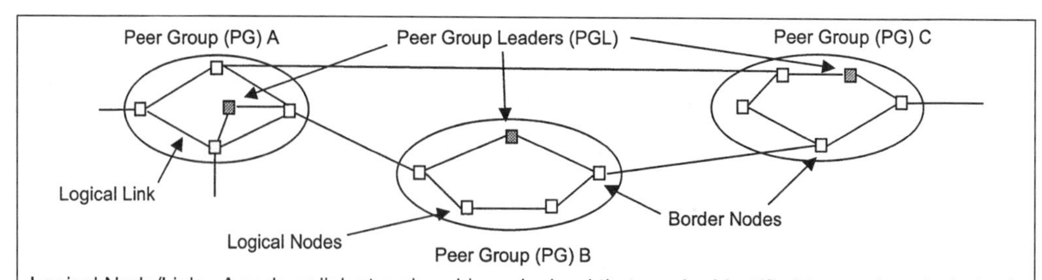

Logical Node/Link—A node or link at a given hierarchy level that can be identified by a unique logical address (Logical Node IDs). Peer Group (PG)—A collection of logical nodes such that all members maintain the same view of the group by exchanging information with the other members of the group. The peer group identifier is specified at network configuration. Peer Group Leader (PGL)—A node that is elected to represent the peer group to the next hierarchical level (logical group node) and coordinate aggregation of PG information to be passed to the next level. The PGL role is passed between the nodes in a PG by an "election" protocol. Apart from aggregation and maintaining PNNI heirarchy, the PGL has no special role. Border Node—A node that has a (logical) link that crosses the peer group boundary. Such links are called "outside links." Logical links inside a peer group are called "horizontal links."

Figure 7-13. Routing definitions and terminology.

members maintain the same (topological and link state) view of the group by exchanging information with the other members of the group. The peer group is characterized by a peer group identifier, which may be specified by the management system when the PNNI network is being configured. Associated with a peer group is a peer group leader (PGL) as noted earlier. The PGL is selected (using an election protocol) to represent the peer group to the next hierarchical level as a so-called logical group node (LGN). It achieves this by coordinating the aggregation of routing information within the peer group and "passing" it up to the next hierarchical level.

It should be noted that this information is not actually physically "passed" to another "node," but simply maintained as a separate logical entity to represent the next "level" of the routing hierarchy within the same node. In this respect, a logical node may effectively represent one or more levels of the routing hierarchy by maintaining the logical separation between the routing information databases at the various hierarchical levels. Again, this concept may be clarified by using the example of the telephony network alluded to earlier. Thus, a local exchange may route a call locally based on interpreting the subscriber number and using its "local" routing database. However, if the number dialed contains a long distance or even an international code, the switch uses its "higher level" routing database to route the call to the appropriate long distance trunk or transit exchange. If an international connection is being requested, the NEs will need to resort to the next "higher" level in the routing database in order to route the call to the appropriate international gateway NE or trunk, and so on. This simple example clearly identifies the advantages of hierarchical routing, since in this model any given NE need not maintain extensive (e.g., global) routing databases in order to provide reachability to every NE attached to the WAN. It simply uses the summarized routing information "exchanged" between the hierarchical levels to route the call to the appropriate domain or border node implied in the addressing hierarchy.

In general terms, the concept of the PNNI routing hierarchy may be represented as shown in Fig. 7-14. As indicated above, the primary reason for adopting such a hierarchical scheme is to achieve improved scalability as well as manageability of the routing databases required for WANs. Thus, if some form of hierarchical routing were not used, each NE in the WAN would be required to maintain a routing database containing the link state and nodal information to enable reachability to all other destinations in the WAN. This requirement may result in unmanageably large routing databases in each NE. However, by introducing some form of hierarchy in

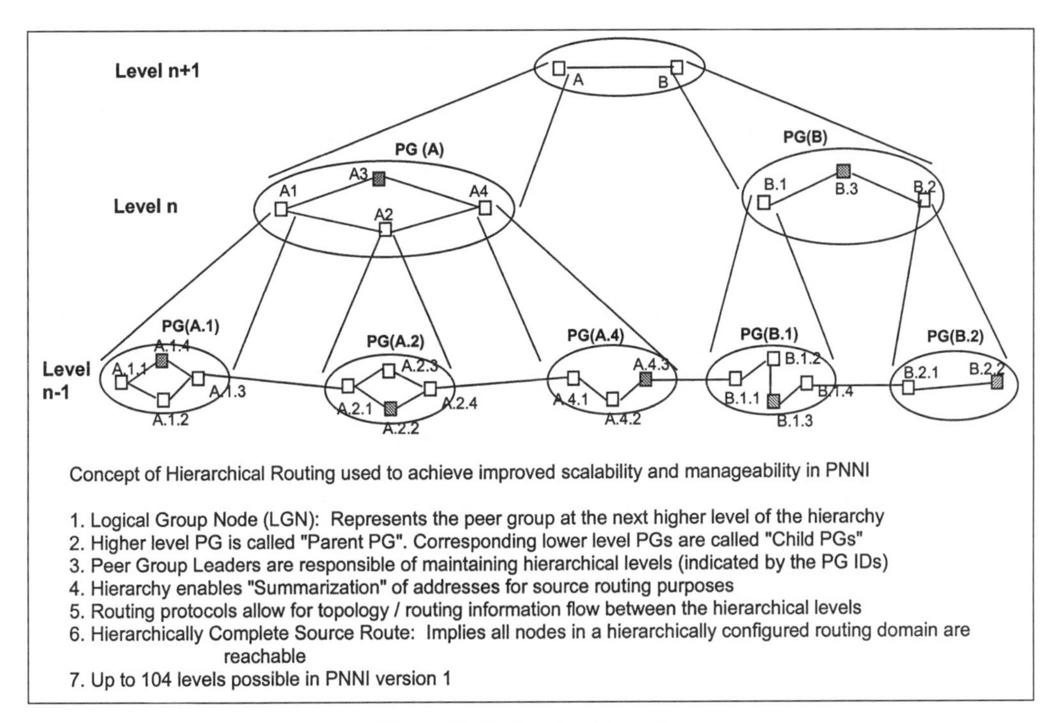

Figure 7-14. Routing hierarchy.

the routing mechanism, either geographic, as in the example considered, or in the form of other "domains," the database need only manage the routing information required to progress the call within its bounded routing domain or to the appropriate border node if the destination address indicates an external endpoint. This results in smaller routing databases if suitably designed, which significantly improves the scalability of the routing protocol. In theory, the PNNI routing hierarchy is capable of supporting up to 104 levels. However, it may be recalled that the DTL IE stack used for PNNI source routing permits only up to 10 levels in the SETUP message (i.e., one DTL IE per level). In practice, the use of 3–4 levels in an efficiently constructed scheme enables extremely large networks to be implemented, e.g., the telephony network.

The generalized terminology adopted for PNNI routing by the ATM Forum is also summarized in the figure and encapsulates the concepts outlined earlier. Thus, the logical group node (LGN) as embodied in the peer group leader node represents a given peer group (PG) at the next higher level of the hierarchy. The higher-level PG is termed a "parent peer group," with the corresponding lower levels referred to as "child PGs." For efficient routing to occur, the hierarchy essentially follows the address summarization inherent in a hierarchical address structure. Referring to the example used earlier, the three-level hierarchy shown in Fig. 7-14 may be used to represent

Level $n + 1$ = country code
Level n = area code
Level $n - 1$ = city code

In this example, the notion of "address summarization" refers to the country code "containing" ("summarizing," as it were) all the addresses in the country. Similarly, the area code corresponding to level n contains or "summarizes" all the address in any given "area" or operator domain.

The city code, corresponding to level $n - 1$, then summarizes all the addresses in the city, and so on. This concept of address summarization is further generalized as described below. However, one way of viewing the "construction" of such a routing hierarchy is to envisage the routing protocols allowing the flow of the "summarized" routing information to be exchanged between the hierarchical levels, as indicated earlier.

7.9 ADDRESSES AND IDENTIFIERS IN PNNI

It has already been stressed earlier that the address structure and routing functions are intimately linked, and the common example of the PSTN described above illustrates the relationship in a typical case. PNNI routing protocols are no exception to this precept, but essentially generalize the address and identifier structures in order that the protocols may be applicable for the wide variety of network configurations anticipated in enterprise or private network deployments. As noted previously, the PNNI signaling and routing protocols assume the use of the NSAP-based ATM End System Address (AESA) format, not only for the source and destination addresses, but also to identify the individual NEs within the network. Although at first glance it may seem paradoxical to specify the use of an "end system" address to identify intermediate systems such as individual ATM NEs, the use of the AESA for this purpose in PNNI routing is in a generalized format, as shown in Fig. 7-15.

The general NSAP AESA structure discussed earlier is 20 octets long, but PNNI routing protocol is specified to operate only on the first 19 octets. The 1 octet "selector" field is not used for routing purposes within the network. Although the internal structure of the other fields are not

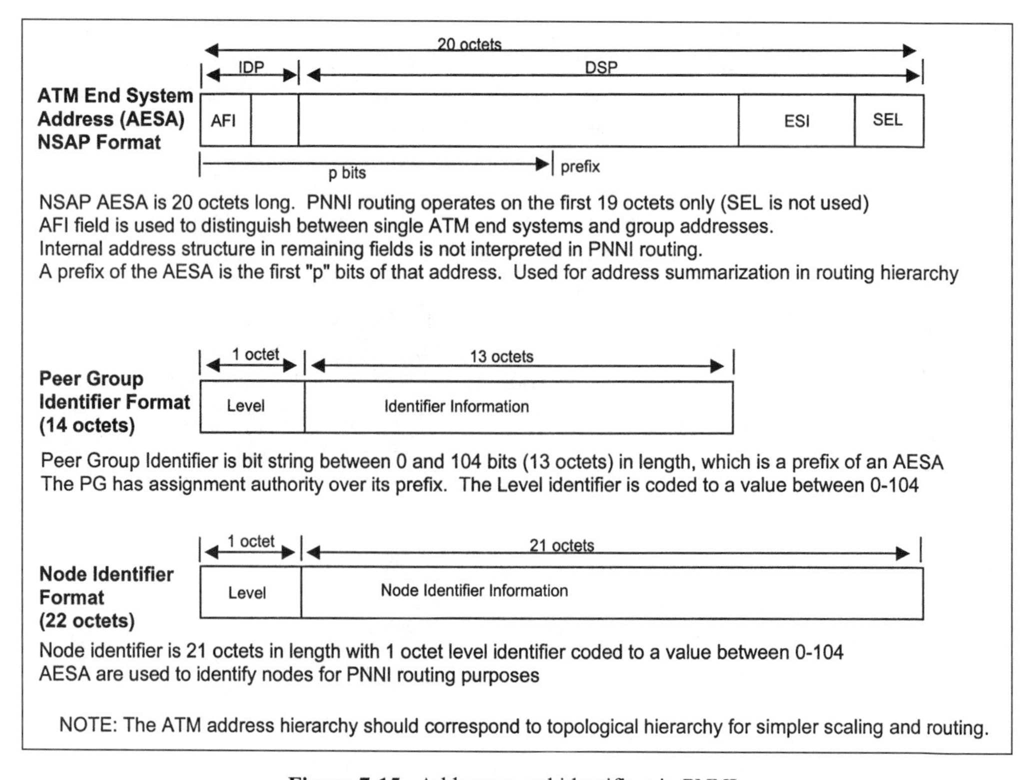

Figure 7-15. Addresses and identifiers in PNNI.

interpreted in any specified way (and depend on the type of address format used as discussed earlier) for PNNI routing, the only exception is the Address and Format Identifier (AFI) field, which is used to distinguish between single ATM end systems and group addresses. The "address summarization" to enable the use of the routing hierarchy is achieved by specifying an address "prefix" as the first "p" bits of the address, as shown. Thus, a prefix of a given value "p" bits may be viewed as "containing," or "summarizing," all the possible address codepoints possible with the remaining bits of the NSAP AESA (not including the selector octet). In the example referred to earlier using the international E.164-based address structure, the address "prefix" = 1 (= country code) summarizes (i.e., contains) all the addresses in the North American Numbering Plan (NANP) domain. Similarly, the address prefix = 44 contains all the addresses in the United Kingdom domain, and so on. The address prefix = 1 613 summarizes all the addresses in the Bell Canada Ottawa/Kingston area, and so on. The PNNI NE interprets the "p" bits of the address prefix to determine how to route the call in any given configuration in an analogous way to the normal telephony case.

Fig. 7-15 also illustrates the format of the "peer group identifier" and the "node identifier" used in the PNNI protocol. The PNNI node identifier is constructed by adding a 1 octet "level identifier" field to a 21 octet field that contains the 20 octet NSAP AESA. The level identifier field is coded to a value between 0 and 104, which signifies the hierarchical level of the node identifier information. The level identifier value is used by the NE to organize the routing databases in accordance with the hierarchical level being utilized. In general, not all level indicator values may be utilized for any given configuration and the values need not be contiguous. This is referred to as the levels not being "densely" assigned. The first (most significant) octet of the node identifier information field is used to distinguish whether the NE represents a peer group or not by encoding a preassigned value (= 160) to designate nodes that are not part of any peer group. If a node is part of a given peer group, the peer group identifier is used to overwrite this field instead. The peer group identifier format also includes the (hierarchy) level Identifier field followed by a 13-octet field, which is a prefix of an AESA, as discussed earlier. The peer group is assumed to maintain assignment authority over its prefix value.

It may be evident from Fig. 7-15 that there is a binding relationship between the NSAP AESA structure and that of the PNNI node identifier and PG identifier, and this information constitutes an essential component of the routing messages exchanged between the NE to construct the hierarchical routing databases required for source routing. In addition to these identifiers, PNNI routing may also use a 32 bit (4 octet) port identifier to uniquely specify a point of attachment of a logical link (e.g., a VP link) to a NE, and a logical link identifier. The logical link identifier is then defined as the node identifier of either NE at the ends of the logical link and the port identifier assigned by the node. In this designation, each logical link may be specified by two logical link identifiers, assuming the nodes at either end of the logical link have different node identifiers. Finally, an identifier termed an "aggregation token" is also defined in PNNI routing as a 32 bit (4 octet) number. The aggregation tokens are used to signify associations between multiple logical links between nodes, as well as associations between the hierarchical level and outside links between peer groups. The use of aggregation token identifiers is outside the scope of this description and details may be found in the PNNI version 1.0 specification [7.3]. The routing protocols also include the concept of an address scope by means of a scope identifier in the routing message. The scope of an address is essentially used to limit the range of reachability by restricting exchange of routing information to specified peer groups (or hierarchical levels). Both address scope and aggregation token identifiers, while useful, are not of primary significance in obtaining a basic understanding of PNNI routing protocols.

7.10 PNNI ROUTING PACKET STRUCTURE AND FUNCTIONS

The five distinct types of messages (or packets) used for PNNI routing information exchange were listed earlier. These were:

1. Hello Packets
2. PNNI Topology State Packets (PTSP)
3. PNNI Topology State Element (PTSE) Acknowledgement Packets
4. Database Summary Packets (DSP)
5. PTSE Request Packets

We now consider in more detail the structure and function of these routing packets and the associated procedures. All the PNNI routing packets include a common 8-octet header as shown in Fig. 7-16. The PNNI routing packets also use the nested type–length–value (TLV) formatting principles analogous to PNNI signaling messages described earlier, so that the common routing packet header encodes the "packet type" and "packet length" as shown. The packet type codepoints were listed earlier and are reproduced in Fig. 7-17 for convenience. The common PNNI routing packet header also includes fields to signify the protocol version for purposes of backward compatibility checking. The description here is based on the PNNI version 1 specification and it may be noted that a version 2 essentially adds some new functions to the routing (and signaling) protocols without changing the basic protocols and need not concern us here. It may also be recalled that an earlier version of the PNNI protocol called PNNI version 0 (or Inter Switch Signaling Protocol, ISSP) did not utilize the automated routing function, but was based on stati-

Common PNNI version 1 Routing Packet Header

Offset	Field Size (octets)	Name	Function / Description
0	2	Packet Type	Packet Type Identifier
2	2	Packet Length	Total length of routing packet (octets)
4	1	Protocol Version	Specifies PNNI Version
5	1	Newest Version Supported	To enable nodes to negotiate common protocol version. Used together with Oldest Version Supported Field
6	1	Oldest version Supported	As above
7	1	Reserved	

1. All PNNI Routing Packets have the above common PNNI Packet Header (8 octets)
2. The routing packets use nested Type-Length-Value (TLV) format principles (analogous to signaling messages)
3. The information in the routing packets are described in terms of "Information Groups" (IG)
4. Analogous to PNNI signaling, there are a large number of Information Groups defined for PNNI routing protocol
5. The IGs also include "Tags" and "Flags" to indicate processing and state requirements
6. All routing packets are encapsulated in AAL Type 5 CPCS in Message Mode (1 packet per AAL 5 SDU)
7. All routing packets are transported on the PNNI Routing Control Channel (RCC) (default VCI=18, or SVCC)

Figure 7-16. PNNI routing packets.

Type Code	Packet Name	Information Groups Contained
1	Hello	Aggregation Tokens, Nodal Hierarchy List, System Capabilities, Uplink Information Attribute (ULIA), Outgoing Resource Avail. (RAIG) LGN Horizontal Link Ext., Optional GCAC Parameters
2	PTSP	PTSE, Nodal State Parameters, Nodal IG, Outgoing and Incoming RAIGs, Next Higher Level Binding, Optional GCAC Parameters, Internal reachable ATM Address, Exterior Reachable ATM Address, Horizontal Links, Uplinks, Transit Network ID, System Capabilities
3	PTSE Ack	Nodal PTSE Acks., System Capabilities
4	DataBase Summary	Nodal PTSE Summaries, System Capabilities
5	PTSE Request	Requested PTSE Header, System Capabilities

Figure 7-17. Information groups and packet types. 1. All PNNI IG types are identified by codepoints in the 2-octet IG type field. 2. The first four bits (MSB) of the IG type field encode the IG tages, indicating mandatory/optional, etc., handling.

cally preconfigured routing and is considered as generally incompatible with version 1. However, the intent of the three version control fields in the common routing packet header is to enable the nodes to negotiate a common protocol version if supported.

The information in the PNNI routing packet is organized into so-called "information groups" (IGs), which may be viewed as analogous to the concept of the "information elements" (IEs) used in the signaling packets (or the "parameters" in the B-ISUP signaling message set). Also in analogy with the signaling protocols, the PNNI routing protocols utilize a large number of different information groups, which are themselves formatted in accordance with the nested TLV concept and identified by individual IG type codepoints. The allocation of the various IG types to the five PNNI routing packets is summarized in Fig. 7-17. It should be recognized that not all the routing IGs need to be present in all the routing packets, as is also the case for the IEs in the signaling messages. However, certain IGs are mandatory and also include a set of "tags" and "flags," which are used to indicate the specific processing requirements, as well as whether the contained information should be handled as mandatory or optional. A detailed description of the use of these tags and flags is outside the scope of this account, as is the description of all the various routing IGs listed in Fig. 7-17. The IG type codepoints are given in Table 7-2, which also indicates the use of the IGs in transferring the routing information up and down the hierarchical levels by referring to the IGs contained either one level up or down (the hierarchy).

The use of a number of the IGs listed in Table 7-2 may generally be inferred from the name, but there are certain key IGs that occur in most packets that may require some additional description. An example is the resource availability IGs (RAIG) which are of two types; the outgoing RAIG (Type = 128) and incoming RAIG (Type = 129). The RAIGs are used to encode the resource-related attributes and metrics of the nodes and links as well as the service categories and QoS parameters required. These include traffic parameters such as peak cell rate, available cell rate etc, and QoS parameters such as CTD, CDV, and CLR, as well as service categories (or ATCs) such as CBR, ABR, etc. As such, the incoming and outgoing RAIGs may contain a number of parameters that may need to be factored into a path computation. The inclusion of the RAIGs in the routing packets provides the capability to take into account a potentially large number of resource and QoS-related parameters in the route calculation, but it must be noted that as the number of such parameters increases, so will the complexity of the algorithms re-

TABLE 7-2. Information group summary

Type	IG name	Contains IGs one level down
32	Aggregation token	
33	Nodal hierarchy list	
34	Uplink information attribute	Outgoing resource availability (128)
35	LGN horizontal link extension	
64	PTSE	Nodal state parameters (96). Nodal information group (97). Internal reachable ATM addresses (224). Exterior reachable ATM addresses (256). Horizontal links (288). Uplinks (289). System capabilities (640)
96	Nodal state parameters	Outgoing resource availability (128)
97	Nodal information group	Next higher level binding information (192)
128	Outgoing resource availability	Optional GCAC parameters (160)
129	Incoming resource availability	Optional GCAC parameters (160)
160	Optional GCAC parameters	
192	Next higher level binding information	
224	Internal reachable ATM addresses	Outgoing resource availability (128). Incoming resource availability (129)
256	Exterior reachable ATM addresses	Outgoing resource availability (128). Incoming resource availability (129). Transit network ID (304)
288	Horizontal links	Outgoing resource availability (128)
289	Uplinks	Uplink information attribute (34). Outgoing resource availability (128)
304	Transit network ID	
384	Nodal PTSE ack	
512	Nodal PTSE summaries	
513	Requested PTSE header	
640	System capabilities	

Type	IG name	Contained in IGs one level up	Contained in packets
32	Aggregation token		Hello (1)
33	Nodal hierarchy list		Hello (1)
34	Uplink information attribute	Uplinks (289)	Hello (1)
35	LGN horizontal link extension		Hello (1) for LGN horizontal Hello
64	PTSE		PTSP (2)
96	Nodal state parameters	PTSE—restricted IG	PTSP (2)
97	Nodal information group	PTSE—restricted IG	PTSP (2)
128	Outgoing resource availability	Uplink information attribute (34). Nodal state parameters (96). Internal Reachable ATM Address (224). Exterior reachable ATM addresses (256). Horizontal links (288). Uplinks (289)	Hello (1). PTSP (2)
129	Incoming resource availability	Internal reachable ATM address (224). Exterior reachable ATM addresses (256)	PTSP (2)

TABLE 7-2. *Continued*

Type	IG name	Contained in IGs one level up	Contained in packets
160	Optional GCAC parameters	Outgoing resource availability (128). Incoming resource availability (129)	Hello (1). PTSP (2)
192	Next higher level binding information	Nodal information group (97)	PTSP (2)
224	Internal reachable ATM addresses	PTSE—restricted IG	PTSP (2)
256	Exterior reachable ATM addresses	PTSE—restricted IG	PTSP (2)
288	Horizontal links	PTSE—restricted IG	PTSP (2)
289	Uplinks	PTSE—restricted IG	PTSP (2)
304	Transit network ID	Exterior reachable ATM addresses (256)	PTSP (2)
384	Nodal PTSE ack		PTSE Ack (3)
512	Nodal PTSE summaries		DBSummary (4)
513	Requested PTSE header		PTSE Request (5)
640	System capabilities	PTSE	All packets

quired. As a consequence, the probability of optimizing all the required parameters will decrease, so that some form of prioritization may be necessary to obtain reasonable convergence time for a path computation. To assist this, the RAIG includes an "administrative weight" parameter, which may be utilized to prioritize a preferred parameter for route selection. In practice, routing computations based only on a limited set of link state metrics are generally used to minimize call set-up latency and reduce call blocking probability.

The Uplink Information Attribute (ULIA) IG is used to associate IGs related to the links between the peer group border nodes (the so-called outside links) at the next higher level in the routing hierarchy. Although the abstraction of "uplinks" has not been explicitly described here, the PNNI routing hierarchy introduces this concept to associate the outside links between the peer groups at one level of the hierarchy to the logical links between the logical group nodes (LGNs) that represent the PG at the next higher level of the hierarchy. The ULIA IG is used to identify the IGs that may be associated with such unlinks. The Transit Network Identifier IG is used to specify the preferred network to be used for routing in analogous fashion to the use in the signaling message of a transit network selector IE. It may also be noted that all the routing messages may contain the so-called System Capabilities IG. The intent of the System Capabilities IG is to enable the transfer of proprietary or "system-specific" routing information between the NEs if required for some purposes by a given vendor or operator. The System Capabilities IG uses the so-called organizational unique identifier (OUI) specified by the IEEE to identify the proprietary information in the IG as belonging to a given vendor or operator.

The Interior Reachable Address IG is used to indicate the AESAs of directly attached endpoints to the PNNI network, whereas the Exterior Reachable Address IG refers to ATM addresses that maybe reached through an "exterior" network that is not part of the PNNI network in question. Both IG types may be present multiple times in the routing messages. The PNNI routing messages also provide the optional capability to exchange so-called Generic Connection Admission Control (GCAC) parameters between nodes. The concept of a Generic Connection Admission Control algorithm was introduced into PNNI routing based on the recognition that the actual CAC function in any given node is not standardized and hence may use any appropriate proprietary algorithms in deciding whether or not a connection request may be accepted. However, the routing computations based on a set of link state and nodal attributes and metrics may require assumptions about the connection acceptance probability within a given NE in order to compute an optimum path. Consequently, the PNNI routing procedures have specified (optional) reference or generic CAC algorithms and associated parameters that may be used to emulate an actual CAC decision process during connection set-up. The fact that the actual (proprietary) CAC algorithms used in any given NE may be different will inevitably lead to a discrepancy in the predicted connection acceptance probability. However, the differences may be kept reasonably small by selecting a reference algorithm and parameters that approximate well to typical CAC implementations. A detailed description of the GCAC procedures is outside the scope of this text, but may be found in the PNNI v.1 routing specification.

Even without delving into the details of each of the listed IGs and the numerous nodal and link state parameters, attributes, and metrics included within them, it may become evident to the reader that the support of the complete PNNI routing procedures and capabilities may involve the processing of large amounts of information within the NEs, particularly for large multihierarchical networks. In addition, it should be noted that a number of the key parameters in the IG are subject to an "aging" process and will need to be updated on a periodic or event-driven basis in order to ensure that correct up-to-date information is available for the routing computations at any given time. The aging requirements imply the use of timers for the topology database entries, as well as the need to advertise "significant" changes in attributes and metrics that are involved in routing computations. In general terms, significant changes are defined to mean any change that may affect a routing decision. In practice, these significant changes will need to be prioritized in order to contain the resulting information exchange to within manageable bounds. The processing and maintenance of the (potentially) large amount of topology database information together with need to generate/process periodic and event-driven routing messages implies that PNNI routing software is likely to be complex. Against this, the designer has to balance the need for reasonably rapid and efficient routing computation to ensure high connection set-up performance for realistic network applications. Such considerations inevitably involve difficult design trade-offs between flexibility (or capability) and performance. This in turn may result in some of the related information in the IGs being assigned a low priority for use in actual routing software implementations.

We now turn from this brief summary of the IGs defined for PNNI routing to the structure of the routing message set used to exchange the IGs between the NEs. As noted earlier, these utilize the nested TLV formatting principles also used for all signaling messages, and the individual IGs may be repeated multiple times in a given routing message. The PNNI hello message structure is shown in Fig. 7-18. In common with all the routing message types, the hello message begins with the 8-octet PNNI common header described earlier. The common header includes the message type, length and protocol version information for each message. It should be noted that the "offset" value, which denotes the beginning (or most significant) octet number of each of the individual fields comprising the message, starts at zero. This is in contrast to the description of the (PNNI and UNI) signaling message format, where it may be recalled the beginning (or most significant) octet is numbered from 1. This minor difference aside, the methodology of using an octet offset value and field size together with the field description or function provides a general and convenient way to describe the PNNI routing packet structure.

The PNNI hello protocol is used to periodically exchange the hello packets over all terminating physical transmission paths between neighboring nodes, and over all VPCs as well as dedicated signaling VCCs (SVCCs) between the nodes. The frequency at which the hello packets are

Offset	Size (octets)	Name	Function / Description
0	8	PNNI Header	PNNI common header with Type = 1
8	2	Flags	Reserved
10	22	Node ID	Node ID of originating node
32	20	AESA	ATM End System Address of orig. node
52	14	Peer Group ID	Peer group ID of orig. node
66	22	Remote Node ID	Node ID of remote node (set when available)
88	4	Port ID	Port ID of orig. node (may be 0 or default value)
92	4	Remote Port ID	Port ID of remote node (set when available, or =0)
96	2	Hello Interval	Indicates frequency of Hello packets (If timed out assumes link failure)
98	2	Reserved	
100 -	Variable	Other IGs	

Notes. 1. The PNNI hello protocol is used to exchange "hello" packets between neighboring nodes over all physical links (transmission paths) and VPCs to discover and verify the identity of the neighbor nodes and determine the status of the links. 2. Hello packets sent over outside links also include other IGs such as Aggregation Token, ULIA, etc.

Figure 7-18. Hello protocol.

generated is determined by the hello interval value, and is generally configurable. Consequently, if no hello packets are detected on any given physical link, VPC or SVCC after a (configurable) timer interval determined by the hello interval multiplied by an inactivity factor, the corresponding link is assumed to have failed and the routing database is modified accordingly. Thus, the periodic exchange of the hello packets provides a powerful (although processing intensive) mechanism to detect link failures, functioning as a "keep-alive" signal (also known as a "heartbeat") analogous to similar mechanisms used in transmission systems at the physical layer. In addition, as will be described later, an analogous "heartbeat" mechanism termed ATM layer "continuity check" (CC) has also been defined within the ATM layer management OAM functions. However, it may not be generally assumed that these other underlying failure detection mechanisms are always present, particularly in private or enterprise networks, so that the PNNI hello protocol is intended to provide direct link status information for the routing databases in each NE in addition to neighbor discovery.

It may also be noted that the hello packet includes both the node identifier and the AESA of the originating NE as well as the peer group identifier. For an efficiently constructed routing hierarchy as described earlier, there will be redundant information in these fields, since the node ID is based on the AESA. This realization may be used to simplify the processing of the hello packets for such configurations. In addition to the fields listed in Fig. 7-18, the hello packet may also contain other IGs, as listed in Fig. 7-17. For outside links (i.e., links between PGs) these may include the Aggregation Token, Nodal Hierarchy, and UpLink Information Attribute (ULIA) IGs, as well as the Resource Availability IG (RAIG). The processing of each of these IGs on a periodic basis enables each NE to acquire and maintain an up-to-date topology database and account for significant changes such as link failures. Since the periodic processing of the hello packets on multiple links may involve significant processing overhead, the role of the hello interval timer in minimizing this overhead needs to be noted. Clearly, for stable network configurations based on reliable links, the hello interval may be relatively long and vice versa, thus involving a trade-off between the processing overhead and the level of flexibility and reliability required for a given network configuration. Such trade-offs need to be carefully considered in designing networks using PNNI routing procedures in general.

After the periodic exchange of the hello packets has resulted in neighbor discovery and the construction of the topology database, the PNNI routing procedures employ the PNNI topology state packets (PTSPs) comprising one or multiple PNNI topology state elements (PTSEs) to distribute (or flood) this information throughout each peer group. As indicated previously, the intent of the flooding mechanism is to ensure that each node within a given PG maintains the same topology database as the other nodes to facilitate call routing through the peer group. The structure of the PTSP and PTSE headers is shown in Fig. 7-19. The PTSP header essentially identifies the originating node and its peer group, and each PTSE comprising the PTSP may be viewed as an IG (with IG type identifier = 64 as noted earlier) that includes other IGs, such as nodal state parameters, internal and external reachable ATM addresses, etc., as listed in Table 7-2. However, it is important to recognize that each PTSE is constructed as an independently exchangeable unit of routing information with its own PTSE identifier, sequence number, and checksum values. In addition, since the PTSEs are subject to aging, the PTSE header also includes a "PTSE remaining life" value to allow the information to be deleted or reflooded in the event of expiry. The checksum calculation is carried out over the entire content of the packet excluding the (variable) remaining life field to detect errorred PTSEs. The PTSE sequence number is used in conjunction with the PTSE acknowledgment and database summary packets as described below.

PTSP Header

Offset	Size (octets)	Name	Function / Description
0	8	PNNI Header	Common PNNI Routing Packet Header
8	22	Node ID	Node ID of originating node
30	14	Peer Group ID	Peer Group ID of originating node

PTSE Header

Offset	Size (octets)	Name	Function / Description
0	2	Type	IG Type identifier (= 64 for PTSE)
2	2	Length	
4	2	PTSE Type	Indicates which restricted IGs are allowed
6	2	Reserved	
8	4	PTSE ID	Identifies one of multiple PTSEs from a node (value = 1 reserved for Nodal IG Type)
12	4	PTSE Sequence No.	
16	2	PTSE Checksum	Checksum over all packet excluding Rem Life
18	2	PTSE Remaining Life	
20 -	var	Other IGs	

Figure 7-19. PTSP and PTSE headers.

The PTSE acknowledgment packets are used to acknowledge the correct receipt of any PTSEs sent from a neighboring NE. The general structure of the PTSE acknowledgment packet is shown in Fig. 7-20, where it may be noted that acknowledgment for multiple PTSEs, each identified by the PTSE ID and PTSE sequence number, may be included. The PTSE acknowledgment packets are sent in response to a PTSE request packet, and this in turn is used for the database synchronization process using the database summary packets shown in Fig. 7-21.

The general procedure is as follows: each NE floods (i.e., distributes or advertises) its topology information using the PTSPs, which, as indicated previously, comprise multiple PTSEs. In the case where the NE generates multiple PTSEs, each describing different attributes or characteristics, each PTSE is identified by a different PTSE ID value assigned by the originating NE. In the case where multiple instances of the same PTSE ID are received, the PTSE sequence number value is used to determine which PTSE is the most recent (i.e., higher in sequence number value). The PTSP flooding procedure enables each node to build a topology database using the PTSE IGs, and identify its nearest neighbors. The PTSEs are acknowledged by sending the PTSE acknowledgment packets. Each NE then initiates the database synchronization process with its neighbors, using the database summary packets shown in Fig. 7-21.

The database summary packets (DSPs) are exchanged in a "lock-step" procedure to ensure that only one outstanding DSP is being processed at any given time. In this process, the NEs operate in a "master–slave" mode, such that the (initiating) "master" NE sends a DSP to the corresponding "slave" NE, which responds by returning a DSP, thereby also implicitly acknowledging receipt of the initial DSP. Each NE then checks that each PTSE in the received DSP corresponds to its topology database. In the event that a PTSE is missing, or if the neighboring NE has indicated the occurrence of a "more recent" PTSE, as derived from the PTSE sequence number or remaining lifetime, the NE generates a PTSE request packet, as shown in Fig. 7-22, to obtain the new (or updated) PTSE from the neighboring NE.

This step-by-step database synchronization procedure is repeated by all NEs within each PG

Offset	Size (octets)	Name	Function /Description
0	8	PNNI Header	Common PNNI Routing Header with Type = 3
Repeat for each set of PTSE Acknowledgment about one node:			
	2	Type	Type = 384 (for Nodal PTSE Ack)
	2	Length	
	22	Node ID	
	2	AckCount	The number of Acks for this node
Repeat for AckCount times:			
	4	PTSE ID	
	4	PTSE Sequence Number	
	2	PTSE Checksum	
	2	PTSE Remaining Life	

Figure 7-20. PTSE acknowledgment packets. 1. PTSE Acknowledgment packets are used to acknowledge receipt of PTSEs from neighboring nodes. 2. Each PTSE Ack may contain multiple PTSE IGs.

Offset	Size (octets)	Name	Function / Description
0	8	PNNI Header	PNNI Routing Packet Header with Type =4
8	2	DS Flags	Flags for synchronization / more bit/ master-slave
10	2	Reserved	
12	4	DS Sequence Number	
Repeat for each set of PTSEs in the topology database			
	2	Type	Type = 512 (Nodal PTSE Summaries)
	2	Length	
	22	Node ID	Node ID of originating node
	14	Peer Group ID	Peer Group ID of originating node
	2	Reserved	
	2	PTSE Summary Count	The number of PTSE summaries for this orig node
Repeat the following structure for PTSE summary count times			
	2	PTSE Type	
	2	Reserved	
	4	PTSE Identifier	
	4	PTSE Sequence Number	
	2	PTSE Checksum	
	2	PTSE Remaining Lifetime	

Notes. 1. Database summary (DS) packets are used during routing database synchronizing process between peers. 2. Procedure based on "master–slave" exchange to ensure step-by-step (one packet at a time) protocol. 3. Database summary procedures should be invoked only when PTSP flooding has occurred (stabilized).

Figure 7-21. Database summary packets.

until alignment of the topology databases is achieved, whereupon the stepwise exchange of the DSPs is terminated. However, it should be borne in mind that the flooding of updated PTSEs (in the PTSPs) will be required in the event of (routing) significant changes, whereupon the database synchronization process should be repeated again in the stepwise fashion described above. The frequency of the exchange of the PTSE request (and corresponding PTSE acknowledgment) packets and the database summary packets depends on the specifics of the network configuration, as well as the thresholds signifying what constitutes a "significant" change. However, based on experience with the analogous routing protocols for the IP-based Internet, it may be anticipated that a substantial processing overhead will also be incurred in the case of PNNI routing for networks of any significant complexity.

The database summary procedures are invoked only after the PTSP flooding process has been completed, to allow the individual NE topology databases to stabilize within any given PG. Assuming that some level of stability has been attained in the organization of the PTSE-contained topological information, the step-wise database summary packet exchange provides for confirmation of the flooded information within each node of the peer group. For PNNI networks where there are frequent significant changes that may affect routing decisions within the nodes, it is clear that such stability may be difficult to attain. In such cases, the database synchronization process may need to be operated frequently with consequent processing overhead, which may be undesirable. In addition, it may be noted that the inherent "aging" of the PTSEs implied by the "PTSE remaining lifetime" parameter included in Figs. 7-19, 7-20, and 7-21 will in any case result in periodic refreshing of the "aged" topology information (either nodal or link state) even for the most stably configured networks. The selection of the aging period as a configurable parameter thus, to some extent, may result in a lower limit of the flooding and subsequent database synchronization processing required.

Offset	Size (octets)	Name	Function / Description
0	8	PNNI Header	Common PNNI Packet Header with Type = 5
Repeat for each set of PTSEs requested			
	2	Type	Type = 513 (Nodal PTSE Request List)
	22	Node ID	Node ID of originating node
	2	PTSE Request Count	Number of PTSEs requested for this node ID
Repeat PTSE Request Count times			
	4	PTSE ID	

Figure 7-22. PTSE request packets.

The procedures associated with the exchange of the PNNI routing packets and their structure as described above primarily serves to show how the topology information is transferred between the nodes within a given peer group and between peer groups. It is important to recognize that these procedures do not in any way constrain or indicate how the individual NEs process or organize the routing information within the nodal databases. The actual processing of the (potentially) large amount of routing information, the associated state machines, and the efficient organization of the topology databases within the software of the node, remain largely implementation-, and hence, vendor-, specific. In this sense, the routing protocols described here are no different from the signaling protocols discussed earlier. The signaling messages, with their contained information elements and associated procedures, simply provide a precisely standardized mechanism to exchange the connection-related information between the NEs. Precisely how this information is processed by the software within a given node depends largely on the implementor and need not be standardized, thereby providing scope for performance enhancement as well as product differentiation. The same principles may be applied to the related routing protocols, whose primary function is to select an "optimum" path for the VCC or VPC from source to destination. The routing packets provide a precisely standardized mechanism to exchange the topology (nodal and link state) information between the NEs contained in the individual information groups. However, the handling of this information within any given NE, as well as the actual algorithms used to select a suitable path, are essentially the choice of the implementor, and hence may vary considerably in terms of efficiency and performance. It may also be anticipated that not all the numerous options and potential functions specified by the protocols may be supported in any given implementation, in order to limit complexity and hence cost, Here also, as with all engineering decisions, there will be trade-offs between flexibility, performance, and costs to be considered.

7.11 GENERAL COMMENTS ON PNNI ROUTING AND SIGNALING

The brief description of PNNI signaling and routing protocols given here is intended to convey an indication of the capabilities and flexibility possible for any type of network configuration.

As may be envisaged, these capabilities are continuously being extended as new applications and functions are proposed. Such new functionality can readily be incorporated, either by defining new messages or, more frequently, by new information elements (IEs) or information groups (IGs) designed to carry the relevant signaling or routing information. This aspect highlights the incredible flexibility inherent in the common channel, nested TLV, message-based signaling and routing protocols used for ATM (as well as for the 64 kbit/sec ISDN, for that matter). On the other hand, this very flexibility and ease of extension can result in excessive complexity, as more and more information in the signaling or routing messages needs to be processed and stored with every new function added. As always, there will be a need for engineering trade-offs between complexity and performance that need to be considered in any given network implementation. In many cases, the initial deployments may only implement the basic capabilities viewed as necessary, with subsequent upgrades for new features performed as and when necessary. A key performance criterion that needs to be considered is the call handling rate required for any given deployment. For typical enterprise network applications, this parameter may be relatively small—on the order of several tens of calls per second per interface. However, for large public carrier applications, call handling rates of several hundreds per second per interface may be required.

It is evident that the PNNI signaling protocols are essentially straightforward extensions of the UNI signaling protocols to the NNI, with enhancements such as crankback and source routing. While extending UNI signaling procedures to NNI may be advantageous for enterprise network applications, which do not require the complexities of SS7-based signaling, the use of source routing for large public carrier network scenarios is as yet unproven. It may be recognized that the use of source routing in PNNI is a key point of difference between PNNI and the B-ISUP/SS7 protocols intended for general NNI signaling. The other major point of difference is the use of the MTP 3 (involving connectionless message transfer) protocol in the B-ISUP/SS7 protocol architecture, which has no counterpart in PNNI signaling. For large networks, the use of a hierarchically complete source routing mechanism may result in relatively poorer performance in comparison to the processing of the B-ISUP initial address message (IAM) for a given level of call processing capability. Whether this factor alone may limit the use of source routing in large networks is as yet unclear. In other respects, despite the differences in terminology and detailed coding, there is actually little functional difference between B-ISUP and PNNI signaling protocols. Moreover, many of the capabilities developed for PNNI, such as soft PVCs, have been incorporated in the evolving B-ISUP signaling procedures, further narrowing the functional differences.

For PNNI routing protocols, the situation is quite different, since it is clear there are fundamental differences between the approach adopted by the B-ISUP/SS7 protocols and the "automated" dynamic routing concepts underlying PNNI routing procedures. As noted earlier, the latter are derived from the routing philosophy employed for IP-based networks, such as the Internet, rather than the administratively configured and managed telephony-oriented SS7 protocol architecture, which essentially relies on statically configured routing. The advantages of automated dynamic routing mechanism to eliminate the considerable management overhead of statically configuring NEs is clear, but it is also evident that the PNNI routing protocols are in general an extremely complex way to provide the desired flexibility. As such, it must be admitted that their overall interoperability, robustness, and stability for use in large-scale public carrier networks is relatively unproven and insufficiently tested at this stage. Despite this, it must be pointed out that a number of network operators have deployed PNNI signaling and routing within their networks quite successfully. With continuing improvements in the underlying technology it seems clear that both the PNNI and the B-ISUP/SS7 routing (and signaling) approaches will compete for acceptance in the public carrier network environment. Given that PNNI protocols are currently widely used in enterprise networks, and available from the majority of ATM switch vendors, whereas B-ISUP protocols are currently only deployed in a limited number of cases, it would be tempting to predict that PNNI routing will prevail. However, this may be far from a forgone conclusion, since the value of providing complex routing protocols in the relatively stable, closely managed, core or backbone ATM networks is unclear. For such networks, which need to be highly robust and protected against physical and other failures, the periodic exchange of hello, PTSP, and database summary routing packets containing large amounts of essentially redundant information simply adds processing overhead to the NEs without really contributing to routing efficiency significantly.

On the other hand, it may also be argued that since the periodicity of the routing information exchanges can be programmed over wide limits to suit the network configuration, the processing overhead in each NE for PNNI routing packets can be substantially reduced for stable topologies. Moreover, with advances in processor speed and capability, software processing costs may continue to decrease at a rate that may render such considerations as irrelevant with respect to other factors important to network design. For example, it may be argued that the robustness and service flexibility designed into the B-ISUP/SS7 architecture, coupled with its natural evolution from existing backbone SS7 networks, makes this approach more appropriate to large-scale carrier backbone networks, where stability and reliable performance are of prime importance. However, it must be admitted that definitive answers as to whether to select the PNNI or B-ISUP/SS7 approaches may be difficult, since many complex factors may need to be considered. Both approaches have advantages and disadvantages, and hence their respective advocates and detractors, which may need to be taken into consideration in any given network design. In such a comparison, the relative complexity of either protocol architecture may not be a significant deteminant. A detailed analysis of both the signaling and routing aspects of both PNNI and B-ISUP/SS7 protocols would suggest that the complexity of either approach is likely to be of the same order.

A number of significant enhancements to the PNNI protocols have been proposed and will likely be incorporated into the next version of the specification (PNNI version 2) issued by the ATM Forum. The key functional enhancements include:

1. Fault-tolerant rerouting capability. This functionality enables the network to recover more efficiently in the event of faults or for traffic load balancing requirements.
2. Closed user group (CUG) capability. The CUG protocols enable the network to support virtual private networking (VPN) services, if required.
3. Network call correlation identifier (NCCI) capability. The NCCI function enables end-to-end information relating to call performance or accounting to be transported when required, by providing a global call identifier mechanism in the Generic Identifier Transport (GIT) IE.

These and other enhancements to the signaling and routing protocols are intended to improve the robustness and range of applicability of these protocols. Another important consideration is the need to be able to interwork between networks employing B-ISUP/SS7 signaling and PNNI-based signaling and routing, since, as noted earlier, these are likely to coexist in the foreseeable future. The interworking requirements between the B-ISUP (or equivalently a B-ICI) network and a PNNI based network have been incorporated in the ATM Inter Network Interface (AINI) specification being developed in the ATM Forum [7.1]. The AINI is essentially the PNNI signaling protocol without the routing capability, mapped into the corresponding B-ICI signaling protocols (B-ISUP), in essentially the same way as is done for UNI (i.e., DSS2) signaling to SS7 parameters. The PNNI routing protocols may either be transmitted at the gateway (interworking)

NEs or be "tunneled" transparently through the B-ICI-based network over a dedicated VCC established between the gateway nodes. In this model, the B-ICI network may simply be viewed as a statically routed "peer group" within an overall PNNI hierarchy.

The above simplified considerations should not be taken to mean that all aspects of interworking between the PNNI and B-ICI-based signaling protocols have been laid to rest. Although the overall interworking framework may be relatively simple, there remain numerous detailed issues that continue to receive extensive debate. Significant among these remains the question of address interworking, as discussed previously. In addition, the mechanisms for ensuring efficient routing when nontopological addressing schemes are deployed in the PNNI network are not well understood and will require more study. There also remain questions about the large number of parameters that may need to be considered in a routing computation, as pointed out earlier, and their impact on the overall performance that may be expected for call handling rates. For routing decisions based on simple binary states (such as if a given link has failed or not), convergence may be rapid. However, if other link (or nodal) metrics are factored into the decision thresholds, the resulting routing algorithms may be complex and relatively slow. As an increasing number of ATM networks deploy the full suite of PNNI routing mechanisms, it may be envisaged that the behavior and performance of the protocols will be more extensively studied and improved.

CHAPTER 8

ATM Layer Operations and Maintenance (OAM) Functions

8.1 INTRODUCTION

The ATM layer operations and maintenance (OAM) functionality that has been designed to support the management of ATM networks is described in detail in this chapter. There is little doubt that in respect to OAM capabilities, ATM technology arguably provides the most capability of any of the various packet mode transfer technologies in use so far. The comprehensive range of management tools potentially available to the ATM network operator was deliberately designed with the intent that ATM should provide the most robust and reliable means for transfer of any type of data in any foreseen network application. Clearly, this is most important in the case of large-scale public carrier networks, whose customers have come to expect the very high levels of dependability and robustness provided by existing telephony networks. Anything less for a new technology such as ATM may be perceived as a step backward. Consequently, there has been extensive work in both the ITU-T and the ATM Forum as well as other bodies to develop and specify not only the tools for both in-service and out-of-service OAM and testing functions, but also for the overall network management systems that utilize these tools to provide the network operator with the information necessary to manage the network. Despite this, it should not be assumed that all aspects of this complex multifaceted problem have been resolved to achieve network management goals. As will be seen, much still remains to be done, and as for the case of other aspects such as signaling and traffic control, the description below should be viewed as a summary of progress to date, with much work still in progress.

8.2 OAM PRINCIPLES AND GENERAL NETWORK MANAGEMENT ARCHITECTURE

Not surprisingly, the underlying principles and network architecture model used as a basis for ATM OAM derive from the approach that has been generally adopted for legacy telecommunications networks. This model, together with the associated management principles and procedures, has been formalized by the ITU-T into what has come to be known as the Telecommunications Management Network (TMN) OAM model. For the ATM case, this model is described in recommendation M.3010 [8.1], and other related recommendations developed by the ITU-T. A detailed description of the TMN model and its associated protocols is outside the scope of this text, but it is useful to summarize the basic underlying principles in order to understand the relationship between ATM layer OAM and the network management (or operations) systems. It should also be pointed out the TMN model is not the only network management model being considered. Other approaches for network management, notably those based on open distributed processing (ODP) concepts (i.e., distributed computing platforms), have also been extensively developed and proposed. However, it is sufficient to consider only a TMN-based model here, since the differences between the various approaches relate more to the detailed protocols for processing, storage, and communication of the management information resulting from the underlying ATM layer OAM mechanisms, and do not significantly impact the actual mechanisms.

In the TMN model, the network-management-related functions are classified into five broad management categories:

1. Fault Management (FM). The functions in this category relate to the detection and localization of defects and failures that may affect the transport of user information. The fault management procedures may rely on continuous and/or periodic monitoring of the connections to generate and communicate alarms in the event of failures or defects observed along the connection path (this is also known as "alarm surveillance"). As a result of such actions, the management system may log the maintenance events and alarms, and the NEs may initiate automatic protection switching actions (if available) to enable rapid recovery from a failure.
2. Performance Management (PM). The functions in this category include the monitoring and analysis of the performance of the network (and of individual connections) to detect any degradations of the performance parameters that may adversely affect the Quality of Service (QoS) commitments to customers. As a result of such performance parameter monitoring/analysis, the management system may log performance degradation data and initiate actions to ensure that the committed QoS can be met.
3. Configuration Management (CM). This category includes all the functions and procedures related to the provisioning of connections and the allocation of the related network resources such as bandwidth, buffer, and processing capacity in the NEs. The routing and configuration of semipermanent virtual circuits (PVCs) is also included in this category of management functions.
4. Accounting Management (AM). These functions are associated with resource usage monitoring and the collection and transfer of accounting information to enable billing for services. As may be expected, accounting management systems are highly network-operator-dependent, and may be viewed as more of an administrative procedure as opposed to an operational one. As such, these functions are outside the scope of this text.
5. Security Management (SM). These functions, which may also be highly network-operator-dependant, relate to the security of all network resources from external infiltration or corruption of sensitive data. Although of growing importance and therefore the subject of extensive study, SM is outside the scope of this discussion.

Of the five general TMN functional categories, the focus here will be on fault and performance management aspects, although it may be assumed that configuration management functions are inevitably invoked in the setting up of any semipermanent virtual channel connections under management (or operational) system control. The broad TMN categories of management functions serve as a useful conceptual framework for more detailed description of individual functions, but it is important to recognize there will be relationships between functions assigned to any of these categories. For example, there may be relationships between PM or FM and AM, in the sense that fault- or performance-related events may result in accounting/billing information if QoS commitments cannot be met. There may also be a relationship between FM and CM if, for example, automatic protection switching schemes are invoked in the event of faults. Irrespective of how the network management functions are grouped, it should be evident that, in practice, management system software will likely be extremely complex but essential for the proper operation of any network designed for dependable and high performance targets.

For the OAM purposes considered here, a generalized network management reference architecture is shown in Fig. 8-1, which also depicts a simplified representation of the TMN layered model. Strictly speaking, the TMN model consists of five layers, but here we combine the (higher) service management and business management layers into one for convenience. The functions and procedures in these higher layers relate to network operator business, financial and service level negotiations, or service contracts, and need not be considered for the purposes of this description. The functions and associated procedures for the network management layer (NML) relate to the management of the network (or subnetwork) as a whole, i.e., for a number of NEs. The network element management layer (NEML) is responsible for the management of an individual NE, or possibly for a small cluster of NEs. Finally, the network element layer (NEL) includes the basic layer management functions of the individual NE, as described in more detail below. Each layer of the TMN model may include some (or all) aspects of the five management functional categories defined above. However, the fact that each of the TMN layers may contain aspects of these management categories does not mean that the functions are in any way redundant or duplicated at every layer. The underlying concept here is that the management information relating to any given category (e.g., FM) that is generated (or processed) at a layer, may be communicated to the next layer after some degree of filtering or abstraction. In this way, the higher TMN layer may receive a suitable relevant subset (or abstraction) of the lower-layer management information that is required for its role in the overall management scheme.

As an example of this management information abstraction process inherent to the TMN model, consider the case where the network element layer experiences a defect of some kind, such as a failure of a component in a circuit pack. The NE may generate a series of alarms as a consequence of this event at the NE layer. If judged sufficiently serious and persistent (i.e., if maintenance actions at the NEL are unable to rectify the defect), the NEL may indicate the alarm state to the NEML. The FM function at the NEML, which may be either within an individual NE or within a controller for a group of Nes, will process and log the alarm information,

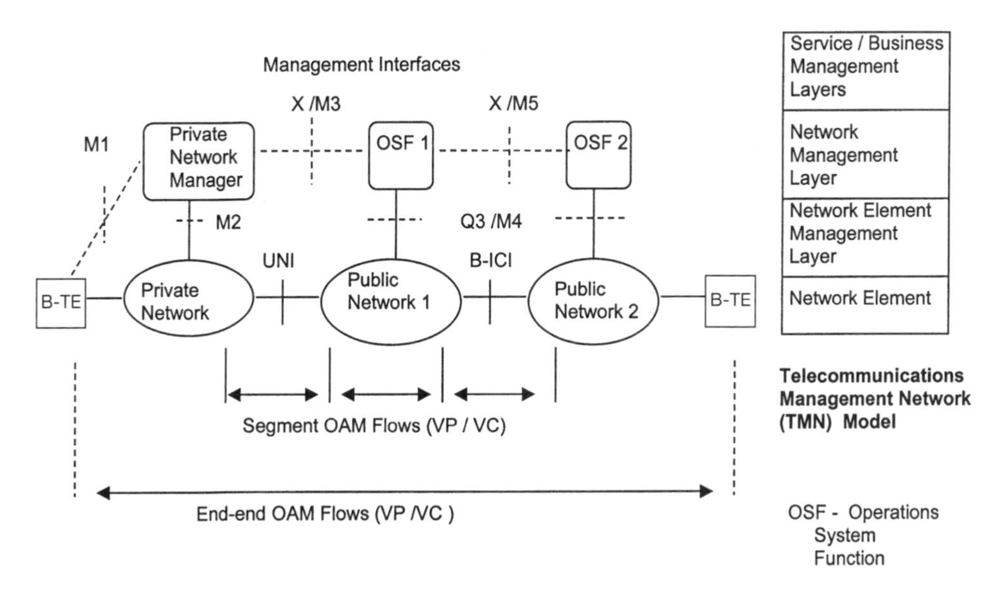

Figure 8-1. General network management architecture.

and may further filter or abstract the information before passing it up to the NML, if required. The FM function in the NML, which is responsible for the network (or subnetwork, depending on configuration), thus obtains an indication of the fault condition in an abstracted or filtered form, and may initiate actions to localize and remedy the fault as necessary. If the fault is sufficiently serious to affect the service contract and has financial implications, the NML may pass the relevant information up to the service or business management layers for appropriate action. In this way, it may be seen that the TMN protocol architecture is essentially a layered model for the ordered handling and processing of management information. This is why the TMN procedures are sometimes referred to as "information modeling." The rationale behind the abstraction or filtering process is also based on the simple requirement that it is not advisable to "flood" the higher layers with all the potential lower-layer management information that may be generated within the myriad of components in any given NE. Only the necessary elements need to be communicated to avoid swamping the management system with unnecessary detail. In any case, for sophisticated modern management systems, the relevant detail may be readily made available if required at any layer.

Figure 8-1 also shows a highly schematic model of the possible management interfaces corresponding to the TMN model outlined above. Considering first the case of a typical public carrier network, the management interface between the operations system functions (OSF) and individual NEs (or groups of NEs) is termed the Q3 interface in the TMN model. An alternative, but broadly equivalent, terminology developed by the ATM Forum refers to this interface between the OSF and NE as an M4 interface. Traditionally, this interface is also called an "OS/NE Interface." Ignoring for now such purely terminological differences, the Q3 interface is intended to provide an essentially open (i.e., standardized) interface for the communication of all management-related information between the NE layers (however configured), the network management functions, and the layers above, if necessary. Although a detailed description of the protocols used across the Q3 interface is outside the scope of this text, we will later consider briefly an example of particular types of management information models. More detailed descriptions may be found in texts dealing with network management and the TMN, specifically [8.1–8.3].

For the exchange of management information between different OSFs, the TMN model identifies an X interface. In the ATM Forum's management specifications, this interface is referred to as an M3 interface if between a "private" network management system and a public network OSF, and as an M5 interface if between two public network OSFs, as shown in Fig. 8-1. In the ATM Forum's network management reference model, two other management interfaces have also been specified that have no direct counterpart in the TMN model as defined by the ITU-T. These are the so-called M1 interface between an ATM terminal equipment (B-TE) and the customer premises (private) network management system, and the M2 interface between a private NE (e.g., an ATM PABX or LAN hub) and its network management system. It is important to recognize that all the management interfaces identified above are so-called "out-of-band" or dedicated management information interfaces between the respective management entities, which operate independently of the user traffic along any given ATM VPCs or VCCs. The management interfaces may use ATM VPCs or VCCs to transport the management information between the OSFs (or management system entities), but these are dedicated or preassigned VCCs for the management protocol and are independent of user VPCs or VCCs, analogous to the case of common channel signaling, where the signaling messages are carried over VCCs dedicated to signaling between the NEs. For some network scenarios, the management interfaces may not even use an underlying ATM VCC, and may utilize other transfer protocols such as X.25, Frame Relay, IEEE 802.x LANs such as Ethernet, etc. to transport the requisite management information.

In contrast to the above out-of-band management information exchanges between management systems, ATM also uses "in-band" dedicated OAM cells that provide specific management functions related to fault and performance management. These dedicated ATM layer OAM cells flows are also represented in Fig. 8-1 in order to show their relationship with the management information exchanged across the out-of-band TMN interfaces. The ATM OAM cells are in-band in the sense that, if present, these cells are interspersed together with normal user data cells along the VPC or VCC, and may be inserted or extracted at specific VPL or VCL termination points along the VPC or VCC in question. As will be seen in detail below, these in-band OAM cells can be distinguished from the user data cells by preassigned codepoints in the ATM cell header. However, since the OAM cell flows must traverse the same connection path as the user cells on that connection, they must have the same VPI and VCI values as the user cells of that connection (VPC or VCC). As shown in Fig. 8.1, the OAM cells may exist at both the VP and the VC levels, depending on whether they relate to a VPC or VCC. In addition, two kinds of OAM cell flows have been defined for ATM, as shown in Fig. 8-1. First, there may be an "end-to-end OAM" cell flow, which pertains to the endpoints of any given VPC or VCC. The end-to-end OAM cells (e-e OAM) may only be inserted or extracted at the endpoints (terminations) of any given VPC or VCC. Intermediate NEs may monitor these e-e OAM flows, but should not terminate or modify them. Second, there may be dedicated OAM cells that relate to a "segment" of the ATM VPC or VCC, which are known as "segment OAM" flows (seg. OAM).

The concept of a connection "segment" as a maintenance entity needs to be clearly understood, not only in the context of the related segment OAM cell flows, but also as an essential network management tool within the TMN framework. The primary need for a connection segment as a distinct maintenance entity derives from the observation that an ATM VPP or VPC may span several different administrative (or operational) domains, and each domain may wish to manage (or maintain) its part of the overall connection. A convenient way to achieve this is for any administrative domain to be able to set up by management action through the appropriate interface (e.g., the Q3 or M4 interfaces mentioned earlier) a connection segment over its domain as a distinct maintenance entity. Consequently, by inserting and/or extracting the relevant OAM cells at the endpoints of its connection segment, the domain network operator may be able to manage its part of the overall connection, without affecting upstream or downstream domains. In practical terms, from an ATM segment OAM cell flow viewpoint, the "segment" is simply a concatenation of VPLs or VCLs whose endpoints are set up by the network management system, and which form a part of the overall ATM VPC or VCC. The segment OAM cell flows can only exist over the segment of the overall connection defined by the network management interfaces and hence must be extracted at the endpoints of the segment. Since the network management system has control over the extent of the maintenance segment it sets up through the TMN interfaces to the NE, the network operator may change the segment endpoints as required. In principle, any VPL or VCL termination point in the NEs along a connection path may be configured as a segment endpoint, although in practice these may typically be set up at the edges of an administrative (or operational) domain.

Before discussing the details of the in-band OAM cell flows, it is important to clearly understand the relationship between these and the management information flow across the network management interfaces shown in Fig. 8-1. The dedicated OAM cells may flow along a connection path (VPC or VCC) carrying some form of OAM information, typically in relation to fault or performance management functions. This information may pertain to the overall VPC or VCC if these are end-to-end OAM cells, or may relate to any given connection segment along the connection path if they are segment OAM cells. Any given NE along the connection path may be configured to process the OAM cells on a VPC or VCC in a number of ways (described in more detail below). The NE may then (if configured) "filter" the resulting OAM information and communicate this information, not the actual OAM cell, to the management system (at either the NEML or NML levels, depending on the specific configuration) via the TMN management interfaces, using one of a number of protocols designed to exchange management information. The OAM cell payloads themselves do not traverse the TMN management interfaces since these are essentially "out-of-band" dedicated communications channels, which may or may not in general be based on ATM VPCs or VCCs. If the NE in question terminates a given VPC or VCC, or is configured as a segment endpoint for the VPC or VCC, it may terminate OAM cells on that connection, process the contained information, and choose to pass the resulting event to the management system if necessary. If the NE in question is simply an intermediate node along the connection path, it may be configured (by the network operator via the management system) to "monitor" any passing OAM cells on the VPC or VCC, without modifying them in any way, or the NE may simply ignore the transiting OAM cells if it is not configured to monitor them by the network operator.

Since the OAM cells travel along the connection path interspersed with the user data cells, if present, they may be used for "in-service" monitoring or maintenance purposes if required. In addition, the monitoring information in the OAM cell payload may be rapidly available, in the order of the propagation delays, along the connection path. On the other hand, the resulting (processed) out-of-band management information that may be exchanged across the TMN interfaces may be relatively slower, particularly if significant (software-based) processing is involved. Although the use of dedicated in-band OAM cells enables the high-speed transfer of maintenance information along the connection paths, the processing and use of this information by the relatively slower management system should not be allowed to overwhelm it. In any case, not all management information such as accounting or configuration functions, needs to be processed in real time.

8.3 OAM FLOWS

Having described in somewhat general terms the relationships between the per-connection in-band OAM cell flows and the out-of-band management information exchange across the TMN interfaces, we turn now to a more detailed description of the ATM layer OAM cell functionality. In fact, the concepts underlying the use of in-band (and in-service) OAM mechanisms derives from analogous functions widely used in transmission systems, notably those based on the synchronous digital hierarchy (SDH) technology developed for modern optical fiber systems. Since such systems are capable of transporting vast amounts of critical user data, it is imperative that the network operations system are able to instantaneously monitor (and react to) the state of the transmission networks in respect to potential defects and performance degradations. To achieve this, transmission systems have developed powerful in-service mechanisms to monitor error conditions, defect states, and performance degradations, and report these to the operations system function (OSF) for timely action. In its role as both a transport and a switching technology, as noted earlier, it is therefore not surprising that ATM has adopted and extended a number of these mechanisms for its own use. In fact, the relationship between the OAM "flows" at the physical layer and those at the ATM layer may be categorized in terms of the general B-ISDN protocol reference model (PRM) as depicted in Fig. 8-2. In this model, which has been used as the basis for the detailed description of the ATM layer OAM functions in ITU-T's Recommendation I.610 [8.5], the OAM flow (F) at any given level in the PRM is denoted by a number, e.g., F1 to F5 from the (lowest) regenerator section (F1) upward.

In Fig. 8-2 it may be seen that three levels of OAM flow, F1 to F3, have been defined in general at the physical layer, and two levels, F4 and F5, at the ATM layer of the PRM. The possibility of OAM flows at higher layers (e.g., an F6 or F7 flow corresponding to the AAL) has also been conjectured, but these will be considered later; here we confine the discussion to

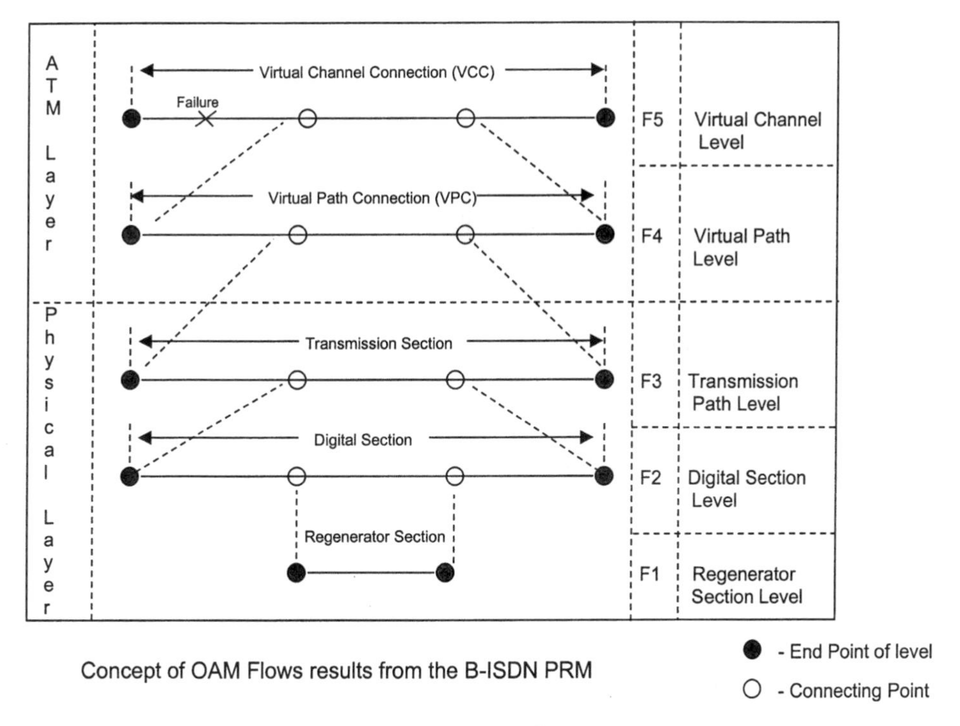

Figure 8-2. OAM flows.

the ATM layer. The definition of the individual OAM flows at the various levels are shown in Fig. 8-2.

The lowest level of OAM Flow (F1) corresponds to the regenerator section, which is essentially a maintenance span of control between any two (optical) regenerator nodes and typically forms a subentity within a so-called digital section. One or more regenerator sections may exist between the digital sections. The digital section OAM flow (F2) corresponds to the span covered by the digital section overhead (SOH) in SDH transmission systems, but may not have a counterpart in other transmission hierarchies. In the SDH system there may be one or more digital sections within the end-to-end transmission path. As noted earlier, the transmission path essentially corresponds to the points where the transmission payload is inserted (or extracted) from the transmission systems. The OAM flow corresponding to this level [e.g., carried in the path overhead (POH) of an SDH-based transmission system] is termed the F3 level OAM flow, and relates to the maintenance of the entire transmission path.

Within the ATM layer itself, separate OAM flows will need to exist, depending on whether the entity that is to be maintained corresponds to the VP level or the VC level. The dedicated OAM cell flows at VP level (i.e., relating to VPCs or paths thereof) are generically termed the F4 level flow, whereas the OAM cell flows relating to the VC level (i.e., VCCs) are termed F5 OAM flows. As noted earlier, OAM flows relating to the maintenance of the layers above the ATM layer have also been proposed, particularly in relation to the AAL, where F6 and F7 flows are conceptually feasible, corresponding to sublayers within the AAL, as will be discussed later. However, these are generally not well defined at this stage and hence will not be discussed in any detail here. In keeping with the principles underlying the layering of protocols inherent in the PRM, the OAM functions relating to a given OAM level are independent of the OAM functions of other layers and therefore have to be provided at each layer. The OAM functions provided within each layer of the PRM are performed by the layer management (LM) entities, which are essentially responsible for processing the OAM information. As noted earlier, this processed OAM information may be provided to the plane management entity (i.e., the system management functions at the node level), and to the adjacent higher layer if required. As a general guideline, it should be noted that OAM related information is typically passed upward in the protocol stack, and it should not be necessary for the higher-layer OAM functions to support any OAM capability at the lower layer. These guidelines for handling of operational information are not specific to ATM, but may apply to other networking technologies as well.

For the physical layer, the OAM flows F1 to F3 and the related OAM functions depend largely on the transport mechanisms of the given transmission system as well as the (operational) supervision functions associated with the physical layer termination functions in the transmission equipment. The OAM capabilities of the various transmission systems in itself constitutes a large field of study that is essentially outside the scope of this text [4.2]. However, it is useful here to summarize some key points relating to the OAM flows for the physical layers of interest to ATM. As noted previously, there are three main categories of physical layer transmission systems of interest to ATM. These are:

1. SDH-based transmission systems
2. PDH-based transmission systems
3. Cell-based transmission systems

In general, not all the physical layer transmission systems support all the possible OAM flow levels termed F1 to F3 above. However, for the case of SDH-based physical layer transmission systems, all three levels of OAM flows may be used. In this case, the F1 and F2 OAM flows are carried in dedicated octets defined in the section overhead (SOH) of the SDH frame. The F3 OAM flow corresponding to the transmission path level is carried in dedicated octets in the higher-order path overhead (POH) of the SDH frame. These SOH and POH OAM functions essentially monitor for loss of signal (LOS) or loss of frame (LOF) conditions (fault management), or for unacceptable error performance (performance management) based on bit interleaved parity (BIP) 8 error computation. The detailed mechanism and codings for these OAM flows can be found in ITU-T Recommendations G.707 [4.5], G.782 [4.8], and G.783 [4.9].

For the case of PDH based physical layer transmission systems, the digital section level F2 OAM flow is not supported by the transmission technology explicitly. However, the regenerator section OAM flow (F1) and the transmission path OAM flow (F3) are provided by dedicated bits in the PDH frame section overhead. The detailed codings and error detection mechanisms are described in ITU-T Recommendations G.702 [4.10], G.804 [4.11], and G.832 [4.12]. For both the SDH and (to a lesser extent) the PDH transmission systems, the supervisory OAM functions (or flows) are intended to support monitoring for fault or performance management conditions, such as "signal fail" (SF) or "signal degrade," etc. If a defect is detected at any given level, the consequent action is generally to propagate a so-called alarm indication signal (AIS), also equivalently known as alarm inhibition signal (AIS), at that or a higher level downstream of the point at which the defect (or fault) was detected. For bidirectional transmission systems, the downstream transmission terminal responds to the AIS flow by generating a so-called remote defect indication (RDI), upstream. Note that this signal used to be called a far-end receive failure (FERF) in earlier terminology. In this way, both downstream and upstream NEs would be aware of the defect condition along the transmission system. In general, error (performance) monitoring is typically carried out by performing a BIP calculation over the transmission frame (or multiframe) and comparing results at either ends of the physical link. If a given error threshold is exceeded, the management system may initiate appropriate action such as automatic protection switching of the physical link. Of the commonly used transmission systems, it is clear that the more recent SDH- (or SONET-) based transmission systems provide by far the most comprehensive OAM capabilities, by design.

Both the widely used PDH- and SDH-based physical layer transmission systems are frame oriented, as described previously, so that the ATM cells are "mapped" into the transmission payload (also called a "virtual container," or VC in the case of SDH), delineated by the specific framing structure. However, for the case of the cell-based transmission system, no framing is used as the ATM cells are directly transmitted onto the physical media interspersed with dedicated physical layer cells, which perform rate adaptation (the idle cells) and OAM functions. In this case, the physical layer OAM (PL-OAM) cells are used to transport the F1 and F3 level OAM flows. Here also, as for the PDH case, the F2 level (i.e., digital section) flow is not provided for. Although the cell-based transmission system has not been commercially deployed to any significant extent at the time of writing, a number of experimental trials have been performed, particularly in the European Union administrations. The cell-based system treats ATM as both the switching and the transmission technology, and is clearly an elegant solution for packet mode transmission systems, with the added advantage of enhanced efficiency, since the "overhead" associated with framing may be eliminated. For the cell-based transmission, the F1 and F3 level OAM flows are carried in the dedicated PL-OAM cells defined for that purpose. It is important to clearly distinguish between these PL-OAM cells, which exist only within the physical layer functions, and the normal ATM Layer cells (including the OAM cells), which exist within the ATM layer at both VP and VC levels.

The PL-OAM cells (and the idle cells considered earlier for rate adaptation purposes) are only generated, processed, and terminated within the physical layer functionality and must not be passed up to the ATM layer under any circumstances. If this were to happen, say as a result of a protocol error, the PL-OAM cells should be discarded as invalid cells. The detailed structure and coding of the PL-OAM cells is given in ITU-T Recommendation I.432 [4.6], but it may be noted here that the essential fault and performance monitoring functions available in the frame-based SDH and PDH systems have also been incorporated in the PL-OAM cells. For error detection, the PL-OAM cells include a BIP-8 calculation over a block of cells called the monitoring block size (MBS), which may be selected up to a maximum of 64 cells. The BIP-8 computation may be performed over a number of blocks which are specified in the PL-OAM cells. If errors are detected in transmission, these are encoded as parity violations in designated fields in the PL-OAM cell as far-end block errors (FEBE), which may be reported to the management system. For defect (fault) detection and alarm surveillance, the PL-OAM cells also include the alarm indication signal (AIS) and remote defect indication (RDI) functions in a completely analogous manner to that in the frame-based transmission systems. To ensure efficient fault and performance monitoring of the transmission cell stream, the frequency of the PL-OAM cells needs to be at least one PL-OAM cell every 512 cells, although higher rates are feasible if required. Consequently, it may be seen that in terms of the F1 and F3 OAM capability, the PL-OAM cell flows in the cell-based transmission system can provide the same essential functionality available in the frame-based systems, the only difference being that the OAM overhead is encoded in the dedicated PL-OAM cells rather than in the transmission frame as in SDH.

8.4 ATM LAYER OAM FLOWS AND FUNCTIONS

At the ATM layer, two OAM flows are defined, as noted earlier. At the virtual path (VP) level, this flow of dedicated OAM cells is called the F4 OAM flow, whereas at the virtual channel (VC) level, the flow of OAM cells is called an F5 OAM flow. These OAM cells are dedicated solely to providing the OAM functionality for either VPCs or VCCs or both, and should not be confused with the physical layer OAM (PL-OAM) cells described above for the physical layer F1 and F3 flows defined for the cell-based transmission systems. In the sections below, we describe in detail the functions of the F4 and F5 OAM flows specifically in relation to fault and performance management categories.

The OAM cells in the F4 flow must have the same VPI values as the user data cells on every VP link (VPL) along the connection path (VPC), since the flow is considered "in-band" to the VPC. The F4 OAM cells are distinguished from the user data cells in the cell stream by one or more preassigned VCI values. The OAM flow may be bidirectional, in which case they have the same preassigned VCI value for both directions. In addition, the OAM cells for both directions of the F4 flow must follow the same physical (transmission) path so that any intermediate connecting points (NEs) may readily correlate the fault- and performance-related information from both directions of the connection. Two kinds of F4 OAM flow can simultaneously exist in a VPC, as noted earlier. These relate to the end-to-end F4 flow and to the segment F4 flow. The end-to-end F4 OAM cells are identified by a preassigned VCI value of 4, and are intended for end-to-end operations and maintenance communications on VPCs. The segment F4 flow is identified by the preassigned VCI value of 3 for the OAM cells. The segment F4 OAM flow is intended for communicating operations and maintenance information within a connection segment as a maintenance entity defined by the network management system (or operations system function) as described earlier. The segment may be viewed as a concatenation of VP links (VPLs) grouped for maintenance purposes. It is also important to note that although segments may be defined contiguously along a connection path, overlapping or nesting (embedding) of segments is not permitted, since the OAM cells belonging to one segment cannot enter another (overlapping) segment.

The same general requirements described above for the F4 OAM cells also apply to the F5 OAM flows used for communicating operations and maintenance information along VCCs. However, since the F5 OAM cells must have the same VPI and VCI values as the user data cells on every VC link (VCL), they cannot be distinguished from the user cells by preassigned VCI values as for the F4 cells. Consequently, the F5 OAM cells are identified by preassigned codepoints in the ATM header payload type (PT) field. The (binary) codepoint PTI = 100 identifies a segment F5 OAM cell (and flow), and the (binary) codepoint PTI = 101 identifies an end-to-end F5 OAM cell (and flow). These codings of the PT field in the ATM cell header were noted previously in the detailed description of the ATM cell header functions in Chapter 3 (see Fig. 3-4 and Table 3-3). The provision of the preassigned VCI values for the F4 OAM cells and the PT codepoints for the segment and end-to-end F5 OAM cells enables any ATM NE (which by definition must process the ATM cell headers) to identify immediately whether the OAM cells present on any given VPC or VCC are a segment or end-to-end F4 or F5 OAM cell flow. If the NE is provisioned as a segment endpoint through the management interfaces (Q3 or M4), then it may extract and process the segment OAM F4 or F5 cells, otherwise, it may simply monitor the flows, generally without modifying them.

The ATM layer OAM functions defined so far are listed in Fig. 8-3, which also summarizes the main application for each of the OAM functions. These functions are defined in detail in ITU-T Recommendation I.610 [8.5], which is used as a basis for this description. The in-band OAM functions defined for fault management are:

1. Alarm Indication Signal (AIS)
2. Remote Defect Indication (RDI)
3. Continuity Check (CC)
4. Loopback (LB)

OAM Function	Main Application
Alarm Indication Signal (AIS)	For reporting defect indications in the forward direction
Remote Defect Indication (RDI)	For reporting remote defect indications in the backward direction
Continuity Check (CC)	For continously monitoring connection continuity
Loopback (LB)	For on-demand connectivity monitoring For fault localization For preservice connectivity verification
Forward Performance Monitoring	For estimation performance on a connection
Backward Performance Monitoring	For reporting performance estimations in the backward direction
Activation/Deactivation	For activation and deactiviation performance monitoring and CC
System Management	For use by end systems only fo user specific management

Figure 8-3. ATM layer OAM functions. (From ITU-T Recommendation 1.610, 1999.)

Each of the above fault management (FM) OAM functions are described in detail below. The OAM cells that provide these functions are generically referred to as fault management OAM cell flows. The in-band OAM functions defined for performance management (PM) are:

1. Forward Performance Monitoring (FPM)
2. Backward Reporting (BR)

The dedicated OAM cells defined to support performance management are sometimes generically referred to as performance monitoring (PM) OAM cells, and are described in more detail below. Finally, two other types of in-band OAM functions (with their associated OAM cells) have also been defined as shown in Fig. 8-3. These are:

1. Activation/deactivation cells, which are intended as one way of "activating" or initiating the performance monitoring (PM) and the continuity check (CC) on any given VCC or VPC.
2. System management OAM cells, which were originally intended for the transfer of systems-specific or security-related information between connection endpoints (or end systems). Currently, the system management OAM cells are used for security applications such as encryption or authentication functions that have been defined by the ATM Forum. The description of these special security-related OAM functions is outside the scope of this text, and the interested reader may refer to the ATM Forum, Security Management Specification [8.6].

8.5 FAULT MANAGEMENT FUNCTIONS—AIS/RDI

The ATM layer alarm indication signal (AIS) and the associated remote defect indication (RDI) functions are essentially an extension of the same concepts used at the physical layer, i.e., within the transmission system overhead. The essential difference is that whereas in most frame-based transmission systems such as SDH or PDH, the AIS and RDI information is transported in dedicated bits or octet codepoints within the transmission overhead, at the ATM layer dedicated

OAM cells are used to transport the AIS and RDI information in-band along the affected VPCs and VCCs.

The basic procedures underlying use of the AIS and RDI FM OAM mechanisms are summarized in Fig. 8-4 Part a, which depicts a given ATM VPC or VCC traversing a number of NEs (sometimes called "connection points" in relation to OAM terminology). If a defect occurs at some point in the connection path as shown, the NE immediately downstream of the detected defect should generate and send AIS cells downstream on all affected active VPCs and VCCs. Considering first the end-to-end AIS and RDI functions, the end-to-end AIS cells thus generated by the NE that detects a defect are transmitted downstream to the connection endpoints (of the VPC or VCC). The AIS OAM cells should be sent as soon as possible after detecting a defect condition and are sent periodically at nominally one AIS cell per second. The AIS OAM cells are generated if the defect occurs within the ATM layer functions either at VP or VC levels, or within the physical layer functions (i.e., due to transmission path defects or failures). In response to the end-to-end AIS cells received at the (VP or VC) connection endpoint, the OAM procedures require that the equipment or NE at the connection termination should generate remote defect indication (RDI) OAM cells upstream along the affected (bidirectional) VPCs or VCCs. The RDI OAM cells are also generated at a nominal rate of one cell per second.

It should be noted that the rate of the AIS OAM cells downstream of the detected defect and the corresponding RDI OAM cells upstream along the bidirectional connection is independent of the user data cell rate on the VPCs or VCCs. It is periodic at nominally one AIS and RDI cell per second on each affected VPC or VCC. However, depending on the type of defect, if it results in the generation of the AIS and RDI OAM cells, there will be no user data cells flowing on the

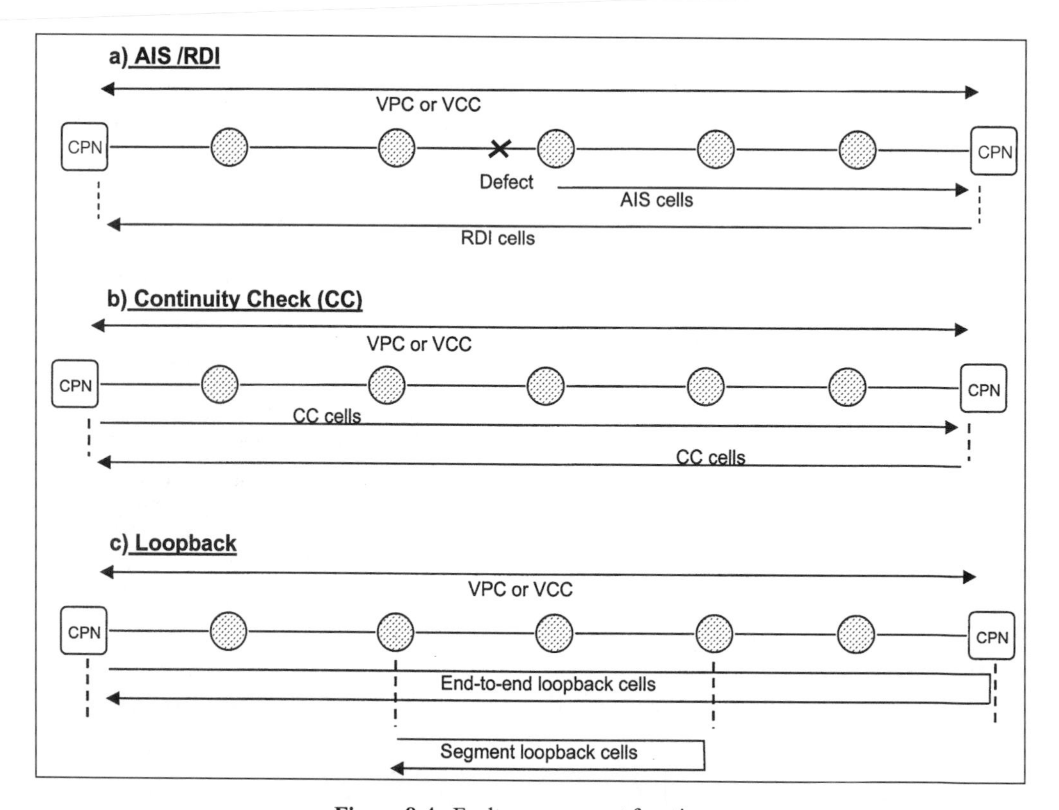

Figure 8-4. Fault management functions.

affected VPC or VCC in the downstream direction. Thus, the only cells flowing downstream in the presence of the defect will be the AIS cells. In the upstream direction, however, user cells may be present together with the RDI cells. For example, if the defect in question is a loss of continuity (LOC) condition due to a component failure or malfunction within the physical layer functions (i.e., transmission system) or within the ATM Layer functions, it is likely that no user data cells will flow downstream of the defect and only AIS cells will be detected at the connection termination point.

Under these conditions, it is customary to specify the onset (and removal) of an "AIS state" at the connection termination point as soon as an AIS cell is received, or a transmission path (physical layer) AIS or LOC condition is detected. The AIS state at the downstream connection termination point is maintained as long as AIS OAM cells are received (at nominally one cell per second). The AIS state is considered to be released the instant the user data cell or a continuity check (CC) OAM cell is detected at the connection end point. The AIS state is also released if no AIS cells are received for a period of nominally 2.5 ± 0.5 seconds (assuming no user data cells or CC cells are present in that interval). Similar considerations apply for the case of the RDI state at the upstream connection endpoint, which declares the RDI state on detecting the RDI cell. The RDI state is maintained as long as RDI cells are detected at nominally one cell per second. The RDI state is released if no RDI OAM cells are detected for a period of nominally 2.5 ± 0.5 seconds. All the above procedures for end-to-end AIS and RDI cells assume that the NEs stop generating AIS (or RDI) OAM cells on all affected VPCs or VCCs as soon as the defect indication is removed or the fault is cleared.

So far we have only considered the end-to-end AIS- and RDI-related actions of the NE that detects the defect and the terminal equipment that declare the (downstream) AIS and (upstream) RDI states. For the intermediate NEs along the (bidirectional) connection path as shown in Fig. 8-4 Part a, two options are possible. The intermediate NEs may be configured (via the management interface to the network management or operations system function) to "nonintrusively" monitor the passing AIS and RDI cell or simply to pass the AIS/RDI cells without monitoring. In either case, it is important to recognize that NEs downstream of the node that generates the end-to-end AIS cells should not themselves generate AIS cells but simply pass the end-to-end AIS (or RDI) cells. This requirement is in keeping with the alarm inhibition (signal) function that underlies the concept of AIS and RDI in transport networks. The presence of the AIS cells every second at any given downstream node indicates that a defect has occurred somewhere upstream of that node, but since the AIS cell is already present, it "inhibits" the generation of alarms from the nodes downstream of the initial NE that detected the defect and generated the AIS OAM cells. The inhibition of AIS and RDI cell generation also prevents the downstream and upstream endpoints being "flooded" with AIS and RDI cells generated by the intermediate nodes.

It is also important to recognize the significance of the term "nonintrusive monitoring" of the OAM cells traversing a given NE. Nonintrusive monitoring of an OAM cell flow implies that both the traffic characteristics and the content of the OAM cells should not be changed in any way by the NE the cell passes through, other than the inevitable (but bounded) cell delay variation (CDV) induced by normal buffering of all cells in the VPC or VCC. However, nonintrusive monitoring does not necessarily mean that the NE can not process the information carried within the OAM cell, as long as it can do this without delaying or changing the OAM cell contents. Typically, nonintrusive monitoring of the OAM cells can be performed by copying the OAM cells in question into a dedicated OAM processor while allowing the OAM cell itself to traverse the NE in the normal way. The information in the copied OAM cell may then be processed within the layer management plane without interfering with the transfer of cells. If the NE is not configured to nonintrusively monitor the OAM cells which may traverse the node, these cells are simply switched through the NE along with the user data cells in the cell stream of any given VPC or VCC and not copied (and thereafter processed) into the OAM processing functions of the NE. The capability to configure any given NE to either nonintrusively monitor the OAM cells or not enables considerable flexibility from the overall network management perspective. It enables the network operator to configure any of the NEs within its domain to monitor and troubleshoot for failures that may occur on the VPCs and VCCs that traverse the given NE. If a given intermediate NE is configured to monitor the OAM flows, it may then inform the NE management system of the occurrence of end-to-end AIS (or RDI) cells on any given VPCs or VCCs. This process will be described in further detail later when we discuss the detailed functional modeling of the ATM NE. The fault management functions in the NE may then alert the OSF through the TMN management interface that a defect or failure condition has been detected by the NE on certain VPCs or VCCs.

Although in Fig. 8-4 part a only one direction of the associated AIS and RDI flows has been shown for simplicity, and a defect occurring on one direction of a typically bidirectional connection is assumed, it is important to recognize that in the case of a bidirectional defect, the NEs on both sides of the defect will generate AIS cells (in either direction of the connection). In that case, the RDI cells will be sent from both ends of the VPCs or VCCs. If the defect condition involves a loss of continuity (LOC), the RDI cells will not be able to be transmitted to the connection endpoints, and are terminated at the defect location. However, even for this case, each NE along the connection path will be traversed by an AIS or RDI cell, and hence will be in a position to monitor for the occurrence of defects if configured by the OSF. Consequently, the simple end-to-end AIS and RDI mechanisms may be used to initiate alarms for either unidirectional or bidirectional defect conditions, directly analogous to the case of their use in conventional transmission system fault management.

So far we have only described the end-to-end AIS and RDI mechanisms relating to the entire span of the VPC or VCC. However, the concept of AIS and RDI OAM cells has also recently been extended for use within the connection segment maintenance entity. In this case, the segment AIS and RDI OAM cells are terminated at the segment boundaries, which, as noted earlier, are configured by the network management system (or OSF). It may be recalled that any given NE can distinguish between a VP level (or F4) segment and end-to-end OAM cell, since these have different preassigned VCI values. For the VC level (or F5) case, the segment and end-to-end OAM cells are distinguished by different codepoints of the PT field in the ATM cell headers, as defined previously. Since the segment AIS and RDI cells are always terminated at segment boundaries (and hence only exist within the OSF-defined connection segments), use of the segment AIS and RDI OAM function enables the network operator to immediately determine whether the defect (or failure) occurred within its defined segment or not. This segment-specific fault "localization" capability enabled by the introduction of the segment AIS and RDI functions was one of the main reasons for introducing segment AIS and RDI functionality in addition to the initially defined end-to-end AIS and RDI OAM cells. Other reasons related to the support of ATM layer protection switching functionality will be described below and need not concern us here. Since many network operators may be primarily concerned with the fault management of their particular administrative domain, as delineated by a maintenance segment, the segment AIS and RDI cells provide a useful tool for such configurations.

However, despite the above rationales, it may be noted that the addition of the segment AIS and RDI capability to the (already present) end-to-end AIS and RDI functionality was not without considerable controversy. Not only was there extensive discussion regarding the detailed procedures for incorporating segment AIS and RDI, but the effective duplication of the alarm inhibition aspects of the AIS function led some experts to question even the need for segment AIS and RDI functionality. From this perspective, it may be recognized that in principle a defect condition that occurs anywhere along a given VPC or VCC (or in the transmission path that contains them) affects the VPC or VCC as a whole, irrespective of how many concatenated connection segments or administrative domains it comprises. The fact that the defect condition occurs in a particular segment, thereby generating segment AIS and RDI cells within that segment, is of interest to the operator of the segment, primarily to indicate whether the defect is within or external to the segment in question. This information is essentially a fault localization attribute, whereas the primary function of AIS and RDI cells is alarm indication and inhibition, as noted earlier. As we will see later when describing the structure of the AIS and RDI cells, it is possible to include defect location (and type) information within the AIS and RDI cell payload as an optional feature. This possibility, although optional and generally not widely used to date, further weakens the case for the additional segment AIS and RDI functions, since the location of the defect indication could be more accurately communicated in this way, if required.

In addition to these more general architectural considerations regarding the need for segment-specific AIS and RDI functionality, the actual mechanisms relating to the segment AIS and RDI OAM cells also proved contentious. The underlying reasons for this discussion resulted from the recognition that for a NE to generate segment AIS cells upon receiving a defect indication, it must be configured as being part of the connection segment by the OSF. The necessity of configuring the NEs to be "segment-aware" was considered an unnecessary burden for the OSF by some OAM researchers. A number of techniques to avoid this segment configuration requirement were extensively studied but since these generally required the setting of codepoints within the AIS cell payload at either the egress or ingress of the segment boundaries on the fly, this approach introduced additional difficulties and was rejected. With the recognition that NEs could more likely be within a segment, it was assumed that the default configuration of the NE could be to generate segment AIS and RDI cells, thereby reducing the burden on the OSF. The default configuration of setting the NE within a segment may in any case be readily reset by the management system or OSF if required and generally does not constitute additional burden, since it is only one of numerous provisional parameters under management control.

It is also important to recognize that the use of segment AIS and RDI capability does not eliminate the need to generate end-to-end AIS and RDI cells as well, so that NEs outside a given segment may also be notified of the presence of a defect on the VPC or VCC. This requirement is a consequence of the architectural consideration mentioned earlier and derives from the alarm inhibition property of the AIS function. Thus, although the segment AIS (and RDI) cells resulting from a defect indication within a segment are terminated at the segment boundaries, it is necessary that an end-to-end AIS (and RDI) cell continue to the VPC or VCC endpoints to enable alarm inhibition (or indication) to occur. Consequently, it is accepted that the NE must generate both an end-to-end and a segment AIS OAM cell on detecting a defect on every affected VPC or VCC. In principle, the segment AIS and RDI functions may be viewed as "superimposed" on the conventional end-to-end AIS and RDI OAM cell mechanisms to provide the additional segment-specific AIS/RDI-based fault management capability to the OSF. In this superposition model, a given NE will generate end-to-end AIS cells (at a nominal rate of one per second) on all affected VPCs and VCCs upon receiving a defect indication. If the NE in question is configured as part of a segment by the OSF, it will also generate segment AIS cells (at a nominal rate of one per second) that will terminate at the segment boundary, whereas the end-to-end AIS cells will continue to the connection endpoint. The segment RDI cells will be generated upstream (at a nominal rate of one per second) and terminated at the upstream segment boundary. However, the end-to-end RDI cells will traverse the segment to the upstream connection termination point. Although easier to visualize for a unidirectional defect condition, the same mechanisms apply for the case of a bidirectional defect condition, for which both end-to-end and segment AIS cells will be sent downstream on either side of the defect location. It should also be noted that the general procedures described earlier relating to AIS and RDI states remain the same for both segment and end-to-end AIS and RDI mechanisms.

In the description of the fault management procedures for end-to-end and segment AIS and RDI cells above, the terms "defect" and "failure" conditions have been used somewhat interchangeably. However, in the TMN operations terminology a distinction is often made between the three types of fault management conditions termed "anomaly," "defect," and "failure." The distinction essentially depends on the duration, or persistence, of the given fault condition, and whether or not an alarm signal is triggered by the condition either to the management interface or in-band, as in the case of ATM OAM cells. An anomaly is considered to be a relatively short-lived fault condition that in general does not initiate the generation of alarms but may be logged by the management system depending upon the type of anomaly. Typically, an anomaly does not result in the loss of service to the higher layers. For the case of ATM, examples of an anomaly may be fault conditions such as loss of cell delineation (LCD), which may persist for many tens of cell times before cell delineation is recovered by the header error control (HEC) function described previously. In the SDH case, a temporary loss of frame (LOF) condition may be considered an anomaly. Such fault conditions may be relatively short-lived (e.g., several tens of microseconds for an LCD) and do not result in any serious interruption of service, and hence do not warrant the setting up of alarms in the management system. However, if the anomaly persists for a significant period of time, the fault condition may be referred to as a "defect," which may or may not result in the generation of an alarm state, depending upon the type of defect as well as the requirements imposed by the OSF. Although no precise durations are specified, the defect conditions may be considered to last in the order of several tens of milliseconds to seconds and may result in a significant interruption to the service. For the ATM case, a defect state may result in the initiation of AIS and RDI OAM cells and/or indication to the NE management system or OSF of the defect state. Finally, if the defect condition persists for a considerable duration, on the order of tens of seconds to many minutes or hours, it may be described as a "failure," which will necessitate alarm states and subsequent management system action to rectify the fault condition.

It may be noted that the duration "thresholds" between the anomaly, defect, and failure conditions are somewhat (deliberately) imprecise, and not generally subject to standardization. This is because it is generally the network operator (and OSF) that must decide upon the relative significance of these conditions in any given network application, and adjust the need for alarm surveillance and the threshold between anomaly, defect, and failure states accordingly. These will also depend on the type and quality of service being offered by the ATM network. In practice, since the majority of fault conditions are caused by physical factors such as cable cuts or disconnection of line cords on NEs and so on, the fine distinction between the defect and failure states is often blurred and the terms are used interchangeably. It is also worth recalling that in accordance with conventional OAM methodology, the occurrence of a lower-layer defect or failure gives rise to a defect indication from that particular layer (if present) and from the layer above, and so on. Thus, the defect states are propagated "upward" along the protocol stack, as it were, with each layer generating its own fault indication function and indicating the defect condition to the layer above. The detailed mechanisms whereby this process may be described functionally is discussed in subsequent chapters when considering the functional modeling for a NE. Here an example may best illustrate this general OAM guideline. If we assume that a defect such as a loss of signal (LOS) occurs at the transmission path (TP) level within the physical layer, then a TP AIS will be sent downstream, corresponding to an F3 flow at the TP level, as described earlier. A defect indication will also be communicated by the protocol stack to the VP level, in the ATM layer. As a result, end-to-end VP AIS cells (and segment VP AIS cells if this function is enabled) will be sent on all the VPCs within the affected transmission path. If any of these VPCs terminate within the network, then at the termination point of the VPC, the defect indication will be communicated up to the VC level, so that end-to-end VC AIS cells (and segment VC AIS if this function is enabled) will be sent on all the VCCs within the terminated VPC.

Focusing on the ATM layer for the moment, the process described above for the propagation of the F4 and F5 level end-to-end (and segment, if present) AIS cells on all affected VPCs and VCCs demonstrates how the AIS cells (together with the corresponding RDI cells sent upstream) enable each VPC and contained VCC connection endpoints (and segment endpoints) to become aware of a defect or failure condition somewhere along its path. If intermediate NEs along the path are configured to perform nonintrusive monitoring of OAM flows as described earlier, these NEs will also be aware of the defect state of the VPCs or VCCs traversing them. It may thus be seen that the use of the in-band (end-to-end and segment) AIS and RDI OAM cells provides an extremely powerful and rapid means of fault monitoring within the network. This capability is also sometimes referred to as alarm surveillance. On the other hand, it is also important to recognize that from the perspective of the network management system or OSF, it is not necessary to report the occurrence of each and every AIS and RDI cell event in each of the affected VCCs or VPCs resulting from a defect, since this would simply flood the OSF with excessive and redundant fault information. To avoid flooding the management system with excessive information via the NE management system interfaces, it is desirable that some level of filtering or "correlation" of the defect (or failure) states should be performed within the NE management function. In many cases, for example, it may be sufficient to inform the management system (or OSF) of the lowest-level defect (or failure) indication that caused the generation of the AIS and RDI cells on any given VPC or VCC. The detailed mechanisms whereby such a "root cause analysis" or correlation function may be performed by the management system within the NE or by the TMN is outside the scope of this text. Such "alarm filtering" or correlation mechanisms are essentially within the domain of the TMN capabilities. However, it is important in this context to visualize the relationship between the "filtered" fault information provided to the network management system by a given NE, which may indicate the lowest-level "root cause" defect or failure, and the AIS and RDI OAM cells generated periodically on all affected VPCs and VCCs. These AIS and RDI cells serve to inform the connection (and segment) endpoints, as well as any monitoring NEs, of the occurrence of a failure while inhibiting downstream and upstream alarms.

It is also of interest to note that provision of the (end-to-end and segment) AIS and RDI functionality in ATM constitutes another example indicating that ATM may be considered as both a "transmission " and a "switching" technology. As noted earlier, the concept of the AIS and RDI mechanism derives from its widespread use for fault management in conventional transmission technologies such as PDH and SDH. The extension of the AIS/RDI concepts to the ATM layer management by defining AIS and RDI OAM cells provides a packet switching technology such as ATM a powerful and flexible means of fault management, available to no other packet switching technology in use to date. In combination with the other ATM layer OAM fault management mechanisms described below, it is clear that the range of fault management tools available to ATM should enable it to provide robust transport and switching networks for the most stringent applications envisaged.

8.6 FAULT MANAGEMENT FUNCTIONS—CONTINUITY CHECK (CC)

The concept of the continuity check (CC) function at the ATM layer is somewhat analogous to (and derives from) the notion of a "keep-alive" signal used in some transmission systems. For the case of the ATM CC function, the primary objective is to enable monitoring the continuity of a given VPC or VCC continuously in order to be able to detect a loss of continuity (LOC) condition due to a defect or failure within the ATM layer functions. The basic CC OAM cell mechanism is described in Fig. 8-4 part b, which depicts a VPC or VCC traversing a number of NEs (or connecting points) as before. Dedicated CC OAM cells, whose structure will be described in further detail later, are inserted periodically into the connection and may be monitored on a continuous basis in order to ascertain whether the VPC or VCC is operating as it should, or a LOC condition has occurred by detecting the absence of the CC OAM cells on a given VPC or VCC. Although the CC OAM cells have been defined for both end-to-end and segment level continuity monitoring purposes at both the VP and VC levels, we consider here the end-to-end case, since the basic CC procedures are essentially similar for both cases.

As noted in Fig. 8-4 part b, there are two distinctly different mechanisms that have been defined, as options, for the generation of the CC OAM cells on any selected VPC or VCC. In option 1, a CC OAM cell is only sent by the source when no user data cell has been sent for a period of nominally 1 second, whereupon the CC cells are sent at a nominal period of one CC cell per second until the user data cells are available to be sent by the traffic source. Thus, in this mechanism of CC cell generation (so-called option 1 CC) the cell stream in the VPC or VCC will consist of either user data cells or, in periods of inactivity longer than 1 second, of CC OAM cells at a rate of 1 CC cell per second. For the so-called option 2 of CC cell generation, the CC cells are inserted into the VPC or VCC cell stream at a nominal rate of one CC cell per second, independent of whether or not user data cells are present on the connection. Thus, in the so-called option 2 mechanism of CC cell generation, the cell stream on the selected VPC or VCC will comprise the user data cells interspersed with the CC cells at a nominal rate of 1 CC cell per second. It should be noted that in this case, the independent insertion of the CC OAM cells needs to be performed without in any way violating the traffic contract for the VPC or VCC in question. Thus, for example the intercell spacing between the adjacent user data cell and the CC OAM cell should not be less than that allowed by the peak cell rate (PCR) and the associated cell delay variation tolerance (CDVT) specified by the traffic contract to avoid the inadvertent discard of a user data cell by the UPC or NPC function. Despite this traffic scheduling aspect, most current implementations of the CC cell generation mechanism use the option 2 method, since it effectively decouples the CC cell generation and insertion from the user traffic state.

Whichever option is chosen for the generation of the CC OAM cells by the traffic source, if the VPC or VCC endpoint does not receive any user data cells or CC cells for an interval of 3.5 $\pm$ 0.5 sec, it declares an AIS state due to a loss of continuity (LOC) defect condition. For the case of the segment CC cells, if the segment termination point (e.g., the NE) does not receive any user data cells or CC cells for an interval of 3.5 $\pm$ 0.5 sec, it declares a LOC defect and consequently transmits AIS cells downstream toward the connection endpoint, as described previously. In this case, the VPC or VCC endpoints then transmit RDI cells upstream along the bidirectional connection, as described previously. It should be noted from Fig. 8-4 part b that for a bidirectional VPC or VCC, the CC cells are sent in both directions of the connection independently of each other, in order to detect unidirectional defects or failures. As for the case of the AIS and RDI cells, the intermediate NEs along a connection path may (or may not) be configured to nonintrusively monitor the CC cells flowing in the VPC or VCC in order to detect the LOC state of the VPC or VCC selected. It is important to note the difference in behavior between the VPC or VCC endpoint and the segment endpoint with respect to the segment CC cells. If the segment endpoint detects a LOC condition due to absence of both user cells and segment CC cells (for the interval given above), it must generate AIS cells in accordance with the AIS and RDI procedures described previously. Similarly for the end-to-end case, the LOC initiated AIS states requires the generation of RDI cells upstream.

The difference in interval between the release of an AIS state after the absence of AIS cells (2.5 ± 0.5 sec) noted earlier, and the declaration of AIS state due to absence of CC cells (3.5 ± 0.5 sec) results from the requirement in option 1 of the 1 sec interval before sending CC cells. The actual interval values specified by the OAM standards [8.5] reflects a broad compromise between the need to establish the persistence of a fault condition and the need to indicate the fault condition as early as possible. Since both the AIS/RDI and CC OAM cells are required to be generated at one cell per second per connection, it is clear that the temporal granularity is of the order of 1 second. This is itself a tradeoff based on traditional circuit availability considerations, whereas the tolerance of ± 0.5 sec for the intervals is included to allow for the OAM cell insertion process in (potentially) a large number of VPCs and VCCs, which may depend on specific implementations.

The number of VPCs or VCCs on which the end-to-end or segment CC function may be performed depends on the network operator requirements and the network applications being supported by the VPCs or VCCs. It was initially suggested by some operations specialists that the CC function should be used on all the ATM VPCs and VCCs, or at least on all the permanent virtual circuits (PVCs) within the network. However, the resulting complexity and OAM burden this would entail led to the recognition that the CC function should be activated only on a selected number of VPCs or VCCs, depending on the network operator's choice. For this purpose, a number of options are available for implementation of CC function activation and deactivation. One option is to activate (or deactivate) the CC function on the selected VPCs or VCCs as required by means of the appropriate command via the management interface to the NE, i.e., by means of the TMN procedures. This approach is generally the preferred mechanism to date. A variation on this option (which depends on the specific implementation of the CC mechanism) is to automatically initiate the CC cell generation process upon (typically PVC) connection establishment, essentially as part of the connection establishment procedures for the selected PVCs. The other option for the activation/deactivation of the CC function on a selected number of VPCs or VCCs is to use the dedicated (in-band) activation/deactivation OAM cells listed in Fig. 8-3 and further discussed below. However, the use of the activation/deactivation OAM cells, although intended for this purpose, involves the additional complexity of incorporating this function as well, and hence has not seen widespread support to date.

The use of the CC OAM cells, whether on some selected or on all connections, clearly provides a powerful fault management capability for network applications in which it is required to continuously monitor the connectivity state of the VPC or VCC. Despite this seemingly obvious benefit, it must be admitted that the CC OAM function has been somewhat controversial and has yet to find widespread use within existing ATM networks. Initially introduced as an optional function within the "OAM tool kit" of fault management functions defined in ITU-T Recommendation I.610 (1995 version), the CC OAM function was not adopted by the ATM Forum's Implementation Agreements Specifications on OAM capabilities (UNI [6.2] and B-ICI [6.1], as well as the NM M4 specification [8.8] on management requirements). It was generally argued that the majority of defects or failures (typically up to 90% or more) occurred as a result of gross physical failures such as inadvertent cable cuts, circuit pack disconnections, etc., which would in any case result in the sending of AIS and RDI OAM cells. For such events, the CC cells were viewed as redundant, since the AIS and RDI states could be used as indication of the LOC condition. Thus, in the majority of fault situations, the CC function was viewed as unnecessary OAM overhead that involved additional implementation complexity for little perceived added value. It will be recognized that the sending of the periodic CC cells on any given VPC or VCC essentially only allows the detection on a continuous basis of ATM layer function defects that could result in the LOC condition. Underlying physical layer defects will not be detected, unless they lead to an ATM layer failure.

More recently, however, there has been a resurgence of interest in the CC function, driven primarily by the need to define more comprehensive "in-service" OAM tools for estimating the availability state of "mission critical" or "premium grade" PVCs. Although this aspect of the use of the fault management OAM functions need not concern us at this stage and will be discussed below, it is clear that the use of the (segment and end-to-end) CC function is likely to be more widely adopted as ATM service providers evolve to support more robust and mission-critical network applications. In some existing ATM networks, notably those based on ETSI specifications, the CC function is used for continuous monitoring of VPC-based PVCs for so-called premium customers, thereby providing a form of "value added" capability. In addition, as will be pointed out later, the CC function may also be used on selected VPCs in conjunction with continuous performance monitoring (PM) capability to provide maximum supervision of the critical resources of the network.

8.7 FAULT MANAGEMENT FUNCTIONS—LOOPBACK (LB)

The ATM layer loopback (LB) OAM function shown schematically in Fig. 8-4 part c is essentially analogous to the existing out of service loopback fault-finding techniques widely used in telephony and data communications networks. As such, ATM LB OAM function constitutes the most popular fault management function for both out-of service and in-service use. The LB function utilizes dedicated loopback OAM cells (whose structure and semantics are described in detail later) for both VP (F4) as well as VC (F5) levels. LB OAM cells may be used for both end-to-end as well as segment use by inserting the end-to-end or segment LB OAM cell into the cell stream, which is then "looped back" at either the (VP or VC) connection or segment endpoint as appropriate. This procedure enables the operations system to obtain a simple and rapid "on-demand" verification of the connectivity of the VPC or VCC. It should be noted that unlike the CC OAM cell, which is generated periodically (at 1 cell per second), the LB OAM cell is only sent under the control of the NE management system, or by the end user management system, on an "on-demand" basis. If the LB cell is not returned to the originating point in (nominally) 5 seconds, the loopback test is considered to be unsuccessful. The result of the LB test may then be reported to the network management system for further action. It may be noted that the loopback procedure may be repeated if required in order to confirm the initial diagnosis. The interval between the successive loopback tests should be at least 5 seconds to allow for the result of the previous LB test to be determined.

As listed earlier in Fig. 8-3, the LB function may be used for a number of fault management capabilities including:

1. In-service connectivity monitoring of any given VPC or VCC under management system control on an on-demand basis.
2. Out-of-service, or preservice, connectivity verification of any given VPC or VCC under management system control. Thus, the LB procedures may be used to confirm the correct connectivity state of a PVC before it is released for service to a customer. This function is probably the most widely used aspect of the LB function in practice.
3. For fault localization purposes, by sending LB OAM cells repeatedly to different NEs along a connection path in order to localize the defect. As will be described in detail below, the LB OAM cells include a loopback location identifier (LLID) capability as an option, which allows for the looping back of the LB cell from any given NE identified by the particular LLID value (if this function is enabled by the management system).

All the above applications of the LB OAM cells may be performed for either end-to-end or segment (if defined) configurations in which the distinction between the end-to-end and segment LB OAM cells is made in the same way as for the other OAM cells, as described earlier. It may also be observed that although the LB procedures are generally intended to verify connectivity on demand by management action, i.e., typically a "one-shot" process, it may also be used on a periodic basis if required to monitor continuity of a VPC or VCC continuously. The period may be programmable, depending on the network application. In this sense the LB OAM function may be made to emulate the CC function described earlier, but since the LB process is significantly more complex than the relatively simple CC OAM function, periodic use of LB cells may become processing-intensive if not constrained to long periods and relatively few connections per interface.

A number of different scenarios for the network applications of LB OAM function have been identified in ITU-T Recommendation I.610 [8.5] and are shown schematically in Fig. 8-5. These indicate the potential flexibility and power of the LB process and explain the popularity of this OAM tool for fault management purposes. It is also of interest to note that the ATM Forum

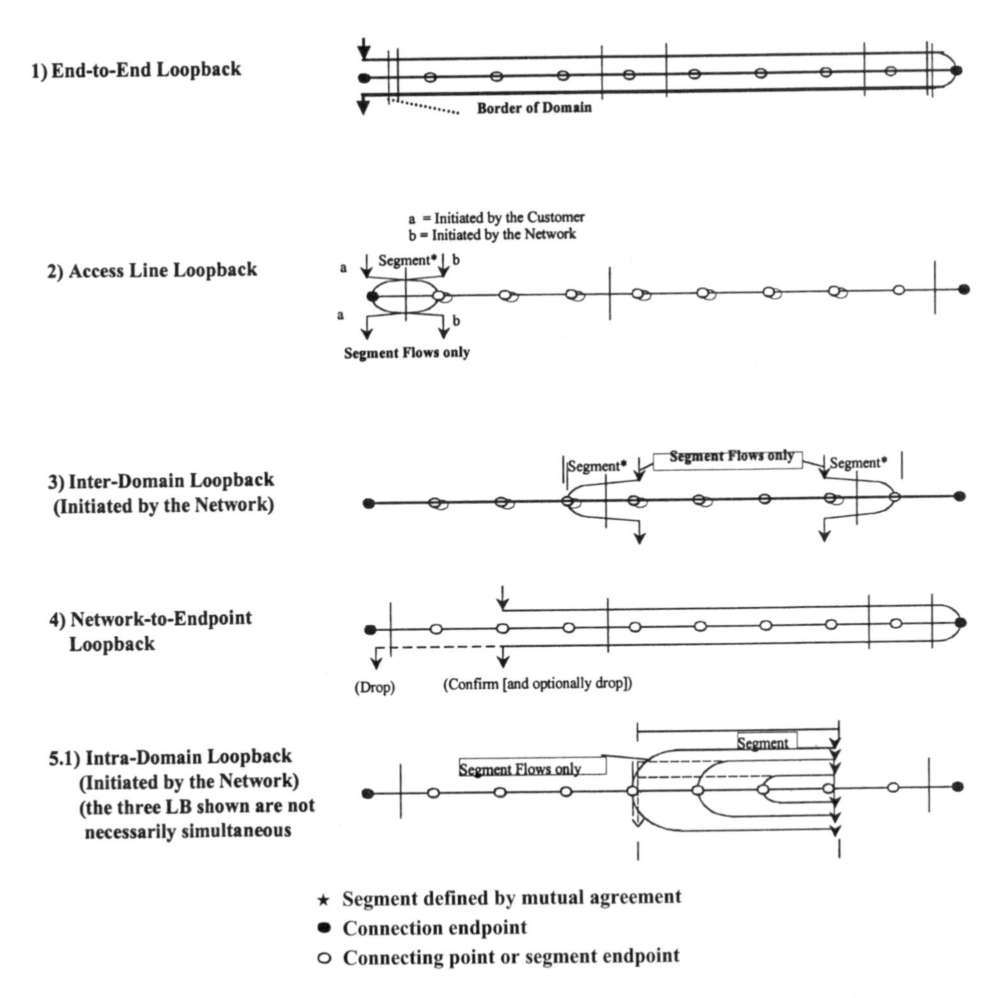

Figure 8-5. Network applications of loopback. (Derived from ITU-T Recommendation I.610)

specifications adopt the LB function as the essential OAM tool for ATM networks. As shown in Fig. 8-5, five general categories of network applications of loopback procedures have been identified. These include:

1. End-to-End Loopback. In this simplest of cases, end-to-end LB OAM cells are inserted at the (VP or VC) connection endpoints (in user or network domain) and looped back from the remote endpoint to verify connectivity as required. This scenario is completely analogous to the familiar so-called "ping" test widely used in IP-based data networking.
2. Access Line Loopback. In this application, segment LB cells may be inserted by either the end user or the network and are looped back by the first ATM NE (or customer equipment) in order to ascertain the status of the access VP or VC link. It should be noted that in order to perform this test, an access segment will need to set up first by the (user or network) management system by mutual agreement across the two administrative domains. This may not always be possible.
3. Interdomain Loopback. In this application, segment LB cells are inserted into selected VP orVC connection cell stream by the (management system) of one network operator and looped back by the first ATM NE in the adjacent network operator domain. In this case, the segment across the administrative domain boundaries will also need to be set up prior to the test by mutual agreement between the respective operators.
4. Network-to-Endpoint Loopback. In this application, end-to-end LB cells are inserted by a network operator into selected VPCs or VCCs and are looped back at the remote connected endpoint to verify connectivity. It should be noted that there are two options in the treatment of the looped-back OAM cells at the originating NE. In one option, the originating NE may simply confirm that the LB cell has been returned while allowing the cell to continue on to the endpoint, where it will be discarded. In the other option, the originating NE may discard the returned LB cell after confirming that it is the LB cell it generated for the test. The confirmation mechanism is described below when we consider the LB cell structure and semantics.
5. Intradomain Loopback. In this application, two possible cases may be considered, as shown in Fig. 8-5, depending on where the segment LB cells are inserted. The intradomain LB application uses segment LB cells that may be inserted in the cell stream of selected VPCs or VCCs either at segment endpoints or at intermediate NEs, and are looped back at the segment (or connection) endpoints. For fault localization, the LB cells may be looped back from the various NEs within the segment in repeated tests, using the loopback location identifier (LLID) option, as noted earlier, to identify the target NEs for each test.

The LB OAM procedures and their network applications described above may be viewed as simple "one-shot" tests to confirm connectivity. In order to perform fault localization, the LB OAM procedure has to be performed repeatedly by the management system by inserting LB OAM cells with different LLID values to identify the target NEs along the VP or VC connection path. Since this process may be relatively time-consuming and somewhat cumbersome to perform, the basic LB procedures have recently been extended to enable the so-called multiple loopback technique (MLT), which significantly simplifies fault localization as well as provides a rudimentary path trace capability if required. The multiple loopback technique is depicted in Fig. 8-6 and has been described in [8.5]. The MLT procedures assume that the network management system (or OSF) has provisioned each NE along the connection path in question with its so-called "connection point identifier" (CPID) value, which is essentially the same as the LLID value used in the case of the "simple" LB procedures described above. However, when invoking

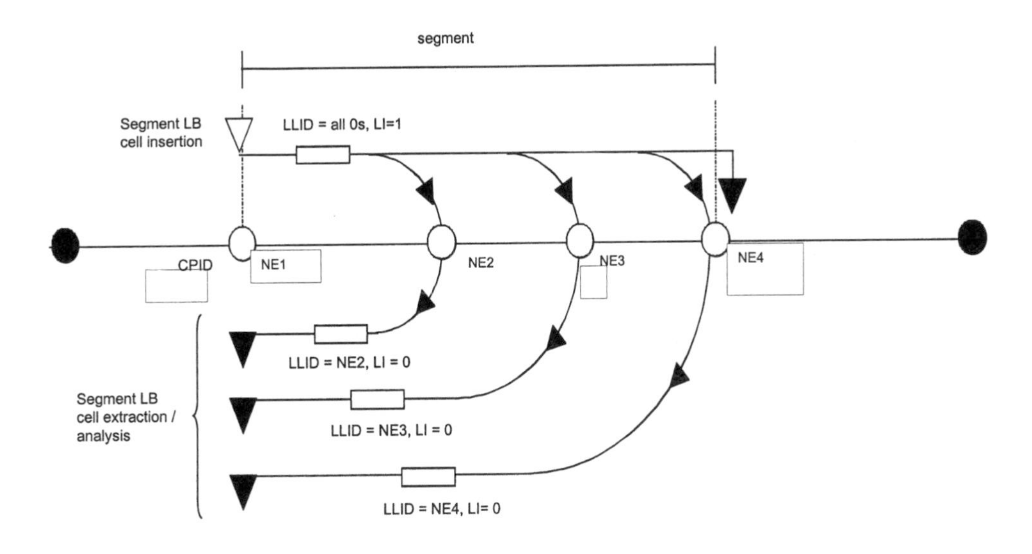

Figure 8-6. Multiple loopback technique.

the MLT procedures, the OSF initiates the insertion of a (segment or end-to-end) LB cell, which is looped back (or returned) from each MLT enabled NE along the connection path. Before returning the MLT OAM cell, each NE along the (VP or VC) connection path "writes" its CPID value into the LLID field of the LB OAM cell. Thus, the NE that originates the MLT LB OAM process receives a number of LB cells returned to it, each containing the LLID (= CPID) value of the NE that looped back the MLT OAM cell.

If we assume that the OSF or network management system knows the CPID values corresponding to the given VPC or VCC along which the MLT procedure has been performed, the returned MLT LB OAM cell contents may be analyzed and compared with the expected NE identifier to verify the connection path. In this way, the MLT procedures provide the OSF with a basic "path trace" or path verification capability, which may be very useful in configuring PVCs in large networks. In addition, fault localization may be rapidly and automatically performed by searching for missing (or wrong) LLID values in the returned MLT LB OAM cell contents. From the perspective of fault localization, it may thus be seen that the MLT extension essentially eliminates the need to perform multiple "single-shot" LB tests to individual target NEs, with a subsequent wait of at least 5 secs before declaring whether the LB test was successful or not. By sending one MLT LB OAM cell and analyzing the contents of the multiple returned LB cells for the LLID values, extremely rapid fault localization (and path verification) may be performed by the OSF or management system, assuming, of course, that it implements the required MLT software. Although the need for the additional management system (or TMN) involvement in the MLT process, as for the simple LB procedures, has been criticized by some as adding unnecessary complexity to the LB function, it seems clear that the resulting automation of the fault localization and path verification capability is useful for configuration management purposes in public carrier networks. It may also be noted that whereas in principle the MLT procedures may be used for both the end-to-end and segment LB OAM cells, in practice they will likely find use in segment-oriented network applications, since many network operators may not welcome the path trace aspects of the MLT being activated by end users.

As pointed out earlier, the simple LB OAM procedures are widely utilized in existing ATM networks for both in-service and out-of-service connectivity diagnostic purposes and have been embedded in the ATM Forum implementation specifications. The more recently standardized MLT procedures have so far only been adopted by the ETSI specifications and may become more widely deployed as more sophisticated ATM network management capabilities are developed for the next generation of ATM equipment. In either case, the close relationship between the capabilities of the in-band OAM cell flows and the NE management system as well as the OSF should be noted. The use of one without the other does not optimize the inherent extensive OAM capabilities designed into ATM technology from the outset.

8.8 PERFORMANCE MANAGEMENT OAM FUNCTION

The use of dedicated performance management (or monitoring) OAM cells (PM OAM cells) allows for in-service and in-band performance monitoring of any given ATM VPC or VCC. Performance monitoring (PM) of the selected VPC or VCC refers to the capability for estimating specific ATM layer performance parameters such as the cell loss ratio (CLR), cell transfer delay (CTD), or cell delay variation (CDV), as well as errored cells caused by transmission error bursts or other causes. The underlying concepts used for the performance management of ATM cell streams derives from those used for performance monitoring in conventional transmission systems, although there are inevitable differences due to the packet transfer mode nature of ATM. In conventional transmission systems, the transmission frame provides a natural unit over which error measurements may be made using mechanisms such as bit interleaved parity (BIP) or cyclic redundancy check (CRC) for bit error detection. However, for cell error measurements within the ATM layer, there is no concept of a transmission frame, so it is necessary to specify a "block" of user data cells, termed a performance monitoring block (PM block), over which an error detection mechanism may be used to estimate cell errors, or cell counts may be used to determine cell losses. As shown schematically in Fig. 8-7 part a, the PM block is delineated by inserting dedicated PM OAM cells after every N user data cells, where N denotes the number of cells in the performance monitoring block. The PM OAM cells contain the performance management information calculated over the block of user data cells. The detailed structure and semantics of the PM OAM cells will be described later; here we focus on the essential PM procedures.

The actual size of the PM block, N, may be selected from a range of values varying from 128, 256, 512, 1024 . . . up to 32,768 cells (in multiples of 2). These block sizes denote the nominal values that may be selected during the PM activation process, since a wide margin of up to 50% of the value of N is also permissible for the actual block size between any two PM OAM cells delimiting a block. The original intention of allowing a relatively wide margin in the actual value of the PM block size N was to avoid the "forced" insertion of the PM OAM cells after, say, exactly N = 256 user data cells had been passed and there was no available slot for the insertion of the PM OAM cell in the cell stream. The forced insertion of a PM OAM cell under these conditions would result in perturbation of the user cell traffic characteristics by introducing some additional cell delay variation (CDV). However, subsequent experience has shown that, in general, the additional CDV introduced by the PM process is relatively insignificant, and that the PM OAM procedure may be somewhat simplified by simply inserting the PM OAM cells after exactly N = nominal block size value has been counted, irrespective of whether a cell slot is available or not. Consequently, the PM OAM cell insertion procedures now allow for the forced insertion of the PM OAM cells after exactly N = nominal block size as an option. The selection of the nominal block size in the range 128 up to 32,768 depends on the peak cell rate (PCR) of the VPC or VCC undergoing the performance measurements.

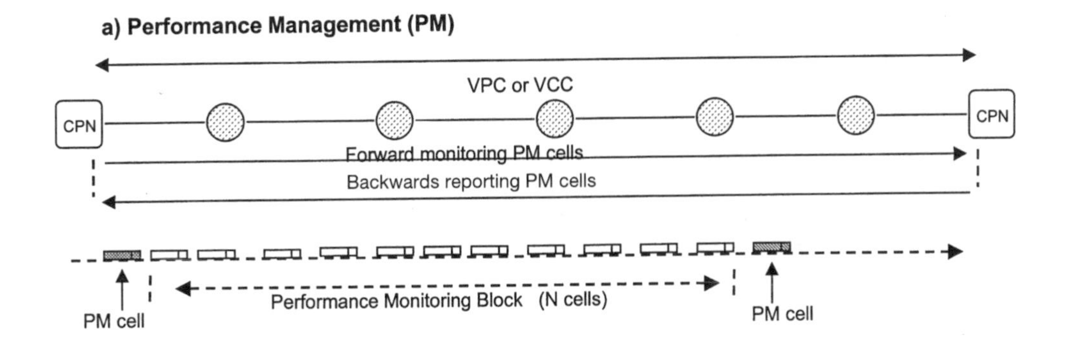

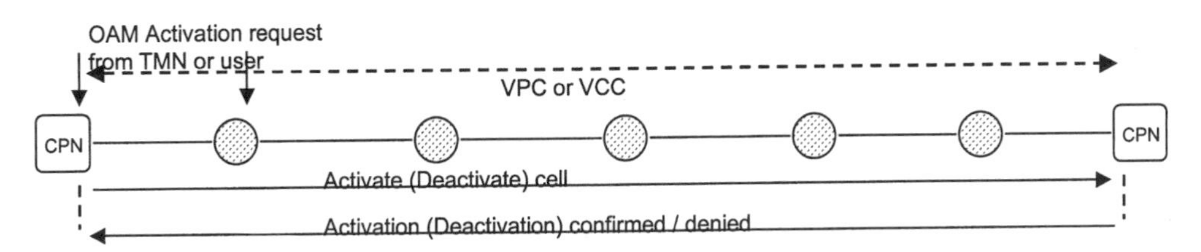

Figure 8-7. Performance management and activation/deactivation functions.

As described in ITU-T Recommendation I.356, which deals with ATM layer performance aspects, the value of N is chosen based on the broad criterion that between 12 and 25 blocks should be monitored per second to provide a reasonable measure of the connection performance.

Two types of PM OAM cells have been defined as part of the overall PM OAM procedures. These are the forward monitoring PM cells (FPM) and the backward reporting PM cells (BR), corresponding to the two directions of a typical bidirectional ATM VPC or VCC. When the forward monitoring and the backward reporting PM OAM cells are used together as a related pair, the procedures enable the collation of the performance monitoring data at a single NE along the VPC or VCC, as may be inferred from Fig. 8-7 part a. In this case, the "paired" BR PM OAM cell carries the results of the FPM measurements to the originating NE, which may constitute a segment endpoint (for segment PM procedures) or connection endpoint (for end-to-end PM procedures). Since in principle either end may initiate the paired PM protocol, involving both the forward monitoring PM OAM cells together with their paired backward reporting PM OAM cells, it may be seen that both directions of a bidirectional VPC or VCC may have FPM and BR cells flowing in this case. However, if it is not required to collate the results at one NE, or the management system may be used to relate the performance measurements in either direction, then only the forward monitoring PM OAM cells may be used to measure either direction of the VPC or VCC, thereby simplifying the overall PM process.

The PM OAM cells may be used to detect:

1. Errored cell blocks
2. Loss or misinsertion of cells
3. Cell delay (as an option)

The detection of the errored cell blocks is obtained by performing a BIP-16 calculation over all the cells in the PM block and comparing the result with a similar computation at the receiver end, analogous to the concept of detecting bit errors in conventional transmission systems. The details of the BIP-16 semantics are given later in describing the PM OAM cell structure. Cell loss or misinsertion may be detected by including a count of the cells transmitted within a given block in the PM OAM cell, followed by comparison with the number of cells received per block at the (segment or connection) endpoints. It may be recognized that cell loss (or misinsertion) has no direct counterpart in transmission systems performance management and is a specific attribute of the packet mode nature of ATM. The same is also true of performance parameters such as CTD and CDV, which may be estimated by including a time stamp value within the PM OAM cells. However, the use of a time stamp is considered an option in the PM OAM procedures at this stage. Consequently, delay measurements may be typically performed by out-of-service test procedures involving the insertion of "test" cells generated by external test equipment. Although such external test procedures are always possible, and the use of standardized test cells has been defined in ITU-T Recommendation O.191 [8.9], we will focus here on the in-band, in-service capabilities afforded by the ATM OAM protocols. It is useful to recall that the results of such in-band, in-service PM measurements will need to be processed and logged by the NE management system, and may then be reported to the (external) network management system or OSF, using either the standard TMN interfaces or proprietary management interfaces.

It may be recognized, even after the brief description above, that a significant amount of per-VPC or -VCC processing is required by the PM OAM procedures. Consequently, it is not envisaged that performance monitoring is continuously enabled on all the active VPCs or VCCs in the network, since this may potentially require unrealistically large amounts of processor capability within the NEs as well as the supporting OSF. In practice, it is generally assumed that a network operator will select a (relatively small) number of VPCs or VCCs on which to activate the PM OAM procedures for a limited duration, or initiate the mechanism in the event of customer complaints on specific PVCs. Thus, it is assumed that the PM OAM protocols may be activated and deactivated on a selected number of VPCs and VCCs, generally under management system control. For segment PM OAM procedures, the activation and deactivation of the PM OAM cell flows and their processing will occur at the NEs that delimit the OAM segment. Intermediate NEs may nonintrusively monitor PM cell flow, if required. For the end-to-end case, the end-to-end PM OAM cells will be generated and processed at the VPC or VCC endpoints. In either case, it is important to note that sufficient bandwidth is allocated on the VPCs and VCCs to accommodate the additional segment and end-to-end PM OAM cell flows without violating the traffic contracts in effect on the selected connections. The maximum number of VPCs or VCCs on which the PM OAM procedures may be simultaneously activated is highly implementation-, as well as network-application-dependent, and may typically range between zero and several tens of connections per physical interface.

8.9 ACTIVATION AND DEACTIVATION PROCEDURES

It was pointed out above that the performance management OAM procedures would need to be activated and deactivated on selected VPCs or VCCs in order to enable the network operator to estimate the performance of a given connection. In addition, activation and deactivation procedures may also be used for the continuity check (CC) fault management function described earlier. Consequently, a common set of activation and deactivation procedures, as well as a dedicated OAM cell for activation/deactivation initialization, has been specified by the ITU-T in Recommendation I.610 [8.5] for use with either PM or CC or both OAM functions. The activation and deactivation procedures are summarized in Fig. 8-7 part b, which also indicates the use of the dedicated activation/deactivation OAM cells between connection (or segment) endpoints in order to initialize the activation process for either or both PM and CC OAM functions. The PM and CC OAM functions may be activated on selected VPCs and VCCs at the connection establishment or at any time after that. It is assumed here that the OAM process being activated is primarily for the case of semipermanent virtual circuits (PVCs), since the use of such OAM functions for the (relatively) short-lived, on-demand switched virtual circuits (SVCs) set up by signaling procedures remains questionable, although in principle they are not precluded.

There are two basic options available for the activation and deactivation procedures for either PM or CC OAM functions. The activation may be initiated entirely by the network management system (the TMN) or OSF via the management interfaces to the NEs in question, or for the end-to-end PM or CC case, entirely by the end user of the network. An initialization procedure needs to be used in order to coordinate the beginning or end of transmission of the PM or CC (or both) OAM cell flows as well as the selected PM block sizes and direction of the PM measurement process. The two options available for this initialization procedure are:

1. The use of the dedicated activation/deactivation OAM cells
2. Entirely through the management system of the NEs involved by using the management interface (TMN-related or proprietary)

If the option using the activation/deactivation OAM cells is used for the initialization process, an "activate" OAM cell is transmitted from the initiating (segment or connection) endpoint to the far end to condition that NE for start of the PM or CC OAM cell arrivals. The remote NE then acknowledges the initiation process by either confirming or denying the required OAM function (PM or CC), by returning an activation/deactivation OAM cell with the appropriate message type (i.e., activation confirmed or denied). This use of the activation/deactivation OAM cells for initialization of the PM or CC function may be viewed as somewhat analogous to the "handshake" procedure commonly used in data communications. However, it will be noted that even with the use of the activation/deactivation OAM cells, the network management system or OSF is required to initiate (and process) the PM OAM cell mechanism via the management interface to any given NE.

This inevitable requirement for the management system involvement in the PM process implies that the second option for the initialization procedures, namely, that of using the management interfaces entirely for the initiation, as well as initialization of PM and CC, remains the more generally preferred option. It is thus reasoned that since activation/deactivation of the PM and CC OAM functions requires suitable messaging via the NE management interfaces to the OSF, the additional complexity involved in utilizing dedicated activation/deactivation OAM cells just for the initialization process is not justified for typical applications. It is primarily for this reason that the activation/deactivation OAM cells protocols, although available as an option in the ITU-T standards, are generally not used for either PM or CC activation. The more generally preferred procedures in current use utilize the NE management interfaces entirely for the whole PM or CC activation and deactivation process, thereby eliminating the need for the NEs to implement the activation/deactivation OAM cell generation and processing. Although this does simplify the overall performance management (and CC) functionality in terms of the OAM cell processing, it has been argued by some network designers that the use of an in-band activation/de-activation capability may have advantages in network scenarios involving multiple administrative domains between which there are no external management interfaces (e.g., the so-called TMN "X" interface between two OSFs). In such configurations, the use of OAM cells for rapid in-band initialization of PM or CC functions across multiple domains may prove advantageous. However, despite such potential uses of the activation/deactivation OAM cell function, it must be admitted that few implementations exist. It may also be noted that the activation/deactivation OAM cells have so far not been adopted in the ATM Forum Network Management (NM) specifications [8.8].

8.10 OAM CELL FORMATS AND CODINGS

The per-VPC or -VCC ATM layer OAM procedures described so far have all assumed the use of function-specific dedicated OAM cells that flow along the connection path together with the user cells to communicate the OAM information between NEs. We now describe the detailed structure and semantics of these OAM cells. The OAM cell structure is the same for both segment and end-to-end OAM cells, and it may be recalled that segment and end-to-end OAM cells are distinguished by the use of different preassigned VCI values (for F4 OAM flows at the VP level). In fact, these preassigned ATM header codepoints enable any NE along the connection path to distinguish the OAM cells from the user data cells on any given VPC or VCC. As shown in Fig. 8-8, all the OAM cell types are based on a common OAM cell format. It will be noted that the first octet after the normal (five octet) ATM cell header is split into two fields of four bits each. The first field specifies the "OAM type" according to the codepoints listed in Fig. 8-8. Thus, the OAM type field coding of 0001 identifies the OAM cell as a fault management OAM cell, whereas the coding 0010 indicates a performance management OAM cell, and so on. The coding of the second field, termed "function type," then indicates the specific function within the OAM type category. For example, the function type coding "1000" with the fault management OAM type (0001) signifies that the cell is a loopback OAM cell, and so on, as listed in Fig.

Common OAM cell Format

ATM Header (5 octets)	OAM Type (4 bits)	Function Type (4 bits)	Function Specific Fields (45 octets)	Reserved (6 bits)	EDC (CRC-10) (10 bits)

OAM Type and Function Type Identifiers

OAM Type	Coding	Function Type	Coding
Fault Management	0001 0001 0001 0001	AIS RDI Continuity Check Loopback	0000 0001 0100 1000
Performance Management	0010 0010	Forward Monitoring (FPM) Backwards Reporting (BR)	0000 0001
APS Coordination Protocol	0101	Group Protection Individual protection	0000 0001
Activation / Deactivation	1000 1000	Performance monitoring Continuity Check	0000 0001
System Management	1111	Only defined by ATMF Security Spec.	

Figure 8-8. OAM cell format and coding.

8-8. The so-called system management OAM type also listed here will be described later and need not concern us at this stage. A dedicated OAM cell type for automatic protection switching (APS) coordination protocol (coded 0101) is also defined for the exchange of protection switching information. Use of this OAM cell type is described in Section 8.17 dealing with ATM layer protection switching.

It will be recognized that the combination of the OAM type and function type fields corresponds essentially to a "message type" function (or protocol control information, PCI) for the OAM cells, analogous to that of signaling or routing messages discussed earlier. However, it is important to note that unlike signaling or routing message sets, the ATM layer OAM cells do not utilize a type–length–value (TLV) coding structure. Instead, the OAM cell structure is based on a fixed-field coding methodology, which although not as flexible as the TLV method, lends itself more readily to very high speed processing in hardware implementations. Since the OAM cells may need to be processed within the NE as part of a very high speed cell stream, OAM cell processing may often need to be implemented in hardware. Following the function type field, the so-called "function-specific" fields include the PCI specific to the information the OAM cell is transporting, and these are described below for the various OAM function types such as loopback, AIS/RDI or forward monitoring, etc. It will be noted that 45 octets are available for such function-specific protocol control information. The last two octets of the common OAM cell format are divided into two fields: a "reserved" field of 6 bits and a 10 bit field which includes an error detection code (EDC) that operates over the whole OAM cell payload excluding the EDC field. The selected error detection algorithm is a cyclic redundancy code (CRC) of 10 bits based on the generator polynomial of $G(x) = 1 + x + x^4 + x^5 + x^9 + x^{10}$. The reserved field is coded as all zeros, and is intended for additional functions that may be defined in the future. The similarity between the common OAM cell format described above and the resource management (RM) cell format described earlier should also be noted. In effect, both the RM cells and the OAM cells considered here may be viewed as specific cases of dedicated management cells to be used for the in-band, per-VPC or -VCC, transport of the relevant management information. These should be contrasted with the "common channel," or out-of-band, signaling messages (or routing information messages) that were described earlier for the exchange of signaling and routing information between NEs. In addition, as described in further detail subsequently, it is also possible to define out-of-band, dedicated VPCs or VCCs to exchange network management information between NEs, corresponding to the network management interfaces either between NE and OSFs, or between OSFs, or between NEs.

8.11 FUNCTIONS SPECIFIC TO THE AIS/RDI AND CC OAM CELL

Having described the common OAM cell functions and format, we turn now to consider the function-specific fields for the OAM cells, starting with the AIS and RDI OAM cells. The purpose of the AIS and RDI cells was described earlier, and it may be recalled that such cells may occur for either end-to-end or segment F4 and F5 flows. As shown in Fig. 8-9, there are two fields currently specified for the AIS and RDI cells, and it should be noted that both are specified as optional functions (in [8.5]). The one octet "defect type" field may be coded to indicate the cause or nature of the defect that initiated the generation of the AIS cell, thereby enabling downstream NEs or the connection endpoint to monitor the cause of the defect. An example may be to indicate whether the defect occurred within the physical layer functions or within the ATM layer functions. For this purpose, a number of coding schemes to catalog defect types have been proposed and discussed as to their validity and usefulness, but it must be admitted that no standardized codepoints for the defect type option have been defined to date. This is primarily be-

Function-Specific Fields for AIS/RDI FM Cell

Defect Type (optional) 1 octet	Defect Location (optional) 16 octets	Unused octets (coded 6AH) 28 octets

The AIS/RDI cell function-specific fields are:

1. Defect Type (1 octet): As an option this is used to indicate the nature (cause) of the defect (e.g., defect in VP or VC level, or lower-layer defect, etc.).
2. Defect Location (16 octets): As an option this field may be used to indicate the location of the defect. For AIS, the location identifier corresponds to generation point. For RDI, the identifier is same as that for the initialing AIS cell.

Note: When not used the fields are coded 6AH.

Figure 8-9. AIS/RDI cell.

cause the overall utility of the defect type function has been questioned, since it has been pointed out that the defect cause information may be made available to the management system via the management interface of the NE in question. In addition, it may not be clear how the defect type information may be of use to the downstream NEs or the segment or connection endpoints of the affected VPCs and VCCs. As a consequence of these unresolved questions, the defect type option in the AIS/RDI mechanism has been limited to essentially proprietary implementations, and is not in widespread use in existing ATM networks.

The 16-octet "defect location" field, whose use is also indicated as optional, is intended to convey information as to the identity, or location, of the NE that generated the AIS OAM cell on detecting a defect or fault. The use of the defect location function enables a downstream monitoring NE to obtain information regarding the localization of a fault along the connection path. Similarly, upstream NEs which may also be (nonintrusively) monitoring the OAM flows may be able to localize the fault, since the returning RDI OAM cell would also include the same value of the defect location identifier as in the AIS cells. The use of the optional defect location function for such fault localization purposes has also been somewhat controversial, although it has received a greater measure of support recently, allowing partial standardization of the location identifier coding, as will be described below. However, as for the defect type function considered earlier, a number of questions have been raised concerning the need for the defect location function. It is argued that the management system or OSF may obtain this information directly through the NE management interface or via the TMN, so that the transport of the defect location in-band along affected VPCs or VCCs is redundant. In addition, some network operators are concerned about the security aspects of sending defect location information along connections that may traverse several administrative domains. It is generally regarded that such information is not of relevance, or may even be misused, by neighboring (and often competitive) administrative domains, and hence should only be restricted to the domain within which the fault is detected. Since in any case the actions required subsequent to fault detection and localization are under the control of the OSF or the network management system of any given domain or maintenance segment, differences in operational procedures are only to be expected.

Despite these questions, a location identifier format that may be used in conjunction with the optional defect location function has been defined in Recommendation I.610 [8.5], although its primary intent is for use with the multiple loopback technique (MLT) described earlier. This location identifier format includes fields for country code and operator code, as well as an operator's specific or proprietary code to enable an element of security in respect to fault localization across multiple administrative domains. It may be noted that, as for the case of the optional defect type field coding, the defect location coding in the RDI cell should be the same as for the AIS cell causing the generation of the RDI cell. If the defect type or defect location functions are not being used, the corresponding fields are coded in the default Hexadecimal 6A (also-called 6AH) to denote that the field is unused, in keeping with the general coding rules for all OAM cells. These coding rules require that unused octets are coded as Hexadecimal 6A, whereas unused bits in incomplete octets are coded as all zeros (e.g., the reserved field in the common OAM cell structure).

For the case of the continuity check (CC) OAM cell, the function-specific fields are null, since this cell is essentially intended only as an indicator that connection continuity is present. Thus, the format of the CC OAM cell is essentially identical to the common OAM cell structure shown in Fig. 8-8, with the function-specific fields coded as Hexadecimal 6A (6AH). In effect, the CC OAM cell may be viewed as the "simplest" of the OAM cells, since no "additional" information is contained in the cell that may require processing. However, assuming that the optional defect type and defect location functions are not being used in the AIS and RDI cells, this is also true for these OAM cells as well. In such cases, the only OAM-related "information" that is conveyed by the AIS/RDI or the CC cells is the presence (or absence) of these cells at any monitoring point in the network. As noted previously, this information may be used by the NE management system to infer the status of the VPC or VCC in question on a continuous basis. More recently, there have been proposals to add functionality to the CC OAM cell, with a view to enhancing its capability to provide additional information for specific network fault and performance monitoring conditions. Thus, it has been suggested that including a source address/identifier field in the CC OAM cells would allow the detection of misrouting faults. In addition, by including cell count fields in the CC cell, the means could be provided for detecting cell losses on a per-second basis to improve performance management capabilities and thereby availability estimation. However, it has also been argued that addition of these functions to the inherently simple CC OAM mechanism would significantly complicate the processing of the CC cells and thereby detract from their utility as a generic means for continuity monitoring on large numbers of VPCs and VCCs. Although in the future it will always be possible to add specific functions to the OAM cell types, the inherent trade-offs between the inevitable added complexity and the relative value of the specific OAM functions need to be carefully balanced. This is often difficult to achieve, given the large differences in network operations and management approaches and possible scenarios for which ATM may be employed.

8.12 LOOPBACK OAM CELL FORMAT AND FUNCTIONS

It was noted earlier that the loopback (LB) OAM procedures constitute one of the most widely used OAM functions for both in-service or out-of-service diagnosis of connectivity. The format and function-specific fields for the LB OAM cell are shown in Fig. 8-10, which also summarizes the semantics of each field. The least significant bit (LSB) of the one-octet loopback indication field is used to indicate whether or not the cell has been looped back. The LB cell source NE encodes this field as 0000 0001, whereas the NE at which the cell is looped back changes the loopback indication coding to 0000 0000 (all zeros), thereby indicating a "returned" LB OAM cell. The use of the Loopback Indication field also prevents the problem of infinite loopbacks occurring when the default values of loopback location identifier field (e.g., all ones) is used to denote the LB endpoint. The four-octet correlation tag field is used to correlate the transmitted LB OAM cell with the corresponding received LB OAM cell. To achieve this, the LB source NE encodes the correlation tag field with a different value for each generated LB OAM cell, which is

Function Specific Fields for Loopback

Loopback Indication (1 octet)	Correlation Tag (4 octets)	Loopback Location Identifier * (16 octets)	Source Identifier * (16 octets)	Unused (coded 6AH) (8 octets)

Unused 0000000	0/1

* - default value = all 1s

Description of loopback cell fields:

1. Loopback indication field (1 octet): The LSB of this field provides indication if cell has been looped back
2. Correlation Tag (4 octets): This field is used to correlate transmitted OAM LB cell with received OAM LB cell
3. Loopback Location ID (16 octets): As an option, this field may be used to identify loopback point (NE) along the connection (e.g. for fault localization). Default coding is all ones.
4. Source ID (16 octets): As an option this field may be used to identify the originating loopback point (NE). Default coding is all ones.

Figure 8-10. Loopback cell format.

then compared with the value in the returned LB OAM cell for positive confirmation of the associated LB process. The correlation tag process is necessary to enable a source to generate several consecutive LB OAM cells (as long as the minimum interval between each is at least 5 seconds, in accordance with the LB procedures).

The 16-octet loopback location identifier (LLID) field is used to identify the NE (also called a connecting point) at which the cell should be looped back (i.e., the LB point along the virtual connection or connection segment). In the event that specific values of the LLID field are not used or known, default values of the LLID field have been specified. The default value of "all ones" is used to identify an endpoint, which may be either a connection endpoint for end-to-end LB OAM cells or a segment endpoint for segment LB OAM cells. The "all ones" default LLID value is widely used for general continuity verification purposes, when no additional fault localization procedures need to be followed. In addition, the default value of "all zeros" has also recently been defined (see [8.5]) to enable the extension to the multiple loopback technique (MLT) described earlier. The "all zeros" default value is used to indicate all the NEs (or connecting points) along the VPC or VCC for which the MLT procedures have been enabled by the management system (or OSF) and hence which have been provisioned with an LLID value. Each of these NEs then inserts its provisioned LLID value into the LLID field and loops back the LB cell to the originating NE. In this way, as noted earlier, the originating NE receives a sequence of LB OAM cells with the LLID values of the participating MLT NEs along the connection path. Analysis of these LLID values by the NE management system or OSF may then enable rapid fault localization or path verification. Even if the MLT process is not available, it will be evident that use of specific "target" LLID values may be used for fault localization purposes, as described earlier.

The structure and coding of the LLID field is analogous to the address structure for identifying ATM NEs for management functions. The first octet (octet 1) of the LLID field is a location (or address) identifier type field, whose coding allows for the use of different address (location) formats in the remaining octets of the LLID field. Thus, the function of the location identifier type field is analogous to that of the authority and format identifier (AFI) field described earlier for the case of ATM end system address (AESA) NSAP-based formats. However, for the LLID field, the defined codings are as follows:

Location identifier type (octet 1 of LLID field)	Coding structure (octet 2 to 16 of LLID field)
0000 0000	No coding structure defined, octets coded all zeros.
0000 0001	Octets 2 to 5 carry E.164 Country Code + Network ID code in BCD coding. Octets 6 to 16 are binary coded to carry operator-specific information.
0000 0010	Octets 2 to 5 carry E.164 Country Code + Network ID code in BCD coding. Octets 6 to 16 are coded as Hexadecimal 6A (6AH).
0000 0011	Partial NSAP-based coding structure to be defined.
1111 1111	No coding structure defined, remaining octets all coded all ones.
6AH	No specific coding structure defined, remaining octets all coded 6AH.
Other codepoints	Reserved for future use.

8.13 FUNCTIONS AND FORMATS OF PM OAM CELLS

In describing the procedures associated with the performance management (PM) OAM cells earlier, it may be recalled that two types of PM OAM cells were described: the forward monitoring PM cells (FPM cells) and the backward reporting (BR) OAM cells. For normal ATM bidirectional VPCs or VCCs these may be used as "paired" FPM and BR cell flows if it is desired to collate the PM measurements at a given NE (e.g., a segment endpoint) along the connection path. Alternatively, if the network management system or OSF can coordinate the PM results obtained at the two connection (or segment) endpoints from the FPM flows via the external management system interfaces, it may be possible to utilize only the FPM cells to measure each direction of the VPC or VCC. The selection of which particular option to use largely depends on the specific network application envisaged, the capabilities of the NE management system, and the OSF, but it should be recognized that both "paired" FPM and BR cell flows, as well as FPM only techniques may be in use.

The function-specific fields and formats of the FPM and BR OAM cells are shown in Fig. 8-11, together with a summary of the semantics of each field. It will be evident that, at least compared to the other OAM functions described above, the PM functions are significantly more processing-intensive. However, this is not surprising in view of the complex performance management parameters required for the per-connection measurements. Consider first the forward performance monitoring function-specific fields shown in Fig. 8-11. This includes the following functions starting from the left:

1. Monitoring cell sequence number for FPM (MSCN/FPM). This one-octet field contains a sequence number modulo 256 for the forward monitoring OAM cell sequence, generated by a running counter at the source NE sending the FPM cell. For the case of the paired FPM and BR cells, it should be noted that independent counters are used at either end of the PM process. The MCSN/FPM values may be used to detect whether FPM cells are lost, thereby invalidating specific block measurements during the PM test process. In addition, as noted below, the MCSN/FPM value enables correlation between

Function Specific Fields for Forward Performance Monitoring (FPM) OAM Cell

MSCN / FPM (8 bits)	TUC_{0+1} (16 bits)	BEDC 0+1 (16 bits)	TUC_0 (16 bits)	TSTP (optional) (32 bits)	Reserved (34 octets)

Function Specific Fields for Backward Reporting (BR) OAM Cell

MSCN / BR (8 bits)	TUC_{0+1} (16 bits)	Reserved (16 bits)	TUC_0 (16 bits)	TSTP (optional) (32 bits)	Reserved (27 octets)	RMCSN (8 bits)	SECBC (8 bits)	$TRCC_0$ (16 bits)	BLER 0+1 (8 bits)	$TRCC_{0+1}$ (16 bits)

Description of the fields:

- **Monitoring Cell Sequence Number (MCSN):** Sequence number modulo 256. Independent counters are used at each end for FPM / BR.
- **Total User Cell number of CLP (0+1) cells (TUC(0+1)):** Total number of transmitted user cells when the FPM cell is inserted (modulo 65536). The paired BR cell contains the TUC value from FPM cell.
- **Block Error Detection Code for CLP(0+1) user cell flow (BEDC(0+1)):** Contains BIP-16 error detection code for the block of user cells. Used for FPM cells only.
- **Total User Cell number of CLP(0) cells (TUC(0)):** Count of total number of CLP(0) cells sent (as for TUC(0+1)).
- **Time Stamp (TSTP):** As an option, this field may be used for inserting a timestamp for delay measurements. (The default value is all ones).
- **Reported Monitoring Cell Sequence Number (RMCSN):** This field is present only in the BR cell. It is used to correlate FPM and BR cells by copying the MCSN/FPM value from the paired FPM cell.
- **Severely Errored Cell Block Count (SECBC):** This field is present only in the BR cell. It contains the number (modulo 256) of SECBs by processing the paired FPM cell.
- **Total Received Cell Count for CLP(0) cells (TRCC (0)):** Used for BR cells only to indicate number of received user cells with CLP=0.
- **Block Error Result for CLP(0+1) cells (BLER (0+1)):** Used in BR cells only to indicate number of errored parity bits in the BIP-16 code of the paired FPM cell.
- **Total Received Cell Count for CLP(0+) cells (TRCC (0+1)):** Used in BR OAM cells only to indicate total number of CLP(0+1) cells received (modulo 65536).

Figure 8-11. Performance management cell format.

the paired FPM and BR cells by including this value in the returning BR cell as well for each cell block.

2. Total user cell number of CLP (0 + 1) cells [TUC (0 + 1)]. The value in this two-octet field corresponds to the total number of user data cells transmitted up to the point at which the FPM cell is inserted into the cell stream. The number includes both the CLP = 0 (high-priority) and the CLP = 1 (low-priority) user data cells in the PM cell block between successive FPM cell insertions. This value is generated by a modulo 65,536 (two-octet binary) running counter at the sending end, which counts user cell-generating events per selected VPC or VCC. By comparing the value in the TUC (0 + 1) field with the total number of received CLP = 0 + 1 user data cells, it can be seen that the number of lost (or misinserted, if the number is greater) user data cells within each PM block can be calculated. The number of lost (or misinserted) user cells may then be used to compute the cell loss ratio (CLR) or cell misinsertion ratio (CMR) experienced by the monitored VPC or VCC. In general, the CLR (or CMR) value is taken as an average over an interval that is large compared with the cell block duration.
3. Block error detection code for CLP (0 + 1) user cells [BEDC (0 + 1)]. The value in this two-octet field is the even parity bit interleaved parity (BIP-16) error detection code calculated over the information fields of all the user data cells in the PM block, i.e., all user cells between two consecutive FPM cells. The BEDC (0 + 1) value may be compared to the result of the received BIP-16 computation to measure the number of errored cell blocks or severely errored cell blocks (SECB) performance parameters as required. The criteria for determining errored cell blocks, severely errored cell blocks, and the related severely errored cell block ratio (SECBR) is described in ITU-T Recommendation I.356 [5.9] and is outside the scope of this discussion.

4. Total user cell number of CLP (0) cells [TUC (0)]. The value inserted in this two-octet field is the total number of CLP = 0 (i.e., high priority) user data cells in the PM block, transmitted between consecutive FPM cells in the same way as for the TUC (0 + 1) counts as described above. By comparing this TUC (0) value with the count (modulo 65,536) of the received CLP = 0 user data cells between successive FPM cells, it is possible to compute the number of lost (or misinserted) CLP = 0 user cells and hence the corresponding CLR or CMR values. Note that this computation relates only to the high-priority (CLP = 0) user cell flow, which may typically, but not always, be regarded as the more important component of the data stream. In the event that only one level of CLP (e.g., = 0) is supported by the networks, it is clear that the PM process may be simplified accordingly.
5. TimeStamp (TSTP). The value in the four-octet TimeStamp field may, as an option, include a time stamp in order to enable measurements of cell transfer delay (CTD) and cell delay variation (CDV) performance parameters. If used, the TSTP field includes the time at which the FPM cell is inserted into the cell stream, but detailed procedures for use of time stamps for delay measurements have not been defined to date since this function is considered optional. Nonetheless, it may be recognized that measurements of round-trip delay and particularly CDV are useful for network traffic engineering. Consequently, both proprietary and out-of-service (e.g., based on external test equipment) test procedures are often used for this function. When not used, the default value inserted into the TSTP field is all ones (since the normal hexadecimal 6A may be interpreted as an actual time stamp value).

The function-specific fields and format of the backward reporting (BR) OAM cells are also shown in Fig. 8-11. Again, describing the fields in turn, commencing from the left-most field, these are as follows:

1. Monitoring cell sequence number/backward reporting (MSCN /BR). The value in this one-octet field is a sequence number modulo 256 for the BR cell sequence, as derived from an independent running counter in the NE generating the BR cell. As noted above, both the FPM and BR sequence number counters are considered independent, and enable the detection of lost FPM or BR cells so that the PM processor may be made aware of invalid cell blocks during the test.
2. Total user cell number of CLP (0 + 1) cells [TUC (0 + 1)]. The TUC (0 + 1) number returned in the paired BR cell to the originating NE removes the necessity to store this value for the subsequent computation on receipt of the BR cell.
3. Total user cell number of CLP (0) cells [TUC (0)]. As for the above case, this two-octet field contains the corresponding TUC (0) value copied directly from the paired FPM cell, thereby removing the need to store this value for any subsequent processing.
4. TimeStamp (TSTP). The four-octet TimeStamp field in the BR cell may be used as an optional function, exactly as for the case of the paired FPM cell. If used, its value represents the time at which the BR OAM cell was inserted at the sending end. If not used, this field is coded with the default all ones as for the FPM cell case.
5. Reported monitoring cell sequence number (RMCSN). The monitoring cell sequence number value of the paired FPM cell (MSCN/FPM) as described above is copied directly into this one-octet field so that it may be used to correlate the paired BR cell with its FPM cell at the originating NE.
6. Severely errored cell block count (SECBC). The value in this one-octet field represents the number of severely errored cell blocks measured on the VPC or VCC after processing

the BIP-16 error detection code in the paired FPM cell, as described earlier. The criteria for determining a severely errored cell block has been defined in ITU-T Recommendation I.356. The SECB count may be logged by an independent running counter modulo 256 that stores the current SECB count.

7. Total received cell count of CLP = 0 cells [TRCC (0)]. The value in this two-octet field represents the total number of CLP = 0 (high-priority) user data cells received between consecutive paired FPM cells. The total received cell count may be generated by an independently running modulo 65,536 counter at the receiver end. By comparing the value of the TRCC (0) field and the corresponding TUC (0) field per PM block, it may be seen that the number of lost (or misinserted) user data cells can be computed at the originating NE. It may also be noted that in the event that CLP = 0 cells are tagged as CLP = 1 cells by the UPC or NPC (i.e., the user data stream does not conform to the traffic contract, as described in Chapter 5), the tagged cells will appear as effectively "lost" to the TRCC (0) count, which may be misleading. However, if only tagging has occurred, the total CLP = 0 + 1 cell counts will not be affected, so that in this case the management system may be notified accordingly.
8. Block error result for total (CLP = 0 + 1) cell flow [BLER (0 + 1)]. The value in this one-octet field represents the number of errored parity bits detected by the BIP-16 code contained in the paired FPM cell, as described earlier. The result of the BIP-16 calculation is only inserted in the BLER (0 + 1) field of the BR cell if both the following conditions are met: (a) the MCSN values of the consecutive FPM cells are sequential, signifying block integrity has been preserved; and (b) the total number of user data cells (CLP = 0 + 1) received in the PM block corresponds to the value in the TUC (0 + 1) count, i.e., no cells have been lost or misinserted for a given block. In this case, it is assumed that cell loss rather than bit errors are the determining performance metric to be considered. In the case when the above criteria are not met, the BLER (0 + 1) field is coded as all ones.
9. Total received cell count for CLP = 0 + 1 cell [TRCC (0 + 1)]. The value in this two-octet field represents the total number of CLP = 0 + 1 user data cells received (at the NE that generates the BR cell) between two consecutive paired FPM cells, i.e., within a given PM cell block. This value, as for the case of the TRCC (0) field, may be generated by an independent running counter modulo 65,536, and is used in the same way as the TRCC (0) number, namely to compute the number of CLP = 0 + 1 cells that are lost or misinserted per PM block. As indicated previously, these values may then be statistically processed to derive the cell loss ratio (CLR) or cell misinsertion ratio (CMR), which are used to characterize the QoS of the VPCs and VCCs being monitored by the PM process.

It is evident from even this brief description of the PM function-specific fields that the PM procedures are by far the most processing intensive requirements on the NE management system in terms of both dedicated hardware and software, when compared to the other ATM Layer OAM functions. The hardware complexity generally results from the number of counters and associated logic that must be provided for the functions described above, assuming that all are supported for both FPM and BR OAM cells. The PM related management system software is also necessary to process and log the per-block results. Consequently, it is not surprising that the ATM PM function has been criticized as being overly complex, to the extent that some initial implementations have not supported these functions, relying instead on out-of-service performance characterization using external test equipment. More recently, however, the availability of powerful OAM processor VLSI hardware has resulted in increased use of the in-service PM functions described here. It must also be borne in mind that the PM OAM cell is complex because it is attempting to measure several different performance parameters, each of which requires complex processing. Thus, the TUC coupled with the TRCC fields for both total (CLP = 0 + 1) and high-priority (CLP = 0) cell flows are required for cell loss or misinsertion measurements to estimate CLR values. The BEDC coupled with BLER BIP-16 function is necessary to detect block errors and hence SECB ratios, since this parameter is used as a criteria for availability of the VPC or VCC. Finally, the TimeStamp function option may be utilized if delay parameters such as CTD and CDV need to be characterized.

However, it must be stressed that, in general, it should not be necessary to continuously perform all the above measurements on all connections on every interface in the network. Aside from constituting engineering "overkill" and waste of resources, this would result in swamping the management system and OSF with excessive performance data. In practice, the PM OAM functions may only be activated on a relatively small number of connections at any given time for diagnostic or troubleshooting purposes under OSF control. Used in this way, the PM processing requirements in any given NE may be reasonably met by providing sufficient capability to support simultaneous PM OAM operation on a relatively small number of VPCs or VCCs per transmission path (or interface). Typically these may range from a few to several tens of VPCs or VCCs per transmission path simultaneously in the NE.

8.14 FUNCTIONS AND FORMAT OF ACTIVATION/DEACTIVATION OAM CELL

In describing the activation and deactivation procedures for the performance management and continuity check OAM functions earlier, it was noted that this could be initiated entirely by the NE management system and OSF, or by a combination of the management system coupled with the in-band activation/deactivation OAM cells. Although this latter option has not been widely used to date, it is still useful to understand, if only for completeness, the functions required in the dedicated activation/deactivation OAM cells.

The function-specific fields and format of the activation/OAM cell is depicted in Fig. 8-12 together with a summary of the semantics. Starting from the left-most field, these include:

Function Specific Fields for Activation /Deactivation OAM Cell

Message Identifier (6 bits)	Direction of Action (2 bits)	Correlation Tag (1 octet)	PM Block Size A-B (4 bits)	PM Block Size B-A (4 bits)	Unused (coded 6AH) (42 octets)

Description of fields:

1. Message Identifier (6 bits): Indicates the message type for activation or deactivation and command or response.

2. Direction of Action (2 bits): Identifies direction of transmission of the cell. The notation A-B or B-A is used to indicate direction of activation / deactivation messages.

3. Correlation Tag (1 octet): Used to correlate commands with responses

4. PM Block Size A-B (4 bits): This field specifies the block size to be used for PM in A-B direction.

5. PM Block Size B-A (4 bits): This field specifies the B-A block size.

Figure 8-12. Activation/deactivation cell format.

1. Message identifier. The codepoints of this six-bit field indicate the activation/deactivation message type as follows:

Message type	Command/response	Coding
Activate	Command	000001
Activation confirmed	Response	000010
Activation request denied	Response	000011
Deactivate	Command	000101
Deactivation confirmed	Response	000110

2. Direction of action. The coding of this two-bit field is used to indicate the direction of transmission of the activation or deactivation OAM cell when used to activate or deactivate the PM or CC process. This field is only used for the ACTIVATE and DEACTIVATE messages, with a default value of 00 when not used. The convention used to define direction is given in ITU-T Recommendation I.610 [8.5] based on a simple source/sink labeling scheme.
3. Correlation tag. The value in the one-octet correlation tag field is used to associate the command and response messages, as described above, during the activation/deactivation cell exchange sequence. This is done by simply matching the correlation tag value in the response message with that in the command message. To enable this, consecutively generated correlation tag values should be different, e.g., they may be sequential modulo 256.
4. PM block size (A to B). The codepoints in this four-bit field is used to indicate the required performance monitoring block size for the forward performance monitoring (FPM) process. The A to B naming convention denotes source (A) to sink (B) for the FPM cells, and is the same convention used for the direction of action field described earlier. As described before, PM block sizes in the range 128, 256, 512 up to 32,768 cells, increasing in factors of two, have been selected with codepoints assigned. These values [see Recommendation I.610 (1999) for the list of codepoints] are used in the "activate" and "activation confirmed" messages. In other cases, this field is coded with the default 0000, even when used for the CC activation process.
5. PM block size (B to A). The value in this four-bit field indicates the required PM block size in the B to A direction, and is coded and used in the same manner as for the A to B direction.

It may be useful to recall that (as in Fig. 8-8) the coding of the four-bit function type field in the common OAM cell header indicates whether the activation/deactivation cell is being used for PM activation (0000) or CC activation (0001).

8.15 OVERVIEW OF THE OAM CELL TYPES

The relatively detailed description of the various ATM Layer Management OAM cell types for fault and performance management purposes provides an overview of the capabilities available for these functions. It is also useful to view all the OAM cell structures together, as shown in Fig 8-13, to obtain an overview of the relative complexity and overall formats. It may be evident from this comparison that the CC cell, with no function-specific fields specified to date, may be considered the simplest, although it should be borne in mind that CC needs to be generated at

Common OAM Cell (and Continuity Check cell)

5 octets	4 bits	4 bits	45 octets	6 bits	10 bits
ATM cell header	OAM Type	Function Type	Function specific Fields	reserved	CRC 10

AIS / RDI cell

1 octet	16 octets	28 octets
Defect Type(opt)	Defect Location (opt)	Reserved

Loopback cell

1 octet	4 octets	16 octets	16 octets	8 octets
Loopback ID	Correlation Tag	LB Location ID (Opt.)	Source ID (Opt.)	Reserved

Forward Performance Monitoring cell

1 oct	2 oct	2 oct	2 oct	4 oct	34 octets
MCSN	TUC (0+1)	BEDC (0+1)	TUC(0)	TSTP (Opt)	Reserved

Backward Reporting Cell

1 oct	2 oct	2 oct	2 oct	4 oct	27 oct	1 oct	1 oct	2 oct	1 oct	2 oct
MCSN	TUC (0+1)	Resv.	TUC(0)	TSTP (Opt)	Resv.	RMCSN	SECBC	TRCC (0)	BLER (0+1)	TRCC (0+1)

Activation / Deactivation cell

6 bits	2 bits	1 oct	4 bits	4 bits	42 octets
Message ID	Direction of Action	Correlation Tag	PM Block Size (A-B)	PM Block Size (B-A)	Reserved

NOTE: Field sizes are not to scale

Figure 8-13. OAM cell types overview.

nominally one cell per second on a (relatively) large number of connections. Their format is identical to that of the common OAM cell format, as are that of the AIS and RDI cells if the optional defect type and defect location functions are not used. However, it may be recalled that the AIS and RDI cells are only generated in the event of a defect or fault on all the affected connections. The loopback OAM cell clearly requires additional processing capability, particularly if the option of the loopback location identifier is used, as with the multiple loopback technique described earlier. Since, in general, loopback tests are on-demand, under management system or OSF control, the actual processing overhead due to LB OAM cells is relatively small when averaged over time. As pointed out earlier, the PM OAM processes, whether only using FPM or the paired FPM-BR OAM cells, clearly require the most processing in operation, but may only be used on a small number of connections for limited durations. In this sense, the OAM functions provided for the ATM network may be viewed as a powerful set of "OAM tools," which may be made available for the proper operation and maintenance of highly reliable, robust ATM networks.

8.16 RELATIONSHIP BETWEEN OAM FUNCTIONS AND CONNECTION AVAILABILITY

From the description of the various ATM layer fault and performance management functions given in the previous sections, the use of these functions for the normal operational aspects of an ATM network may be immediately obvious to the reader. There is, however, another important aspect of the use of the OAM tools that may not be immediately obvious but needs also to be considered. This aspect of OAM capabilities relates to the concept of connection availability and unavailability within the network. In general, network operators and service providers strive to deliver very high availability and reliability to users of a network. Broadly, the availability of a connection or service may be characterized by the percentage of time the connection is available to support the service or application that has been negotiated with the customer, for instance, as part of a service contract. In this sense, the concept of connection availability (or its converse, unavailability) is not restricted to any given technology, such as ATM, but is an aspect of more general network performance characterization studies. Traditionally, telecommunications networks have been engineered to provide very high levels of network availability, as evidenced by the ubiquitous telephony services. Increasingly, data communication networks are also required to provide high reliability and availability, given the limitations of the particular technology used. The criteria for characterizing connection availability has been developed over the years and is essentially based on the performance parameters widely used to measure conventional transmission and switching equipment performance. The parameter used in the formal definition of network (or connection) availability is the so-called severely errored second (SES). This concept of using the SES parameter has also been extended to the case of defining the availability of ATM networks.

In the case of ATM, it may be recognized that connection performance degradation may be characterized by unacceptable cell losses or by severely errored cell blocks. Consequently, the criteria for declaring a severely errored second in ATM (SES_{ATM}) has been defined in terms of both the cell loss ratio (CLR) and severely errored cell block ratio (SECBR) performance parameters. Formally, a SES_{ATM} is defined to occur either if the CLR exceeds 1/1024 or if the SECBR exceeds 1/32. The VPC or VCC is then defined to be "unavailable" if 10 or more consecutive SES_{ATM} occur. It may be noted that the above formal definitions of connection availability/unavailability is essentially independent of the specific bit rates or the traffic characteristics of the VPC or VCC involved, as well as the underlying causes of the cell losses, bit errors, and defects involved. In addition, when the availability threshold of 10 consecutive SES has been exceeded, the 10 second period is considered as part of the unavailable time, and may be logged as such within the NE management system or OSF. The number of such "unavailable" periods is then accumulated by the OSF or management system over the connection lifetime and may be compared with the value negotiated in the service level agreement (SLA) with any given network user. For a bidirectional VPC or VCC, an unavailable state in either direction of the connection is considered to render the VPC or VCC unavailable as a whole. It is also of interest to note that since the underlying QoS performance parameters, such as CLR or error rates, are considered to be valid only during the (normal) available periods of any given VPC or VCC and not during the unavailable periods, measurements of these parameters will need to be suppressed during the declared unavailable periods.

The formal definition of the availability/unavailability criteria in terms of ten consecutive SES measures described above assumes that the underlying performance parameters, CLR or SECBR, are being monitored continuously. However, it was pointed out earlier that the continuous use of the PM OAM process on all or even a large fraction of the connections may pose an unacceptable processing overload on both the NE management system and the OSF. Typically, the PM process used to measure CLR or SECBR parameters may only be available for a small fraction of the VPCs or VCCs at any given time. This would imply that simultaneous availability determination on the remaining connections would not be possible. Recognizing this limitation, the use of other OAM functions in "estimating" the in-service availability state of any given VPC or VCC is also important. In effect, the OAM functions other than the PM function are considered to provide an in-service estimate of the availability state of a given VPC or VCC in the sense that they approximate to the formal definition described above. Four cases may be

TABLE 8-1. Relationship of OAM functions to connection availability

Performance monitoring (forward)	Continuity check	SES_{ATM} criteria
Yes	Yes	CLR $\geq$ 1/1024 or SECBR $\geq$ 1/32 or No CC cell received or $\geq$ 1 AIS cell
Yes	No	CLR $\geq$ 1/1024 or SECBR $\geq$ 1/32 or $\geq$ 1 AIS cell
No	Yes	No CC cell received or $\geq$ 1 AIS cell
No	No	$\geq$ 1 AIS cell

considered, depending on the OAM functions activated on the VPC or VCC in question. These are:

1. PM and CC functions activated
2. PM function only
3. CC function only activated
4. No PM or CC functions available

Each of the above cases may apply either for the segment or the end-to-end flows, enabling the in-service estimation of the availability to a certain level of accuracy. The criteria that may be used in each of the cases may be summarized as shown in Table 8-1. In this classification, it is assumed that as a minimum the AIS/RDI mechanism is implemented.

The table indicates the close relationship that exists between the OAM tools specified for the ATM layer and the methodology that may be used for the in-service estimation of SES_{ATM} and hence the availability threshold. Since connection availability duration may constitute an important element in a service level agreement between network operators or service providers and their customers, the provision of suitable mechanisms for its measurement, whether in-service or out-of-service, will likely play an increasing role in overall TMN-based network management systems.

8.17 PROTECTION SWITCHING AT THE ATM LAYER

The desire to provide high levels of connection availability in ATM networks has also resulted in the inclusion of automatic protection switching functionality in the ATM layer. The concept of automatic protection switching mechanisms to ensure a high level of network survivability in the event of failures, such as inadvertent cable cuts or NE breakdowns, is widely used in transmission systems, particularly those based on the SDH hierarchy. The same basic concepts may be extended to the ATM layer by providing a capability to allow for the automatic switching of any given VPC or VCC in the event that a defect or failure is detected by the OAM mechanisms described earlier. The architectural principles on which ATM protection switching is based are analogous to those used in conventional transmission systems, as shown in Fig. 8-14. The domain over which a given VPC or VCC may be protected may extend between the connection endpoints (termed end-to-end protection switching), or between connection segment endpoints

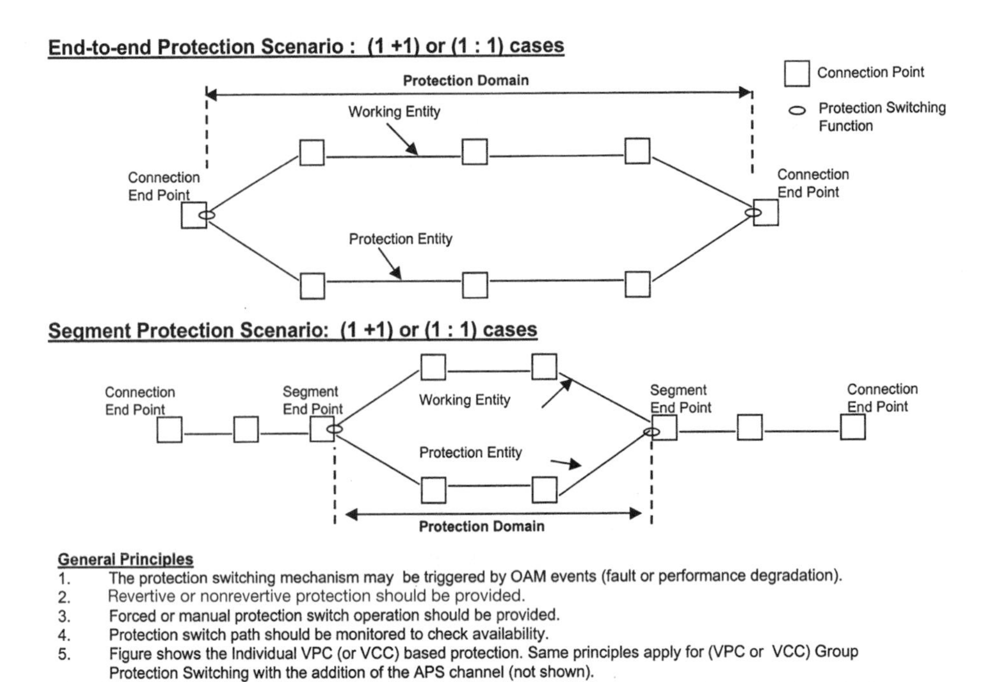

Figure 8-14. ATM layer protection switching principles.

as shown. In principle, combinations of the above types are also possible and extensions to more complex configurations may be envisaged as indicated later. Here we focus on the basic configurations shown to describe the underlying mechanisms used to trigger and coordinate the ATM layer protection switching actions.

In addition to the end-to-end and segment protection switching scenarios, two types of protection switching mechanisms need also to be considered. In the so-called "1+1" protection configuration, the traffic is routed along both the "working" connection path and the (alternative) "protection" connection path simultaneously by means of a bridge or copy function at the source point (of either direction) of the protection switching domain. At the sink point (of either direction) of the protection switching domain, a so-called "selector" function normally only selects the traffic on the working path. In the event that a defect or failure is detected on the working path, the selector function switches over to the protection path to ensure that the traffic flow may continue with minimal disturbance and cell loss experienced due to the fault. Consequently, in the 1+1 protection configuration, the protection VPC or VCC constitutes a dedicated resource, and implies that for every "working" VPC or VCC there is a dedicated protection VPC or VCC available at all times that carries the same traffic. In the alternative so-called "1 for 1" (abbreviated as 1:1) configuration, the protection path is assumed to be predetermined but may carry other, presumably lower-priority, traffic under normal working conditions. In this 1:1 protection configuration, it is assumed that in the event that a defect or failure is detected on the working VPC or VCC, the source point (of either direction) switches the traffic over onto the protection path by preempting the (lower-priority) traffic that may exist on that path. Thus, in the 1:1 configuration, the protection VPC or VCC is not dedicated to the working VPC or VCC as in the 1+1 mechanisms, but may be used to transport other (preemptable) traffic, as long as it is not needed to function as a protection path for any given working VPC or VCC.

In principle, the protection concepts underlying the 1:1 configuration may be generalized to describe a so-called "m : n" protection configuration, where m $\leq$ n. In this generalized case, the "n" working connections may be protected by sharing of the "m" protection connections, such that a failure in any of the n working paths may be protected by assigning its traffic to at least one of the m protection paths. Since the resource assignment mechanisms in such a generalized protection configuration are clearly much more complex and currently not well defined, we will focus here on the basic 1+1 and 1:1 configurations, although it will be clear that some of the procedures may be extended to the generalized m:n configuration.

In addition to the basic per VPC or VCC 1+1 and 1:1 protection configurations described above, the concept of VP or VC "Group" protection mechanisms has also been defined in ITU-T Recommendation I.630 (1999), essentially to enable more rapid protection switching of a "group" of VPCs or VCCs between the endpoints of a protection domain. In the VP group (VPG) or VC group (VCG) protection configuration, an arbitrary group or bundle of VPCs or VCCs are selected by the management system (or OSF) for protection purposes between the endpoints of a group protection switching domain. The VPCs (or VCCs) comprising the VPG (or VCG) share the same transmission path within the protection domain. A separate, dedicated so-called automatic protection switching (APS) VPC or VCC is also established between the end points of the protection switching domain in the same transmission path to serve as a "monitor channel" for the VPG or VCG that needs to be protected. In the event that a failure or defect is detected on the monitored APS VPC (or VCC), the associated working VPG (or VCG) is switched over the protection VPG (or VCG) as a whole group. For the group protection scenario, it should be noted that either of the 1+1 or 1:1 configurations depend primarily on whether the protection entities (in this case a VPG or VCG) are configured to provide a dedicated resource carrying simultaneous traffic (the 1+1 case) or a dedicated path with other preemptible traffic (the 1:1 case).

The provision of any of the above APS functions, whether for individual VPC or VCC or for VP or VC group protection, may be questioned by the reader at this stage. After all, it may be argued that since the vast majority of failures occur as a result of physical layer events such as inadvertent cable cuts or line card disconnections, protection switching functionality should be provided within the physical layer functions, i.e., associated with the transmission path. For many network deployments, this will typically be the case, particularly for SDH- (or SONET-) based transmission systems, which are generally provided with powerful protection switching capabilities. In such cases, the need for additional protection switching capabilities at the ATM layer may appear somewhat redundant. Moreover, in describing ATM signaling protocols earlier, the capability to provide VP or VC connection restoration by means of the so-called soft PVC (SPVC) procedures in both PNNI- and B-ISUP-based networks was also discussed. It may be argued that the soft PVC procedures are also directed to the same end as the OAM-triggered PVC protection switching mechanisms being described in this section, namely the automatic restoration of the VPC or VCC in question to provide high network availability. Here again, the ATM layer protection switching capability appears redundant.

While these observations are clearly valid for many networks, it should be borne in mind that ATM protection switching capability, whether for individual VPCs or VCCs or for group protected VPGs or VCGs, together with the other capabilities, such as SPVCs and transmission system automatic protection switching, may all be considered as useful tools designed to maximize overall network survivability. The concept of network survivability may be viewed as a superset, wherein capabilities such as transmission path and ATM layer protection switching, restoration by means of SPVC signaling procedures, as well as provision of redundant subsystems in NEs, are all coordinated to play a part. An essential concept in designing network survivability is the overall coordination, both from a temporal and spatial perspective, of all the restoration mechanisms that may be deployed in any given network implementation. As an example of the importance of temporal coordination, consider the case where both physical layer (i.e., transmission path) and ATM layer VPC automatic protection switching, as well as SPVC-based restoration, are available over a given administrative domain or subnetwork. In the event of a failure such as a cable cut, it is evident that all three mechanisms may attempt to restore connectivity over a physically diverse path. However, unless a coordinated timing sequence for restoration actions across the protocol layers is implemented, instability may result as, say, the physical layer attempts to switch after or during an SPVC reroute attempt or VPC protection switch. To avoid this indeterminacy, the provision of so-called "hold-off" timers may be arranged at each layer such that a higher layer only attempts to restore the connection if the layer below is unable to successfully complete the restoration of continuity in a predetermined time interval. Thus, it may be arranged that the transmission path will attempt to perform a protection switch first (e.g., SDH-based transmission systems can typically perform a switch in less than 50 msec) and, failing that, the ATM layer VPC (or VPG) will perform a protection switch after, say, 100–500 msec. If these mechanisms are unsuccessful or not available, then SPVC procedures may be used to reroute a connection in the order of , say, 1–2 sec, and so on. For this purpose, the ATM layer APS mechanisms incorporate a hold-off timer that is provisionable in the range 0 to 10 sec, with a granularity of 500 msec.

In such an example of layered protection or, stated more generically, survivability mechanisms, it may be evident that a network operator can arrange that a lower layer (or level) survivability mechanism operates prior to the level above it in the protocol stack by incorporating programmable hold-off timing at each level. This concept of "staggered timing" in maintaining overall network survivability has been termed "escalation" and may be extended to include both the optical networking levels and the fault-tolerant rerouting procedures that have been proposed for some dynamic routing protocols, including PNNI routing, as discussed earlier. However, it will be recognized that, in general, the survivability capabilities available to a network designer will, to a large extent, depend on the level of sophistication of the NEs and the particular network applications intended. For example, in a private network deployment, it may not prove cost-effective to implement several levels of survivability mechanisms or redundancy, whereas for a public carrier backbone network, multiple levels of protection switching capability may be considered practically imperative. In practice, not all NEs are likely to implement multiple protection switching capabilities and it generally remains the network operator's choice as to which survivability mechanism is used in various parts of its administrative domain. Whichever survivability mechanism is employed, the primary intent remains essentially the same—to ensure that network availability targets are achieved. It will be recalled that since the formal definition of the availability threshold is 10 consecutive severely errored seconds (SECs), the restoration of suitable connectivity to meet this threshold remains well within the capabilities of typical ATM protection switching (APS) mechanisms.

Given the relationship between the ATM layer APS mechanisms and the other protection or restoration capabilities available in the context of an overall network survivability strategy, it should also be noted that use of ATM APS configurations enables protection at a finer level of granularity than is typically possible with conventional transmission-path-oriented APS mechanisms. In the case of physical layer (transmission) protection switching, the bandwidth granularity for the protection switch typically corresponds to the entire transmission path bandwidth, which is switched over to a physically diverse path in the event of failure detection. However, for protection switching at the ATM layer, it is feasible to provide 1+1 or 1:1 protection switching for individual VPCs or VCCs within a transmission path, if required. If the group protection switching mechanism is selected (either for VPGs or VCGs), the bandwidth granularity is determined by the number of VPCs or VCCs comprising the group, which is essentially provisioned by the network management system for any given configuration.

It may be noted that the number of VPCs or VCCs comprising the VPG or VCG may vary between one up to the number of simultaneously provisionable PVCs per transmission path, a number that is largely implementation-dependent. For catastrophic failures such as cable cuts, the per-connection (or per-VPG) protection switching granularity available with ATM layer APS may not seem advantageous, particularly if a large number of VPCs or VCCs need to be switched in a short interval. However, the per-VPC (or -VPG) APS mechanisms may also be triggered by defects or failures within the ATM layer functions, which would not normally initiate a physical layer protection switch. Thus, for example, for an ATM layer loss of continuity (LOC) defect within a given VPC or VCC, it would be preferable to invoke a per-VPC or -VCC APS mechanism rather than to switch the entire transmission path, assuming the other connections in that transmission path were operating normally. In this context, it is important to note that for the case of group protection switching (VPG or VCG), triggering due to ATM Layer defects or failures is generally not possible, since in this mechanism it is not individual VPCs or VCCs in the respective VPGs or VCGs that are monitored. As described earlier, the triggering for group protection switching is initiated by monitoring the OAM flows on the separate, dedicated, APS channel set up for any given VPG or VCG. A defect detected on a dedicated APS VPC or VCC does not necessarily imply that there is a corresponding defect in the associated VPG or VCG. Therefore, in general, the group protection mechanisms may only be reliably used for (lower layer) physical layer defects.

The triggering mechanisms for the ATM layer APS are generally initiated by the fault management OAM flows described earlier, particularly by the AIS/RDI cells or where appropriate the CC cells. The particular OAM mechanism that is appropriate to trigger a protection switch depends on the specific network configuration and extent of the protection domain.

Consider first the case of individual VPC or VCC APS scenarios, whether for unidirectional or bidirectional protection or for 1+1 or 1:1 configurations. For cases in which the protected domain is coincident, or coupled, with an OAM segment, the APS mechanisms may be initiated by detection of the segment AIS state, provided the persistence of the segment AIS state exceeds the (provisioned or preassigned) hold-off time, at the sink point of the protected domain. Similarly, for cases where the extent of the protected domain coincides with the end-to-end VPC or VCC, the APS mechanism may be triggered by detection of the end-to-end AIS state, provided its persistence exceeds the relevant hold-off time. For the particular case of the 1+1 configuration, the end-to-end AIS state may be used even when the protection domain is not coupled to the OAM segment, since in this case it is generally possible to distinguish whether the defect (or failure) occurred within the protection domain or external to it, by simultaneously monitoring (i.e., nonintrusively monitoring) for end-to-end AIS cells on both the working and protection VPC or VCC. If the AIS cells are detected only on the working path, it is evident that the defect occurred within the protection domain and therefore a switch is warranted. If, however, AIS cells are observed on both the working and protection VPCs or VCCs, the fault can be assumed to lie outside the protection domain, since in 1+1 configurations, the traffic (including OAM cells) will be "bridged" or copied onto both working and protection paths simultaneously. This of course is not the case for 1:1 configurations, in which the protection VPC or VCC may be used to carry other (preemptible or so-called "extra") traffic under normal conditions. It should be noted that in all cases it is generally wise to monitor the state of the protection path, at least prior to a protection switch, since there is clearly no advantage in switching to a "defective" protection VPC or VCC.

For the case of group protection switching, whether it is for VPGs or VCGs (as well as unidirectional or bidirectional 1+1 or 1:1 configurations), it was noted that an associated, dedicated APS control VPC or VCC is established over the extent of the group protection domain. The APS channel may be monitored for end-to-end AIS cells, and the protection switch is initiated when the end-to-end AIS state persists beyond the provisioned hold-off time at the sink point of the protection domain. It may be noted that for VPG or VCG APS configurations, the protection domain need not coincide with an OAM segment (or connection endpoints), since its extent essentially depends on the associated APS control channel and the requirement that all the VPCs or VCCs comprising the VPG or VCG are contained within a single transmission path (for VPGs) or a single VPC (for VCGs). Despite these basic constraints, the group protection mechanism provides substantial flexibility for most network scenarios envisaged, and together with its potential for rapid switching of the logical group of VPCs has resulted in its widespread acceptance as a simple and effective survivability tool. For either the individual VPC or VCC APS configurations or for the group (VPG or VCG) mechanism, it is evident that the AIS state, whether resulting from segment or end-to-end AIS cells, may be used as a protection trigger. So far, the use of other fault management functions such as CC has not been considered, since this OAM function has been considered as somewhat optional, and hence may not always be available as an APS trigger. In general, the AIS state is assumed to represent a signal fail (SF) condition, but it is also possible to envisage protection switching actions as a result of signal degrade (SD) conditions, analogous to the signal degrade thresholds incorporated in some transmission system APS mechanisms. However, from the perspective of ATM layer APS, the notion of what constitutes a "signal degrade" has not been precisely defined and hence will not be considered further at this stage.

In describing the APS mechanisms so far, it has been tacitly assumed that both ends of the protection domain operate independently of each other during the APS operation. However, for the reliable error-free operation of the APS function for the case of bidirectional connections, it soon becomes clear that coordination between the APS states at either end (bridge or selector) is necessary to prevent misalignment between bridging/selector actions for bidirectional protection switching. These considerations are directly analogous to the APS coordination or control protocols used in SDH-based transmission systems, although, in contrast to the more complex multiphase coordination protocols used there, for ATM layer APS coordination, a simpler single-phase control protocol is used. Also in the ATM case, the APS coordination "request" and "state" information is transferred between the remote ends of the protection domain by means of codepoints specified in a dedicated "APS cell" that is exchanged between the two ends along the protection paths. The APS control protocol is applicable for either 1+1 or 1:1 configurations and for individual or group protection mechanisms. The dedicated APS cell used to exchange the control protocol has the same structure as the common OAM cell format described previously (see Fig. 8-8), with the "OAM type" field coded as 0101 to signify APS coordination protocol. The function type field is coded 0000 to denote group protection and 0001 to denote individual protection mechanisms. The APS control/coordination information is coded into the following so-called K1 and K2 bytes, a terminology also used in transmission systems. The coding of the K1 byte is used to indicate request for switch actions and priority, as well as the identity of the channel (i.e., working or protection) to which the action applies. The K2 is coded to indicate the status of the local bridge or selectior function. Although the details of the APS coordination protocol operation are outside the scope of this general description, it may be evident that the exchange of such request and state information between the remote ends of the protection domain will enable the correct operation of the bridge and selector functions at each end. The interested reader is referred to ITU-T Recommendation I.630 [8.7] for details.

As for the SDH-based transmission system APS mechanisms, multiple levels of switch priority indication are required in order to allow for APS functions such as "forced switch" and "manual switch" "lockout of protection" under management system control, as well as revertive or nonrevertive operation of the protected protocols. An ordered sequence of request priorities for these essentially maintenance and test functions has been established in Recommendation I.630 to enable the APS cell control protocol to coordinate the prioritized states at the remote ends of the protection domain for both directions of the VPC or VCC. The manual override or forced switch capabilities generally provided in APS mechanisms enable the network management systems to periodically check whether the APS operations are functioning as intended. Since there is a (small) probability that one or more of the APS control cells may be lost or corrupted when transmitted on the protection VPC or VCC, the APS coordinating procedures allow for periodic generation of the dedicated APS cells to avoid the incorporation of a more complex "acknowledge–retransmit" protocol. It may be noted that although for the individual VPC or VCC protection configurations the APS control/coordination cells are sent on the protection connection, for the VPG or VCG protection scheme, the APS control/coordination cells are sent on the dedicated APS VPC or VCC established for each group.

The underlying similarities of concepts (and even in some cases, terminology) between the ATM layer protection switching functions described here, whether for individual or group protection configurations, and those provided in conventional transmission systems technologies such as SDH (or SONET), will not have escaped readers familiar with such transmission networks. This aspect provides yet another example of the use of ATM as a "transport"-oriented technology analogous to SDH, as noted earlier. On the other hand, the provision of automatic restoration using the signaling-protocol-based soft PVC (SPVC) procedures described earlier, reflects a view of ATM as a "switching"-oriented technology intended to support on-demand switched VPCs or VCCs using signaling protocols. It is interesting to reflect that both approaches to providing network survivability tools in ATM are equally valid and, in practice, are likely to coexist together for many network applications. In effect, it may be conjectured that a major strength of ATM technology is the fact that it may be treated both as basic network transport and a general purpose switching technology capable of supporting the wide range of services expected for future networking. The rich OAM and survivability mechanisms described above allow ATM-based networks to achieve a level of availability and robustness not possible with any other packet mode networking technology to date. These considerations are clearly of prime importance in the design and deployment of wide-area backbone and carrier networks intended for transport and switching of all "mission critical" traffic types, and explains the choice of ATM as the preferred technology for such purposes.

CHAPTER 9

ATM NE Functional Modeling and Requirements

9.1 THE NEED FOR ATM NE FUNCTIONAL MODEL SPECIFICATIONS

In describing the many functions and protocols related to ATM signaling, OAM, or traffic control functions (among others) so far, no attempt was made to indicate how these various functions related to one another within an ATM NE, or whether coordination between them was required for the harmonious operation of the equipment. In fact, the description adopted was generally from the point of view of network-wide usage, even though it may have been implicit that these ATM functions, whether related to control, traffic, or OAM layer management, always need to be implemented in the hardware and software that is part of any given ATM NE. We now consider how all the requisite ATM functionality considered previously (and in subsequent chapters) may be brought together in a consistent way in a "model" of a generic ATM NE that allows for sufficiently detailed and systematic specification of the ATM functional requirements for any type of NE. In developing such a model, it is important to recognize that the approach needs to be independent of, and thereby not constrain, any given implementation in terms of either hardware or software. However, it should be capable of sufficient detail so as to promote interoperability between different implementations based on the same functional requirements. Having said this, it must immediately be admitted that not only is such a goal somewhat difficult to achieve at best, but that there may be several ways to approach it with no obvious indication as to which may be the best.

Although this problem may be endemic to any generalized modeling representational approach of functional or procedural requirements, in the case of an ATM NE, the functional descriptions appear to fall into one of two broad types of approaches. In the type of approach typified by the ATM Forum family of specifications or the (North American) Bellcore Generic Requirements (GR) series of ATM NE specifications, the methodology adopted is to develop detailed implementation agreements (IAs) based where possible on ITU-T or other standards. These documents generally identify the functional requirements that must be met by the ATM NE for any given application or service and for interoperability purposes. Examples of this approach include the ATM Forum's UNI [6.2] specifications for signaling and traffic control, the PNNI [7.3] or B-ICI [6.1] specifications, Bellcore's GR-1110 [9.5], or GR-1248 [9.6], and so on. In such an approach, the functional descriptions are not necessarily related to a common ATM NE model or representational methodology. Alternatively, in the type of approach adopted in the ITU-T series of equipment recommendations, the functional requirements are based on a common ATM NE functional modeling methodology that may be used for describing any type of transport or switching equipment at various levels of detail. For the case of ATM equipment, this modeling methodology is based (not surprisingly) on the general B-ISDN protocol reference model (PRM) described earlier, and uses the representation methodology previously developed for SDH-based NEs as described in ITU-T Recommendations G.805 [3.2] and I.326 [3.3].

As indicated earlier, either approach to describing NE functional requirements is equally valid and in practice there are no fundamental inconsistencies between the ATM Forum (or Bellcore) approaches and that adopted by the ITU-T modeling representations. It is more useful to view both approaches as essentially different but complimentary, and hence a discussion of the relative merits or demerits of either methodology is unnecessary. However, in general, the more formal functional modeling methodology developed in ITU-T Recommendation G.805 [3.2] and extended to ATM in I.326 and particularly in Recommendations I.731 and I.732 provides a systematic means of NE representation, which may be readily extended as additional functionality is incorporated. For the case of ATM equipment description, in particular, the formal modeling methodology builds on the B-ISDN PRM, thus providing a simple and direct means to establish the relationships and interactions between the control, management, and data transfer functions within any NE. Consequently, the description here will focus on the modeling representation developed in ITU-T Recommendations I.731 [9.1] and I.732 [9.2], while emphasizing the underlying principles rather than details of the inevitably complex representation. In addition, this approach also enables the derivation of a useful classification of the various ATM NE types, which so far have been treated generically.

Although the general B-ISDN protocol reference model (PRM) described earlier is intended as a network model rather than as a specific NE model, its prime result from the NE perspective is that it establishes the basic relationships between ATM functions grouped under the categories of user (or transfer) plane, control plane, and management plane (both layer management and plane management). Where the PRM partitioning is applied to the description of an ATM NE, a general functional architecture of an ATM NE may be developed as shown in Fig. 9-1. This representation is based on the model developed in ITU-T Recommendation I.731. In this

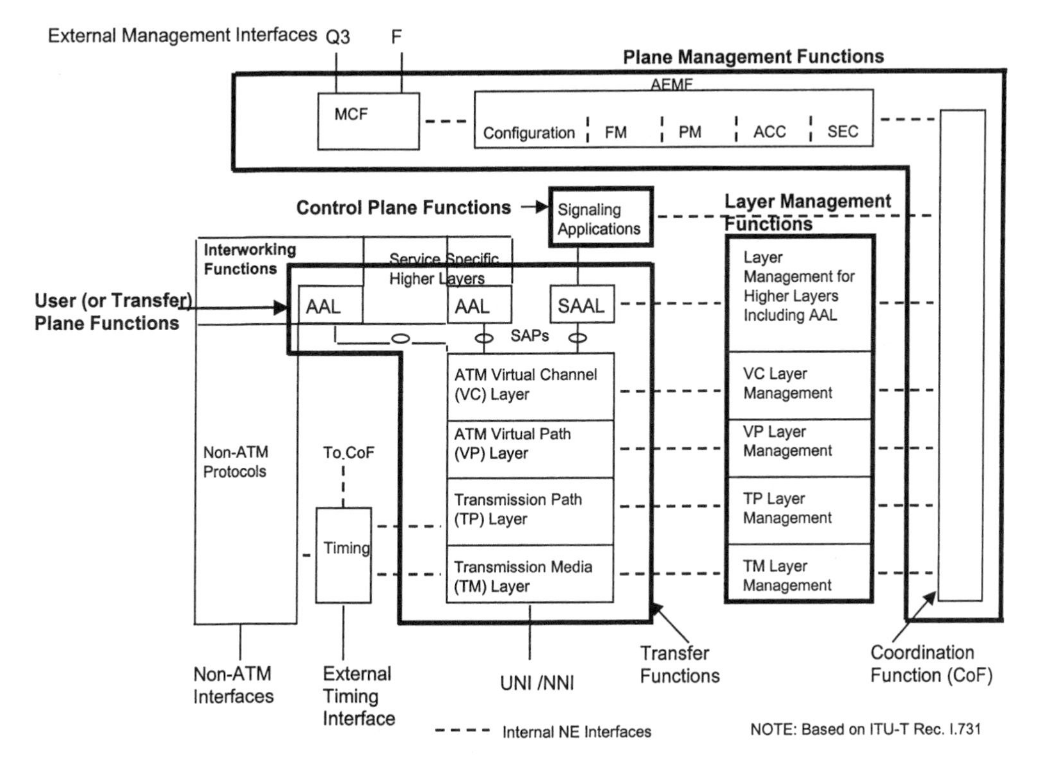

Figure 9-1. General functional architecture of an ATM NE.

model, the partitioning and grouping of the functions into the PRM categories will be evident. The user plane (or transfer) functions are layered according to the transmission (physical) media, virtual path (VP), virtual channel, (VC) and ATM adaptation (AAL) layer, as described earlier. Although the service-specific or higher layers in the user plane are not explicitly detailed here, it will be evident that in the event the ATM NE is a terminal equipment, these functions will need to be included. In general, since the equipment may include non-ATM interfaces (e.g., narrowband ISDN, frame relay, or IP, etc.), interworking functions (IWF) may need to be included as part of the user plane functions. The (external) ATM interfaces are identified as either UNI or NNI as shown. An external timing interface may also be incorporated in cases where timing synchronization is required. In addition to these external interfaces (which may generally be standardized in terms of rates and formats), the model also includes numerous "internal interfaces," notably between the transfer, layer management, and plane management functional blocks. It is important to recognize that such internal so-called "interfaces" do not imply points of access as such, but essentially represent internal communication paths between various functional blocks that may be embedded in the hardware or software of the NE, and as such are not standardized in any way.

Corresponding to the functional layering within the user (or transfer) plane, there is a one-to-one correspondence to the associated layer management functions, such that each layer management block is responsible for its associated layer in the user plane, as described earlier. It will be noted that, in fact, the "ATM layer" of the PRM has been effectively sublayered into the VP and VC layers or, more strictly speaking, the VP and VC levels, since the term "layer" may be confused with the protocol layering defined by the ISO Open System Interface (OSI) protocol model for general data communications. However, in the ATM modeling context, the terms "layer" and "level" are often loosely used interchangeably, but it is useful to keep in mind the formal distinction between these concepts to avoid confusion. Since the layer management functional blocks are assumed to interact through the overall plane management in the PRM representation, an important functional block in the plane management is the so-called coordination function (CoF). As will be noted from Fig. 9-1, the coordination function block not only enables the communication between the layer management blocks but also allows for the interactions between Control Plane functions (e.g., signaling and routing) and the timing function block. A more detailed description of the CoF block will be considered later, but it is important to recognize its role in the overall (i.e., plane management) ATM NE operation even at the level of the rather general NE model described here.

In addition to the overall coordination function (CoF), the plane management function block also includes the so-called ATM equipment management function (AEMF) block and the message communication function (MCF) block, which terminates the (external) management interfaces to the ATM NE. These management interfaces may correspond to the standardized TMN Q3 (or the ATM Forum M4 interface) described earlier, and to the (generally proprietary or local craft) F interfaces for operational control of the NE. A detailed description of these management interfaces and their associated MCF block is outside the scope of this text, since it relates more to the domain of the TMN or operational system function (OSF) in the network. Although the same is essentially true for the AEMF block, it is important to note that the AEMF includes the NE parts of the basic TMN management functions briefly described earlier in relation to the ATM Layer OAM & P capabilities. These general TMN functions comprise configuration management (CM), fault management (FM), performance management (PM), accounting management (AM), and security management (SM). These capabilities are also termed "FCAPS" functions in conventional Network Management terminology (for **F**ault, **C**onfiguration, **A**ccounting, **P**erformance, **S**ecurity). It may thus be envisaged that the AEMF block represents the management "intelligence" of the NE when considered as a system, processing and storing the individual layer management information passed to it through the coordination function and/or control plane. Conversely, the AEMF also serves as the mediator between the external network management system, or OSF, and the NE through the OSF-to-NE management interfaces such as the Q3, M4 or F interfaces shown.

The control plane functions include the signaling applications and routing functions (e.g., PNNI routing), as well as the signaling-specific AAL (SAAL) and the MTP 3 functions as described earlier. It may be noted that the signaling or routing functions also utilize the transfer function block through the appropriate service access point (SAP) since, as will be recalled, both the signaling and the routing messages are exchanged over dedicated (i.e., preassigned) VCCs in any given VPC. Although it is not explicitly shown in this simplified representation, the MCF block responsible for the external management messages may also utilize dedicated ATM VCCs for the exchange of management messages, and hence may also be viewed as using the services of the ATM transfer function block in the same way as the signaling applications.

Even at the somewhat general level described so far, it is evident that the decomposition of the functional blocks of the NE in accordance with the partitioning principles established by the PRM results in a modeling methodology that may be used to simply and clearly describe the interactions between the functions required in an ATM NE. As will be seen below, this general ATM NE functional model may be systematically extended to any required level of detail in order to describe precisely the location and relationships of any function in the NE. It may also be recognized that such a functional representation, in terms of logical blocks or groups of functions, is essentially independent of any implementation considerations, thereby imposing no constraint on equipment design other than those normally required by standardized interfaces or protocols. This aspect is clearly of importance in the acceptance of any general modeling methodology. It is also of interest to note that the explicit decomposition of management functions into the plane management and layer management functional blocks and the resulting explicit identification of the CoF is essentially unique to ATM technology. No other packet mode (or, for that matter, circuit mode) technology explicitly models the role of a coordination function, even though such a function may implicitly be present in actual NEs. While it may rightly be argued that a detailed modeling representation of a NE is not really essential to its proper design or operation, there can be little doubt that a clear functional model representation helps considerably in obtaining a much more precise understanding of NE operation, and thereby more efficient design.

9.2 A GENERAL TAXONOMY OF ATM EQUIPMENT TYPES

Before further developing the general ATM NE functional representation discussed above, it is useful to develop a more precise understanding of the various types of ATM equipment that may be described in terms of the functional blocks constituting the model of the NE. So far, the term "ATM network element" (NE) has been used in a generic sense to somewhat loosely imply specific types of equipment, such as ATM "switches," "cross-connects," or "multiplexers/demultiplexers," and even terminal equipment that may be deployed in any given network scenario. In fact, any equipment that incorporates some aspects of the set of functions we identify with the ATM layer may be considered as an ATM NE. The question arises as to what are the essential functional differences between say, an ATM switch and an ATM cross-connect, or between a cross-connect and a multiplexer, and is it possible to categorize these types in a systematic way? The fundamental functional criteria for distinguishing the equipment types have been described in ITU-T Recommendation I.311 [9.7] and further developed in I.731 to provide a general classification of ATM equipment types. Two basic criteria have been identified:

1. On-demand Signaling Capability. This criteria checks whether the signaling function is present or not in the ATM NE in question.
2. Restricted Connectivity. The concept of "restricted connectivity" implies that the ATM NE has multiple transfer interfaces to one direction (e.g., user side) and one transfer interface towards the other direction (e.g., the network side). In addition, it is required that there be no connectivity between the multiple transfer interfaces, say on the user side. If these conditions do not apply, the equipment is considered as having "unrestricted connectivity."

The two basic criteria above lead to the basic ATM NE types shown in Table 9-1.

TABLE 9-1. Criteria for basic ATM equipment types

Signaling capability	Not present	Present
Unrestricted connectivity	Cross-connect	Switch
Restricted connectivity	Multiplexer	On-demand multiplexer

In addition to the two basic criteria described, which result in the four basic ATM equipment types, there are additional distinguishing characteristics that may be used to provide a further subclassification of each of the basic ATM NE types. The two additional criteria identified in ITU-T Recommendation I.731 are:

1. VPI-based connectivity or VPI/VCI-based connectivity (VP level or VC level connectivity)
2. Whether or not interworking function (IWF) is present for support of non-ATM interfaces

By applying the above two subcriteria to each of the four basic ATM equipment types defined previously, a further, logical subclassification of 4 × 4 = 16 types of so-called "derived" ATM equipment types is possible, as shown in Table 9-2.

It may be noted that for each of the basic ATM NE types (cross-connect, switch, multiplexer, and on-demand multiplexer) there are four possible derived equipment types, depending on the

TABLE 9-2. Derived ATM equipment types

Basic ATM equipment types	Derived ATM equipment types		
Additional criteria	**IWF for non-ATM interfaces**	**VPI connectivity**	**VPI/VCI connectivity**
Cross-connect types	Not present	VP cross-connect	VC cross-connect
	Present	Interworking VP cross-connect	Interworking VC cross-connect
Switch types	Not present	VP switch	VC switch
	Present	Interworking VP switch	Interworking VC switch
Multiplexer types	Not present	VP multiplexer	VC multiplexer
	Present	Interworking VP multiplexer	Interworking VC multiplexer
On-demand multiplexer types	Not present	On-demand VP multiplexer	On-demand VC multiplexer
	Present	Interworking on-demand VP multiplexer	Interworking on-demand VC multiplexer

functionality incorporated in the NE. The terminology used to describe each of the derived equipment types is simply based on the functional criteria used, and it is worth bearing in mind that in some cases this may not correspond to terminology in more common use. As an example, the derived ATM NE type termed "interworking (VP or VC) multiplexer" is sometimes commonly called "service multiplexer" or "service access multiplexer," and so on. It should also be recognized that use of the term "multiplexer" will in general also include the converse "demultiplexer," a combination that is sometimes referred to as a "muldems," by analogy with the more common modem equipment. More commonly, the term "ATM switch" is often loosely used to refer to a wide variety of equipment types, including cross-connects or even multiplexers, regardless of whether or not signaling capability is incorporated in the NE. This, however, is inaccurate, since there is a traditional distinction made between a cross-connect as a NE that is controlled only through the management interfaces (i.e., no on-demand signaling possible), and a switch, which allows for connectivity control through on-demand (i.e., user-initiated) signaling capabilities.

It will be recognized that the formal classification of ATM NE types into the basic and derived equipment categories is simply the logical consequence of applying the selected functional criteria, using the functions identified in the general ATM NE functional architecture described earlier. While these basic and additional functional criteria are reasonably based on the conventional distinctions made between switching and transmission equipment, augmented by the concept of restricted connectivity implicit in a multiplexer/demultiplexer, this does not imply that all 16 derived equipment types necessarily exist or have valid network applications. Thus, it is clear that ATM cross-connects, switches, and multiplexers of various shapes and sizes are commonplace. However, the use, or even meaning, of the "on-demand multiplexer" category, which appears as a logical consequence of applying the two basic criteria, may be questionable, perhaps even meaningless, given the condition of restricted connectivity that typifies a multiplexer. For this case, the use of the "on-demand signaling" function cannot simply mean the setting up and release of switched VPCs or VCCs, since these do not ostensibly exist in a multiplexer. However, if a more "restricted" interpretation of on-demand signaling is permissible for this category, where signaling capability is interpreted to mean the setting up or dynamic modification of bandwidth or resource allocations for the VPCs or VCCs being multiplexed, then this category of equipment may have network application as a "bandwidth-controllable" multiplexer. However, whether the use of such equipment is required in practical ATM network deployments still remains an open question.

It needs also to be recognized that many practical implementations of ATM equipment may combine one or more categories within one framework or physical embodiment. Thus, for example the physical equipment may embody both a VP cross-connect category and a VC switch category in order to make best use of the physical resources involved and thereby minimize costs. Whereas this in itself does not invalidate the above essentially functional (or logical) classification methodology, it may not be clear how one should refer to such a NE, other than simply as a "composite" or "hybrid" NE. Since, in practice, many vendors are likely to "mix and match" functionality to optimize costs, a terminology based on strict functional classification may have limited usefulness in the marketplace. In addition, the question has often been raised as to the classification of a NE that may use signaling procedures in order to set up PVCs, such as "soft PVCs" (SPVCs). However, this issue may readily be resolved by recognizing that the application of the basic criteria is independent of the specific protocols that may be used in the control of ATM NEs. Thus, so long as control of the PVC set up/release is by means of the management interfaces to the NE, the use of "signaling" protocols to route or reroute the PVC does not change the basic classification as a cross-connect. It may be seen that, provided the interpretation of the functional criteria used is kept in context, the general classification of ATM NE types is useful in obtaining a more precise understanding of the rather generic terms "ATM equipment" or "NE" as well as their place in any given network.

9.3 EXAMPLES OF ATM EQUIPMENT TYPES

The basic and additional criteria used in deriving the general (i.e., logical) classification of the ATM equipment types outlined above utilize a "functional" approach that clearly is closely related to the general functional model of an ATM NE that was described previously, based on the B-ISDN PRM. Consequently, it should be relatively simple to model any of the NE types by using the functional "building blocks" that constitute the general ATM NE architecture depicted in Fig. 9-1. This approach has been adopted in ITU-T Recommendation I.731 and some examples of typical equipment models are shown here to demonstrate the usefulness of this simple modeling technique in the description of NEs. In Figs. 9-2, 9-3, and 9-4, three examples of ATM equipment types are modeled using this approach. The examples shown include:

1. VP cross-connect
2. VC switch
3. VP multiplexer

In each of these examples, the external interfaces appear along the bottom of the functional model for convenience. The examples demonstrate the ease with which the general ATM functional model may be used to describe any equipment type by using the basic ATM functional blocks essentially as "building blocks" that may be assembled in a systematic way based on the PRM. It will also be recognized that a number of the functional blocks, particularly those relating to the plane and layer management blocks such as the CoF, AEMF, and MCF, occur in all the NE types shown. However, it is important to note that this does not necessarily imply that the same level of functionality is required in each of these common blocks for all the types of NEs. In practice, the level of sophistication of the control plane or management plane functionality included in the various functional blocks modeled here will clearly vary greatly, depending

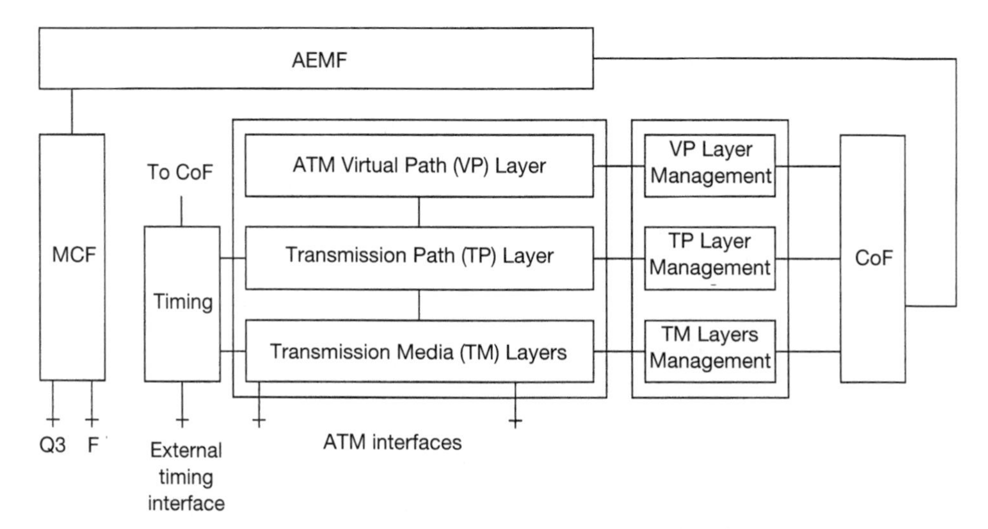

Figure 9-2. Example of a VP cross-connect model. Figure derived from ITU-T Recommendation 1.731.

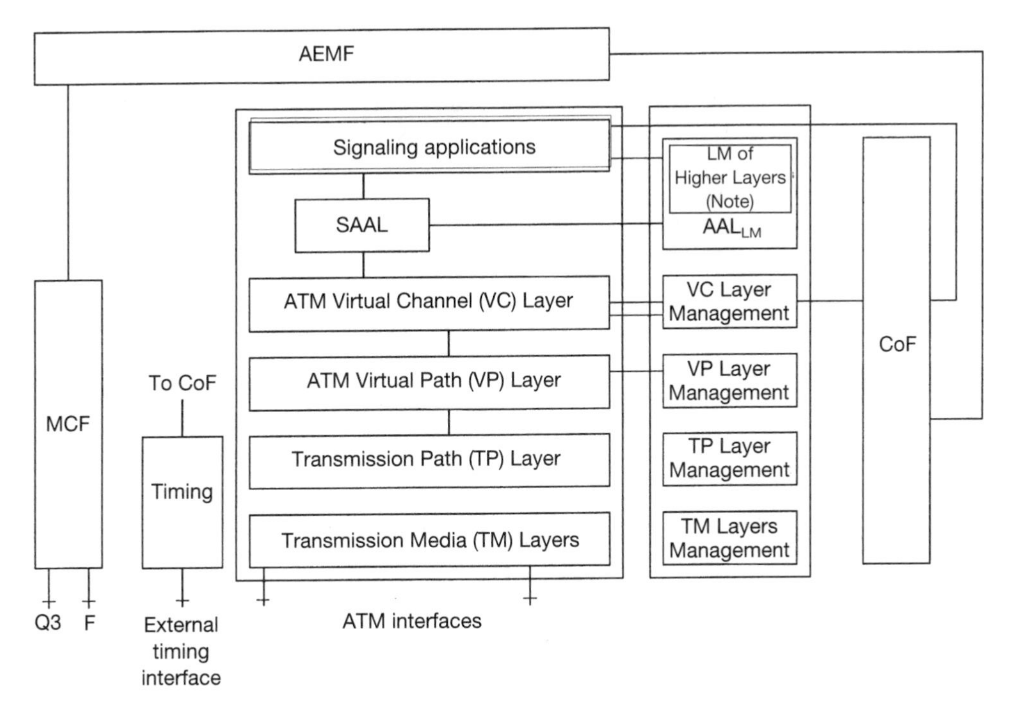

Figure 9-3. Example of a VC switch model. Figure derived from ITU-T Recommendation 1.731.

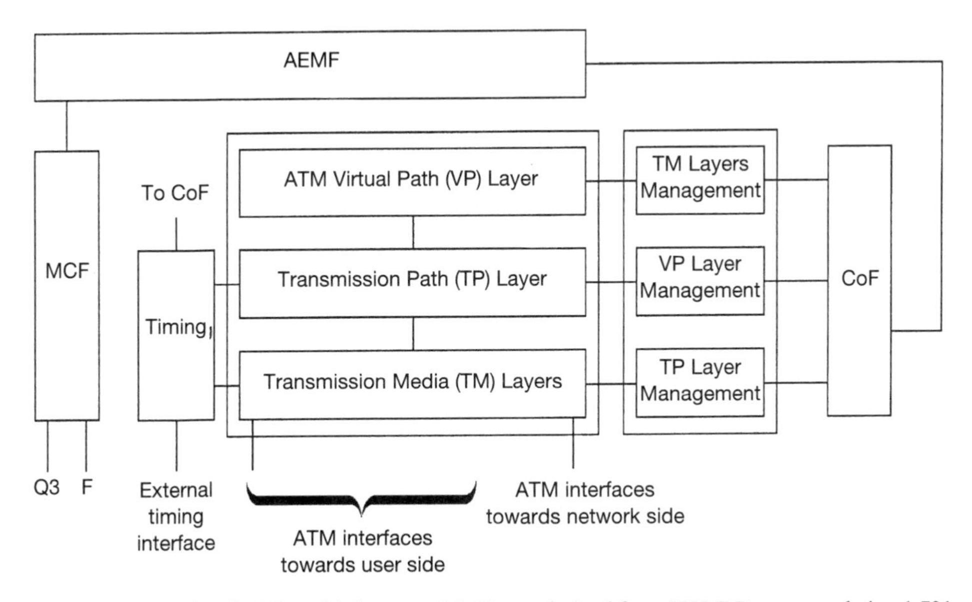

Figure 9-4. Example of a VP multiplexer model. Figure derived from ITU-T Recommendation 1.731.

on the network applications and capabilities that are required of the NE. In effect, the modeling representations described here essentially indicate the incremental functionality required in any given NE type, but the actual functionality provided in any given functional block requires much more detailed description to more fully characterize the capabilities of the NE. In practical implementations, much of the management (AEMF) and control/signaling functionality may typically be provided by software, and the detailed modeling of these capabilities will likely be complex, even for relatively simple NEs.

9.4 THE DETAILED FUNCTIONAL MODELING OF ATM NEs

The general functional architecture of an ATM NE based on the PRM that was described above may be viewed as a useful first step in developing a NE representation methodology convenient for high-level descriptions of ATM equipment, as evident from the examples given. However, it may be obvious that each functional block in this representation comprises a large number of individual ATM-related functions, whether they are transfer functions such as UPC/NPC, or layer-management-related OAM&P functions, and so on. It seems clear, therefore, that in order to develop a more detailed functional representation of ATM NEs, a more precise methodology for describing and grouping these functions needs to be used. Although the B-ISDN PRM clearly establishes the basic relationships between the layers and the control and management functional planes, it does not provide guidance as to the ordering or grouping of the individual functions within any given layer such as the VP or VC layers (or levels). Consequently, in order to describe functions within each level or layer, it is necessary to utilize a more detailed representational methodology that enables a systematic ordering of the functions within any of the layers represented in the general ATM NE architecture. A general methodology for representing so-called "layer transport networks" was developed by ITU-T Recommendations G.805 and I.326 for ATM, and has been adapted for the functional representation of SDH equipment (in ITU-T Recommendations G.782 [4.8] and G.783 [4.9]), and extended to ATM NE in ITU-T Recommendation I.732 [9.2]. The functional representation methodology described in ITU-T Recommendation I.732 enables the description of any ATM NE to a further level of detail and hence provides the basis for the discussion here.

The underlying principle of the modeling techniques used by ITU-T Recommendations G.805 [3.2] and I.326 [3.3] and adopted by ITU-T Recommendation I.732 [9.2] is to systematically group the individual functions within any layer (or level, for the case where sublayering is utilized) into three basic categories, each represented by its own unique symbol. As depicted schematically in Fig. 9-5, the three groups of functions are termed:

1. Termination Functions. These functions relate to the termination of any aspect of the so-called "characteristic information," which in the ATM case essentially corresponds to the relevant protocol control information (PCI) for the layer or level in question. Termination functions are represented as a triangular symbol as shown in Fig. 9-5.
2. Adaptation Functions. The functions that enable conversion of the PCI between the adjacent layers or levels, including any multiplexing or demultiplexing that may be present in order to transform information between the layers, are considered to fall into the category termed adaptation functions. As may be expected, the adaptation functions serve to mediate between any two layers (or levels), and are denoted by a trapezoid symbol in the I.732 and I.326 representations as shown in Fig. 9-5.
3. Connection Functions. The functions associated with providing the connection between any of the VP or VC links are grouped into this category, as suggested by its name. The

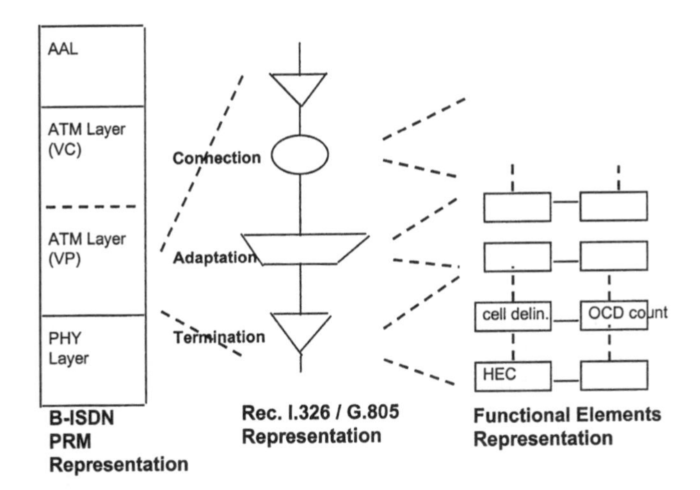

The detailed Functional Modeling of ATM NEs is based on the representation developed in ITU-T Rec. I.326 / G.805 (based on SDH equipment modeling) in which the equipment functions within a layer are grouped into the following functional blocks:

1) Termination
2) Adaptation
3) Connection

Each block may have many individual functional elements to the required degree of detail.

Figure 9-5. Principles of detailed functional modeling of ATM NEs.

connection function (sometimes referred to as a matrix connection, particularly in transmission systems) is typically symbolized by a circle or oval, as shown in Fig. 9-5. The connection function is responsible for the cross-connection or switching between the VPLs or VCLs along the VPC or VCC path, in accordance with the ingress-to-egress VPI/VCI associations established during the connection set-up phase.

The splitting of the ATM functions within any given layers into the groups of termination, adaptation, and connection functional blocks enables a systematic means of grouping of related functions in the representation of any generic ATM NE. The symbols and related methodology for depicting these three functional groups are essentially a convention with no significance other than to provide a convenient way to describe any NE. However, the use of a common set of symbols and terminology, as well as the methodology to describe a range of NE types from Optical SDH to ATM equipment, is convenient from both the design and network planning perspectives. Nonetheless, it must be pointed out that in some cases the terminology used in the G.805- or I.326-based representations is different from that in common usage in ATM and hence may result in confusion. A case in point is the use of the term "adaptation functions," which are clearly not the same as those corresponding to the ATM adaptation layer (AAL) functions described briefly earlier and in further detail in a later chapter. It is clearly important to keep in mind the differences between the AAL, which enables the ATM layer service to be used by a wide range of different applications such as voice, video, or data, etc., and the more narrow use of the term "adaptation functions" in the context of ATM NE functional representation. The confusion over use of the term adaptation is but one instance of the terminological mismatch between words in common usage relating to ATM, and those used in the I.326-based modeling methodology.

Although such differences in terminology are somewhat unfortunate, it is important to recognize that there are no fundamental inconsistencies in the functional representations derived from the I.326-based methodology and an approach based directly on extending the PRM. The one-to-one correspondence between either of the approaches is demonstrated in ITU-T Recommendation I.732, which includes both methodologies while employing a common terminology to describe each functional block in detail. By adopting this "combined" I.732-based approach and decomposing the functions within each layer into the termination, adaptation, and connection functional blocks, a more detailed functional architecture of a generic ATM NE based on the G.805 or I.326 representation may be obtained, as shown in Fig. 9-6. Alternatively, an equivalent representation of an ATM NE based directly on an extension or sublayering of the PRM (i.e., without using the concepts of termination, adaptation, and connection function blocks) may also be derived, as shown in Fig. 9-7. Although these equivalent (but different) representations of an ATM NE appear somewhat complex at first glance, a more careful inspection of either representation (bearing in mind the earlier simple NE model depicted in Fig. 9-1) illustrates that the model simply results from the systematic application of the functional grouping principles described above. Without entering into a detailed description of the functions in each of the blocks shown (this may be found in Recommendation I.732), we now outline the structure of both the equivalent representations shown in Figs. 9-6 and 9-7.

In the first place, a comparison with the earlier simple ATM NE model of Fig. 9-1 shows that the fundamental separation between the transfer, layer management, control plane (signaling/routing), and plane (i.e., overall system) management function blocks is preserved in either representation, as may be expected. The suffix "T" denotes the "transfer" function blocks and the suffix "LM" is used to indicate layer-management-related functions in either representation, as shown. It should also be noted that each functional block in the transfer plane has its associated layer management function block (in either representation) with which it communicates by means of internal "paths"as well as through the coordination function (CoF) block described earlier. These internal communication paths are termed layer management indications (LMI), and will be described in more detail below. The function blocks within the plane (or system) management described previously remain essentially the same as in the simple model of Fig. 9-1, and their potential further decomposition will be considered later. Attention is also drawn to the presence of the symbols "A" and "B" at the top and bottom, respectively, of either representation of the ATM NE model. These symbols are used solely to indicate information flow directionality within the NE, by unambiguously indicating whether the cell flow is in the "B to A" or "A to B" direction. Finally, to avoid repeating the words "in either representation" in the description below, the naming convention for the PRM-based representation is indicated in square brackets [ ] where appropriate, and we consider the functional blocks in the transfer plane, omitting the suffix for simplicity (recall that each block has its associated LM block in any case) starting from the bottom (reference point B) of the NE model.

The transmission media layer (TML) or [Transmission Media, TM] includes all the functions related to the physical transmission system (whether SDH- or PDH-based, or any other as described earlier), that is used in the NE for the transparent transport of the ATM cells. It will be recognized that a detailed account of these functions essentially lies in the domain of transmission systems engineering and will in general be strongly media- (e.g., optical, coaxial cable or twisted-pair copper, etc.) dependent.

The transmission path termination (TP_T) or [Transmission Path, TP] block includes all the functions related to termination of the transmission path such as extraction, processing and/or insertion of the transmission system overhead for either SDH- or PDH-based systems. For SDH-based systems, the handling of the overhead is described in detail in ITU-T Recommendations G.707 [4.5] and G.783 [4.9], whereas for PDH-based systems, this is described in ITU-T Recommendations G.804 [4.11] and G.832 [4.12], for example. Although in general these procedures are independent of the ATM cell payload contained in the selected transmission system,

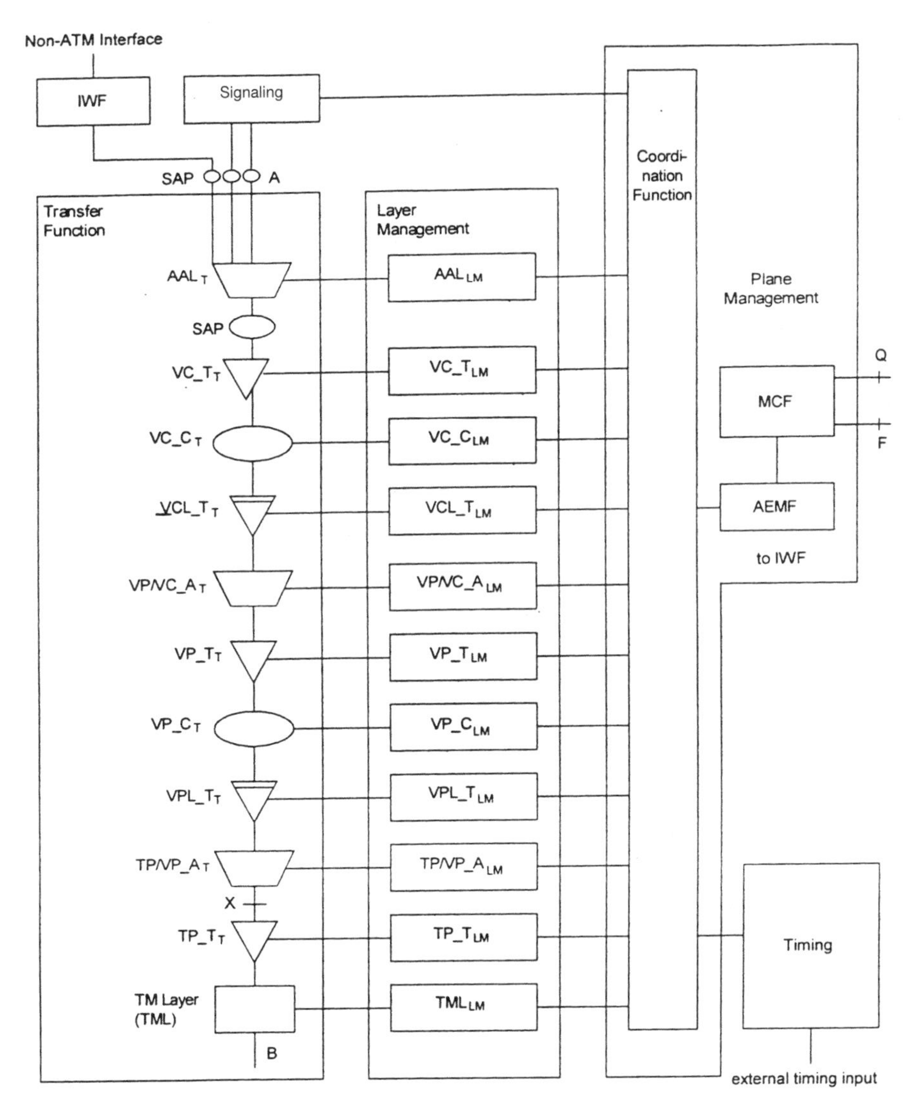

Figure 9-6. Detailed functional architecture of an ATM NE (I.326/G.805 representation). Figure derived from ITU-T Recommendation 1.732.

from the NE perspective, the equipment is required to terminate and possibly process the transmission path overhead involved.

The transmission path/virtual path adaptation (TP/VP_A) or [Virtual Path Multiplexing Entity, VPME] block includes the functions necessary to "adapt' or map the ATM cell streams into the transmission path structure. It is important to note that this functional block includes the multiplexing (in the A to B direction) and the demultiplexing (in the B to A direction) of the VP Links (VPLs) into a given transmission path payload. The TP/VP_A or [VPME] block also includes a number of other functions that are important to ATM NE operation. These are:

1. Mapping (insertion/extraction) of cell stream
2. Cell delineation and processing of header error control (HEC) field

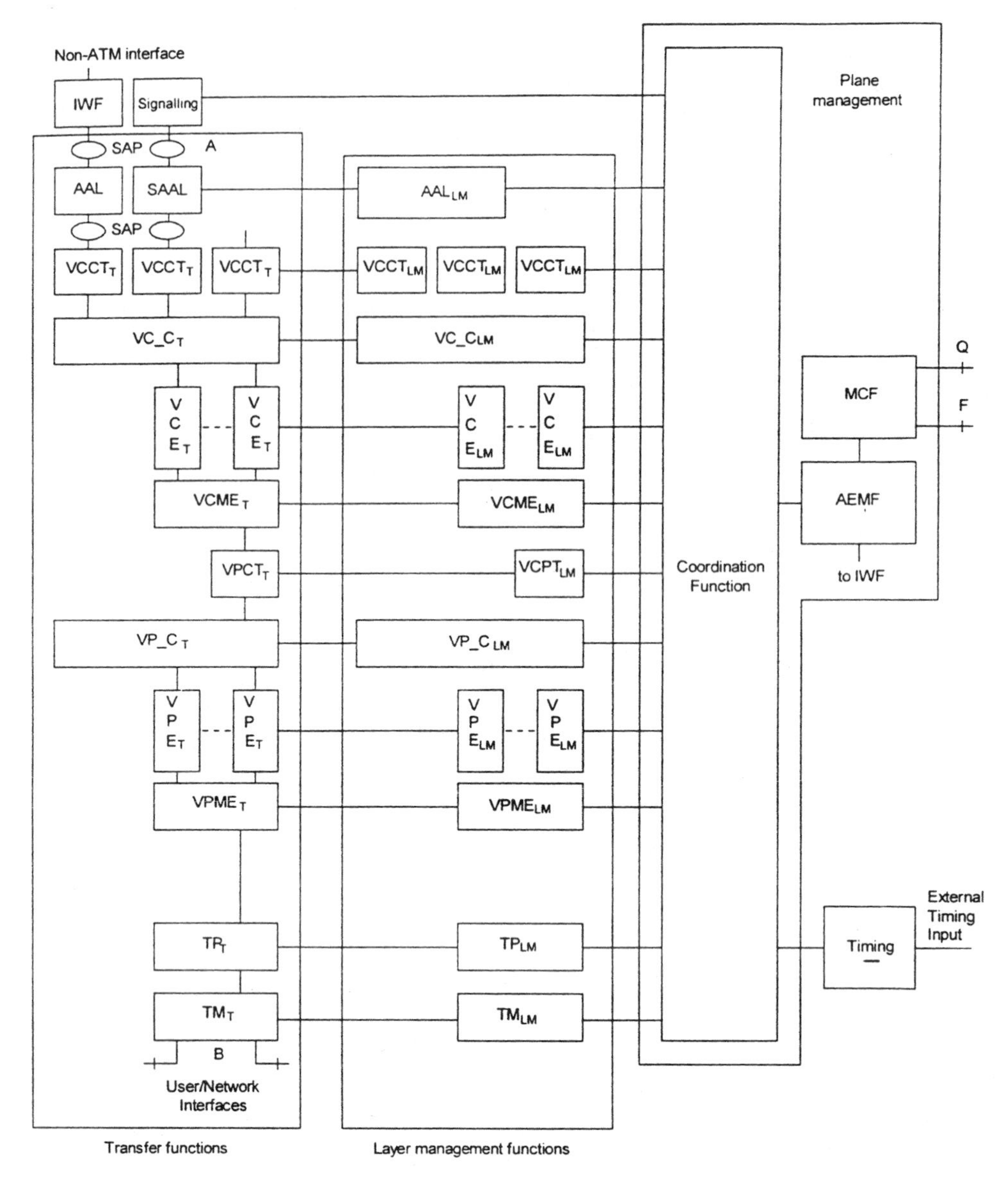

Figure 9-7. Detailed functional architecture of an ATM NE (PRM representation). Figure derived from ITU-T Recommendation 1.732.

3. Scrambling/descrambling of the ATM cell information field
4. Cell rate decoupling by insertion/extraction of idle cells
5. Usage measurements such as counts of incoming/outgoing total cells
6. Cell header verification and count of invalid HEC
7. Generic flow control (GFC) processing when activated
8. Verification of VPI values
9. Congestion control by selective cell discard according to CLP value

From this list of the included functions, it may be seen that the TP/VP_A or [VPME] block constitutes a complex mix of functions that corresponds approximately (but not exactly) to the original grouping of functions in the so-called "transmission convergence" sublayer of the B-ISDN PRM described earlier. To some extent, the allocation of certain functions to the TP/VP_A or [VPME] block is somewhat arbitrary and hence often subject to some controversy, since this block essentially represents a "boundary layer" between the conventional transmission-oriented domain and the packet-oriented ATM domains of modeling methodologies. For example, the allocation of the last two listed functions, namely congestion control by selective cell discard and VPI value verification, may appear as somewhat surprising since these functions would seem to relate more closely to ATM-connection-specific requirements rather than to multiplex-oriented capabilities. While this may be argued from either perspective, it serves as a useful example of the issues that may arise in any representation of disparate technologies in a unified general model. As noted earlier, since ATM technology may be approached from either a general transport (or transmission system) viewpoint or a general packet switching viewpoint, it is not surprising that differences of viewpoint occur in what is effectively a boundary layer functional block such as the TP/VP_A block.

It is also important to recognize that the associated layer management TP/VP_A or [VPME] block is responsible for all the management aspects of the functions listed above. This includes the processing and counting of the events such as out of cell delineation (OCD) or loss of cell delineation (LCD) events, invalid HEC or VPI events, GFC processing and activation/deactivation of functions such as cell discards, and so on. The layer management block is also responsible for communicating the relevant counts or event flags to the AEMF as required by means of the CoF and layer management indications (LMIs). Although the corresponding layer management functions will not always be explicitly described in this necessarily brief account, it is important for the reader to bear in mind that such processes may always be involved in considering each functional block in the general model. However, a detailed listing and systematic description of all the layer management procedures and requirements is available in ITU-T Recommendation I.732, so that the account here may be viewed as a summary review of the underlying principles to assist the interested reader in undertaking a more thorough review of the reference.

Progressing further up the NE model the virtual path link termination (VPL_T) or [Virtual Path Entity, VPE] blocks (i.e., link transfer and layer management) also includes a number of functions related specifically to the individual VPCs. These functions include:

1. Insertion/extraction/copying of end-to-end or segment F4 OAM cells, their processing and nonintrusive monitoring
2. Handling of the VP level resource management (RM) cells where applicable
3. The VP level UPC and/or NPC function for traffic compliance enforcement
4. Shaping function if applicable
5. Setting of the EFCI field if applicable
6. VP level usage measurements such as counts of incoming/outgoing cells per VPL if required
7. Setting (translation) of the VPI values

The VPL_T block also includes the VP level segment termination function if provisioned by the management system.

It may be noted that in the PRM representation depicted in Fig. 9-7 several instances of the [VPE] blocks are shown, essentially representing the individual VPCs that may be multiplexed/demultiplexed into the [VPME] block. These are not explicitly shown in the G.805/I.326

representation of Fig. 9-6 by convention, but inherently may be indicated in the same way. In addition, readers familiar with the G.805/I.326 methodology may be surprised to note the indicated "double triangle" symbol representation of what may be viewed as a normal termination function (triangle symbol). The double triangle symbol for the link termination block is used for the ATM NE model case in ITU-T Recommendation I.732 to indicate that a VPL_T instance may be provisioned as a maintenance segment endpoint by management action if required, as noted earlier in the description of the ATM Layer OAM&P functions. In the strict G.805/I.326 modeling methodology, which is based on the concept of so-called "subnetwork connections" rather than maintenance segments, it is somewhat difficult to represent the concept of provisionable segments. Consequently, a different symbol (i.e., the occluded triangle) was used to indicate the conceptual difference between the conventional link termination symbol and a provisionable link termination such as a VPL_T function block. As will be seen below, the same considerations apply to the virtual channel link termination (VCL_T) functional block within the VC level. It may also be noted that this aspect does not appear in the PRM representation for the [VPE] or [VCE] blocks, since no prior modeling conventions were applicable to this representation.

As suggested by its name, the function of the virtual path connection (VP_C) or [Virtual Path Connection Entity, VP_C] is the interconnection between the VPLs to provide the cross-connection or switching function in the NE. Note that for the connection function, the same terminology applies for both representations. As noted earlier, the label switching function results from the association between the incoming VPI value and port number with the outgoing VPI value and port number and the consequent translation of VPI values from the so-called "look-up" or context tables in the corresponding layer management block. It may also be recalled that for the case of point-to-multipoint (or multicast) connections, the VP_C block is responsible for the generation of cell copies for the various leaves of the point-to-multipoint VPC. For the converse case of a multipoint-to-point VPC requiring the so-called "merge" function, it may be assumed that this capability is also related to the VP_C block, although at present the allocation of the merging function is not well defined in the ATM NE function model.

The virtual path termination (VP_T) or [Virtual Path Connection Termination, VPCT] block includes all the functions associated with the termination or endpoint of the individual VPCs. In addition to the transfer of the cell payload (i.e., information field) and any relevant parameters in the ATM cell header, such as PTI value or CLP indication, to the higher layers (including the AAL) for the case when the NE is limited to the VP level only, the VP_T block also includes for the general case:

1. Insertion/extraction and processing of the F4 OAM cells including loopback
2. Insertion/extraction and processing of the RM cells when applicable

It will be recalled that in respect to the handling of the OAM (F4) cell flows the VP_T block must allow for both the end-to-end and segment OAM functions in the case where these points coincide. The VP_T or [VPCT] block represents the upper functional boundary for the VP level only functions in the NE.

When VC level capability is present in the NE, the next block to be considered is the virtual path/virtual channel adaptation (VP/VC_A), or the [Virtual Channel Multiplexing Entity, VCME] function block. This includes all the functions related to the multiplexing (in the A to B direction) and demultiplexing (in the B to A direction) of the VCLs into the assigned VPLs, in accordance with the VPI/VCI assignments in context (or look-up) tables set up for the cross-connection functions described earlier. In addition to this basic VC to VP multiplexing/demultiplexing function, the VP/VC_A block also includes:

1. Verification of the VCI values and discard of cells with invalid VCI values as well as count of cells with invalid VCI (in layer management)
2. Selective cell discard congestion control for the VC level
3. Extraction/insertion of metasignaling cells if and when appropriate (as noted earlier, the metasignaling capability has not been widely used to date for on-demand signaling applications)

In comparison with the transmission path/virtual path adaptation (TP/VP_A) block described earlier, the virtual path/virtual channel adaptation (VP/VC_A) block would appear to contain somewhat less functionality, as may be expected. However, it may be anticipated that the potential number of VCLs that may require multiplexing (or demultiplexing) within a VPL may be significantly larger than the number of VPLs within a TP, depending on the network application.

The virtual channel link termination (VCL_T), or [Virtual Channel Entity, VCE] block includes all the functions related to the termination of the individual VCLs, performing essentially the same role at the VC level as the VPL_T block described earlier does at the VP level. As for that case, the functions in the VCL_T block include:

1. The insertion, extraction and processing (in layer management) of the F5 OAM cells, including nonintrusive monitoring, as described in more detail earlier
2. Handling of the VC level RM cells where applicable
3. The VC level UPC or NPC function for cell traffic compliance enforcement
4. Traffic shaping function if applicable
5. Setting of the EFCI field if applicable
6. VC level usage measurement such as counts of incoming/outgoing cells per VCL if required
7. Setting (translation) of the VCI values

It is also important to recall that, as for the case of the VPL_T block, the VCL-T block also includes the VC level segment termination function if provisioned by the management interface (through the AEMF block and via the CoF, as may be evident from the model). This aspect of the VCL_T block is represented by the "occluded triangle" symbol as for the VPL_T block described earlier. For both the VCL_T block or the VPL_T block, the segment termination function may be provisioned for either the B to A or the A to B direction in the NE. The "Virtual Channel Connection" (VC_C), or [Virtual Channel Connection Entity, VC_C] block is responsible for the interconnection of the VC links to provide the cross-connection or switching capability for the VC level, as did the corresponding functional block (VP_C) at the VP level. Note that for this connection block, the terminology in both the G.805/I.326 representation and the PRM representation is essentially the same (e.g., VP_C or VC_C). The description of the VC_C block is the same as that given earlier for the VP_C block, by simply replacing reference to the VPI label by the VCI label, as may be expected.

The virtual channel termination (VC_T), or [Virtual Channel Connection Termination, VCCT] block is responsible for all the functions associated with the endpoint or termination of the individual VCCs. As for the VP_T block, in addition to the transfer of the cell payload and any relevant parameters in the ATM cell header such as PTI value or CLP indication to the AAL and higher layers via the service access point (SAP) indicated, the VC_T block also includes:

1. Insertion/extraction and processing of the F5 end-to-end OAM cells
2. Insertion/extraction and processing of the RM cells when applicable

Again, it should be noted that the comments related to the segment and end-to-end OAM flows made in connection with the VP_T block also apply to the VC_T block.

The description above has summarized the decomposition of the transfer and associated layer management function blocks and demonstrates the underlying symmetry in terms of the decomposition of each layer into the adaptation, termination, and connection functional blocks. Although the PRM-based representation does not use the same terminology or symbolism, the one-to-one correspondence between the two representations should be noted carefully, and either representation of the detailed NE model may be used as preferred. It will also be recognized that the other functional blocks relating to control plane (signaling and routing) and management plane are essentially as described previously in the simplified general NE model. A further decomposition of each of these functional blocks may be made but generally these are not represented in terms of the G.805/I.326 or the PRM-based methodology, but described in terms of the signaling or routing protocols considered earlier, or in terms of TMN-related information models [also termed object models or management information base (MIB)] along with their associated procedures. The use of the coordination function to enable the (internal) communications between all the functional blocks should also be noted. It is also important to recognize that for bidirectional connectivity in any given ATM NE, both the B to A and A to B directed blocks need to be present at both "ingress" and "egress" ports of the NE.

The functional decomposition described above is summarized for convenience in Table 9-3, based on the tabulation contained in ITU-T Recommendation I.732. Each of the transfer and associated layer management functional blocks may be described in much further detail, as is done in ITU-T Recommendation I.732. However, further detailed description is outside the scope of this account, although it may readily be followed by an understanding of the principles underlying the functional decomposition methodology described here. In addition, it should also not be assumed that the representations described above constitute the only ways to model an ATM NE. Other methodologies for systematic functional decomposition are possible and, more recently, a modeling methodology developed by ETSI for SDH equipment description has been extended to the representation of ATM NEs as well. This so-called "atomic function" modeling methodology is similar to the G.805/I.326 representation considered above (although, unfortunately, using somewhat different terminology) and attempts to combine both graphical representation as well as logical relationships between blocks to develop a very detailed description of the NE. Whether such an approach will gain wider usage and acceptance in the ATM industry remains an open question at this stage.

9.5 LAYER MANAGEMENT INDICATION (LMI) AND THE CoF

As was briefly mentioned previously, the "internal" communications between the transfer plane function blocks, the layer management functional blocks and the coordination function (CoF) may be represented as "layer management indications" (LMIs). The concept of LMIs, and its extension to the communication paths between the CoF and the AEMF blocks, termed AEMF indications (AEMFI), is illustrated in Fig. 9-8.

The LMIs themselves consist of two types of information flow, depending on the direction. The LMI is called a "Report LMI" if it represents information flow between the transfer block and the corresponding layer management block. The LMI between the layer management block to the corresponding transfer block is termed a "control LMI." The same principle may be extended to the AEMFIs, where a "control AEMFI" passes from AEMF to CoF blocks, and a "report AEMFI" in the opposite direction. The abstract representation of such internal communication paths between the various functional blocks comprising the NE enables a clear visualization

TABLE 9-3. General functional decomposition of an ATM NE

Level			B to A Function description		A to B Function description	
PRM	Rec. I.326	Function	Transfer	Layer management	Transfer	Layer management
AAL	AAL	SAR CPCS SSCS (Rec. I.363)	AAL1 AAL2 AAL3/4 AAL5 SAAL		AAL1 AAL2 AAL3/4 AAL5 SAAL	
VCCT	VC_T	F5 OAM cells	Extraction of F5 OAM cells	F5 OAM cells processing	Insertion of F5 OAM cells	F5 OAM cells processing
VC_C	VC_C	Link interconnection	Interconnection of VC links	Association of VC links	Interconnection of VC links	Association of VC links
VCE	VCL_T	F5 OAM	ins/ext of F5 OAM cells Resource management cells	Restricted rules for cell ins/ext	ins/ext of F5 OAM cells Resource management cells	Restricted rules for cell ins/ext
		Resource management		Resource management cells processing		Resource management cells processing
		F5 OAM nonintrusive monitoring	e-t-e and segment F5 OAM cell copy	e-t-e and segment F5 OAM cell processing	e-t-e and segment F5 OAM cell copy	e-t-e and segment F5 OAM cell processing
		Shaping[a]	VC traffic shaping	Traffic descriptors	VC traffic shaping	Traffic descriptors
		VC UPC/NPC	VC compliance checking and corrective action	Traffic descriptors, discarded cells and tagged cells if necessary	counters	
		VC usage measurement	Detection of cell arrival	Count incoming cells per VC for CLP = 0 + 1 and CLP = 0	Detection of cell arrival	Count outgoing cells per VC for CLP = 0 + 1 and CLP = 0
		EFCI setting			Setting of the EFCI bit of PTI field for user congestion indication	EFCI generation
		VCI setting			VCI field setting	VCI translation
VCME	VP/ VC_A	VC mux	Demultiplexing of the VCs according to the VCI values		VC Multiplexing	
		Congestion control	Selective Cell discard (CLP-based)		Selective Cell discard (CLP based)	
		Metasignaling	Extraction of metasignalling cells	Metasignaling cells processing	Insertion of metasignaling cells	Metasignaling cells processing
		VCI processing	Reading of VCI; discarding of cells with invalid VCI	Verification for invalid VCI cell. Count of cells with invalid VCI cells		
VPCT	VP_T	F4 OAM cells	Extraction of F4 OAM cells	F4 OAM cells processing	Insertion of F4 OAM cells	F4 OAM cells processing
VP_C	VP_C	Link interconnection	Interconnection of VP links	Association of VP links	Interconnection of VP links	Association of VP links

(continued)

TABLE 9-3. *Continued*

Level			B to A Function description		A to B Function description	
PRM	Rec. I.326	Function	Transfer	Layer management	Transfer	Layer management
VPE	VPL_T	F4 OAM	ins/ext of F4 OAM cells	Restricted rules for cell ins/ext	ins/ext of F4 OAM cells	Restricted rules for cell ins/ext
		Resource management	Resource management cells	Resource management cells processing	Resource management cells	Resource management cell processing
		F4 OAM nonintrusive monitoring	e-t-e and segment F4 OAM cell copy	e-t-e and segment F4 OAM cell processing	e-t-e and segment F4 OAM cell copy	e-t-e and segment F4 OAM cell processing
		VP usage measurement		Incoming cells count per VP		Outgoing cells count per VP
		Shaping[a]	VP traffic shaping	Traffic descriptors	VP traffic shaping	Traffic descriptors
		VP UPC/NPC	VP compliance checking and corrective action if activated	Traffic descriptors, discarded cells and tagged cells count		
		EFCI setting			Setting of the EFCI bit of PTI field for user congestion signalling	EFCI generation
		VPI setting			VPI field setting	VPI translation
VPME	TP/ VP_A	VP mux	Demultiplexing of the VPs according to the VPI values		Multiplexing of VPs onto TP	
		Congestion control	Selective cell discard (CLP-based)	Act./deact. of cell discard based on congestion detection	Selective cell discard (CLP-based)	Act./deact. of cell discard based on congestion detection
		VPI processing	Reading of VPI and discarding of unassigned cells and cells with invalid VPI		Verification for invalid VPI cell count of cells with invalid VPI	
		GFC	Read of GFC field (if applicable)	GFC processing	Setting of GFC field in an unassigned cell or insertion of unassigned cells	GFC processing
		Header processing	Read header and discard cells with invalid header pattern			
		TP usage measurement	Detection of cell arrival	Incoming cells count per TP	Detection of cell arrival	Outgoing cells count per TP
		Cell rate decoupling	Idle cell discard		Idle cell insertion	
		HEC processing invalid HEC cells	Header verification, correction (if applicable) and discarding of correction mode	Invalid HEC event. Invalid HEC cell discarded event. Act./deact. of	HEC generation	

TABLE 9-3. ***Continued***

Level			B to A Function description		A to B Function description	
PRM	Rec. I.326	Function	Transfer	Layer management	Transfer	Layer management
VPME	TP/ VP_A	Scrambling	Cell information field descrambling		Cell information field scrambling	
		Cell delineation	Cell delineation	LCD defect detection OCD anomaly event counter and consequent actions		
		Mapping	Cell stream extraction		Cell stream mapping	
TP	TP_T	SDH or PDH transmission path termination	Transmission path overhead extraction	Transmission path overhead processing	Transmission path overhead insertion	Transmission path overhead processing
TM	TML					

[a]The shaping function may not be present in NE. If present, the shaping function can be activated/deactivated in egress side or ingress side per connection. The shaping function should not be simultaneously activated on both B to A and A to B of the same connection.
[b]Not all above functions need to be present in a given network element. Selected link termination functions can become segment termination functions by management action. The extraction of metasignalling could be in the VCL_T or VC_T.

of the mechanisms by which the control, timing, or management-related information necessary for equipment operation passes between the (external) management interfaces (such as the TMN Q3 or F interfaces), through the AEMF, CoF, LM, and transfer function blocks and vice versa. Such a visualization clarifies the relationship between the OAM F4 or F5 functions described earlier to the management information traversing the OSF to NE interfaces such as the Q3 (or M4) and F interfaces.

It should be borne in mind that although the control and report LMIs and AEMFIs provide a simple representation of the internal communication paths between the transfer, layer management, CoF, and AEMF function blocks, the precise information content in any given LMI (or AEMFI) depends on the exact partitioning of the processing contained in these blocks. As was pointed out earlier in relation to the description of OAM&P functions, the question of the extent to which filtering (also sometimes called "defect correlation" function) and processing of the OAM&P (and RM cells) functions are partitioned between the layer management, CoF, and AEMF blocks is still the subject of debate from a NE modeling perspective. However, from an implementation perspective, such questions are largely irrelevant, since the internal functional partitioning within the NE is entirely implementation-specific, so that the "LMIs" are embedded within the software and hardware structures of the equipment. Nonetheless, from the perspective of obtaining a more precise understanding and representation of the functional decomposition of the ATM NE, the location of the OAM filtering and processing capabilities is of importance. In more recent work in this area, it has been generally recognized (if not yet fully described in detail or universally accepted) that the CoF block may provide a component of the OAM "filtering" or so-called defect correlation functionality described earlier, before passing the relevant OAM information to the AEMF for processing. The layer management blocks include the processing/generation of the OAM&P cell fields based on the procedures described earlier (as well as RM cells when used for traffic control).

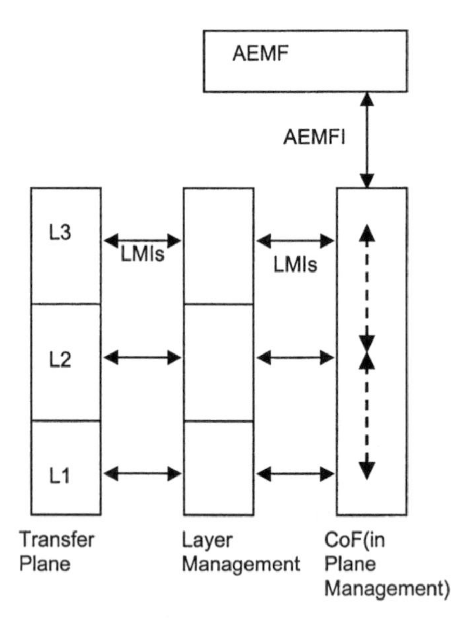

Figure 9-8. LMI and coordination function (CoF).

In the NE model described above, the CoF block may be seen to contain the following functional components:

1. The communications between the layer management blocks as well as the LM and AEMF blocks using the control and report LMIs/AEMFIs described above.
2. The selection and distribution of the timing information from the timing (or synchronization) functional block and/or the external timing interface.
3. The connection admission control (CAC) functional block.
4. When accepted for the enhancement of the current model in ITU-T Recommendations I.731 and I.732, the CoF may also include the relevant aspects of the defect correlations or OAM filtering functions noted earlier.

The inclusion of the CAC function within the CoF block may at first glance appear somewhat surprising, considering the relationship of the CAC function described earlier to the signaling and traffic control or resource allocation functions in the NE. However, it may be recognized that the CAC function essentially performs a "coordination" role between requests for network resources (such as bandwidth on each individual TP, VPL, or VCL) from either the on-demand signaling (control plane), the management system (AEMF), or in-band through the use of RM cells for the available bit rate (ABR) or ATM block transfer (ABT) ATM transfer capabilities (ATCs). In this respect, the CAC function may be represented (or modeled) as shown in Fig. 9-9, and it would consequently appear appropriate that the CAC function should be located within the "coordination" function (CoF) as currently described. The CAC model shows that the requests for transmission path (TP) bandwidth for VPCs or VCCs may originate from the signaling messages for on-demand switched VPCs or VCCs, from the management interface (Q3, M4 or F) for PVCs, and from the RM cells for flow controlled (ABR) or "fast-reservation"-based (ABT) ATCs. As noted earlier, the CAC function includes a CAC algorithm that allocates the resources to the connection requests if sufficient bandwidth is available to satisfy the QoS re-

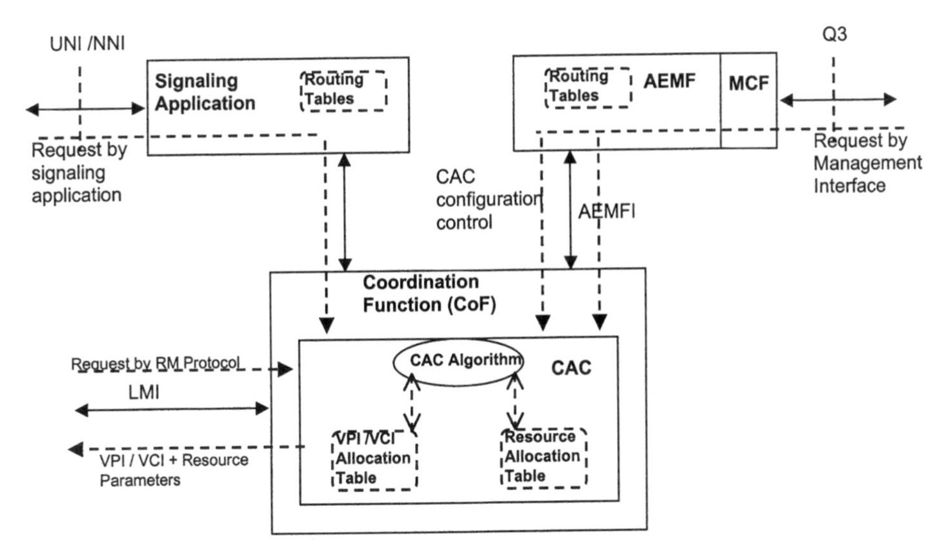

Figure 9-9. Connection admission control (CAC) function model. The CAC function may be separated from the CoF.

quirements of the VPC or VCC request, as determined from the "resource allocation" and "VPI/VCI allocation" tables accessed by the CAC algorithm. Consequently, if insufficient resources are available, the CAC function indicates that the connection request (whether from signaling or management) cannot be accepted.

The relationship between the instantaneous bandwidth requests inherent in the RM cells and the CAC resource allocation table is evidently more complex, since in this case it is not simply a matter of acceptance or rejection of a VPC or VCC during the set-up phase, which is independent of the RM related procedures. The use of RM cells in ABR, for example, implies dynamically changing instantaneous cell rates between some minimum cell rate (MCR) and some PCR value, as indicated in the explicit rate (ER) value in the RM cell, as described earlier. While it may be intuitively evident that the explicit rate algorithm used in the processing of the ABR RM cells will need to access the "bandwidth pool" represented by the "resource allocation tables" in the CAC model, the detailed functional modeling of the procedures involved are not well defined at this stage. However, it has been generally recognized that it will be necessary to represent the RM-cell-related functions in a specific "traffic control" function block that is able to communicate with the CAC function block.

In addition, given the potential complexity and importance of the CAC function block, its inclusion in the CoF block has also been questioned in recent work. As a result, it is now generally accepted that it is preferable to represent the CAC function as a separate functional block within the plane management block, independent of, but connected to, the CoF, analogous to the AEMF or MCF block. This further decomposition of the plane management block and CoF, together with the addition of a dedicated "traffic-control"-related function block will clearly change the representation of the plane management block of the general NE functional model described above in relations to Figs. 9-6 and 9-7. It should also be anticipated that the considerable current work in the ITU-T and ETSI on developing more detailed representation methodologies for the description of ATM NEs will also significantly enhance the basic NE model described here, while maintaining the underlying principles on which the model is based.

Returning to the description of the LMIs and their relationship to the CoF, it should be noted that a complete classification of the LMIs has been provided in ITU-T Recommendation I.732

for each functional block. An example of the I.732-based representation of the control and report LMIs for a particular block, the VPL_T [or VPE] block, is shown in Fig. 9-10 for the fault management (FM) related LMIs. It should be noted that the "control LMI" is defined from the LM to the transfer block, whereas the "report LMI" is for the reverse direction, as indicated earlier. Fig. 9-10 also shows that the transfer block primarily includes the insertion, extraction and copy functions related to the (fault management) VP OAM cells described earlier, whereas the layer management block includes the processing and generation of the specific OAM cell payloads (i.e., the function-specific fields). A similar representation may be found in I.732 for each of the relevant function blocks, and may be envisaged for each of the OAM functions. The complete list of FM-related LMIs which are numbered FM1 to FM32 are listed in Table 9-4, derived from ITU-T Recommendation I.732. It is also useful to visualize the relationship between the FM LMIs in the A to B and B to A directions of OAM cell flow in the NE and the role of the CoF in enabling this relationship between the layers as well as flow directions. A simplified version of this overall relationship is shown in Fig. 9-11 for the FM-related LMIs reproduced from ITU-T Recommendation I.732. Note that, to simplify the representation, the corresponding transfer function blocks, as well as connection-related blocks, are not shown.

In the description of the ATM layer OAM functions earlier, the general OAM rule that defect indications in a given layer (or level) result in the generation of AIS cells in the same layer, as well as the adjacent higher layer, was noted. This general OAM guideline is illustrated in the model shown in Fig. 9-11. For example, referring to the LM blocks in the B to A direction it may be noted that detection of a transmission path (TP) defect indication, such as TP AIS or loss of pointer (LoP) etc., results in the generation of VP AIS cells in the VPL_T or [VPE] block initiated by means of the LMI labeled FM2. Similarly, the detection of a VP loss

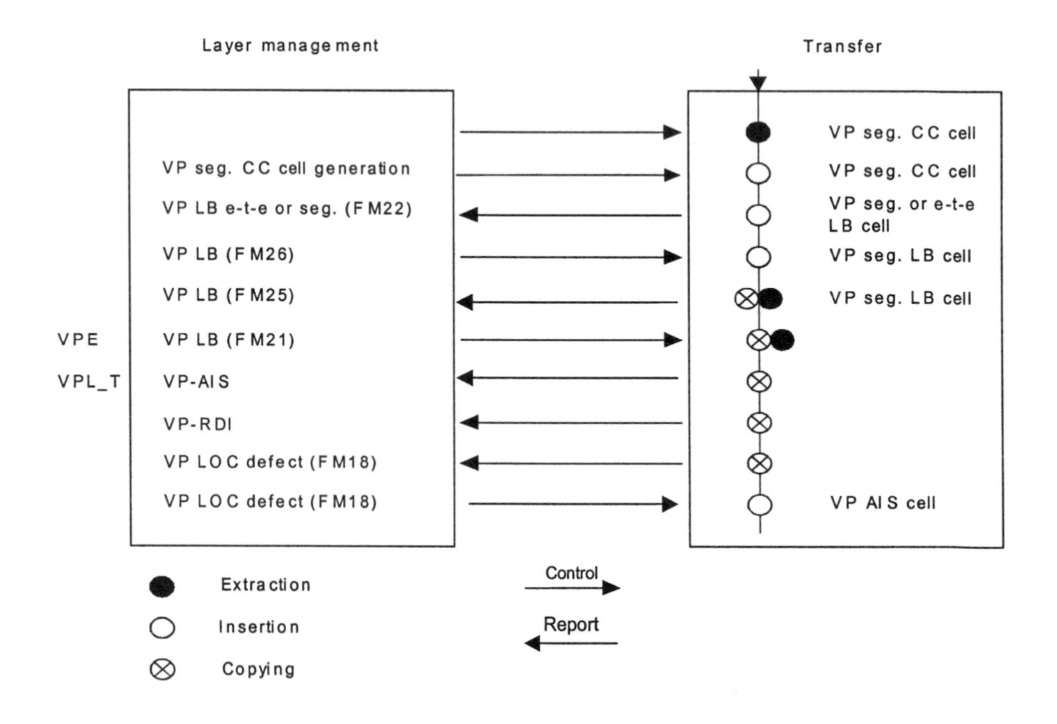

Figure 9-10. Layer management indications (LMIs) for VPL_T/VPE (A to B).

TABLE 9-4. List of LMIs for fault management (derived from ITU-T Recomendation I.732)

LMI	Description	Status
FM1	TP-RDI generation in accordance with Recommendations G.783 and G.784	Present for SDH or PDH interfaces
FM2	VP-AIS generation due to TP defect	Present
FM3	TP-RDI generation due to LCD defect	Present
FM4	VP-AIS generation due to LCD defect	Present
FM5	VP-AIS generation due to VP-LOC defect	Present if continuity check is implemented
FM7	VP-RDI generation due to VP-LOC defect	Present if continuity check is implemented
FM8	VC-AIS generation due to VP-LOC defect	Present if continuity check is implemented
FM9	VP-RDI generation due to VP-AIS detection	Present
FM10	VC-AIS generation due to VP-AIS detection	Present
FM11	VC-AIS generation due to VC-LOC defect	Present if continuity check is implemented
FM13	VC-RDI generation due to VC-LOC defect	Present if continuity check is implemented
FM14	VC-RDI generation due to VC-AIS detection	Present
FM16	VC-AIS generation due to VC-LOC defect	Present if continuity check is implemented
FM18	VP-AIS generation due to VP-LOC defect	Present if continuity check is implemented
FM21	VP LB detection and timer activation in A to B due to VP LB generation in B to A at a source point	Present if LB is implemented
FM22	VP LB detection and timer activation in B to A due to VP LB generation in A to B at a source point	Present if LB is implemented
FM23	VC LB detection and timer activation in A to B due to VC LB generation in B to A at a source point	Present if LB is implemented
FM24	VC LB detection and timer activation in B to A due to VC LB generation in A to B at a source point	Present if LB is implemented
FM25	VP LB generation in B to A due to VP LB detection in A to B at a loopback point	Present if LB is implemented
FM26	VP LB generation in A to B due to VP LB detection in B to A at a loopback point	Present if LB is implemented
FM27	VC LB generation in B to A due to VC LB detection in A to B at a loopback point	Present if LB is implemented
FM28	VC LB generation in A to B due to VC LB detection in B to A at a loopback point	Present if LB is implemented
FM29	VP LB detection and timer activation in B to A due to VP LB generation in A to B at a source point	Present if LB is implemented
FM30	VC LB detection and timer activation in B to A due to VC LB generation in A to B at a source point	Present if LB is implemented
FM31	VP LB generation in A to B due to VP LB detection in B to A at a loopback point	Present if LB is implemented
FM32	VC LB generation in A to B due to VC LB detection in B to A at loopback point	Present if LB is implemented

of continuity (LOC) defect in the VP_T or [VPCT] block results in the generation of VC AIS cells in the VCL_T or [VCE] block initiated by the LMI labeled FM8. The LMIs initiating AIS cell generation in the same layer, such as FM5 for the VPL_T block, and FM11 for the VCL_T block are also shown. The generation of the corresponding RDI cells in the (reverse) A to B direction may also be noted, as initiated by the LMIs labeled FM9 for the VP_T block and FM14 for the VC_T block. An exception to this general role should also be pointed out, since this has generated some controversy in the description of NEs where both ATM and SDH technologies are utilized. The exception relates to the LMI labeled FM3, which initiates the generation of a transmission path RDI (TP RDI) as a consequence of detecting the loss of

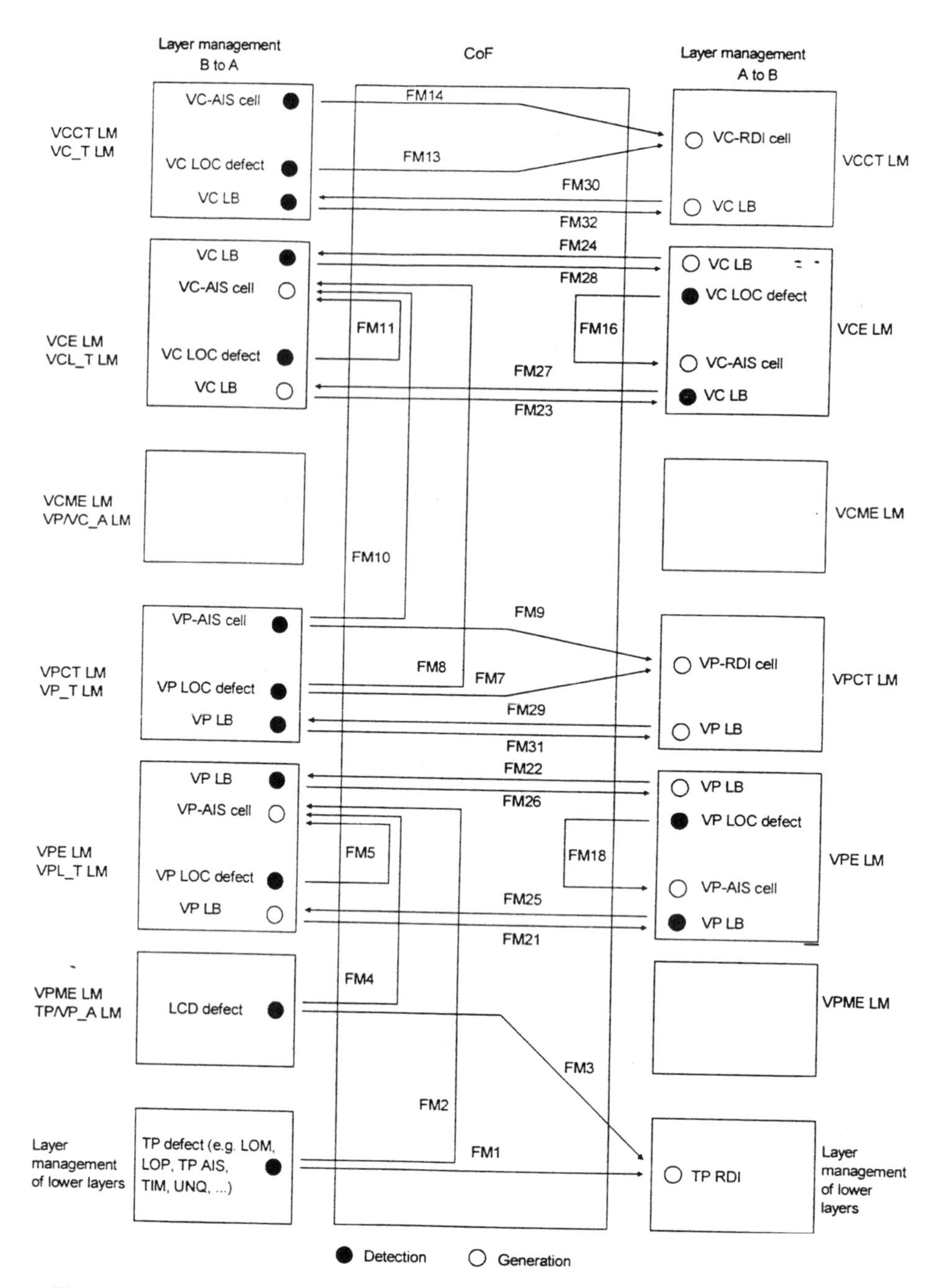

Figure 9-11. LMIs for fault management. Figure derived from ITU-T Recommendation 1.732.

cell delineation (LCD) defect indication in the TP/VP_A or [VPME] block, which lies above the TP-related block. The reason for this apparent breakdown of the general OAM guideline noted above is essentially historical, since as may be recalled from the B-ISDN PRM concepts described earlier, the cell delineation function was included in the transmission convergence sublayer, which was also considered to include the RDI function at that stage. The subsequent evolution of the NE functional modeling methodology allocated these functions to different functional blocks, resulting in an anomaly whereby a defect in a higher-level block initiates an RDI in an apparently lower-level block.

TABLE 9-5. List of LMIs for performance management (derived from ITU-T Recommendation I.732)

LMI	Description	Control or report
PM1	VP segment or end-to-end PM nonintrusive monitoring activation	Control
PM2	VP segment or end-to-end PM result	Report
PM3	VP segment PM activation	Control
PM4	VP segment PM result	Report
PM5	VP segment PM generation	Control
PM6	VP end-to-end PM activation	Control
PM7	VP end-to-end PM result	Report
PM8	VC segment or end-to-end PM nonintrusive monitoring activation	Control
PM9	VC segment or end-to-end PM result	Report
PM10	VC segment PM activation	Control
PM11	VC segment PM result	Report
PM12	VC segment PM generation	Control
PM13	VC end-to-end PM activation	Control
PM14	VC end-to-end PM result	Report
PM15	VC end-to-end PM generation	Control
PM16	VC segment PM generation	Control
PM17	VC segment PM result	Report
PM18	VC segment PM activation	Control
PM19	VC segment or end-to-end PM result	Report
PM20	VC segment or end-to-end PM nonintrusive monitoring activation	Control
PM21	VP end-to-end PM generation	Control
PM22	VP segment PM generation	Control
PM23	VP segment PM result	Report
PM24	VP segment PM activation	Control
PM25	VP segment or end-to-end PM result	Report
PM26	VP segment or end-to-end PM nonintrusive monitoring activation	Control
PM27	VP segment PM activation in B to A due to VP segment PM generation in A to B at a segment endpoint	
PM28	VP segment PM activation in A to B due to VP segment PM generation in B to A at a segment endpoint	
PM29	VP end-to-end PM activation in A to B due to VP end-to-end PM generation in B to A at a connection endpoint	
PM30	VC segment PM activation in B to A due to VC segment PM generation in A to B at a segment endpoint	
PM31	VC segment PM activation in A to B due to VC segment PM generation in B to A at a segment endpoint	
PM32	VC end-to-end PM activation in A to B due to VC end-to-end PM generation in B to A at a connection endpoint	

So far, the detailed description has concentrated on the LMIs related to the OAM fault management processes within the NE. A similar representation of the performance management (PM) related LMIs may also be described, as depicted in Table 9-5, which lists all the PM LMIs illustrated in Fig. 9-12, both taken from ITU-T Recommendation I.732 and reproduced here for convenient reference. Although such a representation of all the PM-related LMIs may appear complex at first glance, the underlying symmetry, as well as independence between the VP- and VC-related blocks for the PM OAM procedures may be exploited to understand the PM LMIs. In addition, the differences between the PM LMIs and the FM LMIs resulting from the different underlying mechanisms need to be borne in mind. These primarily relate to the activation (deac-

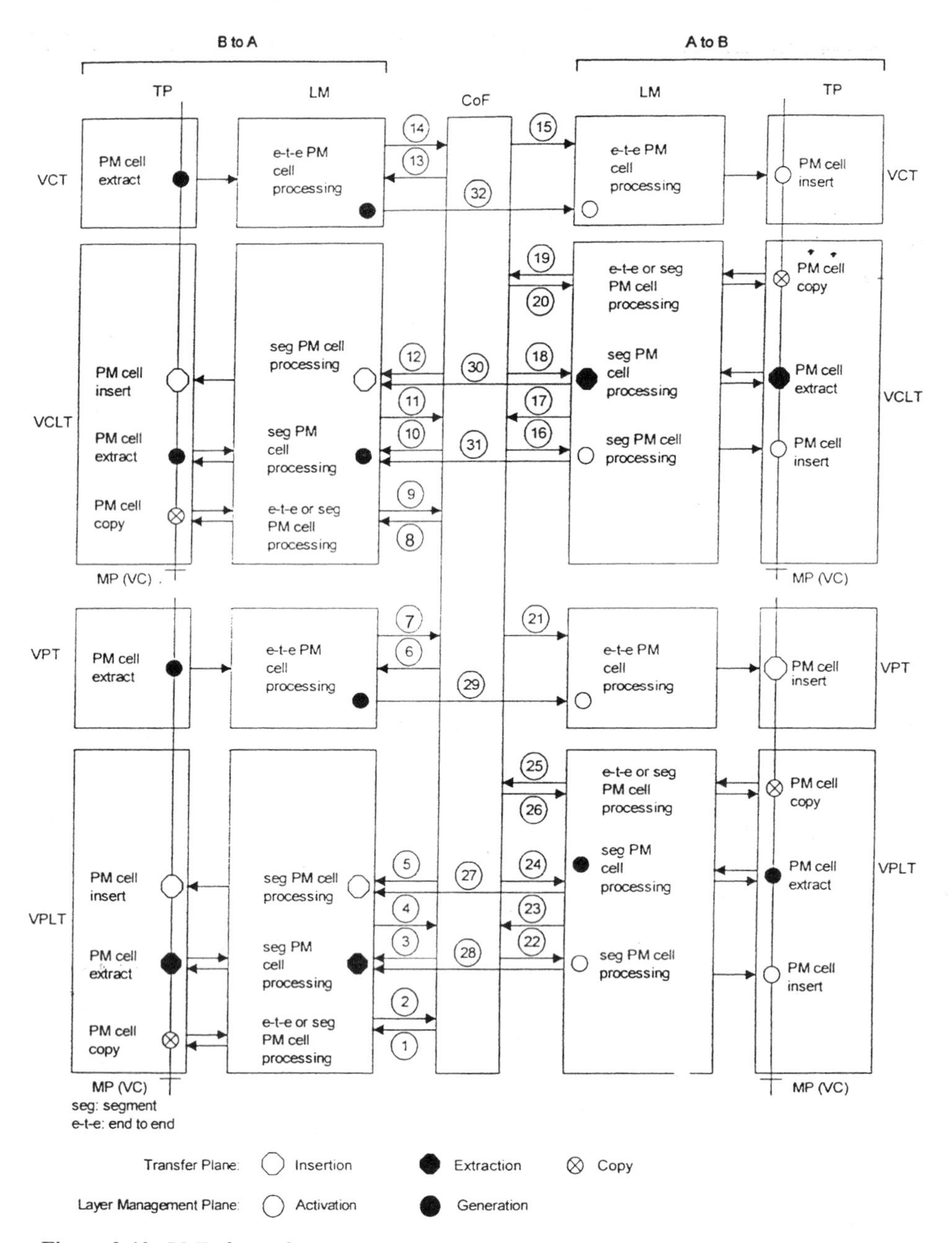

Figure 9-12. LMIs for performance management. (Derived from ITU-T Recommendation I.732.)

tivation) of the PM OAM cell procedures and to the reporting of the results of PM cell processing to the AEMF via the CoF. For simplicity, the corresponding AEMFIs are not shown but it should be assumed that where a "PM result" LMI is represented, the corresponding result is communicated to the AEMF block through the CoF by means of a related AEMFI. It may also be recognized that the relationships between the paired forward monitoring (FPM) and backward reporting (BR) PM OAM cells are represented by the PM LMIs labeled from PM27 up to PM32 for both segment and end-to-end PM cells. Also, as may be expected and unlike the FM case, the PM LMIs do not initiate OAM cell generation/insertion between blocks at different layers, thereby simplifying their representation within each layer.

9.6 EXAMPLES OF THE DETAILED FUNCTIONAL MODEL

The basic ATM NE functional model described earlier provided a simple means to describe any type of ATM NE by using the functional blocks in accordance with the general equipment categories such as switches, cross-connects, or multiplexers. It will also be evident that the detailed ATM NE functional model may also be used in the same way, while serving to represent a further level of detail in the decomposition of the basic functional blocks. In such a description, either the G.805/I.326 or the PRM-based representation may be used. Some examples of applying the detailed modeling representations are shown in Fig. 9-13 and Fig. 9-14, using some of the same equipment types (VC switch and VP Multiplexer) that were described previously for comparison.

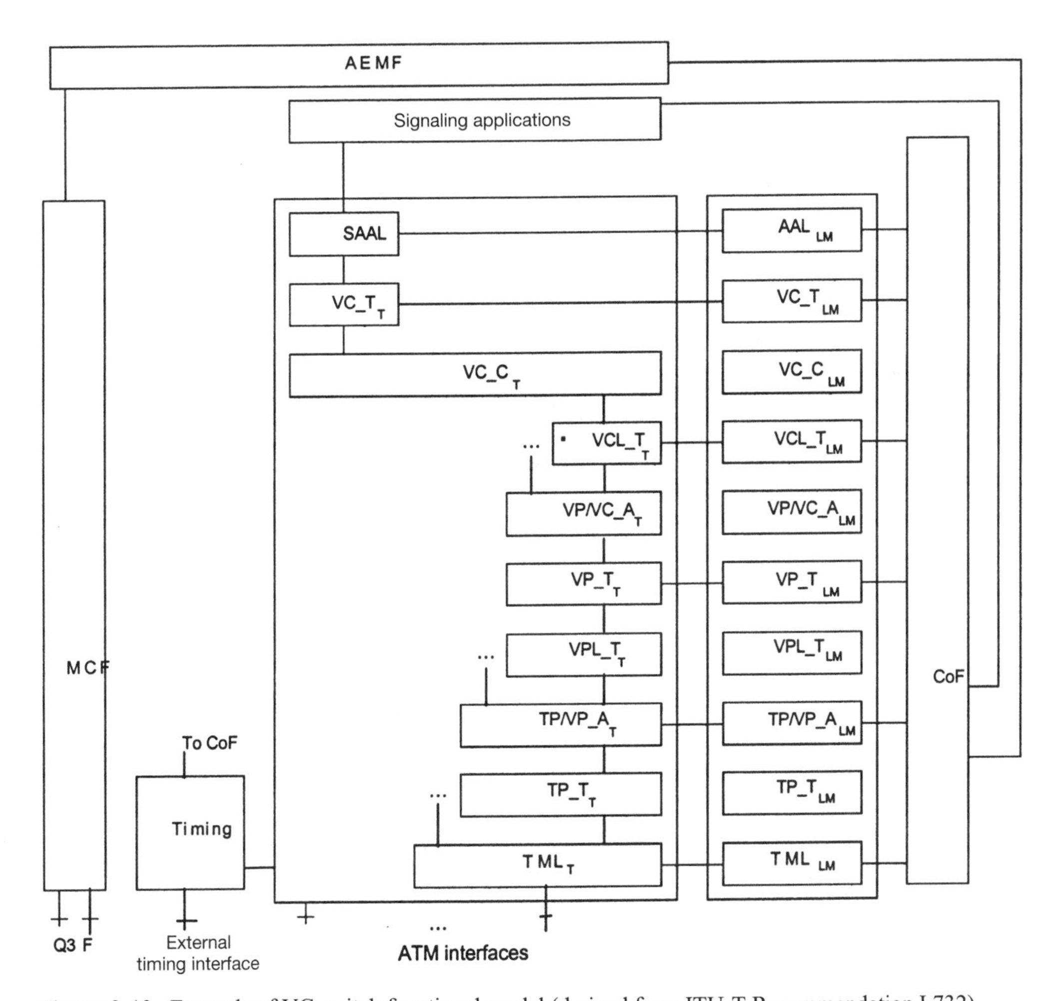

Figure 9-13. Example of VC switch functional model (derived from ITU-T Recommendation I.732).

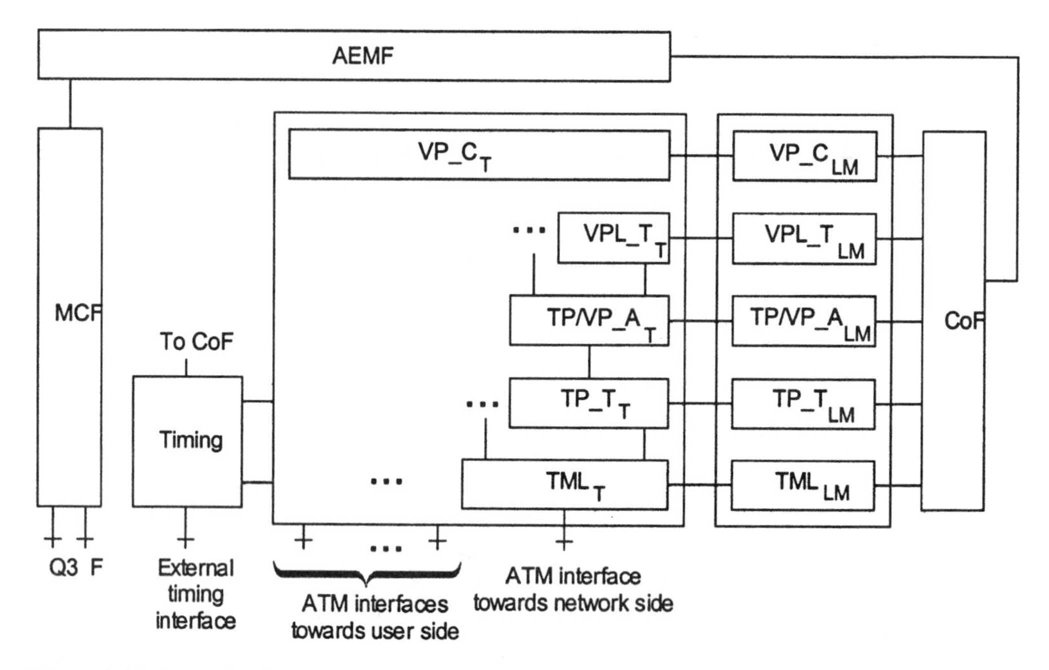

Figure 9-14. Example of VP multiplexer functional model (derived from ITU-T Recommendation I.732).

These examples demonstrate the use of the functional blocks of the detailed NE model to extend the characterization of any type of ATM NE to the level of detail required. It should be stressed that in interpreting such representations of the ATM NE, the general comments made regarding the examples described previously should be borne in mind. Thus, while such representation provides a convenient means of visualizing the main capabilities of the NE in question, the individual blocks do not necessarily imply that all the contained functions are present in the equipment. Additional specification is clearly required to define the selected subset of functions implemented in any particular equipment, based on the intended network applications.

9.7 THE AEMF AND TMN-RELATED INTERFACE PROTOCOLS

In describing the ATM Layer OAM functions, their close relationship with the overall telecommunications management network (TMN) framework [also loosely referred to as the network management system or the operations support functions (OSF)] was stressed. This relationship may now be more clearly seen in modeling the detailed functional decomposition of ATM NEs. The modeling necessitates the introduction of concepts such as the LMIs, CoF, and AEMF, together with the external (open) management interfaces such as the Q3 (or M4 in ATM Forum terminology) interface, which enable the flow of management and control information between the NE and any external management system. In principle, it is important to recognize that the extensive management capabilities designed into ATM technology can only be fully utilized if both the ATM layer OAM&P functions and the associated TMN-related functions work in concert for any given network application. When this occurs, it may be evident that the network operator can utilize any or all of the individual capabilities in the basic TMN management categories represented within the AEMF block. These TMN functional categories include:

1. Fault management (FM)
2. Performance management (PM)
3. Configuration management (CM)
4. Accounting management (AM)
5. Security management (SM)

The study of the TMN related functions, together with the associated protocols, management information base (MIB) structures, and related mechanisms in itself constitutes a vast subject of interest for networking technology. Although this important and specialized area of study is outside the scope of this text, it is useful for the reader to appreciate its relationship to the description of the AEMF block for completeness in understanding of the ATM NE functional modeling. This section provides a summary overview of TMN architectures and procedures, and interested readers may refer to one of a number of specialist accounts for a fuller description of network management information modeling [9.3].

The general TMN protocol architecture model shown in Fig. 9-15 is essentially intended to provide a broad and flexible framework for the description and design of general network management and operations systems. It may be recalled that traditionally telecommunications network management systems were largely developed to meet operator-specific requirements and often tailored around specific NE vendor implementations, thereby precluding interoperability. The increased trend toward so-called "open" or standardized management and control interfaces between NEs and operating systems, as well as between different operating or network management systems, has resulted in numerous architectural approaches to representing network management protocols, including those increasingly based on so-called "open distributed processing" (ODP) architectures. However, for our purposes, the TMN model provides a layered

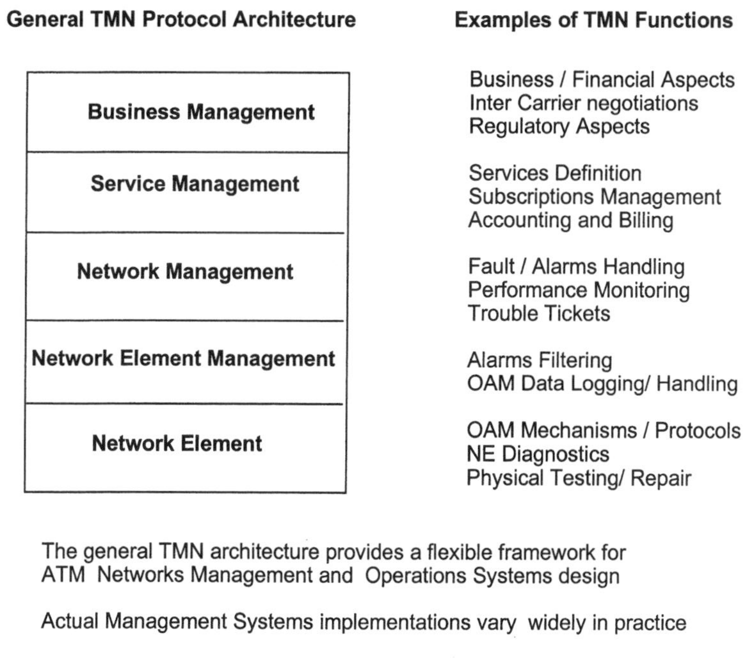

General TMN Protocol Architecture	Examples of TMN Functions
Business Management	Business / Financial Aspects Inter Carrier negotiations Regulatory Aspects
Service Management	Services Definition Subscriptions Management Accounting and Billing
Network Management	Fault / Alarms Handling Performance Monitoring Trouble Tickets
Network Element Management	Alarms Filtering OAM Data Logging/ Handling
Network Element	OAM Mechanisms / Protocols NE Diagnostics Physical Testing/ Repair

The general TMN architecture provides a flexible framework for
ATM Networks Management and Operations Systems design

Actual Management Systems implementations vary widely in practice

Figure 9-15. Telecommunication management networks (TMN) architecture.

approach that may be more easily related to the NE modeling representations considered here. It may also be recognized that the three lower layers are of primary interest from the perspective of the AEMF. These include the network element, network element management, and network management layers, as shown in Fig. 9-15. Considering the general categories of functions listed in these TMN layers, it may be envisaged at first glance that the lower three layers of the TMN framework very broadly encompass the management functions described in the ATM NE layer management block, the AEMF block, and the external network management system. The higher service and business management layers will not be considered here since these relate more to the business than technical aspects of network operations. They are, however, of interest to the configuration, accounting, and security management function categories listed earlier.

For example, the functions listed for the TMN network element layer (NEL), also called the element management layer (or EML), would appear to generally correspond to the physical and ATM layer OAM functions, as well as any equipment-specific diagnostics and/or test capabilities incorporated in the NE. Similarly, examples of the functions listed for the network element management layer (NEML, also referred to as the element management layer, EML) roughly correspond to those falling within the AEMF block, as described earlier. The TMN network management layer (NML) functions broadly correspond to those that may fall within the scope of the network management system or OSF, assuming a management interface (e.g., Q3) between the EML and NML. However, in practice, the actual partitioning of management functions between the various NEs and management-related entities in any given network implementation may vary widely, so that the model represented in Fig. 9-15 should be viewed simply as a general framework. An example of a physical realization of a network management architecture is shown schematically in Fig. 9-16, together with the corresponding TMN layers to indicate a possible mapping. In such an implementation, it may be seen that management interfaces, which may be either open or proprietary, exist between both the network management system or OSF and the element management system (EMS), as well as between the latter and the ATM NE. In this example, the EMS, corresponding to the element management layer (EML) of the TMN

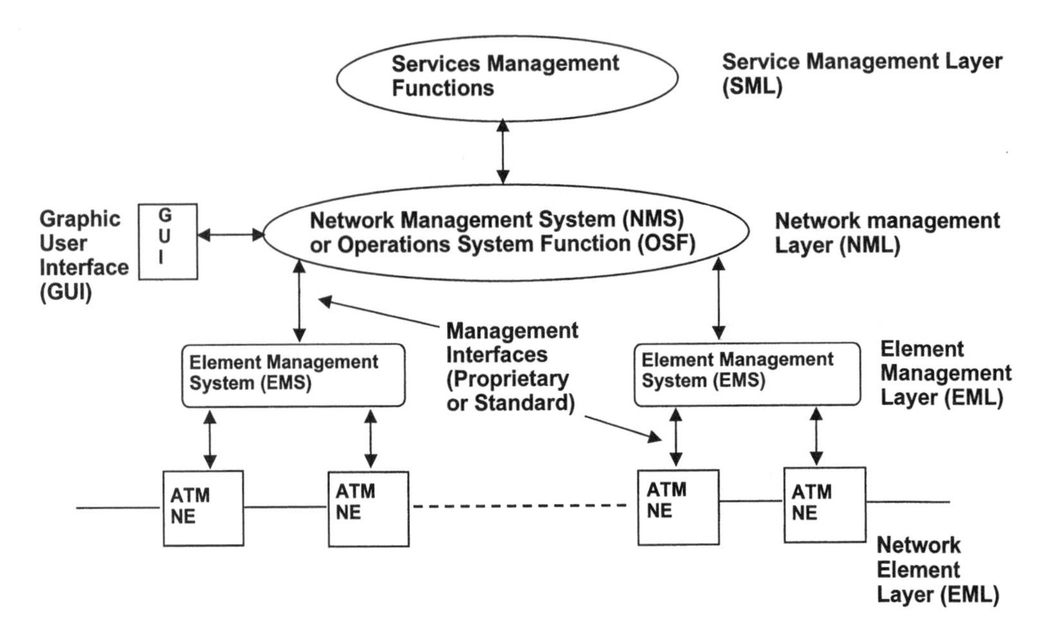

Figure 9-16. Network management system example.

model, is responsible for more than one ATM NE. Also, it may be noted that the overall network management system (NMS) utilizes a so-called graphic user interface (GUI) to facilitate the visual correlation of management information across the network at various levels of detail, if required. Other means of data handling and status monitoring of remote locations may also be provided in practice.

It will be recognized that a substantial part of the network management architecture consists of the processing, communication, and handling of management-related information. This information is often structured in the so-called management information base (MIB) within management entities such as the AEMF block or OSF block. The description of such ordered information structures lends itself readily to object-oriented (OO) modeling approaches and thus is generally followed in the TMN-related information modeling protocols. For the structured communication of management-related information, the TMN management interface protocols are based on a "manager–agent" architecture, as shown in Fig. 9-17. In this model, the concept of a so-called common management information service element (CMISE) is introduced to provide a framework for specifying the procedures enabling exchange of management information between the "manager" and "agent" entities within the operating systems, for example across a typical OS to NE interface. The CMISE entities may be described for each of the groups of management functions that are represented in the system, e.g., for fault management or performance management or subsets thereof, which may be of interest. The exchange of management information in the TMN interfaces utilizes the common management information protocol (CMIP). The CMIP messages that encode the management-related information, may use any one of a number of underlying transfer protocols including preprovisioned ATM VCCs (or X.25 or frame relay services, etc.) to exchange the information between the manager and agent entities.

Within the manager and agent entities, the information is structured in the MIB as groups of so-called managed objects (MOs), which may be manipulated or exchanged across the management interface by means of the CMIP message set. In general terms, a managed object is simply

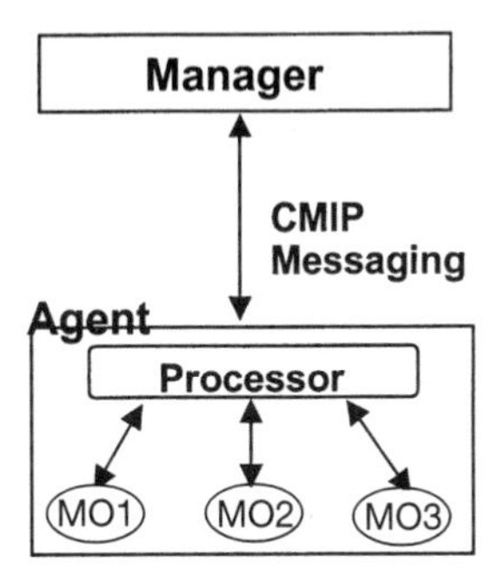

Notes. 1. TMN protocol operations based on "manager–agent" structure. 2. Common management information service element (CMISE) concept provides framework for procedures enabling exchange of management information between managers and agents in operating systems. 3. Common management information protocol (CMIP) is the TMN protocol used to exchange management information across the management interface. 4. Management information base (MIB) is the structured information used in the CMIP/CMISE protocol. 5. MIBs may be proprietary or based on standard (or open) information models that have been defined by many standards/forums (e.g. ATM Forum, IETF, ITU-T SG 4, and SG 15). 6. Standard management interface protocols are commonly based on CMIP or SNMP (simple network management protocol).

Figure 9-17. TMN protocols.

an abstract representation of any management-related aspect of the NE, regardless of whether it is a physical entity such as a specific circuit pack, or a logical entity such as an ATM VCL or an AIS OAM cell event. To facilitate their processing using object-oriented software design technology, as well as to provide a common description methodology for the construction of MIBs, the MOs are generally represented in terms of an abstract syntax notation (ASN) language in accordance with a defined set of rules. The details of this methodology and the complex set of rules and relationships between the managed object classes used to construct the MIBs need not concern us here, but it may be of use to note that, in general, managed objects may include four components. These are classified as:

1. Attributes. The attributes are used to uniquely describe any given MO at the management interface.
2. Operations. The operations are the actions that may be performed on any given MO, such as updates or modifications.
3. Notifications. These are used to signify any notification the MO may make to any management entity, such as a report of a defect indication.
4. Behavior. This component essentially describes the behavior of the MO in response to any management actions.

In addition to these components of the MOs contained in any given MIB, the TMN procedures often distinguish between the "administrative" and "operational" states of the MOs or resources represented in the MIBs. In general terms, administrative states are occasioned by direct action by the network operator through the management interfaces to any given NE, whereas operational states pertain to events that may affect any of the MOs in the MIB. An example of an "administrative" state is the blocking of a VPC or VCC resource by management action for the purpose of testing the resource. An example of the "operational" state is when an AIS or RDI OAM cell event indicates a defect (or fault) condition along any given VPC or VCC. In this case, the operational state is an AIS state that may be reported to the AEMF and possibly to the OSF via the management interface using a CMIP message procedure.

The general TMN-related concepts briefly summarized above enable us to visualize the methodology by which the AEMF may be modeled and the close analogy with the detailed NE functional decomposition described earlier. Thus, the AEMF may be represented as a TMN agent that comprises a MIB (or a set of MIBs, depending on how these are structured), which itself is comprised of ordered groups of managed objects (MOs), each representing a particular management aspect of the relevant functional components of the NE. The AEMF exchanges the (processed) management information represented by the MOs with the TMN manager entity (e.g., the OSF) through the management interface using the CMIP messages that operate between the peer CMISE entities in the manager–agent architecture. Since processing of management information is typically performed by software, the MIBs are structured in terms of object-oriented methodology to facilitate modular software design. Despite this somewhat oversimplified view, however, it must be stressed that the ATM MIBs are typically complex and unfortunately not uniquely defined. In practice, the ATM MIBs may be proprietary to specific implementations, thereby making interoperability between different management systems difficult (or even impossible), whereas different standardized MIBs have been defined by a number of bodies, necessitating choice in the selection of a suitable management interface for any network application. In effect, parallel work on specifying ATM-related MIBs has resulted in somewhat different approaches taken by the ITU-T, the ATM Forum, the TeleManagement Forum, and the IETF. Although there are some underlying similarities between some of these standardized ATM MIBs, it should not be assumed without detailed testing that management systems (e.g., AEMFs) based on these different approaches may interoperate. An added complication is that the TMN-based CMIP approach to exchanging management information between MIBs is not as widely used in existing ATM management systems as is the competing so-called simple network management protocol (SNMP) based interfaces originally developed by the IETF for IP-based network management purposes.

9.8 CMIP/CMISE MESSAGES AND FUNCTIONS

As noted above, the messaging protocol used for the exchange of the management information across the management interfaces of the TMN model is called the common management information protocol (CMIP). These messages are exchanged between the peer common management information service elements (CMISE) in the manager or agent entities. The CMIP procedures use the basic message set listed below:

1. M-Event REPORT. This message is used to report management-related events (e.g., AIS state)
2. M-GET. The GET message is used to obtain any management information from the corresponding peer entity.
3. M-CANCEL-GET. This message may be used to cancel the GET command.
4. M-SET. The SET command is used to modify, update, or "write" the contained management information (or MO) by the peer entity.
5. M-ACTION. This message is used to request a management action by another remote entity on a given managed object.
6. M-CREATE. The CREATE message (or command) is used to instantiate (or create) a managed object. An important example of the use of the CREATE message is the instantiation of a cross-connection between two VPLs or VCLs in setting up a PVC.
7. M-DELETE. The DELETE message (or command) is the opposite of the CREATE command in that it is used to delete a managed object or a parameter that may have been set up by the CREATE command.

It is not necessary here to describe in further detail the CMIP message structures and procedures to recognize that the message set briefly outlined above may be used for any required management function such as reporting of failures to creation of VP or VC Link connections as required. As noted earlier, the CMIP message structures may be carried by a range of transfer protocols, from LAN protocols such as EtherNet to X.25 or frame relay connections dedicated between the management entities. A particular case of interest for ATM networks is to utilize dedicated ATM VPCs or VCCs between the management entities for the transport of CMIP messages, analogous to the use of a dedicated (i.e., common channel) ATM VCC for the transfer of signaling, or for that matter routing messages as described earlier. When an ATM VPC or VCC is used for the exchange of the CMIP messages, they are encapsulated in an AAL Type 5 protocol data unit (PDU), as will be described later. Since the encoding of the ASN-based management information or MOs within the CMIP messages may result in relatively large data frames, the use of a high-speed ATM VCC for the exchange of such data provides significant performance advantages over the use of lower-speed (e.g., X.25-based) data transport mechanisms.

From the perspective of the NE, the CMIP-based messages may be used for configuration management functions. Examples of the use of configuration-related commands such as SET or CREATE include the configuration of physical resources (represented by MOs) such as:

1. CircuitPack
2. EquipmentHolder
3. DS3LineTTPBidirectional
4. TcAdaptorTTBidirectional

Some examples of the configuration of logical resources (represented by MOs) include:

1. UNI (note that this may imply further subgroups of MOs)
2. NNI (as above)
3. atmCrossConnection
4. atmFabric
5. vpCTPBidirectional
6. vcCTPBidirectional

As pointed out earlier, the CMIP procedures (e.g., M-CREATE) may also be used to perform an ATM cross-connect function between any VPLs or VCLs by manipulation of the "atmCross-Connection" managed object listed above.

Note that the terminology used in the above examples of typical MOs is intended to introduce the reader to the kind of terminology used in describing the MO and their classes used in constructing the MIB. This mode of naming the MOs is based on the ASN methodology referred to above. For present purposes it is not necessary to describe all the acronyms used in the naming conventions for the MOs, but it should be recognized that these conventions enable an ordered and unique means of relating the MOs to the relevant individual functional components of the NE.

For the case of performance management (PM) TMN category, the CMIP procedures are used to provide a number of functions such as:

1. Initiating (i.e., activation/deactivation) PM OAM cell flows and assigning the measurement intervals on selected VPCs or VCCs
2. Collection of relevant PM parameters or data collection
3. Storage of PM data including duration, history, and screening functions
4. Thresholding of PM data, including assigning of the parameter thresholds and intervals
5. Reporting of the PM data, including any relevant filtering of the parameters when required

As for the case of the configuration management examples given above, the PM operations may also be carried out on both physical and logical resources (managed objects). Some examples of PM-related physical managed objects include:

1. atmTrafficLoadCurrentData
2. atmTrafficLoadHistoryData
3. atmCongestionCurrentData

Examples of PM-related logical managed objects include:

1. atmConnectionCurrentData
2. upcNpcCurrentData

3. upcNpcHistoryData
4. cellLevelProtocolCurrentData

It may be recognized that due to the potentially large number of PM-related parameters in the ATM network and the resulting data generated by the PM procedures both the physical and ATM layer OAM functions, the PM MIB operations may require complex processing. Nonetheless, this aspect of overall network management remains an essential component in ensuring that the requisite QoS capabilities are maintained.

The latter observation may also be made for the fault management category of TMN functions, in which, as may be anticipated, the CMIP messaging procedures may be used to provide the FM-related capabilities described in some detail earlier. These include the use of the fault management OAM functions for alarm surveillance and testing of VPCs and VCCs. The CMIP procedures are used here for:

1. Reporting of AIS and RDI events using AIS/RDI OAM cell flows
2. Initiation and monitoring of the F4 and F5 level loopback, both for segment and end-to-end for simple and multiple loopback tests
3. Physical layer loopbacks (i.e., facility loopbacks at line level)
4. VP and VC level continuity check (CC) activation and monitoring

The CMIP messages are also used for the purposes of logging of FM events and forwarding of (possibly filtered) fault information such as:

1. Alarm records
2. Results of NE diagnostics
3. Trouble ticket processing

In addition, CMIP messages enable the notification and initiation of the administrative or operational states if and when required.

It should be noted that although the emphasis in the brief summary above has been on the use of CMIP functions from the perspective of the network element (i.e., the AEMF), CMIP procedures are by no means restricted to the OS-to-NE interface. Thus, CMIP may also be used for exchange of management information between the MIB capabilities in different OSFs through the so-called TMN "X" interface, if required. It is clear that the CMIP procedures and functions provide comprehensive capabilities for the overall management of large public domain ATM networks. However, this very comprehensiveness and flexibility has resulted in the general perception that CMIP-based MIBs are unnecessarily complex, particularly for data networking applications. Consequently, the use of CMIP-based management systems has been somewhat limited to date, particularly as an alternative protocol, in the form of SNMP, is widely available and provides similar capabilities.

9.9 SIMPLE NETWORK MANAGEMENT PROTOCOL (SNMP) FUNCTIONS AND MESSAGES

The so-called simple network management protocol (SNMP) was initially developed by the IETF for use in managing IP-based data networks, e.g., those that comprise the Internet. Initially intended for the purposes of exchanging management-related information between the hosts, servers, routers, etc. and the management system of the subnetworks, the IETF defined SNMP as an application layer protocol using connectionless (IP) datagrams for the transfer of SNMP messages. In recent years, the capabilities of SNMP protocols and management systems have been continuously enhanced to accommodate the growing demands of data networking, so that the use of SNMP-based management systems for ATM (as well as IP over ATM) networks is not surprising, despite the parallel work on TMN CMIP procedures. As described further below, SNMP was initially specified by the ATM Forum for use across the UNI as part of the so-called "interim local management interface" (ILMI) for configuration and provisioning of ATM NEs. Subsequently, SNMP-based MIBs have been widely adopted for ATM network management by the IETF, the ATM Forum, as well as the Network Management Forum, resulting in widely available commercial management systems based on these specifications. In addition, there have been recent proposals to introduce SNMP into the ITU-T TMN architecture in parallel with the development of CMIP. However, to date SNMP has not been adopted in the ITU-T TMN work, primarily due to questions of functional overlap and the potential need for interworking between CMIP and SNMP-based management systems and associated MIBs. Since both systems are likely to coexist in the foreseeable future, it is useful to summarize the main features of SNMP in relation to ATM NE management systems.

SNMP uses an asymmetric "request/response" protocol for the exchange of the management information in which the sender is not necessarily acknowledged by the receiver, and where the fault events are notified to the management station on an event-driven basis. However, as shown in 9-18, the basic functional architecture for SNMP is essentially similar to the "manager–agent" entities used in the CMIP framework and is independent of any specific network architecture or topology. The essential functions of the SNMP manager include:

1. Controlling and monitoring the ATM NEs using the Network management system (NMS) processor software.

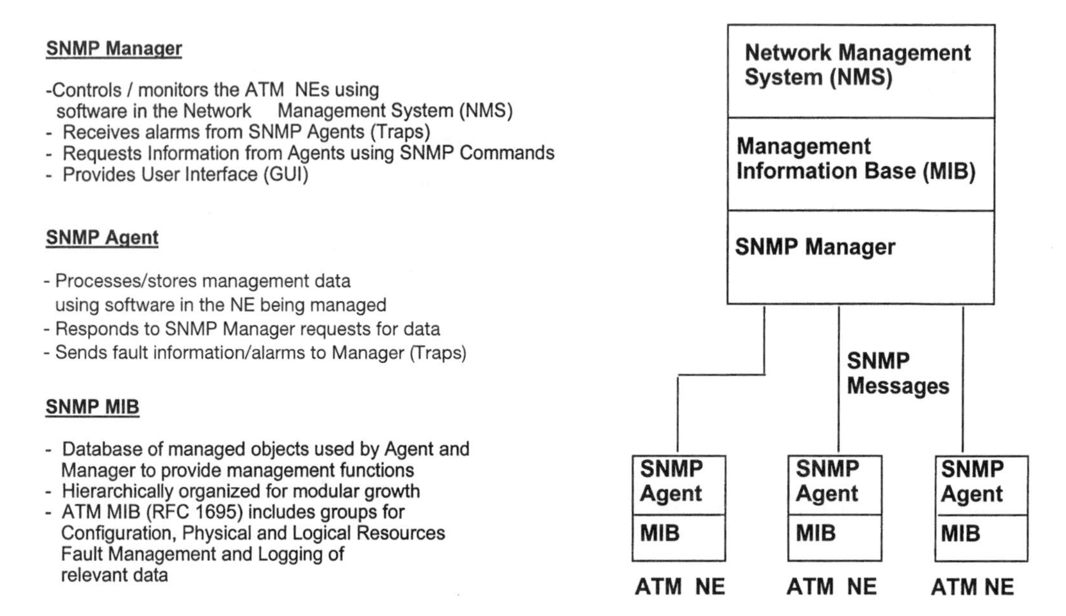

Figure 9-18. SNMP functional architecture.

2. Receives and processes alarms from the SNMP Agents. As seen below, the alarm (i.e., fault) events are contained in SNMP messages known as "traps," a terminology that derives from IP-based data networking use.
3. Requests relevant management information from the SNMP agents using the SNMP messages (commands) described below.
4. Provision of a convenient user interface such as a graphic user interface (GUI) for preservation and manipulation of network management data.

Similarly, the primary functions of the SNMP agent include:

1. Processing and storage of the relevant management data; for example, in the AEMF of the NE being managed.
2. Response to request for information from the SNMP manager.
3. Reporting of the fault (alarm) information on event-driven basis to the SNMP manager by sending "trap" messages.

As for the case of the TMN-based CMIP architecture, the SNMP-based MIBs in the manager and agent entities are an ordered database of managed objects (MO) that are hierarchically organized for modular growth. The SNMP MIBs in general also include MO groups for configuration and fault management for both physical and logical resources (objects), as noted earlier for the CMIP case. Although selected performance parameters may also be logged, performance management capabilities for SNMP MIBs are somewhat less evolved than for the TMN procedures. However, this difference may narrow as SNMP-based systems are more widely used in carrier network, rather than the originally intended enterprise network management applications.

The SNMP procedures utilize a set of four basic message types as described below. These are:

1. GET. This message (or command) is used to request individual management information (managed objects) or the status of any relevant resource. The GET message results in the return of the name and current value of the object instance (or status). In the event that the requested MO is not defined, an error status is returned to the manager entity.
2. GET-NEXT. This message is used to request the name and value of the next managed object instance, and may be used repeatedly by the manager entity to download the status of all relevant MOs if required.
3. SET. This message is used to modify or update the management information or instance of a managed object. It enables the manipulation of the management information in the SNMP MIBs, and may also be used in a "create" mode to set up ATM cross-connections if required, analogous to the use of the "CREATE" command in CMIP.
4. TRAP. This message is used to report the occurrence of any management event such as faults (defects), alarms, and so on, as indicated earlier. A TRAP message is generally sent by an SNMP agent to the manager to notify it of fault conditions on an event-driven basis. Functionally, the TRAP message is similar to the EVENT REPORT message in CMIP.

The underlying functional similarities and differences between the SNMP message set above and the CMIP message set outlined earlier should be noted. Thus, whereas SNMP essentially utilizes four message types in comparison with the seven in CMIP and hence may be considered somewhat simpler, the SNMP SET command is also used functionally in several modes, to duplicate the CMIP "CREATE" and "DELETE" messages that may be used to manipulate ATM cross connections in a NE. The main differences in functionality between the CMIP and SNMP message sets relate to the use of the ACTION message in CMIP, which has no direct counterpart in SNMP, as well as the GET-NEXT message in SNMP, which has no direct counterpart in CMIP. However, the more commonly used SNMP messages such as "TRAP," "GET," and "SET" have direct functional counterparts in the CMIP "M-EVENT REPORT," "M-GET," and "M-SET" messages. It should be stressed, however, that despite the functional similarities between these SNMP and CMIP messages, it should not be assumed that the protocols will automatically interoperate. Detailed differences in the MIB structure, MO specification and coding between the SNMP and CMIP procedures may still require interworking or translation between these different protocols, in scenarios where both SNMP- and CMIP-based management interfaces are utilized. It may also be observed that there is a common perception that SNMP-based systems are "simpler" than the CMIP-based management systems from the perspective of implementation complexity. However, this perception is more likely a result of the historical evolution of SNMP-based systems from data networking applications as opposed to the TMN-oriented CMIP procedures, which are primarily intended for the more demanding requirement and constraints of utility carrier networks. From the strictly functional perspective, assuming that a given set of management capabilities need to be supported in the ATM network (and NEs), it seems unlikely that there will be substantial differences in complexity between CMIP- and SNMP-based implementations, since in any case the complexity of the ATM MIBs implies the need for significant processing irrespective of the information exchange protocol used.

9.10 INTERIM LOCAL MANAGEMENT INTERFACE (ILMI)

The so-called interim (also now called integrated) local management interface (ILMI) was initially defined by the ATM Forum as a "temporary" management interface applicable for both the customer premises network (CPN) and across the UNI. The primary purpose of the ILMI was to provide the essential elements of CPN management until a more comprehensive M2 interface could be specified. Hence the original use of the word "interim" in ILMI. The scope of the ILMI is shown in Fig. 9-19, where it may be noted that ILMI pertains to both the "private" or enterprise network scenarios or to the UNI (i.e., access part) of the public carrier network. The ILMI utilizes the SNMP-based ATM MIB to provide management control for provisioning capabilities such as address registration configuration and link status information using the SNMP messages outlined earlier. The SNMP messages are encapsulated in the AAL Type 5 (described in more detail later) and are transported over a dedicated ATM VCC using the preassigned value of VPI/VCI = 16 on each VPL/VCL. Typically there is one MIB instance per UNI. The ILMI MIB comprises managed objects representing the relevant aspects of:

1. Physical layer configuration parameters
2. ATM layer configuration
3. Numbers of VPC or VCC supported per interface
4. ATM layer performance statistics
5. Address registration per UNI (e.g., AESAs for the interface)
6. Service categories supported, etc.

These basic management capabilities enable the customer- or network-side ATM NEs and the associated links to be configured for operation with minimal management system complexity, if required. However, as additional network applications of ATM such as LAN emulation (LANE)

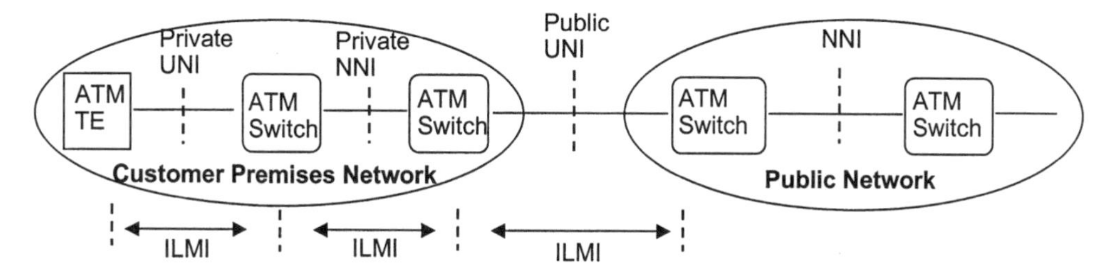

Figure 9-19. Interim local management interface (ILMI).

and MPOA, described in more detail in subsequent chapters, were developed by the ATM Forum, the basic ILMI MIB was enhanced to support these and other capabilities, and it seemed no longer appropriate to regard the ILMI as an "interim" management system. Consequently, it is now more commonly referred to as the "integrated" local management interface, and is widely used in both the ATM enterprise network scenario as well as the access ATM network to provide basic management functionality for provisioning and status monitoring of ATM NEs. It may be recognized that the ILMI protocol is analogous to a (simplified) TMN "X" interface between two network management administrative domains, except that in the case of ILMI, the domains constitute the customer or enterprise network and (public) carrier network management systems. The ILMI in fact provides an interesting example of a basic X interface that uses a dedicated (preassigned) ATM VCC between the management entities in the two domains for the exchange of information by SNMP messages. The use of a relatively high-speed VCC can ensure rapid transfer of large amounts of management (MIB) information if required.

9.11 ATM PERFORMANCE OBJECTIVES AND QoS CLASSES

In describing the ATM layer management and AEMF functions above, the use of the PM OAM cell flows and their processing in the LM and AEMF blocks of the NE to estimate performance measures such as cell loss and delay in service was outlined. Such performance parameters are also of significance in the characterization of the Quality of Service (QoS) of ATM VPCs and VCCs. The QoS in turn may be maintained by the use of the ATM layer traffic control functions described previously, as well as other factors such as network traffic engineering and dimensioning of NE buffers for anticipated traffic profiles. So far, the actual objective values of such QoS parameters was not considered, but it is clear that to obtain reasonable end-to-end service interoperability it is desirable to specify objective values for the relevant ATM performance parameters. The detailed definitions and measurement methodologies for all the ATM layer network performance parameters as well as their objective (target) values is described in ITU-T Recommendation I.356 [9.4] titled "B-ISDN ATM Layer Cell Transfer Performance." This document also describes the procedures by which the relevant performance measures may be allocated to portions of a so-called "international connection" (VPC or VCC), which spans several administrative domains. However, a detailed description of these complex formal definitions and procedures is outside the scope of this account and here we simply summarize the significant target values and the related QoS classes defined by Recommendation I.356 [9.4].

It is important to keep in perspective the relationship of the ATM performance parameters in assessing the overall end-to-end network performance experienced by any end user application. As shown in Fig. 9-20, by considering a layered approach to the characterization of potential performance impairments perceived by an end-user or terminal equipment, it is evident that the

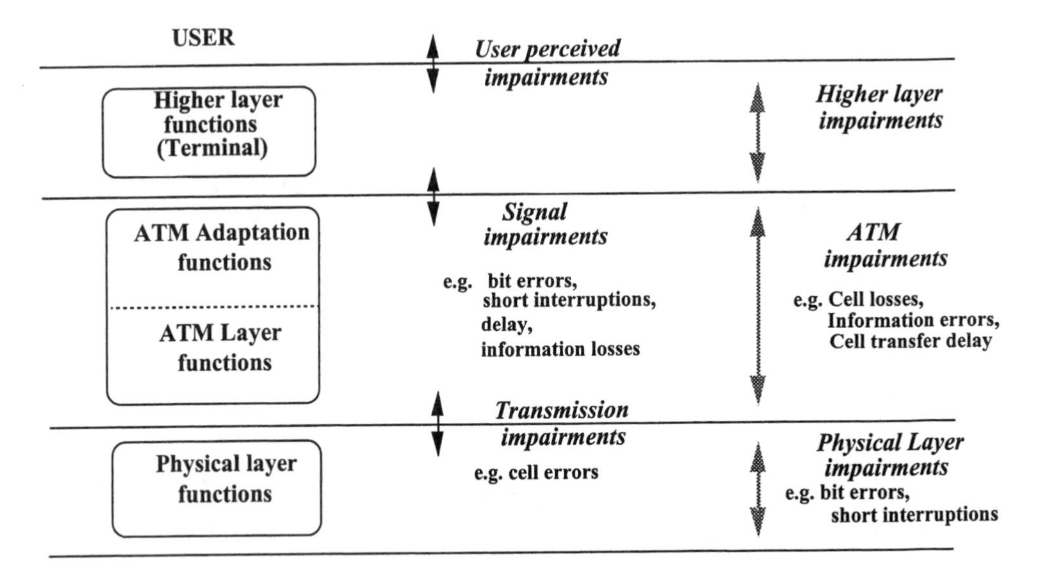

Figure 9-20. B-ISDN performance overview.

ATM layer performance impairments essentially form part of a complex interplay of possible network performance limitations. Thus, for any given end-user application, the user perceived performance impairments may include transmission or physical layer parameters such as bit or burst errors, short interruptions, and ATM layer impairments such as cell loss or delay, errored or misinserted cells, and so on. For the ATM adaptation layer and higher-layer functions, these may be signal impairments such as protocol errors, corruption of information, processing delay, etc. Consequently, it should be of no surprise that the detailed characterization, measurement and statistics of the myriad performance aspects of networking technologies in itself constitutes a vast and complex field of study, of which ATM layer performance is only one aspect. Nonetheless, it is of interest to observe that since ATM is designed as a common underlying switching and transport technology for all foreseen networking applications, its performance, both from the network management and traffic performance (or engineering) perspectives, has received substantial scrutiny. A useful result of this extensive study has been the broad acceptance of the concept of the ATM QoS classes and their relationship to the traffic control and performance management tools described earlier. In fact, it would be fair to say that no other packet mode networking technology in current use is as well characterized in quantitative terms as ATM networks.

Bearing in mind these general observations, we now turn to focus on the ATM layer performance parameters. The ATM network performance objective values for the presently defined QoS classes have been described in Recommendation I.356 and are tabulated in Fig. 9-21. It should be noted that four basic QoS classes have been identified at present (although this does not preclude the specification of additional QoS classes in future, if required). The QoS classes are:

1. Class 1 (Stringent Class). The QoS Class 1 is primarily intended for applications that require stringent bounds on both cell delay (CTD and CDV) and loss (CLR) performance parameters. Examples of such applications include voice, telephony, or video signals over ATM and circuit emulation. This QoS class may be supported by selecting an ATC such

	CTD	2-pt. CDV	CLR(0+1)	CLR (0)	CER	CMR	SECBR
Nature of the Network Performance Objectives	upper bound on the mean CTD	upper bound on difference between upper /lower 10**-8 quantileCTD	upper bound on cell loss probability	upper bound on cell loss probability	upper bound on cell error probability	upper bound on mean cell misinsertion rate	upper bound on the SECB probability
Default Objectives	no default	no default	no default	no default	4* 10**-6	1 per day	10 **-4

QOS Classes

Class 1 (Stringent class)	400 msec.	3 msec.	3* 10**-7	none	default	default	default
Class 2 (Tolerant class)	U	U	10**-5	none	default	default	default
Class 3 (Bi-level class)	U	U	U	10**-5	default	default	default
U Class U - Unspecified or Unbounded	U	U	U	U	U	U	U

Figure 9-21. QoS classes and network performance objectives. Based on ITU-T Recommendation I.356 (1996).

as deterministic Bit Rate (DBR) or statistical bit rate option 1 (SBR1), although the use of any other ATC is not precluded, provided it allows to meet the parameter objective values of this class.

2. Class 2 (Tolerant Class). The QoS Class 2 is primarily intended for applications that require bounded cell loss ratios (CLR) for the total cell stream (both CLP = 0 and CLP = 1 cells) but no strict bounds on cell delay parameters such as CTD and CDV. Examples may include data applications such as file transfer or e-mail, which may be able to tolerate non-real time transfer of information. A typical ATC that may be used for this class may include SBR 1, although it is clear that DBR may also be more than sufficient for this purpose.
3. Class 3 (Bilevel Class). The QoS Class 3 is somewhat similar to the Class 2 (Tolerant Class) except that the CLR upper bound value of 10^{-5} pertains only to the high-priority CLP = 0 cells in the total cell flow. The CLR value for the lower-priority CLP = 1 cells is unspecified (U) and may be determined by traffic loads and network traffic engineering. The bilevel QoS class is intended for data networking applications in which two levels of cell discard priority may be utilized to advantage: for example, for cost reduction purposes. An interesting analogy may be made with the case of the committed information rate (CIR) service using frame relay access devices (FRADs), which utilizes essentially the same underlying concept based on the use of the discard eligibility (DE) field in the frame relay header. In addition to DBR ATC, the QoS Class 3 may also be supported by using SBR options 2 and 3 as well as ABR, as described earlier.
4. Unspecified (U) Class. The QoS parameters in this class are not specified by the standards, and may be left to be determined by the network operator based on general traffic

engineering and dimensioning principles. Such services are also termed "best efforts" services and may provide a low-cost networking capability which, for example, may utilize spare capacity in the network when required.

It should be noted that the primary performance parameters value differentiating the QoS classes described in Fig. 9-21 are the cell transfer delay (CTD), the cell delay variation (CDV) according to a two-point measurement methodology (described in Recommendation I.356 [9.4]), and cell loss ratio (CLR) for either CLP = 0 + 1 or CLP = 0 cell flows. The other ATM performance parameters such as cell error ratio (CER), cell misinsertion rate (CMR), and the severely errored cell block ratio (SECBR) are assigned default values for all classes except the U-Class. This is primarily because it is assumed that these parameters are not generally adjusted on a per-connection basis but may be more a function of general network engineering and technology aspects. It is anticipated that, typically, well-designed and stable networks are likely to meet these default objectives under normal operating circumstances. It should also be stressed that the QoS parameter bounds given in Fig. 9-21 are viewed as provisional in the sense that they may be revised in the light of further empirical evidence obtained in the future from the study of actual network deployments. That such a proviso is explicitly identified in ITU-T Recommendation I.356 [9.4] is an indication of the complexity and uncertainty of estimating suitable performance bounds for a technology capable of supporting a wide range of applications and traffic profiles. It should also be pointed that the parameter values listed in Fig. 9-21 are primarily intended for so-called public carrier networks, and may not be applicable in smaller enterprise networks in which the need for strict performance bounds may be less obvious, but is clearly not precluded.

It is also important to recognize that the parameter values listed for the various QoS classes are end-to-end network performance targets, based on the assumption of a so-called hypothetical reference connection (HRC), which for the case of Fig. 9-21 is assumed to be 27,500 km and consequently may span several administrative domains, including potential satellite links. Moreover, since some performance parameters (such as CTD) may be additive, assumptions are also made concerning the number of ATM NEs in various portions of such a hypothetical reference connection, which may generally differ for VPC and VCCs. As mentioned earlier, the detailed procedures for apportioning the performance parameter values to the various national, international, and satellite portions of an overall end-to-end VPC or VCC, including the number of NEs, are outside the scope of this summary account, but may be found in ITU-T Recommendations I.356 and G.826 [9.8].

In utilizing the rules for allocating portions of the performance allowance to the various domains, it should be recognized that whereas some parameters (e.g., CTD) may be considered additive to a reasonable approximation, others, such as CDV, are not. Since allocation of CDV portions are difficult to derive, block allowances are generally provided for QoS Class 1. As an example, the estimated bound is given as 1.5 msec of CDV for no more than three ATM NEs operating with 34 to 45 Mbit/sec transmission speeds. In general, the CDV value is expected to decrease with increasing link speed for a given level of link utilization.

The overall network performance objectives outlined above based on the concept of the hypothetical reference connection assuming a number of NEs in each portion may be viewed as the design objectives for network deployment. However, individual ATM layer performance parameters are also important as design objectives for an ATM NE. The NE performance objective values for selected parameters are defined in ITU-T Recommendation I.731 [9.1] and are summarized here for completeness and comparison with the network perspective, particularly for support of QoS Class 1 (Stringent Class). The ATM NE performance objectives are specified as follows:

1. For cell loss ratio (CLR) objectives, two possible cases are identified:
   a) For selected "demanding" applications CLR $< 2 \times 10^{-10}$ on specified connections (VPC or VCC) per NE.
   b) For less demanding applications CLR $= 10^{-7}$ on specified VPCs or VCCs per NE.
2. For cell transfer delay (CTD) objectives only the DBR ATC (also called CBR service category in ATM Forum specification terminology) is considered as values for SBR (or VBR) and ABR are unspecified. For DBR, the CTD objectives are:
   a) Maximum CTD (10^{-10} quantile) = 300 microsec.
   b) CTD (99 percentile) = 150 microsec.
   c) Mean CTD = 100 microsec.

   per NE (as measured between any ingress to egress port on the NE).
3. For cell delay variation (CDV) only the DBR ATC is considered (as measured between the ingress and egress port of the NE), where the maximum CDV value should be CDV (10^{-10} quantile) $\leq$ 250 microsec.

It may be noted that the CDV objectives for the NE fall within the block allocation for the QoS Class 1 network performance objectives outlined earlier (i.e., 1.5 msec for three NEs, assuming additive CDV values per node as a worst case), but it may be anticipated that improvements in performance will result with proper engineering and partitioning of NE buffers and scheduling algorithms as the technology improves. In addition, the per-NE CLR objectives for so-called "demanding " applications (e.g., high-quality video streams) also appear to be more stringent than the overall network (HRC-based) CLR (0 + 1) of $\leq 3 \times 10^{-7}$, though, here again, it seems possible that future improvements in design may result in improved CLR performance. In fact, a target of 1×10^{-8} has been suggested although, clearly, further empirical evidence based on wider service deployment is necessary to obtain improved confidence levels in the objective parameter values. However, even with present values, it would be reasonable to assume that the QoS performance of well-engineered ATM networks can more than adequately meet foreseeable application requirements.

CHAPTER 10

ATM Adaptation Layer (AAL) and Interworking

10.1 THE GENERAL AAL FUNCTIONAL MODEL AND AAL TYPES

In the description of the general B-ISDN protocol reference model (PRM) considered earlier (see Fig. 2-2) it will be recalled that the ATM Layer user (or transfer) plane functions are essentially independent of the type of higher-layer applications (or services) that are using the services provided by the ATM layer functions. This aspect of ATM is by design, since the "service independence" at the ATM layer enables the common transport and switching of cells containing any type of higher-layer applications typically exhibiting widely differing characteristics, such as voice, video, or data signals. However, a consequence of obtaining service independence at the ATM layer is that any service specific functions will need to be provided in an intermediate layer between the ATM layer and the specific (higher-layer) service or application. This layer is termed the ATM adaptation layer (AAL) and its primary purpose is to provide the functionality necessary to carry any type of application over ATM. Since a wide variety of applications may be envisaged, it is necessary to define several types of AAL protocols to handle these, and it is interesting to note that essentially four main types of AAL protocols have been defined, as described in further detail below. However, before considering each type, it is useful to examine the general structure of the AAL as well as develop familiarity with the terminology employed in its description.

In general terms, the AAL may be divided into the three "sublayers" shown in Fig. 10-1, although it is important to bear in mind that not all the sublayers need to be present in all cases. The three general AAL sublayers, as shown, include:

1. The Segmentation and Reassembly Sublayer (SAR). As it name implies, the SAR sublayer contains the functions required to segment or reassemble the portions of the higher-layer information into the fixed size ATM cell payload of 48 octets, including any "protocol control information" (PCI) added within the SAR sublayer. Clearly, if the higher-layer information is less than 48 octets, a so-called "padding" function will be needed to fill the ATM cell payload field.
2. The Common Part Convergence Sublayer (CPCS). The "common part" of the AAL refers to those functions (and associated PCI) that are common to a range of, or possibly all, applications or services intended for the given AAL type. The "convergence sublayer" refers to the group of functions required by the application in question in order to ensure the integrity of the data being transferred over the ATM VCC or VPC prior to or after the SAR functions are performed. Typically, the CPCS functions (together with their associated PCI) apply to the data "as a whole," as opposed to portions of it as in the case of the SAR. It is important to note that the CPCS together with the SAR functions constitute the so-called "common part" of any given AAL, since they generally

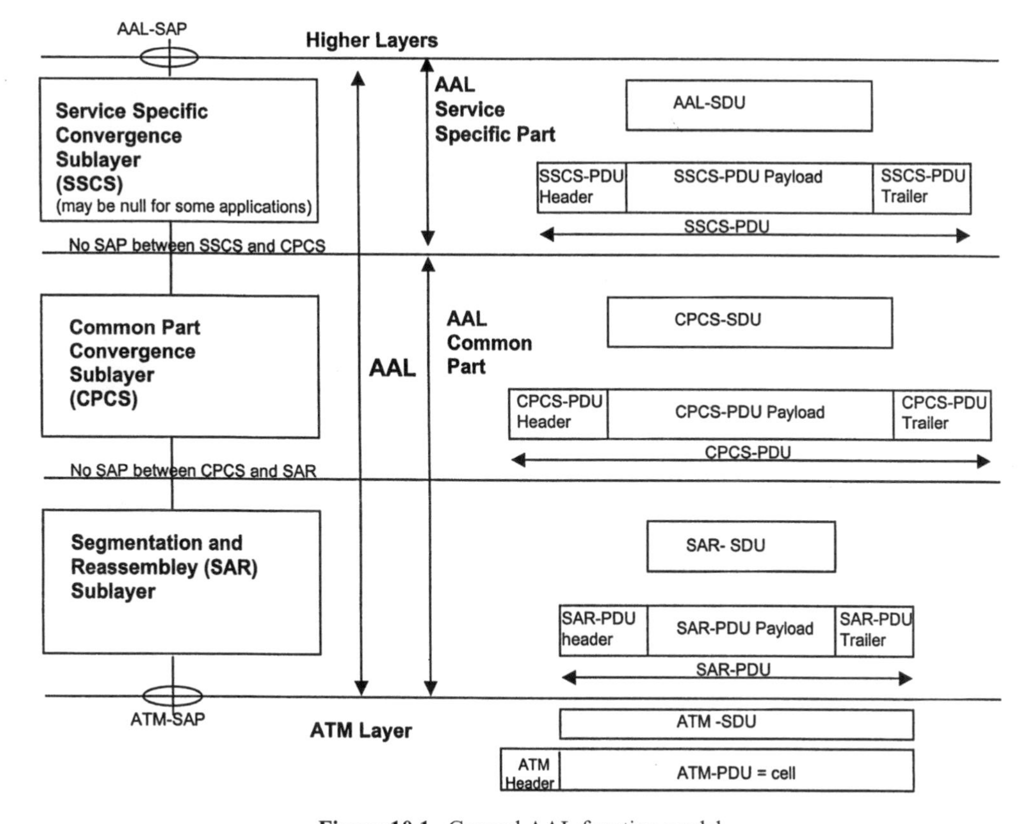

Figure 10.1. General AAL function model.

comprise all the necessary functions needed to support a variety (or all) the applications for which the AAL is intended.

3. The Service-Specific Convergence Sublayer (SSCS). As its name implies, the SSCS includes the functions (with their associated PCI) that are specific to the application or service using the ATM VCC or VPC, as distinct from those functions that are common to a class of applications and would thereby be grouped within a CPCS. In effect, it is useful to view the design of any given convergence sublayer as a grouping of service-specific functions (PCI) and common functions (PCI). In this view, the AAL may be subdivided into the SAR sublayer and the convergence sublayer (CS), the latter being further sublayered into groups of common and service-specific functions. In this way, it is only necessary to define a new SSCS part to accommodate any new application or service if required, while retaining the basic capabilities of the AAL type in the common part of the protocol, which remains unchanged.

It will be recognized that service access points (SAP) are indicated between the AAL and the higher layers as the AAL-SAP and at the interface between the AAL and the ATM layer as the ATM-SAP. In keeping with the general principles of protocol design, a SAP implies the specification of primitives by which information may be exchanged between the layers in a structured and open manner. It may be noted that there are no SAPs defined between the CPCS and SAR sublayers or between the SSCS and CPCS, thereby implying that there is no exchange of defined primitives between these sublayers in general. However, in the event that a particular application or service does not require an SSCS (i.e., a so-called "null" SSCS), the AAL-SAP clearly reduces to the interface between the CPCS and the application or service in question. In principle, since any (or all) of the AAL sublayers may be null, it is also possible to envisage an extreme case wherein the entire AAL is null so that the AAL-SAP coincides with the ATM-SAP. This case has sometimes been referred to as the AAL Type 0, Null AAL, or a "proprietary AAL," since the primitives are essentially undefined.

It may also be noted from Fig. 10-1 that the general structure of the protocol data unit (PDU) within each of the sublayers essentially follows the OSI guidelines in that the PCI required for the given sublayer is contained in the relevant "header" and/or "trailer" parts of the PDU. Thus, at the AAL-SAP the higher-layer information is transferred into (or, conversely, from) the AAL as an "AAL service data unit" (AAL-SDU) in the appropriate primitive. Within the SSCS, the functions of the SSCS are encoded as PCI within the SSCS-PDU header and/or SSCS-PDU trailer to form the complete SSCS-PDU. Similarly, the SSCS-PDU may be viewed as a "CPCS service data unit" (CPCS-SDU) when tranferred to the CPCS, which then adds the appropriate PCI encoded in the CPCS-PDU header and/or trailer to create the CPCS-PDU. The same process may be repeated into the SAR sublayer, bearing in mind a primary constraint here is that each ensuing SAR-PDU corresponds to the 48-octet ATM cell payload. Consequently, the 48-octet SAR-PDU is transferred to the ATM layer as the ATM-SDU. The addition of the ATM-PDU (cell) header to this ATM-SDU creates the ATM-PDU, or cell, as shown. The relationship between the "SDUs" transferred between the layers and the "PDUs" within the layers (or sublayers) should be clear from inspection of Fig. 10-1, which essentially serves as an instance of the general OSI protocol modeling methodology for the case of the AAL. It should be pointed out that although for the general case shown there, the PDUs include both header and trailer fields, however, in practice, as seen below, not all the AAL formats defined so far necessarily include PCI in both header and trailer fields. In such cases the relevant header or trailer "field" may be viewed as null.

The general AAL functional model summarized above provides a useful means for categorizing and reducing the complexity of AAL protocols required by sublayering the service-specific functions, as distinct from the common part functions, to support any given service. However, it was recognized from the outset that it would be desirable to define a limited set of AAL types, despite the wide range of applications or services ATM was envisaged to support, primarily to facilitate interworking between networks providing the different services. To this end, a set of four broad "service classes" was originally identified, based on a set of criteria common to each class. It was then assumed that by designing an "AAL type" capable of supporting each one of these four broad service classes, the majority of network applications would be catered to. Consequently, for the purposes of this simple classification, the four broad service classes were originally termed "Service Classes A, B, C, and D" and a corresponding set of AAL types, termed "AAL Types 1, 2, 3, and 4," respectively, were identified to support these service classes. In general terms, Service Class A included CBR and delay-sensitive services such as voice or video, and AAL Type 1 was designed to accommodate these requirements. Service Class B was intended for variable bit rate (VBR) applications that required bounded delay, such as VBR video signals, whereas Service Classes C and D were intended for VBR data applications in which strict delay bounds were not required. The essential difference between Classes C and D was that Class C included connection oriented data, whereas Class D included connectionless data applications, as noted earlier. Historically, AAL Types 3 and 4 were intended for these Class C and Class D applications, respectively.

As further work on the specification of the AALs continued however, it was realized that the initial, somewhat rigid one-to-one coupling between the service classes and AAL types was unnecessarily restrictive and often resulted in confusion when discussed in relation to the traffic-based service categories (ATCs). The historical association of service classes and AAL types was further complicated by the widespread adoption of a new AAL type called AAL Type 5, also intended for data applications. Consequently, it soon became clear that although a rough grouping of services (or applications) into broad classes may have been useful for the initial classification of the main AAL types, the use of the service classes in themselves were not useful and were often confused with the traffic-control-related service categories or ATCs described earlier. As a result, the use of Service Classes A, B, C, and D was eventually discontinued, and currently has no significance other than a historical relationship with the original classification of the four AAL types. In effect, since the original concept of the service classes is no longer accepted or used, the AAL types as presently defined may be viewed as essentially "decoupled" from any given higher-layer service or application. Whereas, in principle, this implies that any AAL type may be used with any higher-layer service or application, in practice it will be evident that a given AAL type is more suited to a certain set of services or applications than the others. This is simply because the functions required by the service or application have been designed into the AAL protocol in question as a consequence of its historical relationship with those applications. As additional AAL protocols and other applications are developed and brought into focus, it is likely such linkages will weaken eventually, but even now that there is no formal relationship between service classes and AAL types, there is often an association between an AAL type and the applications for which it may be optimized. Clearly, these observations apply to the common part of the AAL, since it may be assumed that the service-specific part (the SSCS) is more closely linked to an individual service or application.

With this historical background to the development of the AAL types in mind, essentially four main AAL types are of interest at this stage for all foreseeable applications. These are listed in Table 10-1, together with their main intended applications. However, as described above, since services are essentially "decoupled" from the AALs (common part), it is not mandated that the listed services should use the AAL type indicated, and any other type may be employed if considered suitable. It may be recalled that the choice of AAL type may be communicated by means of on-demand signaling (for SVCs), or by provisioning in accordance with a service order. The AAL types listed here are described in further detail below, but it is useful to summarize here the general background that has resulted in the present set of AAL types, as well as their primary determining characteristics.

TABLE 10-1. AAL types

AAL type[a]	Main intended applications
AAL Type 1	Circuit emulation, voice on ATM, CBR video (e.g., MPEG2), and any other constant bit rate (CBR) or delay-sensitive services.
AAL Type 2[d]	Variable packet size, delay-sensitive services with multiplexing.
AAL Type 3/4[c]	VBR, delay-insensitive services (e.g., data /file transfer, etc.). Commonly used only for connectionless data service (e.g., SMDS).
AAL Type 5[b]	VBR, delay-insensitive data services (other than connectionless). Also specified by ATM Forum for video (MPEG2) transport on ATM.

[a]The choice of AAL type for any specific service is not unique.

[b]AAL Type 5 is most commonly used for data applications.

[c]Connectionless data services (SMDS and CBDS, etc.) have traditionally used AAL Type 3/4.

[d]New AAL Type 2 (formerly called AAL CU), mainly intended for compressed voice multiplexing on a single ATM VCC (e.g., for wireless/trunking applications), is defined by ITU-T Recommendation I.363.2 [10.3].

1. AAL Type 1. AAL Type 1 is designed for applications or services that require a bounded timing relationship between the source and destination and hence includes the capability for timing recovery (of source clock frequency) at the receiver. It is intended for applications such as voice over ATM, circuit emulation, or CBR video and audio signals (such as provided by MPEG2 encoding). Typically, AAL Type 1 may be considered for delay-sensitive applications. The detailed description of the AAL 1 protocol is given in ITU-T Recommendation I.363.1 [10.2] and includes a range of functionality such as forward error correction (FEC) codes, which may be used for such applications. However, the description of such specialized functions are outside the scope of this account.
2. AAL Type 2. The original purpose of AAL Type 2 was for variable bit rate (VBR) delay-sensitive applications such as VBR video codecs. However, development of this protocol did not progress for many years, primarily because adoption of such applications appeared remote, whereas the other AAL types could be used for typical VBR applications. More recently, there has been renewed interest in developing a specific AAL type for the support of multiplexed, compressed voice samples such as are widely used in wireless (cellular) telephony networks, for example. Initially termed "AAL Composite User" (AAL CU), this protocol was rapidly developed into the present AAL Type 2, which enables the multiplexing of multiple compressed voice (and/or data) channels into one VCC while enabling bounded delay to be maintained. Details of this new AAL 2 protocol are given in ITU-T Recommendation I.363.2 [10.3], and the main functions are described in further detail below.
3. AAL Type 3/4. It is of interest to note that the protocol now called "AAL Type 3/4" (3 and 4) was originally two separate but similar protocols known as "AAL Type 3" and "AAL Type 4." The initial intention was that AAL Type 3 was designed for connection oriented data applications, whereas AAL Type 4 was designed for connectionless (i.e., datagram) data applications, as visualized in the broad (historical) service classification scheme that was intended to categorize the AAL types as described above. In this original manifestation, the primary functional difference between AAL 3 and AAL 4 was the inclusion of a so-called message identification (MID) function in AAL Type 4 to enable it to support the (higher-layer) connectionless data packets. However, with the subsequent introduction of the simpler AAL Type 5 for support of most data types, as described below, it was eventually decided to combine both AAL Type 3 and Type 4 into a single protocol termed AAL Type 3/4, which would be suitable for both connection-oriented as well as connectionless data transport over ATM. As will be seen later, AAL Type 3/4 is capable of robust and flexible data transport but was viewed as overly complex, particularly in comparison with the subsequent AAL Type 5, which may be used for essentially the same types of applications. As a result, AAL Type 3/4 is at present mainly used for the support of certain specific types of connectionless data services such as the North American Switched Multimegabit Data Service (SMDS), and its European equivalent, termed Connectionless Broadband Data Service (CBDS).
4. AAL Type 5. The AAL Type 5 protocol was initially proposed as a simpler alternative mechanism for the support of VBR, delay-insensitive data applications such as IP over ATM, essentially for the same types of application as were intended for the original AAL Type 3. Interestingly, the original protocol proposal was called "simple and efficient adaptation layer" (SEAL), and it generated significant controversy in competition with the relatively established AAL Type 3. However, it was rapidly adopted by the ATM Forum, hence gaining industry wide acceptance. Subsequently, it was also incorporated in the ITU-T, where it is described in detail in ITU-T Recommendation I.363.5 [10.5] as the current AAL Type 5. It may also be noted that although primarily designed for the support of

VBR data applications, the ATM Forum has also specified the use of AAL 5 for support of MPEG 2 encoded video signals as an option. An interesting aspect of the AAL Type 5 protocol that distinguishes it from the other AALs (and has also led to controversy) is its use of information in the underlying ATM layer to delineate the "end of frame" function, as described below. Strictly speaking, this mechanism essentially contravenes the condition of layer independence, since it implies specific AAL protocol control information in the ATM layer PCI (i.e., in the cell header PTI field). Conversely, it also implies that a "service specific" attribute has been introduced into the ATM layer, which is designed to be service-independent, as noted earlier. Nonetheless, despite these formal objections, it would be fair to say that the AAL 5 protocol is arguably the most widely used in ATM networks to date. Moreover, the capability to delineate the (higher-layer) frame structure within the ATM layer has subsequently been shown to result in other applications for (traffic) congestion control and multipoint-to-point connections, which may be useful in some data networking applications.

In describing the various AAL protocols in further detail below, it is worth bearing in mind the above general observations regarding their original intended applications and subsequent evolution. In addition, with more widespread use and new network applications, it is likely that completely new AAL protocols may be developed, and not just new SSCS protocols using established common parts. Although numbered in sequence, the chronological development of the AAL types varies, since the AAL 1 and AAL 3/4 definitions preceded that of AAL Type 5, with the most recent being AAL Type 2. This chronological sequence will be followed in this account.

10.2 AAL TYPE 1—FUNCTIONS AND FORMAT

The AAL Type 1 protocol is intended for the support of delay-sensitive CBR applications so it must be capable of transferring SDUs with constant source bit rate to the destination with the same bit rate, together with any timing information required. If the application utilizes a structured data format, the protocol also provides for transfer of the structured information, as well as indication of errorred information, if required. The functions provided by the AAL1 to achieve these objectives include:

1. Segmentation and reassembly (SAR) of user information.
2. The blocking and deblocking of the user information in the event that the higher-layer information is less than the 47 octets necessary to form the SAR-PDU. Blocking implies the concatenation of the user information stream to form the complete PDU, and deblocking is the reverse process at the receiver to deconstruct the information stream to its original form.
3. Handling of the cell delay variation (CDV) and cell payload assembly delay (packetization delay) by the appropriate dimensioning of the AAL 1 buffers at the transmitter and receiver ends of the connection.
4. The processing of lost or misinserted cells. The indication of lost or misinserted cells and other fault conditions such as loss of timing or synchronization, errored cells, or PCI may be passed to the management system through the appropriate layer management entity using the layer management indication (LMI) described previously, if required.
5. Recovery of the source clock frequency at the receiver.

6. Recovery of the source data structure at the receiver.
7. Monitoring and processing of the AAL PCI errors.
8. Monitoring and processing of information field errors by means of service-specific error detection/correction codes.

The main functions listed above are provided over both the SAR sublayer and the convergence sublayer (CS), with service-specific functions being provided as part of the SSCS, which may be selected for any particular application. It should also be evident that not all of the above functions are provided by codepoints in the PCI. Thus, functions such as handling of the CDV and packetization delay by buffer management partitioning and scheduling are viewed as the responsibility of the designer of the AAL 1 implementation, rather than on any PCI transferred in the AAL 1 header. In some ways, the AAL 1 protocols may be viewed as a "tool-kit" of functions, of which a set may be selected for the support of a particular higher-layer application such as 64 kbit/sec voice signals, high-quality audio or video signals, or circuit emulation. A detailed description of the AAL Type 1 capabilities, some of which are complex and somewhat specialized, is outside the scope of this text. However details may be found in ITU-T Recommendation I.363.1 [10.2]. The intent here is to summarize the basic functions necessary for a general understanding.

The structure and functions of the AAL Type 1 SAR sublayer are shown in Fig. 10-2, which depicts an AAL 1 SAR-PDU consisting of a one-octet SAR-PDU header (AAL Type 1 header) plus a 47-octet SAR-PDU payload. These together constitute the 48-octet SAR-PDU which is exchanged via the primitives at the ATM layer SAP (as an ATM-SDU) to form the payload of the ATM cell. It should be noted that the SAR-PDU header consists of two four-bit fields, a four-bit sequence number (SN) field, and a four-bit sequence number protection (SNP) field. However, each of these four-bit fields is further subdivided into two subfields as described below:

1. Sequence Number Field (four bits). The SN field is made up of a three-bit sequence count, simply indicating the sequence number (modulo 8) of the SAR-PDUs making up the CS information stream. The remaining one-bit field is called the convergence sublayer indication (CSI) bit, which, as its name suggests, is used to indicate the presence of a convergence sublayer. The default value of the CSI field is "0," whereas CSI = 1 indicates the presence of a "structured" CS. This value is set by the CS functions. In addition to indicating the presence of structured data in the CS, the CSI field may also be used to perform another function, which is to encode clock frequency recovery information when using a mechanism known as synchronous residual time stamp (SRTS), an option that is briefly described below. However, it is of interest to note that the single-bit CSI field may be used to support two quite distinct capabilities.

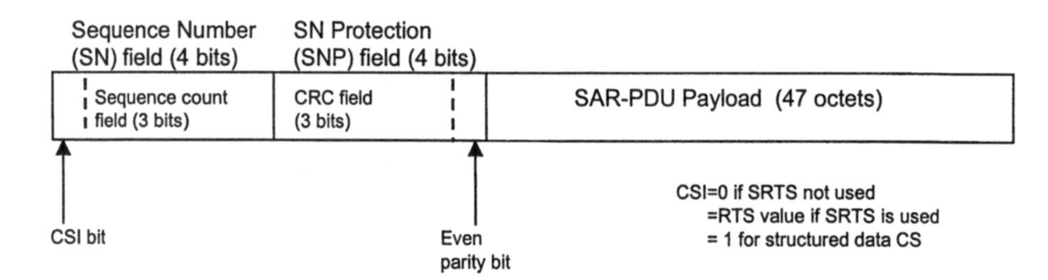

Figure 10.2. AAL Type 1 SAR format.

2. Sequence Number Protection (SNP) Field (four bits). The SNP field is used to provide error detection and correction capability over the SAR-PDU header. As shown, the SNP field consists of two subfields that encode a three-bit CRC-3 and a one-bit even parity field to provide two-stage error protection for the SAR-PDU header. The CRC-3 is performed over the four-bit SN field using the generator polynomial $P(x) = x^3 + x + 1$, and the resulting value is inserted in the CRC field. The parity bit is set to give even parity over the eight-bit SAR-PDU header, thereby providing additional error detection capability. These procedures enable the receiver to perform either single-bit error correction or multiple-bit error detection using the properties of these error protection codes. The AAL 1 receiver processor may operate in either a "correction mode" or a "detection mode," with default to the single-bit correction mode, in much the same way as the ATM header HEC function described earlier.

It is evident that the combination of the sequence numbering and its associated error protection capability is intended to provide a very robust means for the detection of lost or misinserted SAR-PDUs, which correspond to lost or misinserted ATM cells. This information is passed to the CS and hence may be used by the higher-layer application (if required), for example, by inserting "dummy" cells or discarding misinserted cells to maintain bit count integrity.

In addition to the processing of the information on possible lost or misinserted SAR-PDUs and the reporting of this information to the management system, the AAL 1 convergence sublayer may also include other functions such as:

1. Blocking and Deblocking. As noted earlier, this function sequentially concatenates "blocks" of user data (i.e., AAL-SDUs less than 47 octets, e.g., single-octet blocks) to form the 47-octet SAR-PDU payload needed at the SAR sublayer. The deblocking process is simply the reverse of the blocking process, namely, separation of the user data units or retrieval of the original structure at the receiver end.
2. Structured Data Transfer (SDT). This function provides mechanisms for support of any fixed octet-based structure to be transferred in the CS, such as, for example, an 8 kHz based circuit mode service or MPEG 2 encoded video signals. For structure sizes larger than one octet, a pointer mechanism may be used to delineate the frame size. For structured data transfer the CSI bit in the SAR-PDU header is set to CSI = 1 to indicate the presence of structure in the data stream. The use of the pointer mechanism is described in detail in [10.2].
3. Cell Delay Variation Processing. As indicated earlier, the buffer management used to compensate for effects due to CDV at the receiver is considered as part of the CS function, although no specific PCI is required in the CS to accommodate this capability. Removal of any accumulated CDV to meet the required tolerance values for the service in question essentially implies suitable design of buffer scheduling to deliver the AAL-SDUs to the (higher-layer) AAL user at the "constant" bit rate at which they were sent. Since the receiver buffer playout may also relate to recovery of source timing information as described below, both aspects may be coupled in the receiver design to meet overall jitter requirements.
4. Forward Error Correction (FEC). For the support of some CBR services such as broadcast video or high-quality audio, the use of FEC capability to protect against bit errors may be desirable. For such purposes, the AAL 1 CS may incorporate an FEC mechanism based on a Reed–Solomon code with octet interleaving capable of correcting up to two errored octets or four erasures in a block of 128 octets. A detailed description of such complex and specialized error correction mechanisms is outside the scope of this account, but some

details may be found in [10.2], if required. Since not all services require the use of such powerful FEC mechanisms, this capability may be viewed as a service-specific option in AAL Type 1 CS.

5. Partial Cell Fill for Reduction of Packetization Delay. The capability to provide a partial fill of the SAR-PDU payload and thereby reduce the assembly or packetization delay is also provided as an optional function in AAL Type 1. Thus, it will be recognized that for a CBR service such as 64 kbit/sec PCM voice samples, the packetization delay to "fill" the 47 octets in the SAR-PDU payload is given by approximately ($8 \times 47/64$ kbit/sec) 5.9 msec. It has been argued that for certain applications, lower packetization delay would be desirable, such as, for example, when using echo cancellation mechanisms. A convenient means to obtain lower packetization delay is to partially fill the cell, at the expense of some wasted bandwidth. For example, a partial fill of 32 octets in the payload would result in an assembly delay of only 4 msec. If the partial fill option is used, the most significant (i.e., leading) octets are used and the remainder of the payload is filled with dummy octets or padding. The extent of the partial fill needs to be negotiated and provisioned for the specific service during the connection set-up phase.
6. Source Clock Frequency (Timing) Recovery. The AAL Type 1 protocol provides two options for the recovery of timing information in the case of asynchronous CBR services for which the transmitter and receiver clocks are not synchronized with a common network clock. The two mechanisms are termed: (a) adaptive clock method and (b) synchronous residual time stamp (SRTS) method. For the case of synchronous CBR services, for which the transmitter and receiver clocks are fully synchronized with a common network clock, it may be noted that there is no need to use either of the above mechanisms, since timing may be derived directly from the network clock. Hence, the above options are only to be considered for asynchronous CBR applications of AAL Type 1. The adaptive clock method may be used when only the jitter requirements as specified in ITU-T Recommendations G.823 [10.15] and G.824 [10.16] need to be met, whereas the more complex SRTS method may be required if it is necessary to meet both the jitter and wander requirements of [10.15] and [10.16]. A description of these requirements is essentially outside the scope of this text, as is a detailed description of AAL 1 timing recovery mechanisms. However it is useful to summarize the basic underlying concepts here to obtain a general understanding of the required mechanisms.

   The adaptive clock method utilizes the observation that the actual instantaneous amount of data received by the AAL 1 connection endpoint is in itself an indication of the source frequency. Consequently, any variations in the amount of data received may be used by the local clock to adjust its frequency. For example, an increase in the amount of data received, relative to some long-term average value, indicates that the source frequency is too high relative to the local clock, and vice versa. These differences in the instantaneous amounts of received data may be used in a "local" feedback loop to control the frequency of the receiver clock. Although the adaptive clock mechanism may be viewed as a general means to compensate for timing differences between transmitters and receivers and its actual implementation may vary in detail, implementations typically rely on monitoring of the AAL 1 data buffer fill levels between suitable upper and lower thresholds. An example of such an algorithm is described in more detail in [10.2]. It may be observed that the adaptive clock method essentially only requires additional receiver functionality (at both ends of a bidirectional connection), and no additional timing information needs to be transferred through the network. In general, the receiver AAL 1 buffer fill levels will be thresholded to avoid buffer overflow or underflow, so that the monitored fill level may be used for the purposes of the adaptive clock technique. Consequently, this method is widely deployed for support of asynchronous (and plesiochronous) CBR services in which stringent jitter and wander tolerances are not mandated. For more demanding requirements more complex algorithms clearly may be needed, or the alternative option of using the SRTS mechanism described below.

The synchronous residual time stamp (SRTS) mechanism incorporated in AAL Type 1 may be used in either of two modes, depending on whether a common reference network clock may be derived, or, alternatively, a plesiochronous network operation is warranted. For the case of plesiochronous operation, the SRTS mechanism may need to be further enhanced to meet jitter and wander requirements resulting from the lack of synchronization between the plesiochronous network domains. However, for the SRTS mode of operation in which a common derived reference clock may be obtained at both the transmitter and receiver, the mechanism operates by encoding frequency difference information in the conversion sublayer indication (CSI) field of the AAL Type 1 header described earlier. Essentially, the SRTS mechanism is based on the observation that the source clock frequency may be reconstructed as the sum of a constant (or nominal) part and a frequency difference encoded as a residual time stamp (RTS), which is transmitted to the receiver in the AAL 1 header. Since the constant part is assumed to be available at the receiver end (derived from the "common" network clock) only the "residual" frequency difference needs to be made available to the receiver to reconstruct the original source frequency. To enable this, the frequency difference is encoded as an RTS value in a four-bit field that is carried in the CSI bit of the successive AAL 1 headers. Strictly speaking, the modulo 8 sequence count of the AAL Type 1 header enables the CSI bit to be used to provide an eight-bit frame structure over consecutive AAL Type 1 SAR-PDU headers. The CSI field values of SAR-PDU headers corresponding to the odd number sequence count (of 1, 3, 5, and 7) are used to carry the RTS information, whereas the remaining (even) sequence counts of CSI values may be used for other convergence sublayer purposes, such as identifying the type of CS present.

The SRTS procedures are essentially intended for the accurate recovery of source (or service) clock frequency when AAL Type 1 is used for the transport of conventional (CBR) asynchronous circuits based on the DS-1 (T1) or E1 hierarchies. This application of AAL 1 (and the underlying ATM VCC or VPC) is also commonly referred to as "circuit emulation." In circuit emulation, conventional DS-1 (1.544 Mbit/sec), DS-3 (44.736 Mbit/sec), E1 (2.048 Mbit/sec), or E3 (34.368 Mbit/sec) multiplexed signals are "mapped" into ATM VCCs or VPCs using AAL Type 1, and may be switched within the ATM network. However, circuit emulation (or transport) is not restricted only to the carriage of asynchronous circuits, but may also be used for transport and switching of SDH-based signals or even $n \times 64$ kbit/sec signals. In addition to its application in circuit emulation, the AAL Type 1 protocol may also be used for:

1. Video signal transport, e.g., MPEG 2 encoded signals
2. High-quality audio signals, e.g, CD-quality music signals
3. Voice band signals, such as 64 kbit A-law or μ-law signals

As summarized earlier, the capabilities necessary to support any of these network applications of AAL 1, such as sophisticated forward error correction (FEC) or structured data transfer (SDT) mechanisms, although outside the scope of this account, have been defined and hence may be selected for any specific implementation of the AAL 1. It is nonetheless interesting to note that the AAL 1 protocol has been designed to cater to a wide range of applications with the relatively small overhead of one octet for the SAR-PDU and possibly one octet at the CS for SDT applications. This economy of means has been achieved not only by multiple use of the protocol control information in the header, but also by specifying related procedures at the AAL that do not necessarily require use of the header fields. Moreover, application-specific functions such as FEC or SDT need not be present when not required, thereby saving the associated overhead and processing requirements. It should also be pointed out that the AAL 1 procedures allow for the "partial filling" of the payload field to accommodate lower packetization delay for delay-sensitive applications such as compressed voice encoding. However, this aspect of AAL 1 use has to some extent been superseded by the development of the newer AAL Type 2 protocol, which has been specifically designed for compressed voice applications, as described later.

10.3 AAL TYPE 3/4 FUNCTIONS AND FORMAT

As mentioned earlier, the AAL Type 3/4 protocol (called AAL 3/4 for short) evolved by combining the AAL Type 3 protocol, intended for connection oriented data applications, with the essentially similar AAL Type 4 protocol designed for connectionless data applications. In effect, the AAL 3/4 protocol may be viewed as providing comprehensive capability for the carriage of any type of data application over ATM connections, but this very flexibility has resulted in its relative complexity and subsequent neglect. Despite this, it is useful to describe in some detail the structure and functions designed into AAL 3/4, since this protocol to some extent typifies AAL protocol design principles in general.

The overall AAL type 3/4 protocol is divided into three distinct sublayers in accordance with the general structure of the AAL described earlier in relation to Fig. 10-1. This includes a segmentation and reassembly (SAR) sublayer, a common part convergence sublayer (CPCS), and a service-specific convergence sublayer (SSCS), which may be null, depending on the specific network application. It is important to recall that although no service access points (SAP) are defined between these (functional) sublayers, in principle, internal "primitives" may be envisaged that describe in abstracted terms the exchange of information (as parameters) between the sublayers. Although a detailed description of these internal AAL 3/4 primitives and the associated procedures is not necessary in this introductory account, the nature of these primitives may readily be inferred from the functions and formats of the PDUs within each sublayer as described below. Further details may be found in ITU-T Recommendation I.363.3 [10.4].

The structure of the AAL 3/4 CPCS is shown in Fig. 10-3 a, which illustrates that the CPCS-PDU is constructed by encapsulating the (higher-layer) user information, or so-called CPCS-PDU payload, within a four-octet CPCS-PDU header and a four-octet CPCS-PDU trailer. In principle, the length of the user information field may be anywhere between 0 and 65,535 octets, assuming that the higher-layer information is structured in octet-oriented format, although, in practice, other factors may serve to limit the size of this field to lower values (e.g., buffering restrictions). The services provided by the CPCS may be understood by considering the function of each one of the fields comprising the CPCS-PDU shown in turn, as well as in conjunction with the SAR sublayer functions described below. Starting from the left of the CPCS-PDU header, these functions are as follows:

1. Common Part Indicator (CPI). This one-octet field is used to indicate the message type as well as the interpretation of all subsequent fields in the CPCS-PDU. The "message type" function may be used to distinguish between management (or control) messages at the AAL level from those carrying user data, analogous to the case of the in-band OAM cells at the ATM layer described earlier. At present, no specific layer management functions have been defined for AAL 3/4 use, but the potential for future use exists. The CPI field

a) AAL Type 3/4 CPCS

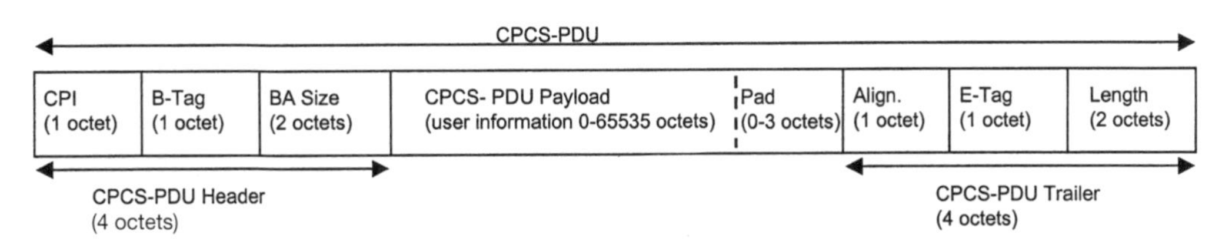

b) AAL Type 3/4 SAR

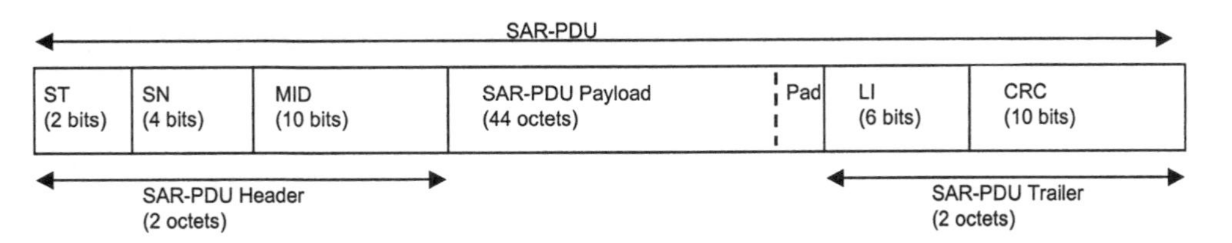

Figure 10.3. AAL Type 3/4 CPCS and SAR functions and format.

also allows for the possibility of using "counting units" or word lengths other than octet-oriented formats within the CPCS. However, at present, the only defined codepoint (CPI = 0000 0000) corresponds to the octet-oriented interpretation of all the fields in the CPCS-PDU.

2. Begining Tag (B-Tag). This one-octet field is used in conjunction with the value in the associated end tag (E-Tag) field in the CPCS-PDU trailer to correlate the CPCS-PDU header and trailer and thereby detect any missassembly or error condition. At the transmitter end, the same value is inserted into both the B-Tag and E-Tag fields for each CPCS-PDU (i.e., message or user data frame). In reassembling the CPCS-PDU, the receiver correlates the B-Tag and E-Tag values to check for correct assembly and may alert the layer management in the event of an error. It may be noted that the B-Tag and E-Tag values need not be in sequence but they need to be the same for each CPCS-PDU or frame.
3. Buffer Allocation Size (BA Size). This two-octet field is used to indicate to the receiver end the maximum buffer size required to enable accommodation of the CPCS-PDU and thereby assist in buffer management while avoiding buffer overflow. The BA size is binary encoded to indicate the number of octets if the CPI value is set to the default or otherwise in the number of counting units (e.g., word length), according to the CPI value. In typical message mode use of AAL 3/4 the BA size value may be the same as the length of the CPCS-PDU payload field. However, in the so-called "streaming mode" use of AAL 3/4, the BA size value may exceed the payload size. The difference between the message and streaming mode operation of AAL 3/4 is discussed below in relation to the overall functions of the CPCS.
4. CPCS-PDU Payload. As noted previously, the CPCS-PDU payload is a variable-length field that incorporates the (higher-layer) AAL-SDU, or user payload. At present, the CPCS-PDU payload is in units of octets, as denoted by the default CPI value (= 0000 0000), although other units may be used in the future if required in any specific application.

5. Padding (PAD) Field. A padding field of between 0–3 octets may be inserted between the end of the CPCS-PDU payload field and the start of the CPCS trailer, as shown in Fig. 10-3 to enable alignment of the AAL 3/4 payload field to a 32-bit (four-octet) word length. The PAD field is simply coded as all zeros and discarded at the receiver end. The 32-bit alignment of protocol structures is desirable from an implementation perspective since typical processing engines operate on either 32-bit- or 64-bit-wide bus architectures. It may also be noted that both the AAL 3/4 header and trailer fields are also consistent with 32-bit-wide alignment for the same reason.

   For the AAL Type 3/4 trailer, the fields include:

6. Alignment (AL) Field. This one-octet field is intended to provide for 32-bit alignment of the AAL 3/4 trailer for the reasons cited above. The alignment field is simply coded as all zeros and is not intended for any other protocol information (PCI).
7. End Tag (E-Tag). As noted earlier, the one-octet end tag field is encoded with the same value as is inserted in the B-Tag field in the AAL 3/4 header for a given CPCS-PDU, to form an association between the CPCS-PDU header and trailer. This correlation assists the AAL 3/4 receiver to detect missassembly of the entire CPCS-PDU in the event of fault conditions. It will be recognized that although in practice the B-Tag and E-Tag values may simply be a sequence count value modulo 256, the AAL 3/4 receiver is only required to check that the values are the same in the B-Tag and E-Tag and not whether they are in sequence.
8. Length Field. The two-octet length field is binary encoded with the length of the CPCS-PDU payload, and provides a means to detect whether any information has been lost (or misinserted) during the data transfer phase. The value of the length field, which is typically in octets unless an alternate word size is specified by the CPI value, is used to delineate the CPCS-PDU payload at the receiver, and, in the event of a mismatch, an error indication may be passed to the AAL management system and the PDU discarded.

The protocol control information in the CPCS-PDU header and trailer as described above is designed to provide a number of functions, which may be summarized as follows:

1. The preservation of the CPCS-SDU between the peer entities, coupled with the delineation of the AAL-SDU.
2. The detection of error conditions that may result in mismatch between B-Tag and E-Tag values, length value mismatch, as well as buffer overflow conditions. Although errored PDUs are typically discarded in most data applications, as an option it may be possible to deliver the errored PDU to the SSCS, provided an error indication is also passed via the management functions.
3. An abort function may also be provided that enables the discard of a partially transmitted CPCS-PDU. This function is only used in the streaming mode operation and requires additional parameters in the internal AAL 3/4 primitives. The abort capability does not require any PCI in the CPCS-PDU header or trailer fields.

Reference has been made above to the operation of the AAL 3/4 protocol in the so-called "message" and "streaming" modes, particularly as some functions, such as abort, relate to the latter mode of operation. In the message mode of operation, the entire CPCS-SDU (i.e., the payload of the CPCS-PDU or the AAL user information in other words) is exchanged with the higher layer as a single unit. In simple terms, the message mode service may be envisaged as the case in which the AAL 3/4 user data frame is exchanged "as a whole" between the AAL and the higher layers (or SSCS, if present) after processing within the AAL 3/4 entity. Although the message mode of operation is the more typical AAL Type 3/4 service, an alternative mode of operation, termed streaming mode service, may also be used for some applications. In streaming mode service, the exchange of information between the CPCS and the higher layers (or SSCS, if present) occurs in blocks separated in time, which may be envisaged as a continuous flow, or stream, of the user data across the CPCS-to-SSCS interface. Such a mechanism has also been referred to as an internal "pipelining" of the CPCS-SDUs, whereby delivery of user data at the receiver AAL-SAP may be initiated before the entire CPCS-SDU is available at the receiver from the transmitter AAL 3/4 peer entity. For streaming mode service, the CPCS-SDU may be viewed as a concatenation of several "blocks" of information of one or more octets long that may be exchanged across the AAL-SAP interface. It may be evident that for such a streaming (or pipelined) mode of operation, the abort function is required to be able to discard partially transferred user data information in the event of error conditions in underlying layers.

Turning now to the AAL Type 3/4 SAR sublayer functions, the structure of the SAR-PDU is shown in Fig. 10-3b. Each AAL Type 3/4 SAR-PDU is made up of a two-octet SAR-PDU header, a SAR-PDU payload (which may be up to 44 octets in length), and a two-octet SAR-PDU trailer. When the SAR-PDU payload is less than the maximum 44 octet, padding octets set to all zeros are inserted in the remaining field as filler. The total length of the SAR-PDU is 48 octets, to coincide with the ATM-SDU (in other words, the ATM cell payload). The operation of the SAR sublayer functions may be readily understood by considering the function of each of the PCI fields in the header and trailer. Starting with the leftmost field in the SAR-PDU header, these functions may be summarized as follows:

1. Segment Type (ST). The two-bit segment type field is coded to indicate whether the SAR-PDU contains the beginning of message (BOM), continuation of message (COM), or end of message (EOM) as a result of the segmentation process on the entire CPCS-PDU (message). In the case in which the entire CPCS-PDU can be contained within a single SAR-PDU payload, the ST is coded to indicate a single segment message (SSM). The ST codepoints are:

Beginning of message (BOM)	10
Continuation of message (COM)	00
End of message (EOM)	01
Single-segment message (SSM)	11

   It will be recognized that the ST field codings enable the reconstruction of the complete message at the receiver end by concatenating the SAR-PDU "segments" and delineating the EOM segment from the ST field codepoint. This process is shown in Fig 10-4.
2. Sequence Number (SN). The four-bit sequence number field enables the allocation of a modulo 16 sequence number to successive SAR-PDUs in the stream, thereby enabling detection of lost or misinserted SAR-PDUs. The sequence number is incremented by one on a per-message basis (i.e., within one CPCS-PDU), and may not be contiguous in successive messages.
3. Multiplexing (or Message) Identification (MID) Field. The 10-bit MID field may be used to multiplex up to 1024 "AAL or SAR" connections on an ATM VCC. The use of the MID values for multiplexing is primarily intended for use with connection-oriented data applications between the endpoints of the ATM VCC under end user control. When the AAL 3/4 is used for connectionless data applications at the SSCS or above, the individual MID values are assigned per (higher-layer) connectionless packet. This enables segments

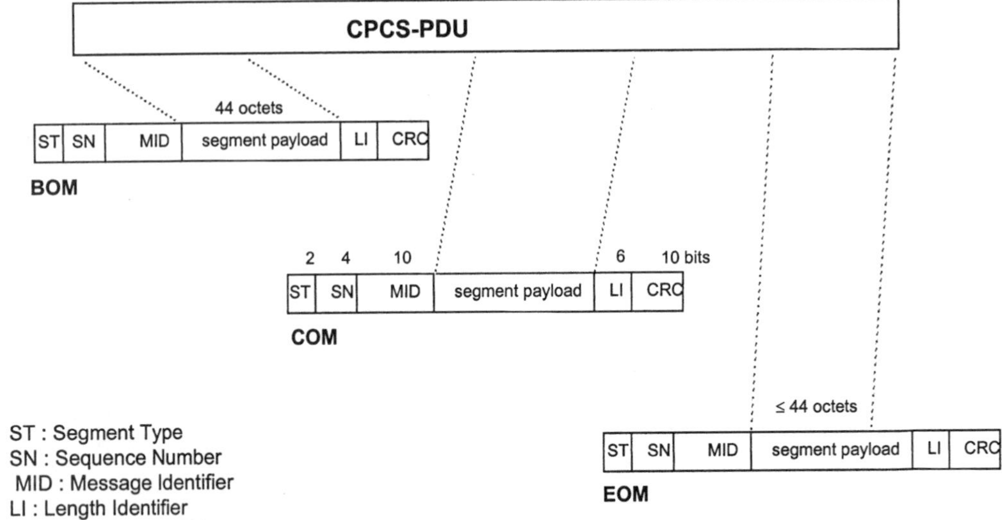

Figure 10.4. AAL Type 3/4 SAR protocol.

of any given packet to be interleaved with segments from other packets within the network while enabling deinterleaving at the receiver end for reassembly of the (connectionless) packet based on the MID values. When not used, the MID value is set to all zeros.

The function of the MID field constitutes the essential point of difference between the use of AAL 3/4 for connection-oriented and connectionless applications and was in fact the key difference between the initially separate AAL Type 3 and AAL Type 4. Initially, the use of the MID function also generated considerable controversy. These aspects of the MID function will be considered further below.

The purpose and structure of the SAR-PDU payload and Pad fields have already been described earlier, so that it remains to describe the functions in the SAR-PDU Trailer.

4. Length Indicator (LI) Field. The 6-bit length indicator field is used to indicate the length (in octets) of the SAR-PDU payload, up to a maximum of 44 octets. Other values of this field are used to indicate specific SAR functions, as described below for the abort function. The remaining octets in the SAR-PDU payload field are then assumed to be padding (or filler) octets that may be discarded on reassembly of the message or user data frame.
5. Cyclic Redundancy Check (CRC). The 10-bit CRC field contains a CRC-10 calculation over the entire SAR-PDU, including the header, payload, and LI field, using the generator polynomial $G = 1 + x + x^4 + x^5 + x^9 + x^{10}$. The CRC-10 values are used to detect errors in the SAR-PDU. Typically, errored SAR-PDUs will be discarded, but as an optional function, the corrupted SAR-PDU may be passed up to the CPCS together with an error indication to the associated management system. This option of a so-called “corrupted data delivery” function is analogous to the case for the CPCS, and has also been incorporated in other AAL protocols such as AAL 5 and AAL 1.

The PCI described above for the SAR-PDU is intended to provide a number of functions at the SAR sublayer. These, as well as other capabilities, are summarized below.

1. Error Detection. The use of the CRC-10 error detection capability, as well as the sequence number, enables the detection of bit errors and lost or misinserted PDUs, as indicated earlier. As a related functional requirement, the sequence integrity of SAR-PDUs must be maintained, as for the case of the ATM layer.
2. Abort Function. The capability to abort the transfer of SAR-PDUs to the CPCS is also provided at the SAR sublayer. The SAR abort function operates by inserting the EOM segment type with a specific value of the length indicator, LI = 63, to indicate to the receiver end that the SAR process needs to be terminated. The payload of the EOM segment initiating the abort procedure may be set to zero, since the receiver will ignore its contents for this specific LI codepoint of 63. The SAR abort procedures may be used in support of the "streaming mode" operation of AAL Type 3/4 outlined previously, or for terminating data transfer in the event of detection of transmission errors or defects.
3. Multiplexing and Demultiplexing. As noted above, the provision of the 10-bit message identifier (MID) field enables the support of multiple (up to 1024) AAL 3/4 connections on a single ATM VCC. Although the inclusion of the MID capability was initially intended to allow for support of connectionless packets over ATM, the multiplexing procedures may be utilized for both connection-oriented and connectionless applications. When used with connectionless applications, each packet from a given source is assigned the same MID value and may be interleaved with other connectionless packets (which will be assigned a different MID value) on the same VCC. At the receiver end, the different packets may thus be deinterleaved and reassembled on the basis of the MID values. The procedures for assigning the MID values depends on the particular connectionless service, but may in general be the responsibility of the management system of the NE, based on a given algorithm. For the case of connection-oriented service support, the MID value may be envisaged as identifying a particular AAL 3/4 connection endpoint (CEP) associated with the AAL 3/4 SAP.

   It may be noted that conceptually, the "interleaving and deinterleaving" of SAR-PDU segments from different (higher-layer) connectionless packets, each corresponding to a particular MID value, is similar to the "multiplexing and demultiplexing" of SAR-PDUs from different AAL CEPs, each corresponding to a different MID value within an ATM VCC. From this perspective, the characterization of the MID function as either a (connectionless) message identifier or as a multiplexing label is essentially similar, even though the procedures for its use may be different for the two cases.

   The MID-based multiplexing capability in AAL 3/4 resulted in considerable controversy when initially proposed, and may be viewed as either a strength or a weakness of the AAL 3/4 protocols. For the case of connectionless applications, the MID function is clearly necessary for the deinterleaving of the connectionless packets at the receiver VCC endpoint in the event that multiple connectionless flows are interleaved within the network. However, its use for connection-oriented applications is questionable, since it may be argued that sufficient multiplexing functionality has already been provided at the ATM layer. In this case, the MID function introduces additional complexity, which may not be warranted. In effect, the MID functionality added to the perception that the AAL 3/4 was unnecessarily complex for many data networking applications and required too much overhead spread over both the CPCS and the SAR sublayers. With the subsequent introduction of the competing AAL Type 5, the use of AAL Type 3/4 has been confined to specific (connectionless) applications such as SMDS and CBDS. Despite this, it may be noted that the design of the AAL Type 3/4 served as a benchmark for the subsequent design of other AALs, and though it is possibly overengineered in a sense, it provides a high level of robustness and functionality for its intended application.

10.4 AAL TYPE 5 FUNCTIONS AND FORMAT

In many respects, the services provided to an AAL Type 5 user (i.e., an application at the higher layer) by the common part of the AAL Type 5 is similar to that provided by the AAL Type 3/4, since both protocols are designed for essentially the same types of applications, namely VBR data frame transport. The primary difference, as discussed above, is the provision of a multiplexing capability in the AAL Type 3/4 common part, which does not exist in the AAL Type 5 common part described below. If necessary, such a multiplexing function could be included in a service-specific convergence sublayer (SSCS) to be used over the AAL Type 5 CPCS, but this has not yet been identified for general use. However, as will be seen later, specific types of multiplexing may be considered to be present in some SSCS protocols intended for the transport of services such as frame relay service (FRS) or for the internetworking protocol (IP).

The AAL Type 5 (often abbreviated to AAL 5) procedures include both the "message mode" and "streaming mode" operation, as described earlier for the case of AAL Type 3/4, together with the "abort" function, which may be used in conjunction with the streaming mode operation option. Other similarities include an option for the delivery of an errored frame (or AAL-SDU) to the higher layer, together with an indication to the management system that the frame was corrupted. Both the AAL 3/4 and AAL 5 common part protocols are considered as providing a nonassured service, in the sense that in the event of loss or corruption of the user data, no capability for the retransmission of the user data frame is provided in the common part of either protocol. When required, it is assumed that such a retransmission capability will be made available at the SSCS or higher layer in order to obtain a so-called assured service for the data transfer. Nonetheless, despite the similarities in the services provided by both AAL 3/4 and AAL 5, it is interesting to note that the functions and format of AAL 5 are very different from that of AAL 3/4, and a significant simplification of the protocol has been obtained by eliminating multiplexing as well as using a capability in the ATM layer to delineate the user frames. A detailed description of the procedures, primitives, and operations for the AAL Type 5 common part is given in ITU-T Recommendation I.363.5 [10.5]. Here we summarize the essential functions and formats.

The structure of the AAL Type 5 CPCS-PDU is shown in Fig. 10-5, which shows that it is made up of a CPCS-PDU payload field of variable length, a padding field, and an eight-octet CPCS-PDU trailer comprising four fields described below. The CPCS-PDU payload length may vary between 1 to 65535 octets and is octet aligned, encapsulating the (higher-layer) user data frame (or AAL SDU). The padding field (PAD) is essentially a filler field used to complete the CPCS-PDU to an integral multiple of 48 octets and may be coded in any way convenient, e.g., all zeros. Thus, the inclusion of the variable length PAD field enables the entire CPCS-PDU to made up to an integral number of 48 octets for segmentation into the ATM cell payload, as shown in Fig. 10-5.

The PCI in the eight-octet CPCS-PDU trailer includes:

1. CPCS User-to-User Indication (CPCS-UU) Field. This one-octet field is used to transparently transfer AAL user-to-AAL user information if required. This information may be required by the SSCS or higher layers, but does not affect the operation of the AAL 5 common part in any way. Consequently, no coding is specified for this field in the AAL 5 common part protocol.
2. Common Part Indicator (CPI) Field. The one-octet CPI field is used in essentially the same way as for the case of the AAL Type 3/4 CPI field described earlier. Although it may be used to distinguish layer management messages from user data messages as a "message type" function, the codepoints for such use are not defined at present. The pri-

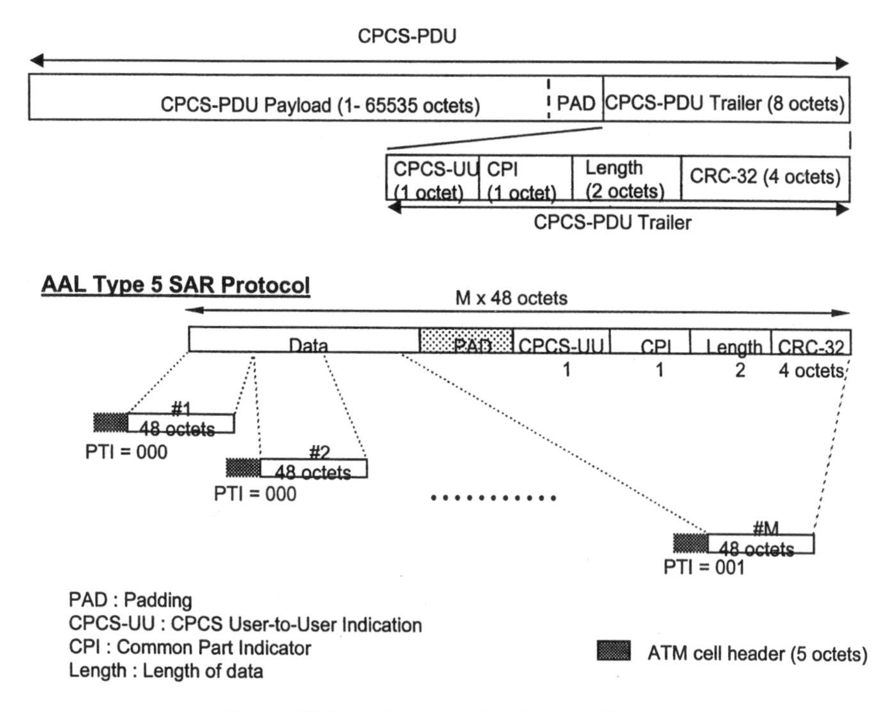

Figure 10.5. AAL Type 5 functions and formats.

mary use of the CPI field is to achieve alignment of the CPCS-PDU trailer to 64 bits to facilitate high-speed implementations commensurate with 64-bit processors. When used only for the 64-bit alignment function, the CPI field is set to all zeros.

3. Length Field. The two-octet length field is used to binary encode the length of the CPCS-PDU payload field in octets. The length function enables the receiver to separate the PAD or filler octets and may also be used to detect whether there has been loss or gain of information during data transfer. When the length field is set to zero, it indicates that the abort function should be initiated at the receiver end for the discard of partially transferred information when streaming mode operation is being used. This function is not used for message mode operation, as noted earlier.
4. Cyclic Redundancy Check (CRC) Field. The four-octet CRC field is used to detect bit errors in the AAL 5 payload by performing a CRC-32 calculation over the CPCS-PDU payload, the PAD field, and the first four octets of the CPCS-PDU trailer. The CRC-32 generator polynomial is:

$$G(x) = x^{32} + x^{26} + x^{23} + x^{22} + x^{16} + x^{12} + x^{11} + x^{10} + x^8 + x^7 + x^5 + x^4 + x^2 + x + 1$$

The detailed algorithm for the calculation of the CRC-32 values is outside the scope of this description, and may be found in ITU-T Recommendation I.363.5 [10.5] on AAL Type 5. Nonetheless, it is sufficient to note here that the provision of the CRC-32 based error detection capability in AAL 5 has been shown to be sufficiently robust for all current data network applications envisaged and has been widely implemented.

Typically, in the event of the detection of a corrupted CPCS-PDU (whether by bit errors or loss of information), the user data will be discarded at the receiver end, which may result in the retransmission of the data initiated by some mechanism at the higher layers, together with a "corrupted data" indication to layer management, signifying that a defect condition has been detected and needs to be taken into account. The corrupted data delivery option may be used for some applications, such as MPEG-2-encoded video signals, where it is deemed preferable to transfer an errored frame rather than no information at all. Typically, in such cases, the retransmission of data is not feasible, and bit errors may only result in a somewhat degraded performance of voice or video information, which may still be subjectively acceptable.

As may be observed from Fig. 10-5, the AAL 5 SAR sublayer function essentially only consists of segmenting (or reassembling) the CPCS-PDU into an integral number of 48-octet SAR-PDUs, which then form the payload of the ATM cells. No additional PCI is required at the SAR sublayer and, therefore, no SAR header or trailer fields are present, in contrast to the case for AAL 3/4. More importantly, the AAL 5 protocol achieves a significant simplification at the SAR sublayer by using the ATM user-to-ATM user (AUU) function in the payload type (PT) field of the ATM cell header to delineate the "end of message" condition. This mechanism is illustrated in Fig. 10-5. It may be recalled from the description of the ATM layer cell header PCI that the three-bit payload type field includes codepoints for the so-called ATM user to ATM user (AUU) indication. The AUU coding was intended to be used by the AAL 5 SAR sublayer to indicate when the SAR-PDU contains the end of the CPCS-PDU (or message) by using the AUU (or PTI) codepoint 001. Thus, when the receiver detects that the PTI (or AUU) value = 000, it assumes that more segments of the overall CPCS-PDU will follow. When the receiver detects the PTI (or AUU) = 001 in the ATM cell header, it assumes that the cell payload (= SAR-PDU) contains the "end" of the CPCS-PDU, including any padding octets and the eight-octet CPCS-PDU trailer. In other words, the PTI coding AUU = 000 denotes a "continuation of message" state, whereas the PTI coding AUU = 001 signifies an "end of message" state. It should be noted that no distinct coding to signify a "beginning of message" state is defined for AAL 5, in contrast to case of AAL 3/4, since it is assumed that the AUU = 001 codepoint signifying the final segment of a CPCS-PDU or frame also implies that the subsequent segment belong to a different user data frame on the same AAL 5 connection.

Although it is clear that the use of the PT AUU function in this way provides a simple and elegant mechanism for the delineation of the AAL 5 user data frames, it may be recognized that, strictly speaking, it violates the "layering" principle inherent in data networking protocols. This is because from the strict perspective of the protocol layering model, AAL 5 uses the PCI at the ATM layer in the PT field in order to perform a function that essentially belongs within the SAR sublayer, namely the delineation of the SAR-SDU (or equivalently the CPCS-PDU). In effect, the use of the PT AUU codepoints to identify whether the cell payload constitutes a continuation or terminating segment of the user data frame or CPCS-PDU creates a dependency between the ATM layer PCI and the SAR sublayer for the case of AAL Type 5, and thereby dilutes the intended "service-independence" of the ATM layer functions. In the initial stages of the development of the AAL Type 5 protocol proposals, these strictly layer modeling considerations resulted in significant controversy and discussion between the proponents and detractors of AAL Type 5. From the formal perspective, it may be argued that, in analogy with the ISO open system interconnection (OSI) seven-layer protocol architecture, both the ATM layer and the AAL correspond to Layer 2 functionality in the ISO model, to which the principle of layer independence applies. Since, from the perspective of the ISO model, the ATM layer and AAL constitute "sublayers" within the ISO Layer 2, layer independence is not actually violated. In addition, there is precedence for the use of a "more" bit to indicate additional segments in data networking protocols, a methodology analogous to the use of the AUU coding for AAL Type 5. However, despite these formal arguments, the practical convenience of the procedure soon resulted in general acceptance of AAL 5, initially by the ATM Forum, then closely followed by standardization in [10.5].

It may also be noted here that the fact that AAL 5 requires an "end of message" indication at the ATM layer, i.e., in the PT field, has resulted in potential benefit to other aspects of the ATM layer that have nothing to do with the AAL, notably for congestion control. Since the PT AUU coding enables the ATM layer to gain "awareness" of the user data frame, it becomes possible to use this information to discard the entire data frame in the event of congestion, rather than discarding "single" cells from many different data frames as a congestion control technique. This congestion control mechanism is referred to as early packet discard (EPD), and variations of it are also known as partial packet discard (PPD). In the event of congestion, it is seen to be preferable to discard the entire data frame, since, in any case, discard of a single cell from a given data frame will result in the discard of the entire data frame at the receiver, and the possible retransmission of the data at the higher layers, assuming that a retransmission capability is provided (e.g., as for the case of TCP/IP over ATM). Thus, the discard of the entire frame in the event of congestion within a NE based on the EPD (or PPD) mechanism, resulting from "frame awareness" at the ATM layer, releases more buffer space to minimize the congestion. If an equivalent number of cells is discarded at random from a number of different VCCs in the event of the congestion, the resulting loss of several frames may initiate the retransmission of the lost data on many VCCs, which may exacerbate the congestion condition in the NE.

The capability to delineate the entire user data frame at the ATM Layer using the PT AUU function may also find applications in other areas, such as enabling a merge function for the support of a multipoint-to-point connection type. As pointed out earlier, the merge capability requires the transfer of cells from several incoming VCLs onto a single outgoing VCL, to instantiate a multipoint-to-point connection type. For the case of AAL 5 connections, the NE operating as the merge point may utilize its "frame awareness" at the ATM layer to effectively merge the incoming cell flows on a frame-by-frame basis onto the outgoing VCL, thereby avoiding the interleaving of cells from the different VCLs onto the outgoing VCL. It has been suggested that the possibility of using multipoint-to-point ATM VCCs may find application in the aggregation of data flows in IP over ATM network applications, as will be discussed subsequently. However, it must be noted that such merging capabilities have so far not been widely implemented in existing ATM networks, and clearly will require some additional complexity in the NE for scalable implementations. It is, however, of interest to note that these applications of the ATM layer frame awareness resulting from the PT AUU function for AAL Type 5 were not envisaged at the time of the initial design of the AAL 5 protocol, and in principle bear no direct relationship to it.

As mentioned previously, the relative (to AAL Type 3/4) simplicity of the AAL 5 protocol and its widespread acceptance has resulted in its use for the majority of data network applications of ATM. As will be described in more detail later, AAL 5 is specified for the support of frame relay over ATM, IP over ATM, and as an option, MPEG 2 video signals over ATM. An increasing number of commercially available ATM multiplexing/switching VLSI circuits incorporate AAL 5 functionality, often on the same IC as the ATM layer functions for cost reduction and convenience. It has also been suggested that AAL 5 may be used for the transfer of voice signals, although this is evidently not optimum and also inefficient in comparison with the use of AAL Type 1 for voice signals. The AAL Type 5 common part is also specified for the carriage of the ATM signaling and routing messages for common channel signaling, as noted earlier. AAL 5 is also generally used for the transport of the out-of-band management information across the TMN Q3 and X interface in the event that ATM VCCs are provisioned to transport such management messages.

10.5 AAL TYPE 2 FUNCTIONS AND FORMAT

The AAL Type 2 (often abbreviated to AAL 2) protocol is primarily intended for the carriage of relatively low bit rate, variable length packets for applications that require bounded delay. An important example of such an application is the transport of compressed voice signals such as are used in wireless cellular telecommunications. In fact, it was the explosive growth in this sector of the communications industry and its potential use of ATM technology for more efficient switching and transport of the growing cellular voice traffic that spurred the development of AAL Type 2 as a protocol optimized for the transport of compressed voice packets. Typically, wireless cellular voice applications result in the generation of relatively short packets of variable length, depending on the type of voice compression algorithm employed across the wireless interface. The details of such higher-layer voice compression algorithms are outside the scope of this discussion, but it may be noted that a number of commonly used compression algorithms generate packets of length shorter than the ATM cell payload, e.g., between 8–20 octets. The actual packet length depends on the bit rate and compression algorithm selected. Consequently, it was recognized that several compressed voice packets from different connections may be "concatenated," in other words, effectively multiplexed, into a single ATM cell payload to enhance both bandwidth efficiency and limit the packetization delay. Thus, a key feature is that the AAL 2 protocol incorporates multiplexing of more than one AAL 2 (user information) connection, up to a maximum of 248, on one ATM VCC.

In addition to its role in the wireless cellular network applications, as shown in Fig. 10-6, the AAL 2 protocol may also be used for network applications involving the "trunking" of multiple voice connections (compressed or otherwise) over a single VCC between NEs in order to obtain relatively bandwidth-efficient transport. This so-called "AAL 2 trunking" application essentially uses the ATM VCC as a "trunk" (in the conventional sense) to transport multiple AAL 2 (voice) connections across a network or part of it, thereby avoiding the set-up of multiple ATM VCCs, which may be considered relatively costly in terms of network resources in certain implementations. The trunking application of AAL 2 utilizes the multiplexing capability incorporated in the AAL 2 "common part sublayer" (CPS), essentially in the same way as the wireless cellular application. It should be noted that for AAL 2, no separate SAR sublayer is required, since by definition there is no segmentation function required and the AAL 2 CPS-PDU is equivalent to the ATM-SDU or cell payload. Within a given AAL 2 CPS-PDU, there are concatenated one or more so-called "CPS-packets" corresponding to the separate AAL 2 user information streams (or AAL 2 connections).

The structure of the CPS-packet and the CPS-PDU are shown in Fig. 10-7, together with the

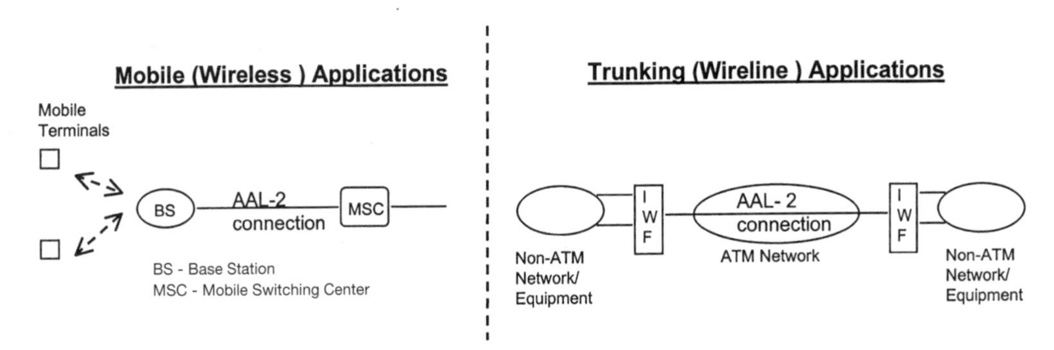

Figure 10.6. AAL Type 2 network applications.

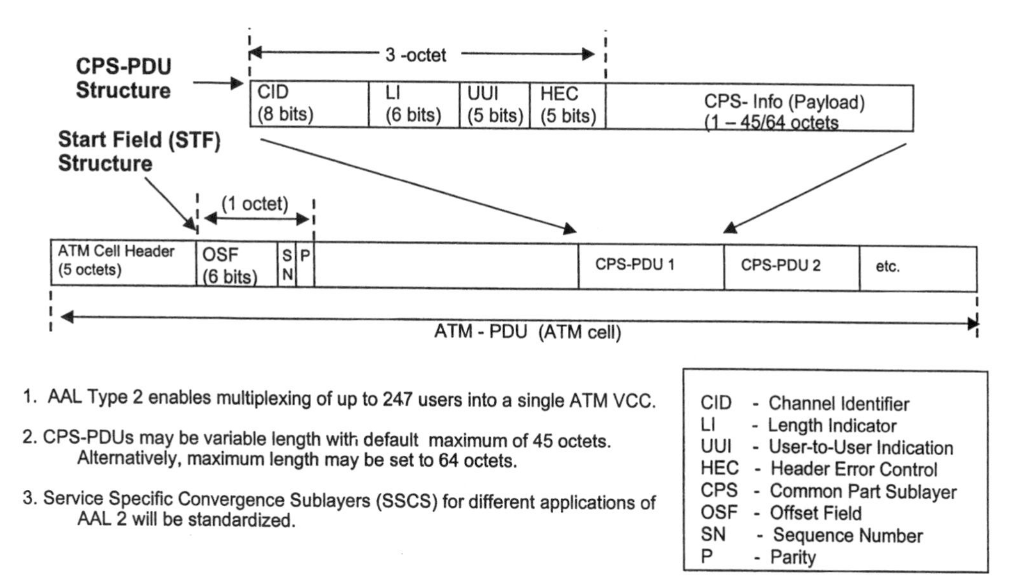

Figure 10.7. AAL Type 2 functions and format.

relationship between these entities. In essence, it should be noted that the CPS-PDU is made up of a one-octet AAL 2 CPS-PDU header (also called the start field), together with a concatenation of CPS-packets, each of which has a three-octet CPS-packet header (CPS-PH). The CPS-PDU may also include a padding (PAD) field of between 0 to 47 octets that acts as a "filler" to make up the (48-octet) CPS-PDU in the event that the CPS-packets do not complete the CPS-PDU payload. As for the case of the other AAL protocols described previously, the operation of the AAL 2 protocol may be clearly understood by considering the structure and function of each individual PCI field, starting with the description of the CPS-packet. The detailed procedures and the associated primitives for the complete AAL Type 2 protocol may be found in [10.3].

As is evident from Fig. 10-7, each CPS-packet consists of a three-octet CPS-packet Header (CPS-PH) and a variable length CPS-packet payload (CPS-PP). The function of each field in this structure is as follows, starting from the leftmost field:

1. Channel Identifier (CID) Field. The one-octet CID value is used to identify the specific AAL 2 user of the multiplexed channel at the AAL 2 level. The binary encoded CID value thus enables the multiplexing of up to potentially 256 separate AAL 2 user channels onto a single ATM VCC, analogous to the use of the multiplexing identifier (MID) function for the case of AAL Type 3/4 described earlier. However, it is important to note that not all the CID values are available for assigning to the user channels. The value CID = 0 (all zeros) is used to activate the padding function by indicating that all subsequent objects are coded as all zeros as filler octets. The value CID = 1 is used to indicate that the CPS-packet is intended for peer-to-peer layer management functions and does not carry any (higher-layer) user information. In addition, the CID values CID = 2 up to 7 are reserved for future enhancements and therefore not available to be allocated to the AAL 2 users at present. The remaining values, CID = 8 up to 255, are allocated to signify the separate AAL 2 user channels to be multiplexed onto the ATM VCC, up to a maximum of 248 channels. The individual AAL 2 channels are bidirectional, so that the same CID value is used to identify both directions of the bidirectional AAL 2 channel, analogous to the case of bidirectional ATM VCCs.

2. Length Indicator (LI) Field. The six-bit length indicator field is used to specify the length of the CPS-packet payload, which, as noted earlier, may vary for each AAL 2 channel. The LI value is binary encoded such that it is equal to 1 minus the number of octets, n, in the CPS-PP. That is, LI = $1 - n$ in octets, where the maximum length may be provisioned to be either 45 octets or 64 octets, with a default maximum value of 45 octets. The significance of the default maximum length is considered later, but it may be noted here that this default results in a maximum size of 48 octets for the CPS-packet, equivalent to that of the ATM cell payload. Clearly, when the default maximum length of 45 is used, LI values between 45 to 63 are not permitted, to avoid ambiguity in the protocol state machine.
3. User-to-User Indication (UUI). The five-bit UUI field may be used to provide a number of functions of which two are currently identified. As for the case of other AALs, the UUI field may be used for the transparent carriage of information between peer AAL 2 users, which may be the SSCS above the AAL 2 CPS, or between the peer layer management entities. In addition, the coding of the UUI field is used to identify specific SSCS protocols that may be used for the various applications of the AAL 2 protocol. Of the 32 (binary) codepoints available for the 5-bit UUI field, the binary values equivalent to 0 to 27 are intended to identify the particular SSCS protocol selected, whereas the binary values equivalent to 28 and 29 are reserved for future enhancements. The binary values equivalent to UUI = 30 and 31 are intended for the layer management functions of AAL Type 2. The main layer management functions identified at present include the allocation of the CID values to the separate user channels, as described above, and the reporting of error conditions to the management system.
4. Header Error Control (HEC) Field. The 5-bit HEC field is intended for error detection of the CPS-packet header, using a particular CRC-5 based algorithm with the generator polynomial $G(x) = x^5 + x^2 + 1$. The detailed algorithm used for the AAL 2 HEC function is outside the scope of this summary but may be found in ITU-T Recommendation I.363.2. When an error is detected, the CPS-packet is discarded and an indication is passed via the layer management function to the management system for the appropriate action.
5. CPS-Packet Payload (CPS-PP) Field. As noted previously, the CPS-PP field is a variable-length field that contains the (binary encoded) AAL 2 user information (i.e., SSCS or higher layer) corresponding to any given AAL 2 channel. The maximum payload size may be set to either 64 or 45 octets, with the latter value providing a default setting.

It is important to recognize that the CPS-packets may "overlap" the CPS-PDU boundaries into the next CPS-PDU (or, equivalently, ATM cell payload). In other words, any given CPS-PDU may include "partial" CPS-packets, with the partitioning occurring anywhere in the CPS-packet, including the CPS-PH. This concept is somewhat analogous to the overlap of ATM cells across the SDH frame boundaries in the packing of the ATM cells into the SDH transmission frame structure. In a similar manner, the packing of the individual CPS-packets may be viewed as "floating" with respect to the boundaries of the CPS-PDU structure. The one-octet CPS-PDU header or start field includes:

1. Offset Field (OSF). The purpose of the six-bit offset field is to indicate the value of the offset in octets between the end of the CPS-PDU header and the beginning of the next CPS-packet header, thereby allowing the overlap of the CPS-packets across CPS-PDU boundaries as described above. The value of the OSF is binary encoded in the number of octets to the start of the nearest CPS-packet or, if one is not present, to the beginning of the PAD field. When it is encoded with the value equivalent to 47, this signifies that the CPS-PDU does not include any start boundary of a CPS-packet.

2. Sequence Number (SN) Field. The one-bit SN field enables a modulo 2 sequence number to be assigned to the consecutive CPS-PDUs to enable detection of lost or misinserted information. If sequence number error is detected, the CPS-PDU may be discarded and the error condition reported via the AAL 2 layer management to the management system of the receiver.
3. Parity (P) Field. The one-bit parity field is used for the detection of errors in the STF of the CPS-PDU. In the event that the (odd) parity violation is detected at the receiver, the CPS-PDU is discarded and the error condition is reported via AAL 2 layer management to the management system as for the other error conditions.

Since the AAL 2 protocols allow for the overlap of the individual CPS-packets across the CPS-PDU boundaries to achieve efficient packing of the multiple AAL 2 channels into a given CPS-PDU, the need for the padding function in the CPS-PDU payload may at first glance appear superfluous. In general, for a relatively large number of AAL 2 "channels," it would appear that there will likely be a sufficient supply of CPS-packets from the various AAL 2 user streams at any given time to be able to "pack" all consecutive CPS-PDUs for transmission. However, in the event that this situation does not always occur and the fact that it is necessary to bound the packetization delay for the intended (real-time) applications for the AAL 2 protocol, the Padding function may be used to limit delay. In such cases, assuming a delay bound may be exceeded due to insufficient availability of CPS-packets to fully pack the CPS-PDU, the transmitter may "fill" the unused part of the CPS-PDU payload with the (all zeros) PAD octets, achieved by the CID = 0 value, and transfer the CPS-PDU to the ATM layer for subsequent transmission, thereby maintaining the delay bound. The actual delay bound, as determined by a timer threshold at the transmitter, is in practice provisionable and largely dependent on the intended applications, as well as implementation of the AAL 2 protocol state machine. Although a detailed description of the timer operations [termed a "combined-use" (CU) timer] is not warranted here, it will be recognized that the use of such a provisionable timer threshold together with the padding function enables delay bounds to be maintained when required for any given application.

10.6 AAL TYPE 2 SWITCHING AND SIGNALING

In the initial development of the AAL Type 2 protocol, the primary applications envisaged were for the multiplexing of multiple (e.g., up to 248) separate, possibly compressed, voice channels over a single ATM VCC to achieve bandwidth efficiency in both the cellular wireless and conventional wireline network scenarios. Although these will remain the main applications for the AAL 2 protocol, it was recognized subsequently that for some cellular wireless access network scenarios there may be performance advantages in carrying out a switching, or relaying, function within the AAL 2 CPS itself. This concept has been developed further by describing a so-called "AAL Type 2 switched network" for cellular radio access networks (RAN). Consequently, the concept of using AAL 2 based "switching" has recently been proposed as a potential candidate architecture to be incorporated as part of the ITU-T's International Mobile Telecommunications 2000 (IMT-2000) vision for the evolution of wireless networks. Although the detailed network architectures and wireless technologies underlying these generic evolution scenarios are outside the scope of this text, the basic concepts of AAL 2 switching may be readily understood as an extension of the ATM layer switching concept to the AAL. In the case of AAL 2 switching, the individual AAL 2 CPS-packets, corresponding to a given CID value, are switched between "AAL 2 links" at an AAL 2 switching (or relay) NE by translation of the CID values, analogous to the relaying of ATM cells by translation of the VPI/VCI values between the

VCLs at the ATM layer. In this case, the AAL 2 CPS-packet CID values perform an analogous function to the ATM cell VPI/VCI values in identifying AAL 2 virtual channel links. The concatenation of the AAL 2 CPS-packet CID links forms the end-to-end AAL 2 "connection," by analogy with the ATM VCC concept described earlier.

The main motivation behind the concept of AAL 2 switching in mobile wireless applications is that the compressed voice packets comprising the CPS-packets may thus be switched (or relayed) between the mobile terminals without the need for transcoding the compressed voice to 64 kbit/sec voice channels. In cellular mobile telephony, it is recognized that the conversion (i.e., transcoding) of compressed voice packets to "normal" 64 kbit/sec voice channels may result in some performance impairments. Consequently, it is believed that by enabling switching of the cellular voice channels from the mobile terminals in the form of the compressed voice packets embedded in the individual CPS-packet payloads, without having first to transcode these packets back to 64 kbit/sec voice channels in the RAN, a gain in performance may be achieved. Clearly at some stage, for example, in order to interwork with the fixed PSTN or ISDN, it will be necessary to transcode the voice signal to the uncompressed format, thereby incurring some level of performance impairment. However, by minimizing the need for transcoding of the AAL 2 CPS-packet payloads switched between mobile terminals, the voice quality of service may be maintained within acceptable bounds. Hence, from this perspective, the fact that the AAL 2 protocol was designed for the multiplexing of the compressed voice channels in the individual CPS-PP concatenated in the CPS-PDU makes it a potential candidate for the switching of the CPS-packets, independently of any normal switching of the underlying ATM cells at either VP or VC levels in the ATM layer. Given this possibility, the significance of selecting the (default) maximum length of the CPS-PP = 45 octets becomes clear. It results in a maximum CPS-packet length of 48 octets, equivalent to the ATM cell payload, and implies that potentially the same switching matrix hardware may be utilized either for switching ATM cells or for AAL 2 CPS-packets.

It will be recognized that for NEs capable of switching (or relaying) AAL 2 CPS-packets, the NE must also terminate the ATM VCCs containing the AAL 2 packets, at both ingress and egress ports. The ATM VCCs between any two consecutive AAL 2 relaying nodes along the entire AAL 2 "connection" may either be set up by management system provisioning (i.e., PVCs as described earlier) or by on-demand ATM signaling procedures. In effect, the ATM VCCs are essentially performing the same role for the AAL 2 CPS-packets as ATM VPCs perform for the ATM VCCs, namely providing "trunks" between the AAL 2 switching nodes for the multiplex of AAL 2 CPS-packets. Purely from the perspective of the AAL 2 switching, it is essentially immaterial as to how the ATM VCCs are set up and managed between the AAL 2 switching nodes, and it may be assumed that any normal procedures may be used. Consequently, it was recognized that independent procedures were required for the set-up and management of the AAL 2 connections in order to achieve AAL 2 level switching functionality. To meet this need, a dedicated signaling protocol has been developed by the ITU-T for AAL 2 switching requirements, in particular for the potential applications foreseen in enhancing future cellular wireless networks based on ATM technology.

Although the AAL 2 signaling protocol bears some similarities to the ATM signaling protocols described in some detail earlier, it is important to note that there are differences between them; the two protocols operate independently and hence should not be confused. Although there may be a number of reasons for the differences between the AAL 2 signaling and ATM signaling protocols, the two main reasons are: 1) the need to keep the AAL 2 signaling as simple as is possible to meet the foreseen requirements for AAL 2 switching applications, and 2) to accommodate the requirements for the basic AAL 2 multiplexing functionality which, as noted earlier, forms the primary rationale for the development of AAL 2. The similarities between the

ATM and AAL 2 signaling procedures include the use of common channel, message-based protocols with the use of function-specific "parameters" for the set-up, release, and management of the AAL 2 connections on a hop-by-hop basis along the connection path. It is not necessary here to undertake a detailed description of the AAL 2 signaling procedures, which the interested reader will find in ITU-T Recommendation Q.2630.1 [10.17]. The AAL 2 signaling procedures are currently (at the time of writing) in the process of being finalized, and whereas their implementation in certain specialized network applications, such as cellular radio access networks (RANs), may be potentially envisaged, more general usage remains an open question at this stage of evolution of ATM networks.

The AAL Type 2 signaling protocols to enable switching of the AAL 2 CPS packets bear some functional similarities to the ATM B-ISUP signaling protocols described previously, although they are designed to be substantially simpler. Thus, the AAL 2 signaling utilizes a set of 11 signaling messages, which may include one or more "parameters" in order to transfer the signaling-related information as for the case of B-ISUP signaling for the ATM layer. The AAL 2 signaling message set and associated message identifiers are:

Message name		Message identifier
1. Block Confirm	(BLC)	0000 0001
2. Block Request	(BLO)	0000 0010
3. Confusion	(CFN)	0000 0011
4. Establish Confirm	(ECF)	0000 0100
5. Establish Request	(ERQ)	0000 0101
6. Release Confirm	(RLC)	0000 0110
7. Release Request	(REL)	0000 0111
8. Reset Confirm	(RSC)	0000 1000
9. Reset Request	(RES)	0000 1001
10. Unblock Confirm	(UBC)	0000 1010
11. Unblock Request	(UBL)	0000 1011

Note that although the terminology used for the message (and parameter) names are different to ATM signaling, the function of each message may readily be inferred from its name in general terms.

It may be noted that the capability to "block," "unblock," or "reset" the AAL 2 connections for maintenance purposes is analogous to the case of B-ISUP procedures intended for the same purpose. The parameters defined for AAL 2 signaling messages are:

1. Cause
2. Connection Element Identifier
3. Destination E.164 Service Endpoint Address
4. Destination NSAP Service Endpoint Address
5. Destination Signaling Association Identifier
6. Link Characteristics
7. Originating Signaling Association Identifier
8. Served User Generated Reference
9. Served User Transport
10. Service-Specific Information (Audio)

11. Service-Specific Information (Multirate)
12. Service-Specific Information (SAR)
13. Test Connection Indicator

As for the case of ATM signaling, not all the listed parameters need to be included in all the AAL 2 signaling messages. The details of precisely which message contains which parameter, together with the associated procedures, need not be described here and may be found in ITU-T Recommendation Q.2630.1. Suffice it to say that in these respects, the operation of the AAL 2 signaling procedures in establishing, releasing, and maintaining the AAL 2 connections is broadly analogous to the procedures used for ATM signaling, with some inevitable differences of detail. Nonetheless, it is important to note that the significant differences between AAL 2 and ATM signaling arise in terms of the structure of the message and parameter formats for the case of AAL 2 signaling. These differences arise in order to obtain some simplification of the AAL 2 signaling procedures, primarily by using some "fixed length" fields where appropriate, thereby limiting the amount of signaling information that needs to be transported and processed at AAL 2 switching nodes. Consequently, the ubiquitous type–length–value (TLV) format common to ATM signaling procedures is not always used of all fields in the AAL 2 signaling messages (and parameters). Both fixed length as well as variable length fields are used; for the case of variable length fields, the first octet signifies the actual length of the field. In addition, compared with the ATM signaling messages, the AAL 2 signaling messages are generally shorter, in keeping with the fewer parameters (and hence functions) required.

Another significant difference between AAL 2 signaling and ATM signaling is that the former may also be used for the case where AAL 2 is providing only a multiplexing capability within an ATM VCC. It will be recalled that the primary intent of the AAL 2 protocol is to provide the capability to multiplex AAL 2-CPS packets based on the CID values allocated to each AAL 2 "channel." In the initial development of the AAL 2 protocol, it was assumed that the individual CID values in the multiplex stream may be assigned either by static provisioning, or dynamically by a simple management procedure termed the AAL 2 negotiation procedure (ANP). Although the ANP was initially conceived as a "management" protocol for the dynamic allocation of the CID values, this function may also be viewed as a subset of the capability needed to perform switching of the AAL 2 CPS packet multiplexed stream. Consequently, with the development of the AAL 2 signaling protocols, the use of a separate ANP for management of the multiplexed CID values appeared redundant, as it was assumed a subset of the AAL 2 signaling procedures could be used to allocate CID values. In purely multiplexing applications of the AAL 2 protocol, for example, for simple voice trunking network scenarios, there is no "relaying" of the AAL 2-CPS packets at intermediate NEs. In this case, the AAL 2 signaling procedures simply enable the assignment, release, and maintenance of the CID values on an end-to-end basis, as opposed to a link-by-link basis for the more general AAL 2 switching application. A simpler subset of the AAL 2 signaling procedures can therefore be used for this "nonswitched," or multiplexing, application, as described in more detail in Annex A of the ITU-T Recommendation Q.2630.1 dealing with the complete AAL 2 signaling procedures.

Irrespective of whether the potential applications of the AAL 2 protocol will occur for simple multiplexing network scenarios or for the more complex AAL 2 switching scenarios envisaged for mobile telephony, the AAL 2 protocol evidently results in enhanced capability, and hence complexity, when compared to AAL Type 1 or Type 5 protocols. It is of interest to observe that the inclusion of a "multiplexing" capability at the AAL was perceived as a disadvantage for the case of the earlier AAL Type 3/4, whereas it is seen to be a strength for the case of AAL Type 2. Although this apparent irony may be explained in terms of the quite different intended network applications for the two AAL types, it cannot be denied that the provision of dynamic multiplex or, more importantly, switching capability at both the ATM and AAL levels may lead to functional redundancy and unnecessary complexity in most cases. These considerations need to be carefully balanced in selecting a protocol architecture optimized for any given network application such as voice or data trunking or wireless access. It may also be borne in mind that in principle, on-demand switched connectivity is possible at both the VPC and VCC levels within the ATM layer; both are capable of supporting a wide range of QoS and traffic management functionality. The provision of on-demand switched AAL 2 connectivity on top of this, may seem like "too much of a good thing," in enabling three levels of switched connectivity in what may essentially be viewed as one layer of the overall (OSI-based) protocol stack.

10.7 THE SERVICE-SPECIFIC CONVERGENCE SUBLAYER (SSCS)

It was pointed out earlier that the AAL protocols are generally sublayered into a so-called "common part" (which typically consists of the common part convergence sublayer, or CPCS, and the Segmentation and reassembly, or SAR, sublayer), and a service-specific convergence sublayer, or SSCS. The description so far has focused on the common parts of the four main AAL types defined to date, although it should be stressed that any given AAL common part may be used by an application or service if appropriate, without the need to interpose an SSCS. In such cases the SSCS may be considered to be null, so that the higher layers (application or service) are able to directly utilize the services provided by the common part of the AAL. However, for some other important applications, such as ATM signaling, for example, the functionality provided by the common part is not sufficient and it is therefore necessary to interpose an SSCS that includes the additional functionality required by the higher-layer applications. This is particularly true for a number of interworking applications, as will be seen later, in which the ATM network is required to interconnect with networks based on other transfer technologies such as frame relay services or networks based on the internetworking protocol (IP) stack.

Since by definition the SSCS protocols contain the functionality required by a particular network application or service, these protocols tend to be designed by specialist groups that intend to use the ATM network for the particular application of interest. In addition, as the number of such applications grows, it is evident that a growing number of SSCS protocols are being developed. Consequently, any listing of the SSCS protocols should be recognized as essentially a snapshot of an evolving set of potential protocols and a comprehensive survey will not be attempted here. Nonetheless, a basic set of SSCS protocols has been developed for the main ATM network applications and examples of some of these are listed together with their essential functions in Table 10-2. The SSCS protocols summarized there include:

1. Frame Relay-SSCS (FR-SSCS). The FR-SSCS protocol is described in detail in ITU-T Recommendation I.365.1 [10.6] and is intended for the transport of frame relay service over ATM and for FR to ATM interworking applications. This aspect will be dealt with in more detail below in considering interworking scenarios. The FR-SSCS enables the multiplexing of FR frames using AAL Type 5 common part while preserving the FR protocol control information. It also enables the interworking of the congestion related functions between the ATM and FR parts of an interworked connection.
2. Service-Specific Connection Oriented Protocol (SSCOP). The SSCOP is described in detail in ITU-T Recommendation Q.2110 [10.8] and is primarily used for the transport of ATM signaling messages across both UNI and NNI, as was indicated earlier in the description of signaling protocols. The SSCOP is intended to provide for the reliable (i.e., assured) transport of connection-oriented data in critical applications such as the carriage

TABLE 10-2. Service-specific convergence sublayers (SSCS)

SSCS type	Service	Essential functions
Frame Relay-SSCS (FR-SSCS). ITU-T Recommendation I.365.1. [10.6]	Transport of FR on ATM FR/ATM interworking.	Multiplexing of FR frames in AAL Type 5. Preservation of FR PCI. Congestion information interworking.
Service-Specific Connection Oriented Protocol (SSCOP). ITU-T Recommendation Q.2110. [10.8]	Assured (reliable) transport of connection-oriented data. Signaling message sets.	Assured mode message transfer with retransmission and flow control capability if error conditions detected.
Service-Specific Coordination Function-connection Oriented network service (SSCF-CONS). ITU-T Recommendation I.365.2. [10.7]	For OSI connection oriented network data services.	Error detection/reporting. Connection setup/reset/release. Assured mode message transfer (uses SSCOP service).
Service-Specific Coordination Function- Connection Oriented Transport Service (SSCF-COTS). ITU-T Recommendation I.365.3. [10.9]	For OSI connection oriented transport data services.	As for above, related to OSI transport layer (uses SSCOP service).
Connectionless Network Access/Interface Protocol* (CLNAP/CLNIP). ITU-T Recommendation I.364. [10.10]	Connectionless data services on ATM.	Uses AAL Type 3/4 common part. Global destination addressing (E.164). Transit network selection. Traffic control and OAM capability.

*The CLNAP/CLNIP procedures are similar to the North American Switched Multimegabit Data Service (SMDS) and ETSI Connectionless Broadband Data Service (CBDS), with some differences in detailed procedures/options. This service is titled "Broadband Connectionless Data Bearer Service" (BCDBS) in the ITU-T documents.

of the signaling messages. In SSCOP, the reliable transfer of the data messages is assured by incorporating the retransmission of any "lost" information, as well as the use of flow control capability to avoid congestion and hence information loss. Although primarily intended for the use of the ATM signaling applications, it is evident that SSCOP may be used for any data application that requires the integrity of information transfer. In general, the SSCOP is used on top of the AAL Type 5 common part.

3. Service-Specific Coordination Function-Connection Oriented Network Service (SSCF-CONS). The SSCF-CONS protocol is described in ITU-T Recommendation I.365.2 [10.7] and is designed for the transport of the OSI connection-oriented network data services on ATM. The SSCS-CONS protocol utilizes the SSCOP functions for the flow control and retransmission of lost data to provide an assured mode message transfer service, and includes mechanisms for the reporting and detection of errors as well as AAL connection set-up, release, and reset. The SSCF-CONS protocol may be used in conjunction with the AAL 5 common part to provide a comprehensive set of functions for the reliable transfer of typical data applications.
4. Service-Specific Coordination Function-Connection-Oriented Transport Service (SSCF-COTS). The SSCF-COTS protocol is described in ITU-T Recommendation I.365.3 [10.9] and is similar to the SSCF-CONS protocol, although intended for the support of the OSI connection-oriented transport data services on ATM. As for the SSCF-CONS protocol,

the SSCF-COTS protocol also utilizes the SSCOP functions for retransmission and flow control, in conjunction with the AAL 5 common part. Both the SSCF-CONS and COTS protocols have not found extensive use in existing ATM data networking applications, primarily due to the limited penetration of the OSI based data services in comparison to these based on the IP protocol stack.

5. Connectionless Network Access/Interface Protocol (CLNAP or CLNIP). The CLNAP, also termed CLNIP in some applications, is described in ITU-T Recommendation I.364 [10.10]. It is intended to provide a general connectionless data service over ATM, similar to the earlier switched multimegabit data service (SMDS) deployed in North America, or its ETSI equivalent, termed connectionless broadband data service (CBDS) in Europe. The CLNAP service utilizes the AAL Type 3/4 common part together with global destination addresses based on the conventional ITU-T Recommendation E.164 [6.15] addresses. Additional capabilities for transit network selection (TNS), traffic control, as well as limited OAM capabilities are also provided to obtain a somewhat comprehensive commercial data internetworking service over ATM networks in both North America and Europe. It is important to bear in mind that although the use of the CLNAP/CLNIP SSCS provides a connectionless service to its user (i.e., the application layers above), this connectionless transport utilizes the underlying ATM connection-oriented transport based on VPCs or VCCs, which may either be provisioned or dynamically established using signaling procedures. In this sense, the CLNAP based data services may potentially take advantage of the full range of performance and QoS capabilities offered by ATM VPCs or VCCs, based on the extensive traffic control and OAM functions described earlier.

The structure of the CLNAP (or CLNIP) SSCS is shown in Fig. 10-8, from which the similarities with the SMDS (or CBDS) Level 3 may readily be observed. The CLNAP header includes two eight-octet fields for the E.164-based destination and source ATM address, on which the CLNAP packets are routed at each CLNAP routing node on a packet-by-packet basis. The CLNAP procedures do not restrict the routing mechanism that may be employed, nor whether the underlying ATM VPC or VCC is provisioned or set up using signaling procedures, so that any combination may be used in practice. Initial deployments were simply based on provisioned VPCs or VCCs between the routing (CLNAP) nodes. The six-bit higher layer protocol identifier (HLPI) is used to indicate the particular type of information in the payload field. In order to obtain 32-bit alignment for processing efficiency a padding field of between 0–3 octets is provided, whose length is encoded (in octets) in the two-bit pad length field. The QoS field (four-bits) may be used to select a particular QoS class or priority level for the CLNAP packet, and its use generally depends on the type of service offered by any given CLNAP service provider. As an option, error detection over the entire payload may be provided by incorporating a CRC-32 error detection code as a trailer. The presence of the CRC-32 field is indicated by the one-bit CRC indication bit (CIB) as shown. Provision for an extended header length to accommodate additional PCI is also made possible by used of the three-bit header extension length (HEL) field, which encodes the additional header extension in 32-bit words. The extended header may be used to accommodate longer addresses (e.g., NSAP-based AESA) or other options.

The use of the CLNAP SSCS over AAL Type 3/4 common part essentially provides the most direct way of implementing a connectionless service based on ATM, but it clearly is not the only way, as will be seen later in describing the use of IP-based services over ATM. It must be admitted that the relatively ubiquitous usage of IP-based protocols in both local area and wide area environments has overshadowed the inherent advantages of the CLNAP-based services, such as CBDS and SMDS, which have achieved limited deployment as a result. The advantages inherent in the use of CLNAP SSCS are not only related to the potential for multiple QoS classes and the

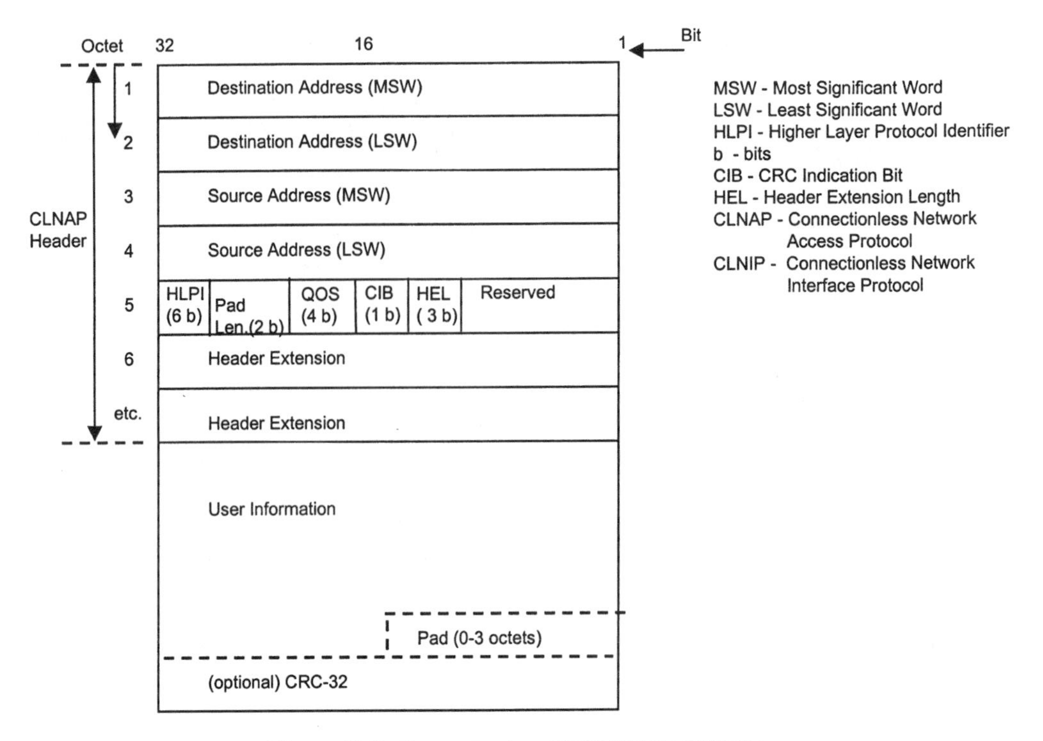

Figure 10.8. Connectionless SSCS-CLNAP/CLNIP.

high-performance and bandwidth capabilities inherent to the ATM infrastructure. The use of E.164-based addressing in the CLNAP SSCS also enables highly efficient routing of the CLNAP packets, resulting from the topological E.164 address structure, as discussed earlier in relation to PNNI routing. In addition, the potential for easier interworking with existing PSTN/ISDN services also exists with the use of E.164-based [6.14] numbering plans, by effectively eliminating the need for address translations between E.164- and IP-based addresses. These aspects will become more evident in describing the more commonly used IP over ATM (IPOA) protocol architectures in subsequent chapters. Regardless of its perceived advantages and drawbacks, the question of whether the CLNAP SSCS-based ATM connectionless data services will ever grow beyond niche network applications remains essentially open. However, at this stage, considering the rapid developments in IPOA based solutions, it seems unlikely that CLNAP-based services will be more widely deployed.

10.8 INTERWORKING

The AAL protocols described above also play an important role in interworking between ATM networks and networks based on alternative transfer technologies such as PSTN/ISDN or IP-based networks. In a number of network applications of ATM, the need for interworking between different protocols is implicit, as will be evident later when considering common ATM networks scenarios. Before describing particular cases in detail, it is useful to examine some general aspects of interworking as it relates to ATM networks. In general terms, the need for interworking arises as a result of ensuring connectivity across networks that may be based on different technologies or protocol architectures (e.g., ITU-T Recommendations I.510 [10.11] and I.520 [10.12]). It is generally assumed that an interworking function (IWF) at the interface between any two networks provides the capability to "convert" between the different protocols and thereby ensure the integrity of the end-to-end user data flow on the interworked connection, despite the different PCI structures on either side of the IWF. It is also important to recognize that in general the interworking requirements do not only relate to the transfer plane procedures. In keeping with the B-ISDN/ATM PRM concept, interworking may also be required between the control plane (i.e., signaling and routing) as well as management plane functions as part of the generic IWF. In many cases, the interworking between control plane or management plane functions in different networks can be more difficult, since these functions generally differ considerably in networking technologies. Clearly, the closer any two protocol structures and functions are, the easier will be the task of the IWF, whether it relates to transfer, control, or management plane interworking requirements.

It can be envisaged that numerous interworking scenarios may be possible, given the number of different technology types, not to mention the detailed differences in network deployments. In general, however, many of these may be categorized into two broad types of interworking scenarios—1) "Network Interworking" and 2) "Service Interworking," using the terminology first introduced in describing interworking between frame relay and ATM networks. However, these categories may be generalized to apply to other technologies when interworked with ATM. The difference between network interworking and service interworking can be described with reference to Fig. 10-9, using an ATM network as the basis.

For the network interworking category, the ATM network is used to transparently transfer the interworked protocol between the IWFs at the interfaces of the two networks as shown in Fig. 10-9. In network interworking applications, the ATM network essentially acts as a "backbone" network for the transfer (including switching and routing) of the interworked protocol, without necessarily modifying the related PCI in any way. Typically, for the case of transfer plane interworking, the IWF may simply "encapsulate" the interworked protocol within the appropriate AAL SSCS and convert it into ATM cells, which are then switched to the remote IWF to be reconstituted to the original protocol structure. It may be noted that the overall (or end-to-end) interworked connection or data path comprises both an end-to-end ATM VPC or VCC, which terminate at the IWFs, and the "other" interworked connections, which also terminate at the IWF,

Network Interworking

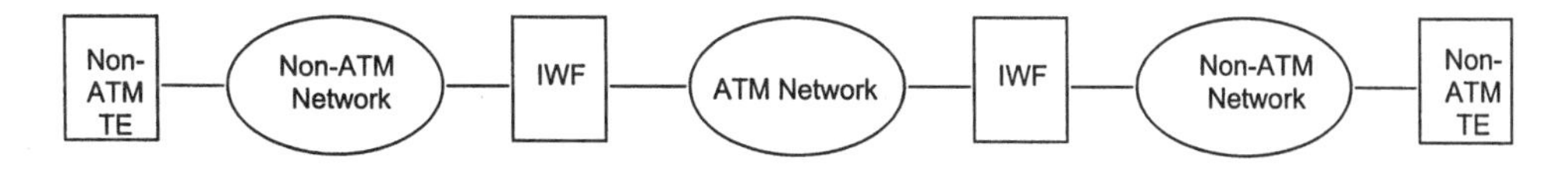

Service Interworking

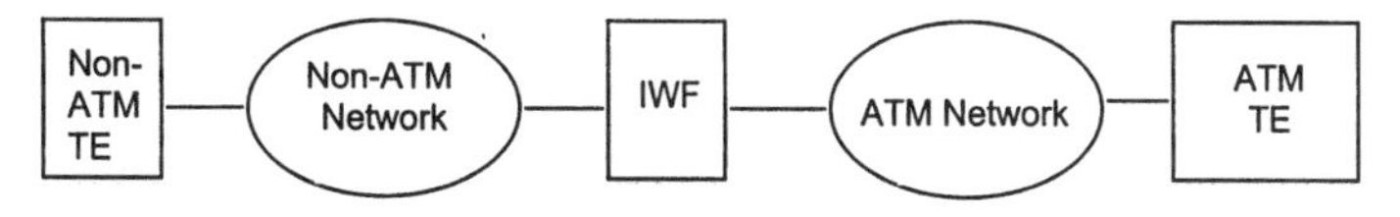

Figure 10.9. General interworking scenarios.

to ensure a continuous data path. Despite the concatenation of the two different types of connection, the IWF must ensure that the higher-layer applications or services are carried "seamlessly" through both types of network. Network interworking also implies the transparent interworking of both control plane and management plane functions where appropriate, although it must be recognized that this may not always be possible due to functional mismatches between control and management capabilities in the two types of network. This will become more evident when we consider specific cases below.

For the case of the service interworking category, it is seen that the IWF essentially terminates the protocols between the two interworked networks, without necessarily reconstituting the PCI at either end. In this case, the ATM network does not act as a backbone or transport underlay for the (other) interworked connection. The IWF may be viewed as "terminating" either PCI and providing a direct translation between the functionalities, where possible. In service interworking scenarios, as for the previous case, it is assumed that the IWF ensures that the higher-layer applications or services that are being interconnected between the end users are also carried seamlessly and without undue modification between the terminals. Thus the "translation" at the IWF essentially refers to the protocol control information and not to the user data payload information. Again, as for the previous case, service interworking also implies that, where possible, interworking between the control and management plane information is also required by termination and translation of these functions at the IWF. However, in the event there is a functional mismatch between the two interworked technologies, it may not be possible to obtain a "direct" translation of the capability between the two networks.

Although the IWF is shown as a separate entity in both the network and service interworking scenarios, it is important to recognize this is purely a functional separation for purposes of illustration. Typically, the IWF may constitute a part of a NE within either one or other of the interworked networks. In addition, although the distinction between network and service interworking is useful from the perspective of understanding the role of the ATM network in an interworked connection, it should be noted that an IWF may be capable of performing both network or service interworking functions in any given implementation. When viewed from the perspective of the IWF, the relationship between network and service interworking applications can be close, since service interworking may be considered to "include" network interworking as a specific instance of protocol translation. In other words, by performing a service interworking translation symmetrically at either end of an ATM VPC (or VCC), it is possible to emulate the network interworking case, as may be apparent from Fig. 10-9, at least for matching transfer plane functions. The close relationship between the AALs and interworking should also be evident, since it is clear that the IWF will by necessity include the appropriate AAL functionality for both SSCS and common part, whether for encapsulation or termination of the PCI.

Bearing in mind the above rather general description of the network and service interworking requirements and the use of the AAL capabilities within specific IWFs between networks, subsequent chapters focus on some important examples of ATM interworking network applications. There are two key areas of interworking network applications in widespread deployment at this stage. These include:

1. Interworking between FR and ATM networks (FR/ATM)
2. Interworking between IP and ATM networks (IPOA)

A third significant application, that of interworking between PSTN (or ISDN) and ATM networks, is of increasing importance as many large network operators continue to expand the capabilities of their core ATM networks to provide services conventionally associated with the existing voice (PSTN) networks [10.13]. The use of AAL Type 1 for interworking between the voice channels and ATM VCCs was already alluded to earlier and forms an essential element of the PSTN to ATM IWF. Call control between the PSTN (or ISDN) and the ATM (B-ISDN) network may be provided in a number of ways, the most direct being the interworking between the ISUP and B-ISUP (SS7-based) signaling procedures described earlier. In addition, the more recent possibility of interworking using the AAL Type 2 protocols for either multiplexing (so-called voice trunking) or AAL 2 switching may also likely be deployed, particularly for mobile (cellular) access networking. In this context, it is also of interest to note in passing that when both AAL Type 1 and AAL Type 2 are utilized (in different parts of a network for example), "interworking" between AAL 1 and AAL 2 procedures will be required. This type of inter-AAL mapping is outside the scope of this text.

CHAPTER 11

Frame Relay and ATM Internetworking

11.1 INTRODUCTION

The previous chapters have surveyed in some detail the essential functions and capabilities designed for ATM-based networks and NEs. In this and subsequent chapters, the focus turns towards utilizing these capabilities in interworking or supporting networks based on other existing networking technologies, such as frame relay (FR) or IP. Internetworking between FR and ATM (often referred to as simply FR/ATM IW) can arguably be considered as the first and most widespread application of ATM networks and will likely continue to play an important role in the development of both technologies. There are a number of reasons for the symbiosis underlying FR/ATM IW requirements, from both the commercial and technical viewpoints. These include the growth of FR-based network services as well as the fundamental underlying similarities between FR and ATM, which can ease interworking, resulting in complementary deployments of the two types of networks in practice. While the commercial aspects of FR/ATM IW are clearly of prime importance in the design of interworked services, the description below addresses the basic protocol interworking aspects of FR and ATM. In order to more fully understand the mechanisms underlying FR/ATM IW, it is useful to summarize the basic principles of the FR protocols, although it should be borne in mind that it is not the intent to describe these in detail here. For a fuller account of FR technology, the reader may refer to [11.1–11.3].

11.2 FRAME RELAY AND ATM COMPARISON

Frame relay services, more formally called frame relay bearer services (FRBS), are based on a frame mode transfer technology that uses the conventional so-called high-level data link control (HDLC) framing mechanisms for the transfer of data packets over connection-oriented virtual channels, which are identified by a logical connection number (LCN), or label, analogous to the VPI/VCI values in ATM cells. However, in contrast to the fixed-size ATM cells, FRBS uses variable length frames, up to a specified maximum length. The maximum frame length may be typically fixed in the range of 1600 to 2000 octets, depending on the intended applications as well as implementation constraints. The HDLC framing technique uses predefined "flags" to delineate the beginning and end of the data frames, and is widely used as the basis for a number of protocols such as X.25, as well as for the carriage of ISUP signaling messages in SS7 systems for 64 kbit/sec-based ISDN services (i.e., the MTP 2 protocol). In fact, FRBS was initially developed as a particular ISDN bearer service intended specifically for data transport and switching (or relaying). In this context, the FRBS was designed to be supported on the ISDN D-channel (i.e., data channel) at bit rates up to 64 kbit/sec, or in clear channel mode up to bit rates of 1.54 Mbit/sec in the North American DS1 (T1) hierarchy, or up to 2.04 Mbit/sec in the European E1 multiplex hierarchy.

From its initial definition as an ISDN-based data service, the FR technology was further developed by both the FR Forum and the ITU-T for use in both enterprise (i.e., private) and public data networks (PDN), triggering its now widespread adoption for general data networking applications in a variety of scenarios. The initial FR services were generally based on permanent virtual channels (PVC), provisioned by the management system interfaces. More recently, the possibility of using on-demand switched virtual channel (SVCs) connectivity is increasingly feasible, as the required FR signaling protocols have been defined in both ITU-T and FR Forum specifications. The FR signaling protocols essentially use extensions of the existing ISDN (i.e., SS7-based) signaling protocols and are described in ITU-T Recommendations Q.933 [11.8], X.36 [11.10], and X.76 [11.11], as well as in corresponding FR Forum specifications. Since ATM signaling procedures are also in essence extensions of the ISDN signaling as described earlier, the underlying compatibility for control plane interworking between FR and ATM may also be noted in this respect as well. However, despite its initial origins in the ISDN concept, it is important to recognize that, in practice, it is not necessary to implement the ISDN services in order to utilize FR service, since the latter can be viewed as an independent, or standalone, technology for data internetworking, and is typically deployed as such.

In chronological terms, the development of FR preceded ATM, although to a large extent the two networking technologies essentially developed in parallel and influenced each other in a number of respects, notably in congestion control functions and some aspects of signaling procedures. From an interworking perspective, it is useful to compare and contrast the key elements of FR and ATM protocols, while bearing in mind that FR was not designed for the support of multimedia or real-time applications as was ATM and therefore may be expected to be considerably simpler in scope and execution than ATM.

As shown in Fig. 11-1 part b, the transfer plane protocol architecture for FR is relatively simple when compared to ATM and in essence the PRM for FR simply utilizes the narrowband (i.e.,

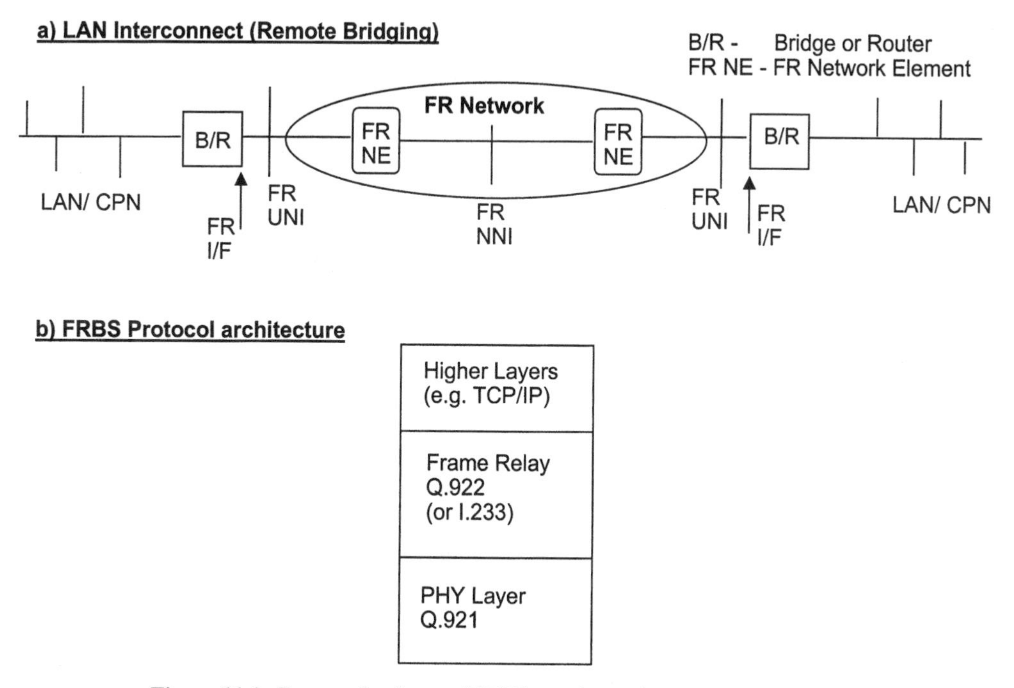

Figure 11.1. Frame relay bearer (FRBS) service and network applications.

64 kbit/sec-based) ISDN protocol architecture, as might be expected. The "higher-layer" applications or services, which typically may be based on the TCP/IP protocol stack, are encapsulated in the FR-PDU as defined in ITU-T Recommendations Q.922 [11.2] and I.233.1 [11.12]. The structure of the FR-PDU is described in further detail below. The physical (i.e., transmission system) layer used for the transport of the FR packets depends on the transmission hierarchy (T1 or E1) upon which the network is designed. More recently, it has been proposed in the Frame Relay Forum to extend the use of FR in conjunction with the SDH (or SONET) transmission system to achieve enhanced transmission throughput and performance. It is of interest to note that no "adaptation layer" is required, since in the case of FR, the PDUs are of variable length and the relaying function is intended primarily for data applications.

As summarized in Table 11-1, FR and ATM have some fundamental underlying similarities, but also some important differences of detail from the protocol perspective. The similarities stem from the use of label multiplexing and the intrinsically connection-oriented basis of the two protocols. In the case of FR, the multiplex label is termed the data link connection identifier (DLCI) and performs essentially the same function as the VPI/VCI labels in ATM. The connection-oriented procedures based on either management-interface-initiated (semi) permanent virtual channels (PVCs) or on-demand signaled switched virtual circuits (SVCs) allows for direct support of QoS classes in both FR and ATM. However, in the case of FR, a set of relative (i.e., nonquantitative) priority classes has been specified as opposed to the fixed QoS classes provided for in ATM. In addition, quantitative measures for the characterization of FR QoS parameters have also been proposed, analogous to the case of ATM. Somewhat analogous to the concept of minimum cell rate (MCR) in ATM, FRBS includes the concept of a "committed information rate" (CIR), which enables the negotiated allocation of sufficient (minimum) bandwidth to a given FR virtual connection to satisfy the QoS requirements.

TABLE 11-1. Functional comparison between frame relay and ATM

Function	Frame relay	ATM
Frame delineation	HDLC flags/zero insertion	Cell delineation/CPCS preservation
Multiplexing	DLCI	VPI/VCI
Packet format	Variable length (up to a maximum value)	Fixed cell size
Error detection	16 bit CRC over frame	Cell header CRC8/AAL 5 CRC 32
Connection types	PVC and SVC bidirectional	PVC and SVC bidirectional
Traffic enforcement	Possible at network ingress	UPC/NPC of traffic descriptors at UNI or NNI
QoS	Relative. Committed information rate (CIR) may be negotiated	Relative or standardized. QoS may be negotiated by specific ATC
Congestion control	FECN, BECN, and DE bits	EFCI, CLP, and RM cells
Traffic parameters	CIR, EIR, access rate, maximum committed burst size (Bc), maximum excess burst size (Be)	PCR, SCR, MBS, CDVT, MCR, ECR, etc.
Layer management	I.620 OAM frames (loopback only)	I.610 OAM cells (F4, F5 flows)
Control plane	Q.933 signaling over DLCI = 0	Q.2931/B-ISUP/PNNI signaling procedures
PVC monitoring	Q.933 Annex A and/or LMI	OAM cells/AEMF
Multicast	Not defined, essentially pt–pt	Unidirectional pt–mpt possible

With respect to traffic and congestion control parameters for use in traffic enforcement or connection admission control to maintain QoS, it will be noted that again there are conceptual similarities between FR and ATM, although here there are some significant differences of detail. Thus, FR provides for the use of the forward explicit congestion notification (FECN) function, which is analogous in principle to the explicit forward congestion indication (EFCI) function described earlier for congestion control in ATM. As noted before, in fact this concept was "borrowed" from FR for the case of ATM congestion control, resulting in some controversy when it was introduced. However, FR also includes a backward explicit congestion notification (BECN) function, which has no direct counterpart in ATM, although, as seen in Chapter 5, there is a "congestion indication" (CI) function in the ATM resource management (RM) cell that may be used to provide an analogous service, but which requires the generation of an RM cell in the backward direction at the congested NE. In addition, FR provides for the possibility of two levels of frame discard priority within a single FR connection, by means of the so-called discard eligibility (DE) function, which is directly analogous in principle to the cell loss priority (CLP) function for ATM, as seen earlier. The traffic parameters used in FR relate to frame rate descriptors such as committed information rate (CIR) and excess information rate (EIR) and related maximum burst sizes B_C, B_E, but it should be stressed that FR does not introduce concepts such as service category (or ATC) as is the case for ATM. As a result, and perhaps because fewer traffic parameters are utilized in general, the traffic and congestion control functions in FR are perceived as significantly simpler than for ATM. It may be also pointed out that FR does not utilize feedback control mechanisms analogous to ABR, or ABT based on RM cells as does ATM, which in itself constitutes a significant simplification.

For the case of per-connection OAM functions, it is interesting to note that there are significant differences in approach between FR and ATM. Although the protocols for using in-band "loopback OAM frames" for in-service or out-of-service continuity verification have been defined in ITU-T Recommendation I.620 [11.4], somewhat analogous to the use of loopback OAM cells in ATM, other techniques have been more widely used in FR, as will be described below in considering management interworking. These techniques have been based on the use of signaling maintenance messages such as STATUS and STATUS ENQUIRY for the monitoring of FR links, rather than in-band OAM frames. In FR, the preassigned DLCI = 0 value is intended for the transfer of FR signaling messages for connection setup, release, and management. However, for the case of FR PVCs, the common DLCI = 0 channel may also be used to transfer STATUS messages to provide information on the status of the FR links and thereby perform an elemental fault management function. Although simple in principle, it may be noted that this approach cannot provide performance management information. The detailed procedures for using the FR signaling messages for this maintenance function are described in ITU-T Recommendation Q.933 Annex A [11.8]. In addition, the FR Forum has also developed other techniques for FR link management based on the simple network management protocol (SNMP), operating over a so-called local management interface (LMI), somewhat analogous to the ILMI procedures outlined earlier for ATM. It is not the intention here to describe these aspects of FR-to-ATM management interworking, but it should be recognized that although FRBS management requirements are somewhat simpler than for ATM, FR-to-ATM interworking for management information may be more complex in nature as a result of the differences in approach.

Apart from the obvious difference of fixed ATM cell size as opposed to the variable length FR frame sizes, Table 11-1 also lists significant differences in respect to frame delineation and error detection mechanisms. For frame delineation, FR utilizes conventional HDLC procedures based on standard HDLC flags and zero insertion mechanisms, in common with other HDLC-based protocols. For the case of ATM, the overall "frame delineation" function must involve both cell delineation using the ATM header error control (HEC) function described earlier, as well as the AAL CPCS preservation mechanism, which differs depending on the selected AAL, as seen earlier. FR and ATM also differ significantly with respect to the approach to error detection procedures. FR utilizes a 16-bit CRC algorithm calculated over the entire FR frame for error detection (typically, frames are discarded in the event of errors), whereas in the case of ATM payload error, detection is performed at the AAL level and hence depends on the AAL type. At the ATM layer (cell level), error detection (and correction) is possible only over the ATM cell header, and not over the cell payload as described in detail earlier. Thus, here again for the case of frame delineation function, the overall error detection function may be viewed as being "spread" over both the ATM layer and the AAL. However, from a FR-to-ATM interworking perspective, these differences in the detailed operation of the two protocols do not pose any significant difficulty, since there is essentially no basic functional mismatch.

Before describing in further detail the FR protocol and its interworking with ATM, it is useful to summarize the typical network applications for the technology and the motivation for interworking with ATM. As shown schematically in Fig. 11-1 part a, the simplest and by far the most popular network application of FR technology is for interconnection between remote local area networks (LANs) or IP-based networks. In general terms, these applications are also loosely referred to as "leased line replacement" or "remote bridging," depending on whether the FR connection is set up between peer LAN Layer 2 bridges or conventional Layer 3 routers. Typically in such applications, the bridge or router NE will be provided with a FR interface card, which is also sometimes called a FR access device (FRAD), coupled with the necessary software upgrade to implement the FR UNI. The FR network comprising the FR NEs (or Switches) provides virtual circuit connectivity (either semi-permanent or on-demand switched) between the remote enterprise customer premises networks (CPNs) or LANs. Since the FR network can provide for statistical sharing of bandwidth resources based on the concept of virtual circuit connectivity, it is feasible to achieve significant cost savings when compared with conventional "leased lines" or nailed-up TDM circuits over the PSTN or ISDN. In essence, the FR virtual circuit simply replaces a TDM leased line which is relatively expensive, since it involves dedicated resources (bandwidth) even when not being used for data transfer. It will be recognized that the same reasoning also applies if the "virtual circuit" connectivity is provided by an ATM network instead of the FR network shown in Fig. 11-1 a. Thus, a similar network application may also be envisaged for ATM, but the relative simplicity and low cost of FRBS, as well as the resulting cost savings to end-users of using virtual circuits, has resulted in rapid growth of FRBS in recent years.

In the FR network application scenarios such as depicted in Fig. 11-1 part a, the data (e.g., IP) packets from the CPN routers are encapsulated directly into the FR payload, to be switched (i.e., relayed) through the FRBS network. The details of the encapsulation protocol has been specified in the IETF RFC 1294 and is referred to as the FR network layer protocol identifier (NLPID) encapsulation. As will be seen later, this encapsulation mechanism is somewhat different to that used in the encapsulation of IP data packets in ATM, which results in the need for further interworking in the case of FR-to-ATM connectivity for IP data packets.

In describing the FR network application for leased line replacement above, the question may have arisen that since FR enables relatively simple internetworking between the CPN routers, why is it even necessary to consider deploying ATM networks and, in particular, why even consider FR-to-ATM interworking? In principle, it is perfectly possible to build relatively large FR networks and many such exist in global deployments. However, as depicted in Fig. 11-2, there are advantages in using an ATM network as a "backbone" for the support of FRBS, particularly for large-scale (e.g., carrier) deployments. The underlying reasons for the utility of the ATM backbone may be understood by recalling that FR interface rates for both UNI and NNI are specified up to typically 1.5 Mbit/sec in the North American transmission hierarchy (DS0/DS1 rates), or up to 2.04 Mbit/sec in the ETSI hierarchy. With these relatively low rates at NNIs, to build large meshed backbone networks using only FR will require significantly more FR links and NEs, in order to maintain reasonable traffic performance for statistically multiplexed

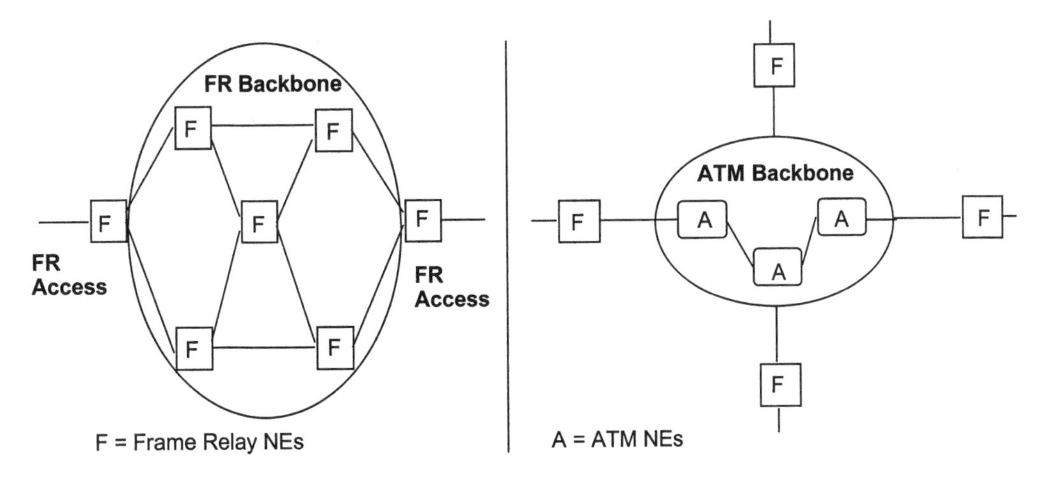

Figure 11.2. FR and ATM backbone networks.

sources. As pointed out earlier in describing ATM traffic performance, the statistical gain depends on the ratio of the total multiplexed rate to the individual source rate that contributes to the overall multiplex. For a given source rate, a high multiplex rate enables a reasonable statistical multiplexing gain to be achieved, assuming the other relevant parameters are held constant. The higher statistical gain translates into relatively lower cost or, equivalently, a larger number of users of the FR virtual circuit services.

These conditions may be more readily realized with fewer links and NEs in the meshed backbone if the relatively higher speed ATM NNIs are utilized in the backbone, as indicated in Fig. 11-2. Since the ATM interfaces readily operate at rates from 45 Mbit/sec (DS3) or 155 Mbit/sec (STM1) to 622 Mbit/sec and above, it is clear that fewer ATM NEs and links can be deployed in the backbone network to build large FR networks while still providing useful FR statistical multiplexing gain. The general recognition among network operators that the relatively high performance and throughput provided by ATM backbone networks in applications such as support of lower-cost FRBS has led to their widespread use in the design of such networks, arguably the most popular network application of ATM networking in initial deployments. As an ancillary benefit, the ATM "backbone" networks may also be used for the transport and switching of other network services such as voice services, circuit emulation or direct IP data applications, as will be discussed in more detail later. In a sense, the typical network scenario involving the use of the (essentially service-independent) ATM backbone network to support a variety of network applications such as FRBS or circuit emulation furnishes the network (or service) provider with a degree of "future proofing" capability, as well as growth potential. In the same sense, the rapid growth of FR- and IP-based networking technologies also results in the concurrent expansion of the underlying ATM backbone networks used for the transport and switching of these services, in a symbiotic relationship between ATM and what are often viewed as "rival" technologies such as FR or IP.

11.3 FRAME RELAY SERVICE-SPECIFIC CONVERGENCE SUBLAYER (FR-SSCS)

The frame relay protocol structure evolved from the basic HDLC technology widely used in other earlier data networking protocols and, as indicated earlier, was initially intended as a specific service within the 64 kbit/sec based ISDN environment. The FR PDU structure is shown in Fig. 11-3, which includes the detailed structure of the FR header format. It will be noted that FR allows the use of three types of header format using either a two-octet, three-octet, or four-octet length. The most commonly used header structure is the (default) two-octet header, although future use of the three or four-octet headers are feasible. The flexibility of the FR header format is an interesting contrast to the fixed five-octet ATM cell header. The FR PDU consists of a one-octet HDLC flag used to delineate the FR frame, followed by the FR header, which can be either two, three, or four octets in length. The variable length FR information field or payload encapsulates the higher-layer protocol (e.g., an IP packet) to be relayed across the network. The two-octet FR trailer consists of a frame check sequence (FCS), which is a 16-bit CRC calculated over the entire frame, excluding the HDLC flags. The FCS is intended for error detection purposes. The frame is delineated by an end HDLC flag, as described in ITU-T Recommendation Q.922 [11.2]. The HDLC flag sequence is 0111 1110. In order to avoid the same sequence appearing within the FR payload, which would result in false delineation, the HDLC protocols include the so-called zero insertion procedures, which are essentially intended to ensure the HDLC flag sequence is not present in the payload. Details of these procedures are outside the scope of this summary but may be found in ITU-T Recommendation Q.922 or other HDLC protocol descriptions [11.1, 11.2].

Considering the most commonly used two-octet (default) FR header structure as the basis of this summary description, it will be seen that this includes a 10-bit data link connection identifier (DLCI) field that provides the logical connection number (LCN) function analogous to the VPI/VCI values in the ATM cell header. As in that case, the DLCI values only have a "local significance" in the sense that each FR switching (or relaying) node may translate (or change) the DLCI values between ingress and egress logical links to perform the switching function. The

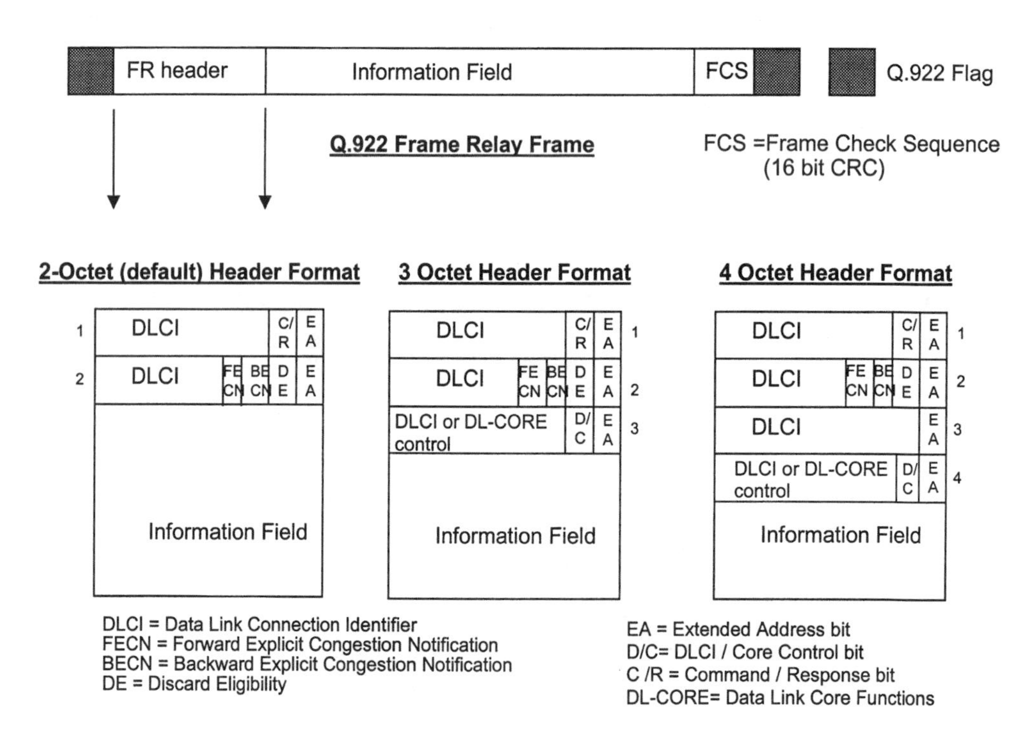

Figure 11.3. Frame relay protocol.

DLCI values = 0 is preassigned for the transport of FR signaling messages. Other reserved and preassigned DLCI values have also been specified for additional and management functions, but these need not be considered here since they do not affect interworking aspects. The so-called command/response (C/R) bit in fact does not carry any FR PCI significance but is used in relation to FR interworking with the earlier X.25 packet transfer mode, which utilizes this function.

The extended address (EA) bit is used to control the FR header length between two-, three-, and four-octet options by indicating the last octet of the header. Thus, the value EA = 0 indicates that additional header octets will follow. The value EA = 1 signifies the final octet in the FR header. For example, the EA bit sequence 0, 1 indicates the two-octet default FR header structure, whereas the EA bit sequence 0, 0, 0, 1 indicates the four-octet FR header format. It will be noted that in the three- and four-octet FR header formats, the last octet may be used either for a DLCI extension function or for the so-called data link CORE (DL CORE) control function. The selection between either of these options is controlled by the value of the DLCI/DL CORE Control (D/C) bit in the last octet of the header. Note that this option does not exist for the default two-octet FR header format, where the DLCI field is always 10 bits in length. Thus, the value D/C = 0 indicates that the other six bits in the last octet should be interpreted as a DLCI extension, whereas the value D/C = 1 signifies the remaining six bits of the last octet carry DL-CORE control information.

In practice, since most of the FR implementations to date only utilize the two-octet header format, the DL CORE capability is not important in existing FRBS deployments. Even in the case of the two- or three-octet FR header structures, typically the DLCI extension capability is of relevance, since the detailed DL-CORE functions and associated procedures have not been defined. In principle, the DL-CORE PCI was initially intended for control and management-related information for frame switching capability, but this possibility was somewhat circumvented by the development of separate FR (common channel) signaling and OAM procedures. However, it may be noted in passing that the four-octet header format with specific DL-CORE coding is specified for use in the case of in-band FR OAM (Loopback) frames, as described in ITU-T Recommendation I.620. Although a detailed description of the FR OAM loopback frames is outside the scope of this overview, it may be pointed out that this capability is analogous to the in-band OAM loopback cells used for continuity verification and fault localization purposes in ATM.

In comparing FR and ATM earlier, it was pointed out that FR incorporates some congestion control functions that are analogous to ATM. These functions include the "forward explicit congestion notification" (FECN) and "backward explicit congestion notification" (BECN) fields, as well as the "discard eligibility" (DE) field in the FR headers depicted in Fig. 11-3. By analogy with the EFCI function in ATM cell header, when the FR NE experiences traffic congestion, it will set the FECN and BECN bits to (binary) 1 to indicate a congestion state to the end user. The intent is that the end-user applications may use this information to reduce the traffic flow and thereby avoid frame loss by relieving the network congestion state. The efficacy of the FECN and BECN approach, as for the case of EFCI in ATM, depends largely on the behavior of the end-user procedures in reducing traffic flow and therefore may not always be relied upon, since, in practice, the detailed procedures for FECN/BECN use in end terminals are not well defined or mandatory. Thus, although several traffic simulation studies have confirmed that, when acted upon appropriately, the use of FECN and BECN feedback information can reduce frame loss levels in the event of network congestion state, use of this capability in practice has not been widespread. The other congestion control function included in the FR header PCI is frame discard eligibility (DE), which is analogous in intent to the cell loss priority (CLP) function in ATM. Thus, in FR, the frames with DE = 1 are subject to discard before frames marked with DE = 0 in the event of traffic congestion. As for the case of CLP use in ATM, the use of the DE bit in FR implies that there are effectively two levels of frame loss priority within a given FR virtual channel. The high-priority (i.e., lower-loss) frames are marked as DE = 0, and the lower-priority (discard eligible) frames are marked as DE = 1.

The FRBS congestion control functions FECN/BECN and DE have been described in some detail above since it will be recognized that in any FR-to-ATM interworking deployment, where these functions are used, the IWF may need to "map," or translate, between the congestion information available on the FR and ATM parts of the overall end-to-end interworked connection. The mapping options between the FR and ATM congestion control functions are described in detail in ITU-T Recommendation I.555, which addresses FR/ATM interworking in detail, as well as in several FR Forum specifications such as FRF 5 and FRF 8, describing FR to ATM interworking scenarios. In a number of FRBS deployments, the use of the DE bit is also closely related to the provision of the so-called "committed information rate" (CIR) service, which essentially allocates a minimum "guaranteed" bandwidth, corresponding to the CIR, to the FRBS user. The CIR concept may be viewed as analogous to the use of the minimum cell rate (MCR) parameter in the relevant ATM ATCs, as described earlier. In FR, the CIR traffic is marked as DE = 0 to indicate that the CIR frames should be treated as having high priority in the event of congestion. For the corresponding FR-to-ATM interworked connection, the CIR traffic will be mapped to ATM cells with CLP = 0 (i.e., high-priority cells) with corresponding bandwidth allocation based on the MCR parameters, if used. In practice, the detailed traffic interworking between the FR network and ATM network depends largely on the specific ATM ATCs (service categories) available, but it will be recognized that the correspondence between the FR and ATM traffic control parameters plays an important role in establishing the interworked connection in any scenario described below.

For FR-to-ATM interworking, the AAL Type 5 protocol is used in the IWF to "convert" between the FR frames and the ATM cells, as shown in Fig. 11-4. The simple procedures whereby the FR-PDU described above is converted into a format to be encapsulated in the AAL Type 5 CPCS payload is called the FR Service Specific Convergence Sublayer (FR-SSCS). The FR-SSCS PDU is obtained by removing the HDLC (i.e., Q.922) flags from the start and end of the FR-PDU, as well as the frame check sequence (FCS) trailer, which includes the 16-bit CRC. In addition, the zero insertion mechanism is removed and the resulting structure, which consists of the FR header and the payload fields, is directly encapsulated in the AAL Type 5 CPCS payload as shown. The FR-SSCS procedures form the basis for the transfer plane protocol interworking between FR and ATM and are described in detail in ITU-T Recommendation I.365.1 [10.6]. The reasons for the removal of the HDLC flags and FCS in the FR-SSCS should be evident, since these functions are redundant in the ATM part of the interworked connection. As described earlier, the AAL Type 5 CPCS includes a 32-bit CRC based error detection capability that obviates the need of the FR FCS in the encapsulated FR-SDU. In addition, the AAL Type 5 CPCS proto-

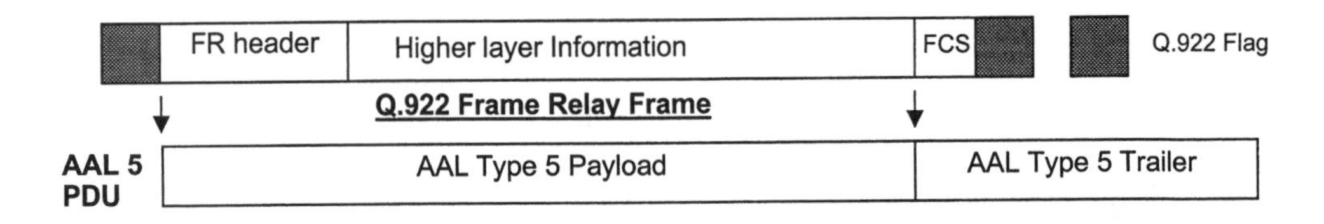

The FR-SSCS- PDU is obtained by removing the Q.922 (Frame Relay) flags, zero bit insertion and Frame Check Sequence (FCS) from the the Q.922 CORE frame and encapsulating the remaining fields in the AAL Type 5 payload

Figure 11.4. FR service-specific convergence sublayer (SSCS).

col utilizes its own frame delineation function based on the ATM user-to-user codepoint in the payload type field of the ATM header. In the reverse direction, that is, when reconstituting the FR PDU, the FR-SSCS simply reinserts the relevant FCS trailer and the beginning and end HDLC flags and zero insertion to reconstruct the entire FR-PDU needed in the FR network side of the IWF. It should also be noted that the FR-SSCS structure is also independent of the FR header format (two, three, or four octets) selected for the interworked connection.

11.4 FR/ATM NETWORK AND SERVICE INTERWORKING

It was pointed out previously in the general description of interworking and AALs that despite the many different specific interworking scenarios possible, typically these may be broadly grouped into two categories known as network and service interworking. The case of FR/ATM interworking is no different and in fact serves as a good example of these general interworking categories. Chronologically, it was the detailed study of FR/ATM interworking aspects that led to the classification of network and service interworking, as described in ITU-T Recommendation I.555 [11.5] and the subsequent FR Forum specifications based on that standard.

The typical FR/ATM network interworking configuration, termed "network interworking-scenario 1" in [11.5], is depicted in Fig. 11-5, together with the transfer plane protocol architecture along the entire interworked connection. The end-to-end interworked connection consists of the concatenation of the FR connections on either side of the ATM VCC (or VPC), each terminating in the IWF. It is important to bear in mind that the "interworking function" as shown is a logical representation, which may in practice reside as part of a NE in either the ATM or the FR networks, depending on the implementation. It is assumed here that the ATM network serves as a "backbone" and that the FR and ATM connections are set up bidirectionally either by provisioning (PVC), or by on-demand signaling (SVCs).

Inspection of the transfer plane protocol architecture (stack) in Fig. 11-5 along the interworked connection shows that the higher layers (in this case, Layer 3 and above) information encapsulated in the FR packet by the FR customer premise equipment (CPE) or terminal is transferred transparently to the remote CPE or terminal through both FR and ATM networks. The so-called Q.922 [11.2] (or equivalent I.233.1 [11.12]) CORE layer corresponds to the FR protocol layer described above. The FR/ATM IWF between the FR and ATM segments of the end-to-end interworked connection utilizes the FR-SSCS protocols described above to encapsulate the FR-SSCS-PDU into the AAL Type 5 CPCS payload, whence it is converted to ATM

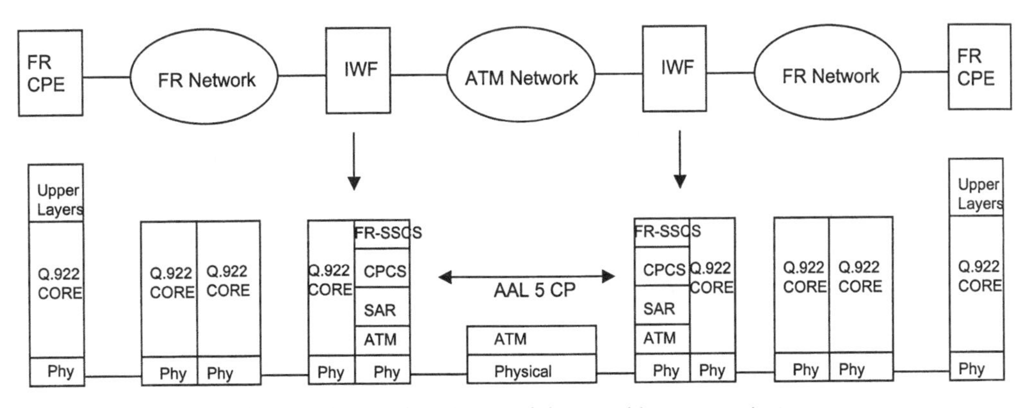

Figure 11.5. FR/ATM network interworking—scenario 1.

cells to be switched across the backbone network. The inverse process occurs at the far end IWF to reconstitute the original FR (or Q.922 CORE) frames, thereby ensuring the transparent transport of the (higher-layer) user information end-to-end across both networks. In the network interworking case, the IWF may be viewed as effectively "mapping" the FR (i.e., Q.922) frames to ATM by means of the FR-SSCS procedures described above. It is evident that on the FR side of the IWF only the Q.922 FR protocols are invoked, for switching of the FR frames through the FR network as shown. Within the ATM network as well, only the ATM layer functions are used (in the transfer plane) to switch the cells in NEs between the IWFs. The combination of the FR-SSCS with the AAL Type 5 CPCS and SAR sublayers in the IWF may thus be viewed as providing the "mapping" of the FRBS to ATM for transport via an ATM network.

In network interworking scenarios, it is sometimes stated that the ATM network provides an "underlying transport" for the FRBS, but it should be clear from Fig. 11-5 that, in fact, the IWF performs a peer-to-peer mapping, since both ATM and FR are essentially Layer 2 protocols. However, the concept of an underlying ATM transport, although not strictly accurate from the protocol perspective, serves as a useful way to envisage network interworking scenarios, particularly as the ATM part of the network will likely also be used for the transport (and switching) of other interworked services, as will be seen later.

An alternative scenario for FR/ATM network interworking, termed scenario 2 in Recommendation I.555 [11.5], has also been identified, as depicted in Fig. 11-6. From the perspective of the IWF, it may be seen that there is no difference between scenario 2 and scenario 1, since both utilize the FR-SSCS + AAL Type 5 protocols for interworking between the Q.922 CORE functions (FR) and ATM. However, FR/ATM network interworking scenario 2 requires the introduction of the concept of the so-called "FR/B-ISDN (or ATM) terminal" device at one end of the end-to-end interworked connection, which incorporates the FR-SSCS protocol to terminate the FR (Q.922) protocol. In a sense, the network interworking scenario 2 may also be viewed as a "special case" of the more general Scenario 1, in which the FR/B-ISDN terminal effectively coalesces the FR network within the device, by including both FR and ATM protocol aspects within its functional architecture. While the hybrid FR/ATM device implied by the scenario 2 may be perfectly feasible from the technical (protocol) perspective, it must be admitted that in most practical FR/ATM interworking deployments, the scenario 1 configuration will be used.

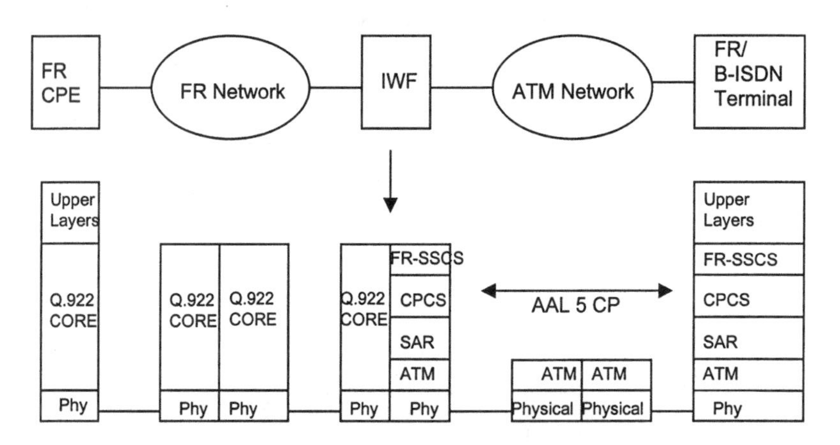

For the FR /ATM Network Interworking scenario 2 it is assumed the terminal attached to the ATM network (FR/B-ISDN TE) implements the FR-SSCS protocol.

Figure 11.6. FR/ATM network interworking—scenario 2.

This is primarily because, as noted earlier, in typical FRBS applications (e.g., for leased line replacement), the CPN or edge router may utilize a FRAD, which will likely only implement the FR (Q.922 CORE functions) protocols and not both ATM and FR protocols, to minimize costs. Consequently, the scenario 2 configuration, althought technically feasible, offers no advantage in practical terms, particularly in view of the FR/ATM service interworking case described below.

Although typical initial FR/ATM network deployments were based on the scenario 1 network interworking configuration, the growth of both FRBS and ATM backbone networks has led to more widespread interest in FR/ATM service interworking, since this enables interworking between both the FRBS and ATM-based services that are increasingly being deployed. The FR/ATM service interworking case is depicted in Fig. 11-7. In keeping with the general concept of service interworking described previously, for FR/ATM service interworking, the FR (i.e., Q.922 CORE functions) protocol is terminated at the IWF and the PCI is mapped (where possible) into the corresponding ATM PCI. Clearly, not all FR PCI will have a one-to-one correspondence with the ATM PCI and vice versa, and it is important to recognize that the service interworking IWF is not "encapsulating" FR PCI as was the case for network interworking based on the FR-SSCS. Consequently, the SSCS functions are different from those described earlier for the FR-SSCS, which essentially encapsulated (omitting the HDLC flags and FCS fields) the basic FR Q.922 CORE functions. In principle, the service-specific convergence sublayer for service interworking between FR and ATM simply terminates the FR PCI while encapsulating the higher-layer (user) information in the FR-PDU payload into the AAL Type 5 CPCS payload field. It will be apparent from the IWF protocol architecture in Fig. 11-7 that, by definition, the IWF terminates both the FR and ATM protocols to enable the transparent transfer of the user information between a purely FR terminal (or CPE) and a purely ATM terminal (or CPC).

The effective termination of both FR and ATM protocols at the service interworking IWF with direct mapping of relevant PCI suggests that the SSCS for service interworking may in fact be a null function. This is the approach adopted by the FR Forum in the FR/ATM Service Interworking Implementation Agreement Specification FRF.8, which simply assumes a null SSCS, but defines the mapping of the relevant PCI between the FR and AAL Type 5 and ATM functions in the interworked connection. It may be noted that this approach provides a pragmatic solution to the problem of service interworking between FRBS and ATM-only networks. From a more formal protocol modeling perspective, however, the mapping functions may be viewed as

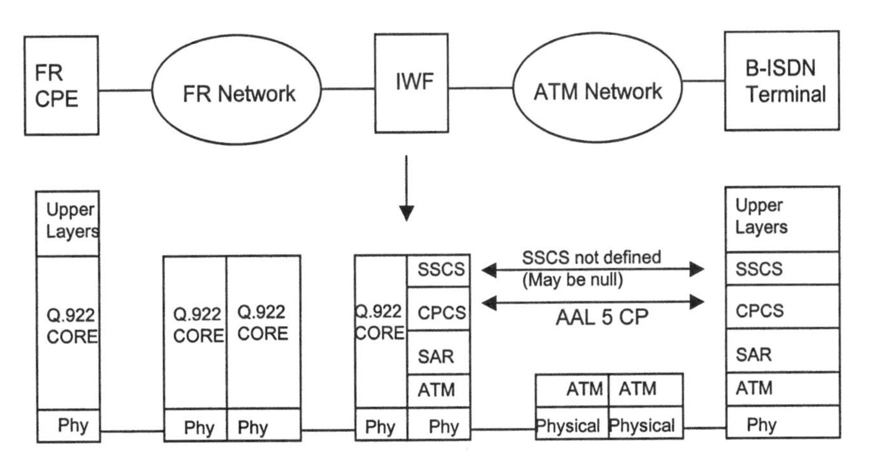

Figure 11.7. FR/ATM service interworking.

constituting the service interworking SSCS capability, an approach assumed in [11.5] for FR/ATM interworking. In this latter description, the SSCS for service interworking, although not explicitly defined in terms of structure and procedures as for the case of the FR-SSCS for network interworking, is nevertheless assumed to represent the termination and mapping of the FR PCI to ATM and vice versa, by analogy with the FR-SSCS. From this perspective, there is no essential inconsistency between the pragmatic approach described in FRF.8 and the more formal protocol interworking representation provided in [11.5]. Nonetheless, it must be admitted that the somewhat different approaches have resulted in some confusion in characterizing FR/ATM service interworking SSCS, particularly in relation to differences with the FR-SSCS, which is more fully specified in Recommendation I.365.1 [10.6] .

In practical terms, however, these differences in approach have not prevented the implementation of FR/ATM IWFs in NEs, which are able to support both network and service interworking capability. The provision of both network and service interworking capability within one FR/ATM IWF implementation is increasingly common, since the growth of both the FRBS and other ATM-based services in actual network deployments makes it clear that both types of interworking will be required in any case. From the perspective of the FR/ATM IWF protocol architecture, it is clear from a comparison of Figs. 11-5, 11-6, and 11-7 that the difference between network and service interworking configurations is the presence or absence of the FR-SSCS, assuming a "null" service interworking SSCS as described above. Equivalently, the service interworking configuration may be viewed as a particular instance of network interworking scenario 2 configuration, in which the FR-SSCS capability is "switched off," i.e., nulled, thereby terminating and mapping the FR PCI to ATM on the interworked connection. It will be recalled that the transfer plane protocol mapping also implies the mapping of the traffic parameters between the FR and ATM parts of the interworked connection, including the congestion control parameters if used. The mapping of the in-band OAM aspects is also of relevance and will be dealt with below.

Despite the relationship between the network and service interworking from the perspective of the FR/ATM IWF protocol architecture, there remains one essential difference between these configurations in respect to the multiplexing of FR connections on ATM. From the earlier description of the FR-SSCS for network interworking and the configuration depicted in Fig. 11-5, it may be recognized that FR/ATM network interworking allows for the possibility of multiplexing multiple FR connections onto a single ATM VPC (or VCC) in the ATM backbone network. The presence of the FR DLCI values in the FR-SSCS PDU may be used as a multiplexing identifier to enable multiple FRBS connections to be transported on a single ATM VPC or VCC set up between the IWFs, analogous to the use of the ATM VCC as a "trunk." The individual DLCI values transported transparently in the FR-SSCS PDU enable the remote end IWF to demultiplex the FR connections according to the DLCI values. This capability of network interworking is termed N-for-1 multiplexing mode and is clearly useful in enabling potential cost savings, since only a single ATM VPC or VCC would be required for the transport of many FRBS users. A special case of the N-for-1 mode is the 1-for-1 network interworking in which a single FR connection is mapped to a corresponding ATM VPC or VCC at both IWFs. The N-for-1 multiplexing mode is applicable for both scenario 1 and scenario 2 network interworking in principle, although for the case of scenario 2 its use will be dependent on the implementation of the FR/ATM terminal device. In the case where such a device represents a single interworked connection endpoint, it will be evident that the N-for-1 multiplexing mode will have limited applicability, such as for the support of different (higher-layer) applications on separate FR connections.

For the case of FR/ATM service interworking, however, it is evident from Fig. 11-7 that only the 1-for-1 mapping case is applicable, since, by definition, the service interworking IWF terminates the FR PCI, including the DLCI values. In this case, each interworked connection consists of a single FR connection part and a single corresponding ATM connection part, since the interworking is between the FR terminal and an ATM (only) terminal, which does not recognize the DLCI syntax. Consequently, the multiplexing of multiple FR connections to one ATM VPC or VCC does not make sense in the service interworking configuration. Since the possibility of multiplexing many FR connections on a single ATM VCC or VPC in network interworking scenario 1 may result in significant cost advantages for FR trunking network applications, the N-for-1 multiplexing mode should be an important consideration in the design of combined network and service interworking FR/ATM IWF in any given NE. In addition, from the traffic control perspective, it will be evident that for the N-for-1 multiplexing mode, the bandwidth allocated for the ATM trunk should be sufficient to accommodate all the FR tributaries while preserving the required QoS end-to-end.

11.5 TRANSLATION AND TRANSPARENT MODE IN SERVICE INTERWORKING

It was mentioned earlier that a major requirement of both the network and service interworking configurations for FR/ATM is to enable the transparent transfer of the "higher-layer" or user data end-to-end along the interworked connection. The higher-layer information here refers to Layer 3 and above, which for the typical case of data networking applications of FR or ATM, involves the transport of IP or other Layer 3 packets. Typically, the FRBS is used as a replacement for leased lines between IP routers located at remote sites of, for example, a large corporate or enterprise data network. In order to allow for the possibility of carriage of the various types of Layer 3 protocols over FR, a flexible encapsulation mechanism has been specified by the IETF in RFC 2427 (formerly RFC 1490) [11.13]), as shown in Fig. 11-8.

This encapsulation mechanism is referred to as the multiprotocol over frame relay protocol and it utilizes the so-called network layer protocol identifier (NLPID) encapsulation procedure, as defined in IETF RFC 2427, to enable multiple Layer 3 protocols to be carried over the FR connection. The example shown in Fig. 11-8 shows an IP packet encapsulated in the NLPID layer, which is then inserted into the payload of the FR-PDU. The NLPID header identifies whether its payload contains an IP, IPX, SNA, or any other type of Layer 3 packet and hence enables the multiplexing of different Layer 3 protocols over FRBS.

In the case of FR/ATM network interworking, the presence of the NLPID layer encapsulation makes no difference to the IWF, since the endpoints of the interworked connection both terminate in peer FR entities. However, for FR/ATM service interworking, two separate interworking modes are possible, since the encapsulation of Layer 3 packets, such as IP over ATM connections, uses an alternative encapsulation called link layer connection/subnetwork attachment point (LLC/SNAP) as specified in IETF RFC 2684 (formerly RFC 1483) [11.14]. A more detailed description of the protocols used for transport of IP packets over ATM is given later, but here for the purposes of FR/ATM service interworking it is sufficient to recognize that ATM uses a different encapsulation protocol for Layer 3 (e.g., IP) data packets than FRBS. As a result of this mismatch between the mechanisms for encapsulation of the Layer 3 data packets in FR and ATM, two distinct service interworking modes have been identified by the FR Forum. These are:

1. Transparent Mode. In the transparent mode, the FR/ATM IWF does not convert (or translate) between the different encapsulation protocols used for the FR and ATM networks. The IWF simply passes the FR (or AAL Type 5) payload transparently between the FR and ATM sides of the interworked connection. In the transparent mode it is assumed the

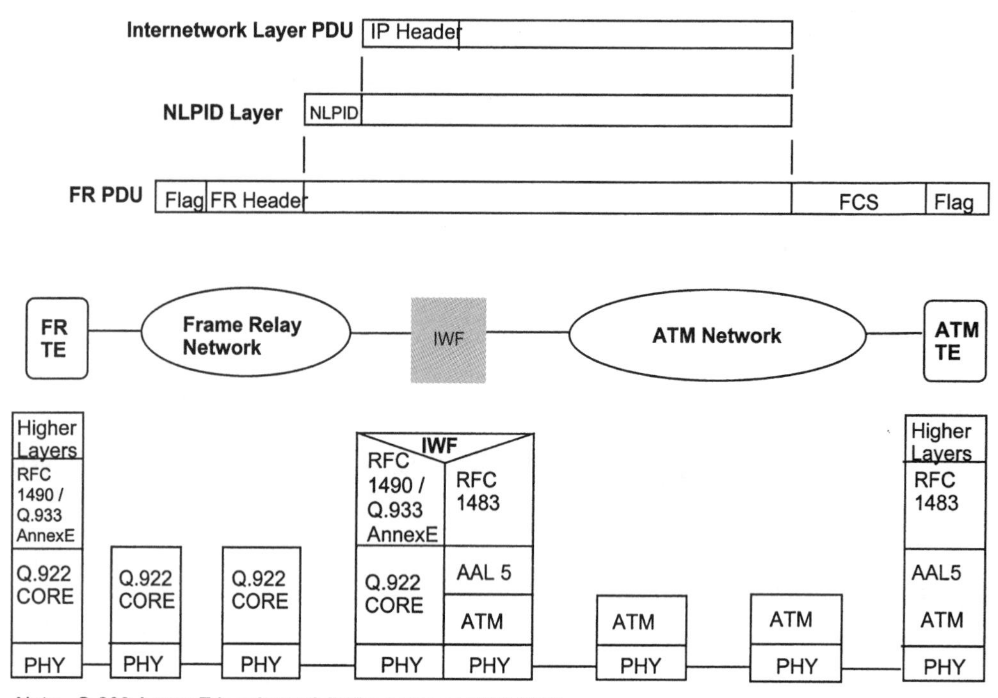

Figure 11.8. Service interworking with translation mode.

conversion between the FR NLPID and the ATM LLC/SNAP Layer 3 encapsulation protocols occurs either within the terminal equipment or elsewhere along the interworked connection.

2. Translation Mode. In the translation mode, the FR/ATM IWF translates (or converts) between the NLPID-based multiprotocol encapsulation used for FR and the LLC/SNAP-based multiprotocol encapsulation mechanism used for ATM. Consequently, there is no conversion required either at the terminals or elsewhere along the interworked connection.

The advantages of using the translation mode in FR/ATM service interworking may be understood from Fig. 11-8, which illustrates the protocol architectures along a FR/ATM interworked connection configured for translation mode service interworking. Since in this case, the IWF is able to perform the "translation" between the NLPID (IETF RFC 2427 [11.13], or its equivalent in ITU-T Recommendation Q.933 Annex E [11.8]) multiprotocol over FR protocol and the LLC/SNAP(IETF RFC 2684 [11.14]) multiprotocol over ATM protocols, no modifications are required at the respective FR and ATM terminals or end systems. In this way, the FR end systems may continue to use the specified NLPID multiprotocol encapsulation, as the FRBS network expands or evolves and the ATM end systems continue unchanged with the LLC/SNAP based multiprotocol encapsulation. If the transparent mode of service interworking had been selected for the IWF, it is clear that one or other of the end systems would have required the addition of a translation capability between the NLPID and the LLC/SNAP multiprotocol encapsulation mechanisms, which is undesirable. Consequently, the translation mode is generally the preferred approach in designing FR/ATM IWF capabilities if the intent is to support service interworking configurations. It is also of interest to note that whereas the approach was initially identified in the FR Forum Service Interworking Specification FRF.8, it was also subsequently adopted in the general FR/ATM interworking standards described in ITU-T Recommendation I.555 [11.5]

11.6 FR/ATM MANAGEMENT AND CONTROL PLANE INTERWORKING

Thus far, the description of FR/ATM interworking has primarily focused on the transfer plane aspects within the FR/ATM IWF for both network and service interworking configurations. However, it was noted earlier that, in general, it is also necessary to consider both the management and, for the case of switched virtual circuit connectivity, the control plane interworking requirements as well. It must be admitted that these aspects of interworking are somewhat more complex and, in some instances, less well developed than the transfer plane functions outlined previously. For the interworking of FR and ATM management-related functions, this is generally because the mechanisms for fault and performance management conventionally used in FR are different to those described earlier for ATM OAM. This is in contrast to the FR and ATM transfer plane functions which, as seen above, are often similar in principle thereby making interworking simpler. Nonetheless, procedures for the end-to-end monitoring of FR/ATM interworked connections have been developed by both the FR Forum and ITU-T to enable network operators to manage interworked PVC connections, as described below. For the case of control plane (i.e., signaling) interworking required for the on-demand set-up or release of end-to-end switched interworked virtual connections, the procedures have been well defined in standards. However these have generally not been widely deployed thus far, since most initial network applications of FR/ATM interworking were based on provisioned semipermanent virtual connections. However, this is likely to change with the growth of on-demand switched virtual circuit based services in both ATM and FRBS networks.

Considering first the case of management interworking, it is necessary to summarize the procedures used for the management of FRBS connections, as depicted in Fig. 11-9 part a. Two distinct methodologies have been defined for the monitoring of FR PVCs based on:

1. Use of the signaling messages, STATUS and STATUS ENQUIRY, sent over the preassigned DLCI = 0 in any given FR transmission path (physical link) across either UNI or NNI. These procedures are described in detail in ITU-T Recommendation Q.933 Annex A, and are often referred to as "link integrity verification using Annex A" procedures.
2. Use of a separate (dedicated) so-called "local management interface" (LMI) based on a management information exchange protocol such as the "simple network management protocol" (SNMP), as outlined previously for ATM. This mechanism is closely analogous to the ILMI procedures defined for ATM network management and, in fact, the ATM ILMI model was initially based on the FRBS LMI approach, both utilizing SNMP.

Since the LMI and the Q.933 Annex A procedures may operate on quite separate FR interfaces, it is in principle possible to use both or either of these management mechanisms across any given FR interface. Typically, the LMI protocols are more commonly used in private (i.e., enterprise or corporate) FR network implementations and for the so-called customer network management (CNM) interface which may be configured across the FR UNI. However, public (or carrier) networks also use LMI procedures in many cases, and the choice is largely a matter of individual implementation and convenience to the service provider. A number of FRBS implementations may support both LMI and Q.933 Annex A protocols to optimize different aspects of management information exchange.

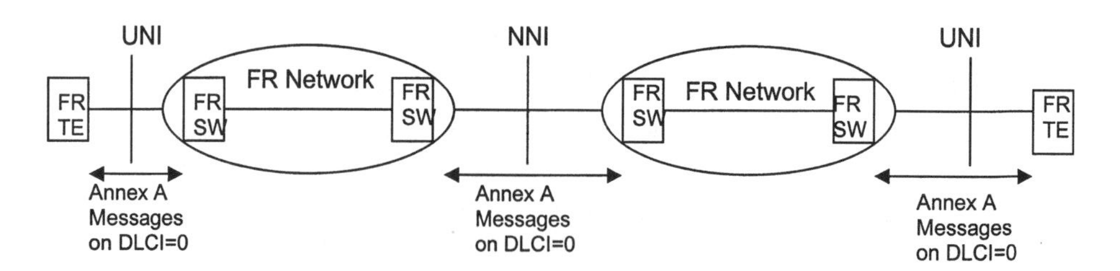

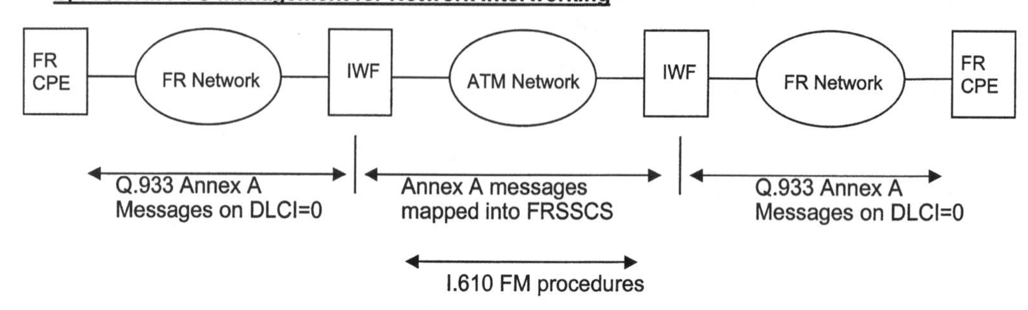

Figure 11.9. PVC management in FRS and network interworking.

From the perspective of FR/ATM interworking, it is of more interest to examine the relationship between the Q.933 Annex A protocols and the ATM OAM protocols described previously, since, in general, the description of (external) management interface interworking may be viewed as a TMN issue. In Fig. 11-9 part a it is seen that the Q.933 Annex A messages, STATUS and STATUS ENQUIRY, are exchanged across either FR UNI or NNI on the preassigned (signaling) DLCI = 0 FR virtual channel. The STATUS ENQUIRY message is sent by the FR NE to request the status of any FR PVC that has been provisioned across the interface. The term "status" or "link integrity verification" (LIV) here simply means a binary indication as to whether the PVC is "up" (i.e., available for user data transfer) or "down" (i.e., unavailable for user data transfer). The STATUS message is sent in response to the STATUS ENQUIRY message, which includes a status field signifying the state of the FR PVC in question. The Q.933 Annex A procedures also allow for an option in which the STATUS message may be sent at any time, independently of the receipt of the STATUS ENQUIRY message, to indicate the status of a single PVC. This option may also be provisioned to operate periodically if it is desired to obtain status information on a continuous basis. It will also be recognized that the STATUS ENQUIRY–STATUS message sequence is not strictly an "in-band" monitoring mechanism, since it is used only on DLCI = 0 within any given FR physical transmission path. However, by specifying the value of any active DLCI on that (or other) transmission path, the status of that specific DLCI link may be conveyed in the STATUS message.

It is of interest to recall from the earlier description of ATM UNI signaling protocols that the initial purpose of the STATUS–STATUS ENQUIRY signaling messages was for the maintenance of the signaling channel only in either the narrowband (64 kbit/sec based) ISDN or ATM signaling network. In effect, the FRBS Q.933 Annex A procedures extend the purpose of these signaling messages to perform a management-plane-related function—that of monitoring the operational state of a FR PVC—thereby blurring the boundary between control and management plane functions assumed in the B-ISDN PRM. From this pragmatic perspective, FRBS does not assume such a clear boundary between signaling (control) and management functionality, despite its initial genesis in the ISDN environment. This is in contrast to the case of ATM, which introduces the concept of in-band OAM cell flows for fault and performance monitoring for either in-service or out-of-service management, and maintains clear separation between control and management functionality, at least from the modeling perspective. Thus, in the case of ATM networks, as seen earlier, the use of signaling messages such as STATUS and STATUS ENQUIRY for monitoring PVCs is not necessary, since this capability can be more directly inferred from the ATM layer OAM flows (AIS/RDI, CC, or loopbacks) and their processing by the ATM NE management system (AEMF) described previously. It may also be recognized from Fig. 11-9 part a that the FR STATUS–STATUS ENQUIRY protocols operate on a link-by-link basis, so that transfer of the PVC status information to a remote part of the end-to-end FR connection requires additional coordination between the management systems of the individual FR NEs, or by means of the separate dedicated external management interfaces (e.g., the TMN interfaces).

As illustrated in Fig. 11-9 part b, for the case of FR/ATM network interworking, the relationship between the Q.933 Annex A messaging and the ATM Layer OAM mechanisms identified in ITU-T Recommendation I.610 [8.5] is relatively straightforward. Since the STATUS–STATUS ENQUIRY messages are transported on DLCI = 0, they may be mapped directly into the FR-SSCS by the IWF to be carried over the backbone ATM VCC or VPC for either the N-for-1 multiplexing, or 1-for-1 mapping configurations. In this way, the FR PVC status information may be transparently transferred to the remote IWF (or FR network) independently of the underlying (backbone) ATM network. However, it is assumed the ATM OAM flows may be used to monitor the status of the ATM part of the interworked connection between the two IWFs as well, independently of the FR Q.933 Annex A status messaging. In order for these two independent PVC monitoring mechanisms to be coupled to provide a coordinated end-to-end fault management capability for the FR/ATM interworked connection, it is clear that the management systems to which the IWFs belong need to externally correlate the FR and ATM PVC status information. In the FR/ATM network interworking configuration, this coordination of the PVC status may only be carried out through the management system interfaces if required, since there are no mechanisms provided to transport the information by in-band messaging. For example, if the FR/ATM IWF detects an AIS/RDI operational state due to a fault in the ATM VCC or VPC, it may select to notify the FR network through the management interface (e.g., a TMN Q3 or SNMP-based external interface), since no mechanism exists to pass this information through the FR STATUS messages. Similarly, if the failure occurs in the FR part of the interworked connection, a STATUS message indicating PVC "down" may be sent transparently through the ATM VCC (or VPC), but cannot be assumed to initiate an AIS cell under normal ATM OAM procedures. However, the management system interfaces on either side may be configured to detect these operational states and exchange the relevant information.

It is important to recognize that for FR/ATM network interworking case, the coordination of the PVC status information by the management systems on the FRBS and ATM sides is not a mandatory requirement, at least as defined in the specifications. However, it is clearly desirable, since it provides the network or service operator with the capability to manage the interworked connection end-to-end from the perspective of fault or availability monitoring. Since the interworking of the FR and ATM fault management information across the management interface of the IWF is not specified explicitly, proprietary implementations in the ATM and FR parts of a multioperator interworked connection may result in interoperability problems unless care is taken in ensuring compatibility in such network interworking configurations.

For the case of FR/ATM Service interworking configurations, the interworking of the in-band ATM OAM flows with the corresponding FRBS Q.933 Annex A procedures is somewhat more complex in practice. The case of fault management interworking is shown in Fig. 11-10, which depicts a FR/ATM service interworked connection utilizing ATM OAM functions for fault management for the ATM part, and FR STATUS–STATUS ENQUIRY messaging on the FR side. Assuming that the end-to-end monitoring of the FR/ATM interworked connection is required, or desirable, it is clear that procedures for the mapping of the ATM OAM fault management functions to the FR Q.933 Annex A STATUS–STATUS ENQUIRY messaging and vice-versa at the IWF need to be defined. These mapping rules have been specified in ITU-T Recommendation I.555 and to some extent in the FR Forum specification FRF.8 [11.7]. In the simplest case, if the STATUS message arriving at the IWF indicates that a given FR PVC link is "inactive" (or down), the IWF will generate an ATM OAM AIS cell downstream on the corresponding ATM VCC to indicate the presence of a defect on the interworked connection, essentially following the normal ATM OAM procedures described previously (see Chapter 8) for in-band alarm surveillance (fault monitoring). Conversely, on receipt of an AIS or RDI OAM cell, the IWF will generate a STATUS message downstream along the FR part with the active/inactive bit set to indicate the PVC is "down" for the corresponding FR DLCI. In this way, the normal FR or ATM OAM procedures may be extended to enable the IWF to transfer end-to-end PVC status information despite the fact that somewhat dissimilar protocols apply in the individual FR or ATM parts of service interworked connection.

Similar OAM interworking procedures are applicable for the case when the ATM continuity check (CC) OAM function or the loopback function are utilized on the ATM side. Thus, a loss of continuity (LOC) state detected by the absence of CC OAM cells at the IWF may be translated on the FR side as a PVC "down" condition in the relevant STATUS message. In all of the above procedures for interworking of the operational states, it will be recalled that the Q.933 Annex A STATUS messages may be generated either in response to a STATUS ENQUIRY message (which is mandatory), or as an option, as an unsolicited message typically under management system control. In either case, the interworking of the operational states is seen to be necessary to enable end-to-end fault management of the service interworked FR/ATM connection.

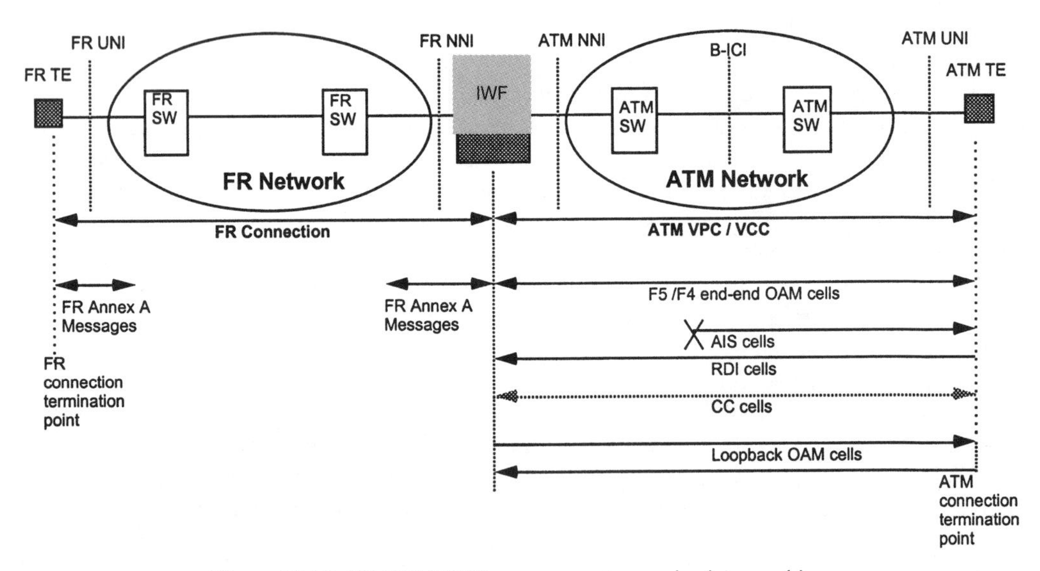

Figure 11.10. FR/ATM PVC management—service interworking.

For the case of the ATM loopback OAM function, it will be recalled that this capability may be used either for in-service, or preservice, continuity verification of any ATM VCC or VPC, as well as fault localization purposes in the general case. Consequently, in the FR/ATM service interworked connection both these aspects of the loopback capability need to be considered. As noted earlier, the loopback OAM mechanisms are particularly useful and hence widely used for the continuity verification of VPCs or VCCs prior to their release in service. This capability may also be extended to the service interworked connection by transferring the "loopback successful" or "loopback unsuccessful" operational state at the IWF to a corresponding indication in the FR STATUS message (i.e., PVC active or PVC inactive). This mapping is applicable for either in-service or out-of-service operation. It may be seen from Fig. 11-10 that the IWF may initiate an ATM OAM loopback test as well as the STATUS ENQUIRY procedures under management system control, in order to verify end-to-end continuity prior to releasing an interworked connection for normal service. This procedure may also, of course, be repeated while the connection is in-service at any subsequent time. However, the potential fault localization capability of the ATM loopback function cannot, at this stage at least, be transferred to the essentially binary Q.933 Annex A PVC status monitoring procedures. In general terms, interworking of fault localization information for the FR/ATM service interworked connection will require the use of the management system across both parts of the connection at the IWF. At least in initial service interworked FR/ATM deployments, fault localization interworking requirements are likely to be somewhat operator- or implementation-dependent.

The procedures for the service interworking of fault management and continuity verification outlined above may be viewed as reasonable extensions of the individual FR and ATM OAM protocols to map operational states across the IWF when possible. However, for performance management (PM) functions such a mapping proves more difficult and is presently not well defined. The primary problem here is that, unlike the case of the in-band PM OAM cell functions for the ATM part, no in-service, in-band PM mechanisms are available for the FR part of an interworked connection. Thus, although PM parameters such as "frame loss ratio" or "frame transfer delay" corresponding to the ATM PM parameters have been defined to characterize the performance and QoS of FR connections, no in-band measurement mechanisms have been specified at this stage. Consequently, the transfer or mapping of the PM-related information may only be possible by means of the management system interfaces, if required. Again, such requirements are likely to be somewhat implementation- or network-operator-specific, and may often be related to the accounting management procedures employed for the service agreements.

Before leaving the question of FR/ATM OAM interworking, it is important to recognize an ambiguity that arises in the currently defined procedures for fault management interworking. As indicated in Fig. 11-10, if a fault occurs in the FR part of the service interworked connection, as indicated by a STATUS message received at the IWF, the AIS OAM cells will be generated by the IWF along the ATM part of the interworked connection. However, in this case, the fault is actually external to the ATM network, but this is not apparent from the normal AIS or RDI cells only, as detection of these by any given downstream NEs may be taken to imply a defect or fault in the ATM network. The same ambiguity arises on the FR side, resulting from a fault or defect in the ATM part of the interworked connection, since the binary indication in the STATUS messages does not identify the location of the fault as external to the FR network. In principle, this ambiguity in locating the fault may be resolved by the transfer of additional (location) information in the network management system across the management interfaces. Since in practice this additional management interface information may not always be present, the inherent ambiguity in the use of normal AIS/RDI procedures for service interworked FR/ATM connections has given rise to controversy over their use, as well as several proposals to avoid the ambiguity. The simplest mechanism to distinguish AIS/RDI cells generated as a result of an "external" fault as opposed to an ATM network fault would be to utilize a suitable codepoint in the (optional) defect location or defect type fields in the AIS/RDI OAM cells, as described earlier. Other mechanisms, such as a dedicated "interworking OAM cell" have also been proposed, but despite extensive discussion no clear mechanism has been adopted to date to resolve the ambiguity. Nonetheless, it is generally recognized that the operational ambiguity resulting from use of a simple AIS/RDI in the FR/ATM service interworked connection is undesirable.

Although the use of the Q.933 Annex A procedures using STATUS–STATUS ENQUIRY messages, or the SNMP-based LMI, are the most commonly used mechanisms to transfer operational status information, it may also be recalled that an in-band OAM Loopback capability has also been specified for FRBS, as described in ITU-T Recommendation I.620 [11.4]. The FR loopback OAM procedures are based on the use of dedicated FR OAM frames with the four-octet header format and are primarily intended for continuity verification or fault localization purposes analogous to the ATM OAM loopback functions. Since the FR OAM loopback frames utilize only the four-octet FR header structure to enable in-band capability, its use in many existing FR network deployments has been limited to date. However, from the perspective of FR/ATM OAM interworking, it will be recognized that the FR loopback functions may be more directly interworked with the corresponding ATM OAM loopback function if it is incorporated in the IWF capability.

Turning to control plane, or signaling, interworking between FR and ATM networks, it may be envisaged that the similarities between the ISDN-based FR as well as ATM signaling protocols will enable relatively straightforward control plane interworking between the FR and ATM parts of an interworked connection. At least conceptually this is the case, as shown in Fig. 11-11, which illustrates the control plane protocol architecture required for on-demand switched virtual circuit (SVC) interworked connectivity between the FR and ATM networks. On the FR

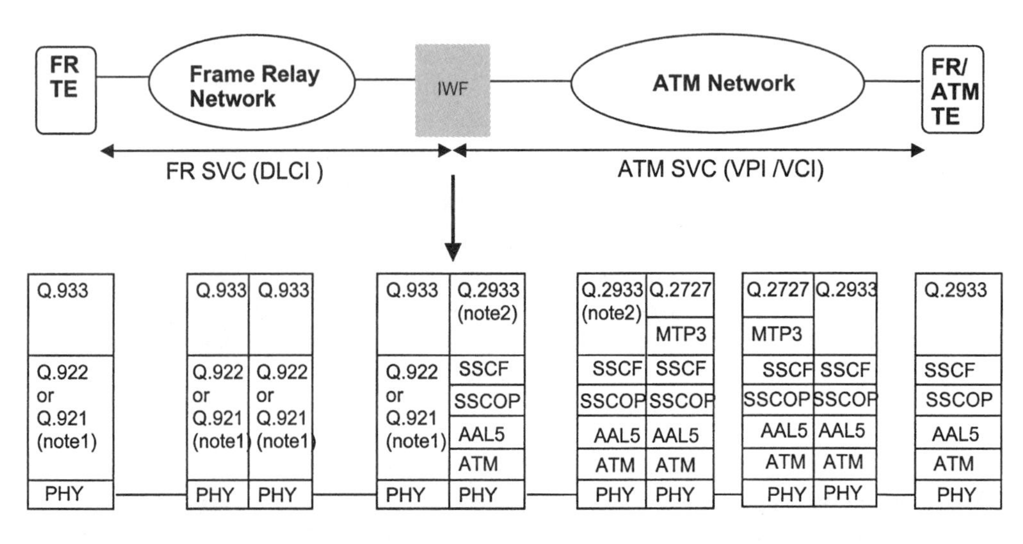

Figure 11.11. FR/ATM call control interworking.

side, the signaling protocols for the set-up, release, and maintenance of the FR connection are described in ITU-T Recommendation Q.933, or equivalently in X.36 (UNI) [11.10] and X.76 (NNI) [11.11]. For the IWF, the control plane interface may either be a UNI or NNI, depending on the network configuration deployed. In the case of a UNI IWF interface, the signaling protocol interworking is described in ITU-T Recommendation Q.2933 [11.15]. This is essentially a variation of the normal ATM UNI signaling protocols described in ITU-T Recommendation Q.2931 [6.8] to accommodate FR parameters. For the case of an NNI IWF, the FR signaling protocols in Q.933 are interworked directly with the B-ISUP signaling procedures, as outlined earlier and described in detail in ITU-T Recommendation Q.2727 [11.16]. In either case, the control plane interworking at the IWF requires the termination of the corresponding FR and ATM signaling messages with subsequent mapping of the corresponding parameters, if present. However, as for the case of the OAM functions, it is clear that not all ATM-related signaling parameters will have a counterpart to FR signaling IEs, for example in the case of traffic management or QoS-related information. In such cases a "best fit" policy is adopted for connection set-up to proceed.

For connection set-up between the FR and ATM networks, two possible modes of operation have been identified, depending on whether the FR part of the connection is set up in one or two stages. In the so-called ITU-T Recommendation Q.933 Case A (also termed "call control with port access"), the FR terminal equipment or end system first establishes an ISDN circuit mode connection to the IWF using normal ISDN signaling procedures as defined in ITU-T Recommendation Q.931 (conventional DSS1 signaling) [6.9]. The FR connection is then established using the FR signaling procedures in Recommendation Q.933 over the ISDN B-channel to the IWF, which then initiates the ATM part of the end-to-end interworked connection. Alternatively, in the so-called Q.933 Case B procedures (also-called "call control without port access"), the FR connection is directly established in a single stage by using the Q.933 signaling procedures (over Q.921 [11.17]) on the ISDN D-channel, using DLCI = 0 for the signaling messages. The IWF initiates the ATM part of the interworked connection as before, using either the Q.2933 procedures [11.15] for an IWF located at the UNI, or B-ISUP procedures for an IWF located at the NNI. It may also be recognized that the same considerations apply if PNNI- (or more strictly, AINI-) based signaling protocols are used on the ATM side instead of the B-ISUP approach, as was discussed previously.

The fact that both FR and ATM signaling (control plane) protocols are based on and may be considered as extensions of conventional ISDN signaling (i.e., existing DSSI + ISUP oriented signaling) clearly eases potential interworking or mapping of signaling information required for the support of on-demand SVCs. Nonetheless, most initial FR/ATM interworking deployments have been based on provisioned (PVC) connectivity, as noted previously, so that use of signaling interworking or SVC applications have been somewhat limited up to this stage. Although the need for on-demand SVC FR connectivity has been questioned, since most FRBS network applications have been used for leased-line replacements, as noted earlier, it is conceivable that with the growth of FR services, the use of on-demand SVCs may increase for economic or operational reasons. Assuming that the same argument applies also to the case of on-demand ATM SVCs, a consequent increase in demand for FR/ATM SVC-based connectivity will become evident in future. In addition, it should also be pointed out that whereas the description here has primarily addressed interworking procedures related to the basic transfer, OAM management, and signaling aspects of the FR/ATM IWF, recent enhancements to the basic FR services will likely result in other interworking requirements. A number of significant enhancements to FRBS are being developed, including the multiplexing of compressed voice over FR, higher-speed FR transport, and additional QoS and service categories for multiprotocol applications. For example, the transport of multiplexed compressed voice signals over FR utilizes a protocol somewhat analogous to the AAL Type 2 protocol described earlier. Thus, interworking between voice over FR and voice over ATM network applications may in future require the use of either AAL Type 2 or AAL Type 1, in addition to the current use of FR-SSCS + AAL Type 5 in the IWF. In the same way, other enhancements to FR/ATM interworking requirements may be foreseen, as additional capabilities are added to these related, interdependent networking technologies.

CHAPTER 12

IP and ATM Internetworking

12.1 INTRODUCTION

The internetworking-protocol (IP) based approach to networking between local area networks (LANs) and attached hosts is currently almost universally used for data networking applications. More recently, its popularity has been further enhanced by the unprecedented growth of the so-called "Internet" and applications such as the Worldwide Web (www), which are based on various versions of the IP stack and allow for fast and convenient exchange of large amounts of data between remote hosts and servers located anywhere on the Global Internet. Despite this relatively recent popular acceptance of IP networking, it is of interest to note that the basic IP approach was developed in the early 1980s, evolving from the earlier data networking protocols such as the "ARPANet," which were initially developed for the robust communication of military information under battlefield conditions. In such environments, it may be anticipated that physical disruption of the communication paths would occur relatively frequently, necessitating automatic rerouting of the data packets to ensure robustness. In addition, it was also clearly highly desirable that the data networking protocols should be essentially independent of the performance, quality, or type of the underlying physical transmission media used, since this could be variable in typical military environments.

The extensive research undertaken to meet these and other broad requirements resulted in the development of a suite of data internetworking protocols for the ARPANet based on a connectionless packet mode transfer, utilizing dynamic routing protocols for automatic path discovery and rerouting of the data packets in the event of transmission path failure. Subsequently, the routing and transfer protocols were further developed into the sophisticated suite of IP-related routing protocols and algorithms utilized in the current global Internet, notably by the Internet Engineering Task Force (IETF) and related bodies. Needless to say, this development of IP-based capabilities continues apace, spurred on by the widely recognized commercial possibilities of the Internet and www applications. As noted above, the IP stack is designed to be transported over any of the existing transmission systems in common use. The potential high performance, quality of service (QoS) capabilities, and flexibility of ATM, both as a transmission and a switching technology, led to the recognition that ATM networks provide the optimum means for the transfer of IP packets or IP-based services. As a result, extensive work in the IETF and the ATM Forum led to the specifications of a number of networking and protocol architectures for the transport of IP over ATM (IPOA), together with related control interworking mechanisms. In subsequent chapters these approaches, still evolving in many aspects, are described in some detail, since the carriage of IP based applications over ATM networks constitutes one of the major uses of ATM backbone networks.

Before entering into the descriptions of the various IP and ATM internetworking architectures, it is useful to obtain a broad understanding of the internetworking protocol architecture and functions, in much the same way as was done for the case of frame relay protocols previously. The brief overview of the IP stack outlined here is not intended to be comprehensive but simply provide sufficient background to enable a clear understanding of IPOA internetworking applications. Readers wishing to obtain a more detailed understanding of IP-related technology may refer to the copious literature that exists for all aspects of IP-based applications [12.1–12.3].

As was seen earlier for the case of FR and ATM interworking, the basic underlying similarities and common roots in the ISDN concept allowed for a relatively brief outline of the FR protocol structure in order to describe FR/ATM interworking. However, for the case of IP and ATM interworking, it will be recognized that fundamental differences of approach exist as a result of the historically different origins of IP in conventional data networking based on a connectionless packet (also referred to as "datagram") paradigm, in contrast to the connection-oriented architecture of ATM (and FR) protocols. These fundamental differences, together with the associated differences in addressing, routing and network management philosophies, make the interworking between IP- and ATM-based networks much more challenging and interesting. It may also be noted as a caveat that IP and ATM internetworking remains a volatile area of intense research, in which today's topical concept may well appear outmoded tomorrow, so that any description may well be viewed as a snapshot in a rapidly evolving story. This danger may be countered by stressing fundamentals, which provides some insight into the likely evolution of the technology. In addition, it is useful to recognize at the outset that in essence both IP and ATM are attempting to achieve the same function—that of transporting information end-to-end between remote locations—by different means. It is therefore not surprising that a healthy tension exists in the relationship between them, a notion that has often been commented on superficially, but which only serves to mask the underlying interdependence between the two internetworking approaches as described below.

12.2 IP INTERNETWORKING OVERVIEW

An elemental IP-based internetworking scenario may be represented schematically as shown in Fig. 12-1 part a. Typically, local area networks (LANs) interconnect workstations (computers or hosts), printers, bridges, etc. by means of a shared physical media access control (MAC) protocol, such as the ubiquitous Ethernet or Token Ring approaches, as described in more detail below. The LANs may be connected to other remote LANs by means of one or more networks of routers using the IP-based routing protocols to forward data packets to the remote destination identified by the IP address. In general, networks using the suite of protocols based on IP are called "internets." The Global Internet is essentially built up by interconnecting edge routers in many such internets by means of a series of backbone internets in a hierarchical structure to enable efficient routing of the data packets (often called IP packets) from source to destination. Physical connectivity between the routers may be provided by a range of point-to-point transmission media based on switched or permanent (conventional) circuits, or virtual circuits (such as frame relay bearer service or ATM VPCs or VCCs). Examples of the group of protocols used for data networking in such scenarios include:

1. Internetworking protocol (IP) corresponding to Layer 3 of the ISO model
2. Transmission control protocol (TCP) at Layer 4
3. User datagram protocol (UDP) at Layer 4
4. File transfer protocol (FTP) at the application layer
5. Many other protocols at Layer 4 and above, some examples of which are listed in Fig. 12.2

In the simplified IP-based internet shown in Fig. 12-2, a user on a workstation (generically termed "host" in internet terminology) may initiate the exchange of e-mail with a user on a remote workstation attached to another LAN at another site. As an example, the so-called

a) Internetworking using IP

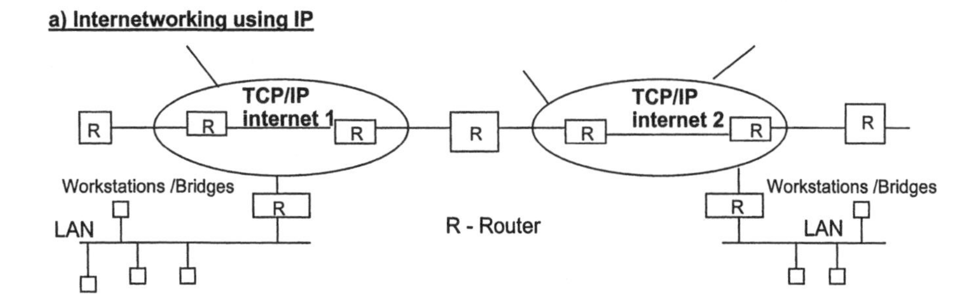

b) TCP/IP Protocol Stack

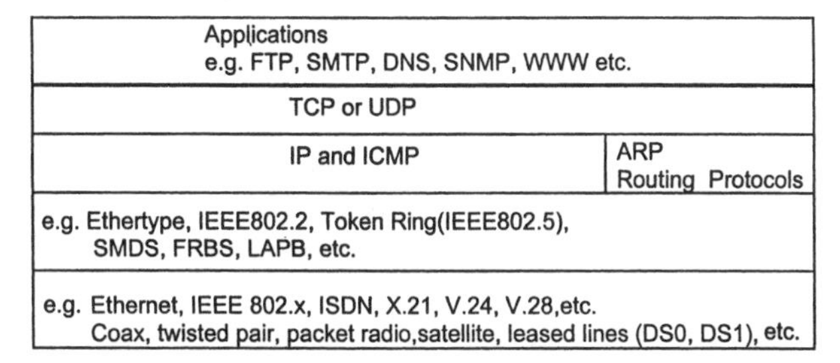

Figure 12-1. Internetworking using IP and TCP/IP architecture.

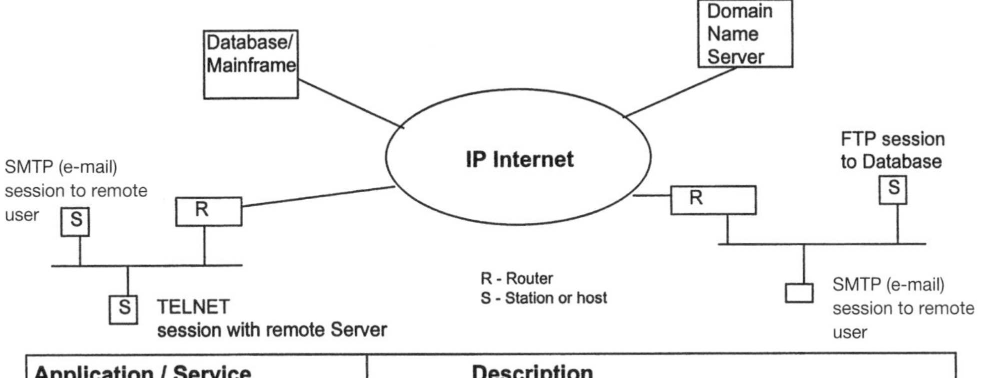

Application / Service	Description
File Transfer Protocol - FTP	Enables transfer of files between the user and remote Computers / Databases
Simple Mail Transfer Protocol - SMTP	Enables electronic mail transfer between users connected to the TCP/IP internet
Terminal Emulation - TELNET	Enables remote login to a host or computer
Domain Name Server - DNS	Provides directory for name translation of network resources / services
Network File System - NFS	Enables access to remote files
Simple Network Management System -SNMP	Enables management of network elements such as Routers, bridges etc.
World Wide Web - WWW	Allows access to global Internet locations to obtain /provide information

Figure 12-2. Examples of internet services.

simple mail transfer protocol (SMTP) may be used for this purpose, whereby the SMTP messages are encapsulated in TCP/IP packets and routed via a series of IP routers comprising the internet to the destination LAN attached to the target workstation. Similarly, a user may invoke the FTP application over TCP/IP to download data files from a remote server to the workstation being used. In all such cases, the connectionless IP packets or datagrams encapsulating the higher-layer messages are forwarded hop by hop by the individual routers comprising the internet, based on the so-called IP address of the destination router port. This process is described in further detail below when considering the structure of the IP address and packet formats. It needs to be borne in mind that the links interconnecting the IP routers may be conventional (e.g., DS0 or DS1) circuits or ATM VCCs or VPCs in the event that the routers are equipped with the appropriate ATM interfaces (often referred to as ATM network interface cards or NICs). Since the Layer 3 IP routing process is essentially independent of the underlying Layer 2 and Layer 1 protocols used for the transport of the IP datagrams, the use of either ATM or FRBS virtual circuits within the Layer 2 functionality provides advantages in terms of potentially lower cost as a consequence of the statistical multiplexing capabilities of virtual circuits.

It is of interest to compare the protocol architecture of the IP suite of protocols to the ISO open systems interconnection (OSI) protocol stack as shown in Fig. 12-3. In order to promote interoperability between the various data networking protocols being developed as a result of the demand for networked computers (hosts), the ISO developed the well-known seven layer protocol reference model depicted in Fig. 12-3. The basic principles underlying this generalized modeling of networking protocols have been well established and extensively described, so that it is only necessary here to summarize briefly the key features of relevance to IP and ATM internetworking. Each layer of the OSI reference stack provides specific functions that may be encoded into the protocol control information (PCI) in the PDU header and/or trailer corresponding to that layer. The key function of each layer may be summarized as:

1. Layer 1: Physical Layer. Provides for the transparent transmission of bit streams and for conversions between the physical media (e.g., electrical to optical, etc.).

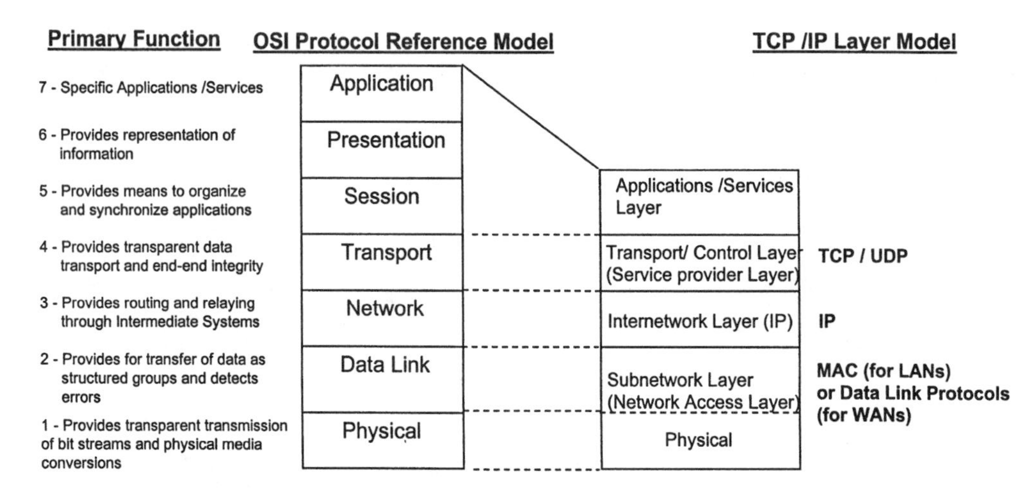

Figure 12-3. OSI and TCP/IP layer models. In general, the TCP/IP protocol layer model has fewer layers than the OSI model. The subnetwork layer includes the data link (for WANs) or MAC (for LANs) and physical layer functions. The session and presentation functions are grouped with the application layer functions. The service provider layer functions are implemented by the TCP or UDP functions. IP-based model retains the protocol control information (PCI) encapsulation methodology of OSI model.

2. Layer 2: Data Link Layer. The primary function is for the transfer of the data packets as structured groups (e.g., defined frames) and for ensuring data integrity by providing for error detection, for example. It is of interest to note that ATM (including the AAL) is generally viewed as corresponding to the Layer 2 functionality, as is frame relay, which is derived from the archetypical HDLC Layer 2 protocol.
3. Layer 3: Network Layer. Provides for the routing and/or switching (relaying) of the data packets through the "intermediate systems" (e.g., the routers or switches) in the network. In the case of a typical Layer 3 protocol such as IP, the routing through the network relies on the IP address (also called the "protocol address," or Layer 3 address). For a connection oriented networking technology such as ATM, the routing or switching is based on analysis of the E.164-based (or AESA) address, which therefore corresponds to the Layer 3 functionality, coupled with signaling.
4. Layer 4: Transport Layer. Transport layer functionality is intended to provide for transparent data transport between the end systems as well as for end-to-end integrity of the data stream. In the case of a typical Layer 4 protocol such as TCP, these ends are achieved by incorporating mechanisms such as flow control and retransmission of data in the event of loss. It is of interest to note that for the case of ATM, these functions only occur in specific AALs such as SSCOP, intended for transport of signaling messages, or assured mode data applications.
5. Layer 5: Session Layer. The functions grouped in the session layer are intended for the synchronization and organization of the specific application, such as the initiation (and termination) of a data transfer session between a host and server, for example.
6. Layer 6: Presentation Layer. The functions in this layer are intended for representation or translation of the information in the application layer, and may often be combined within the application.
7. Layer 7: Application Layer. This generally refers to the specific (data) application or service that is being invoked, such as a file transfer or e-mail, etc.

In the ISO protocol model, the information between layers may be represented as exchanged by means of primitives across a service access point (SAP) between the layers. The PDU within any given layer is formed by adding the PCI in the header or trailer (or both) to the SDU from the layer above (encapsulation). Conversely, the header/trailer from the layer below is stripped (deencapsulation) to create the SDU passed to the layer above. The encapsulation/deencapsulation methodology inherent in the concept of a protocol stack is also evident in the case of the ATM PRM, as well as the IP stack, also shown for comparison with the ISO reference model in Fig. 12-3. However, it will be noted that the IP stack essentially condenses the ISO model into four layers by "combining" the functions of the physical and data link layers into a so-called "subnetwork" or network access layer and those of the session, presentation, and application layers into a single application/services layer. The network (Layer 3) and transport (Layer 4) layer functions remain separate in the IP stack, as may be expected, since these functions essentially constitute the core of the IP and TCP stack. In LAN configurations, the IP stack subnetwork layer (i.e., data link plus physical) uses a so-called media access control (MAC) function such as Ethernet or Token Ring protocols with the appropriate shared (or switched) media physical connectivity. However, in the WAN, the IP stack may utilize typical data link plus physical capabilities such as point-to-point circuits or ATM or FR virtual circuits, as indicated earlier.

Although the primary focus here is on IP and ATM internetworking, on account of its relative ubiquity, it should be noted that similar considerations will apply to other related Layer 3 (and Layer 4) protocols such as IPX, DECNET, etc. In terms of some commonly used protocols, the general IP stack may be represented as shown in Fig. 12-1 part b, which also includes the IP routing protocols, the Internet control message protocol (ICMP), and the address resolution protocol (ARP), the functions of which are described in further detail below.

Since the generalized IP stack shown refers to both LAN and WAN configurations, the subnetwork layer is separated into the data link and physical layers to allow for the WAN configurations. As indicated, the physical media may vary from Ethernet twisted pair to DS0 or DS1 leased lines or wireless, whereas the data link layer functions may be provided by a variety of protocols including FRBS, LAPB, SMDS, ATM, etc., as listed below:

Local Area Networks (LANs)

- Ethernet
- IEEE LANs (e.g., Token Ring)
- Fiber-distributed data interface (FDDI)

Wide Area Networks (WANs)

- Frame relay
- X.25
- ISDN
- Leased lines (point-to-point circuits)

It should also be recognized that not all the listed protocols or applications may be implemented in any given host or device. Typically, IP stack implementations vary widely, tend to be highly dependent on the host operating systems, and may be implemented either as part of the (1) kernel, (2) device drivers, or (3) the applications software.

Before embarking on a description of the packet structure and semantics, it is useful to summarize the main characteristics of the TCP (or UDP) and IP layers as follows:

1. Transmission Control Protocol (TCP)
   - Provides for reliable data transfer using flow control and sequence numbering and acknowledgements of data bytes received.
   - Provides for set up of "virtual connections" similar to connection-oriented (lower-layer) transport.
   - Provides for the multiplexing/demultiplexing of the upper layer protocols by means of "port numbers."
   - Provides for full duplex (bidirectional) data transfer.
2. User Datagram Protocol (UDP)
   - Provides a connectionless transport protocol that is simpler and has less overhead than the TCP.
   - Provides basic unassured data transfer with no reliability mechanisms such as flow control or retransmission.
   - Generally used in LAN configurations and for the transfer of management information such as SNMP messages. May also be used for voice over IP and other real-time applications.
3. Internetworking Protocol (IP)
   - Provides a connectionless networking protocol for the logical routing of packets.
   - Encapsulates the higher-layer TCP or UDP packets.

- Uses a 32-bit IP address (or protocol address) (for IP Version 4) which is globally unique and generally represented in a so-called "dotted decimal" notation, as described below.
- Provides for fragmentation and reassembly of the higher layer packets to meet lower-layer maximum transmission unit (MTU) limitations if required.

12.3 OVERVIEW OF LAN PROTOCOL ARCHITECTURES

The IP architecture and functions evolved from the basic need to internetwork between remote LANs seamlessly, thereby enabling the efficient exchange of data files located on geographically dispersed computers (hosts or servers). Consequently, the mechanisms employed by IP are often extensions of those used in LANs and it is therefore useful to briefly summarize the underlying principles of LAN protocols to obtain further insight into the IP functionality. The two most widely used LAN protocols, Ethernet and Token Ring, are considered below as typical examples of LAN capability, with no attempt being made at a comprehensive coverage of this large area. More detailed accounts may be found in the extensive literature on the subjects, as well as in the detailed IEEE specifications [12.4–12.7].

The basic Ethernet frame structure is shown in Fig. 12-4, indicating a variable length data frame with a maximum payload size of 1500 octets, generally referred to as maximum transmission unit (MTU). Taking into account the header and trailer fields, the maximum Ethernet frame size is 1514 octets without the CRC trailer, or 1518 octets including the CRC 32 function. A minimum frame size of 60 octets (without CRC) or 64 (with CRC) is also specified. In addition, an eight-octet synchronization (or so-called "preamble") field is also required, which contains a predefined sequence of bits to provide a frame delineation (start of frame) function. The Ethernet header includes a six-octet field for each of the destination and source addresses, which are also referred to as "media access control" or MAC addresses. The structure of the MAC includes a three-octet "manufacturer code" and a three-octet "host code," as shown. It may also be noted that the leading bit of the manufacturer code field is set to indicate whether the frame is intended for multicast or for an individual host attached to the LAN. The six-octet MAC address is in-

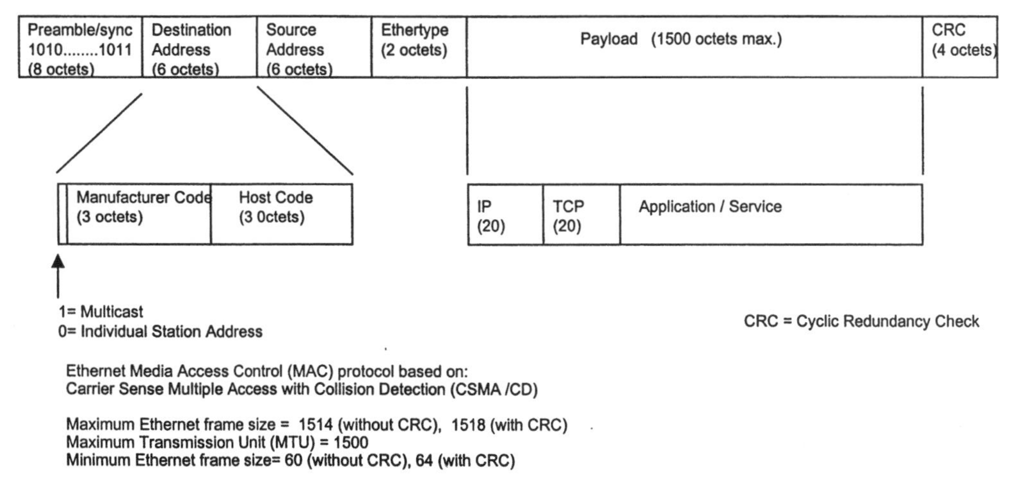

Figure 12-4. Basic ethernet protocol architecture.

tended to uniquely identify each host (or station) that may be attached to any LAN, and hence constitutes a global connectionless address for the delivery of the data frames on the shared physical media interconnecting the individual stations. It will be recognized that the MAC address of the station constitutes a "physical" address, designed into the actual equipment hardware, to identify itself uniquely for reception of the connectionless Ethernet frames that are transmitted onto the physical media by other (source) hosts.

In addition to the fundamentally connectionless mechanism implied by the physical MAC address, LANs such as Ethernet were intended for shared physical media, e.g., simple twisted pair wiring between the stations. The use of shared physical transmission media simplifies broadcast (or multicast) of the Ethernet packets by setting the multicast bit in the destination MAC address as indicated earlier. As the data frame propagates along the shared media, each attached station compares its embedded MAC address with the destination MAC address written in the ethernet frame. If the frame is intended for an individual station (i.e., not multicast) the packet is extracted (or copied) by the station in question. More recently, so-called "switched ethernet" LANs have become more popular due to their significantly enhanced performance over conventional Ethernet protocols. However, their operation remains essentially similar and hence need not be considered further here. In addition to the destination and source (MAC) address fields, the conventional Ethernet header also includes a two-octet Ethertype field, which provides both a message type function as well as a length indicator for the payload field. The four-octet Ethernet trailer only includes a CRC 32 capability for error detection over the entire frame. Corrupted frames are typically discarded. As shown in Fig. 12-4, the payload field encapsulates the IP packet, with TCP encapsulated within that and so on, as required by the ISO OSI layered model.

The apparent simplicity of the basic Ethernet protocol structure outlined above masks a basic function required for any shared media networking technique, namely that of the media access control (MAC) or contention resolution mechanism to avoid the corruption of data in the event of simultaneous transmission by more than one station. In the case of Ethernet, this mechanism is known as carrier sense multiple access with collision detection (CSMA/CD). In the CSMA/CD approach, "collisions" of data frames, which may be caused by more than one station transmitting at any given instant, are avoided by each station sensing if a signal is already present on the shared media prior to transmitting its own data. If the station detects the presence of a signal on the line, transmission is delayed for a preprogrammed period, after which the station attempts to send data again after sensing for potential collisions. In the event that a collision does occur despite the sensing mechanism, the stations in question abort the transmission and attempt retransmission after a random duration of time. The CSMA/CD mechanisms outlined above work well in practice, as may be clear by the ubiquity and low cost of Ethernet LANs in data networks.

The initial Ethernet-based LAN implementations were designed for line transmission rates of up to 10 Mbit/sec, although, typically, actual data transfer rates between the stations are generally lower and dependent on a number of other parameters, including distance, as well as the number of active attached stations. Despite its evident performance limitations, Ethernet's use of relatively simple connectionless physical addressing and CSMA/CD-based MAC protocols allows for low-cost hardware implementations and wide applicability. Inevitably, the growing demand for higher data transfer rates, spurred by high-performance workstations and PCs, has resulted in development of Ethernet LAN transmission rates of 100 Mbit/sec, as well as the more recent so-called "gigabit Ethernet" technology capable of transmission speeds of 1 Gbit/sec or more. Significant performance enhancement may also be achieved by evolving to LAN configurations based on the "switched Ethernet" concept, as noted earlier, which circumvent the limitations of distributed shared media. While these and other continuing developments of the Ethernet-based

LAN technology are outside the scope of this cursory account, it is important to recognize that from its initial, somewhat humble beginnings, Ethernet LAN technology has proved capable of evolution to very high performances, thereby challenging other competitive networking technologies, including ATM.

The proliferation and rapid development of LAN technologies has prompted the IEEE to categorize the various types into a well-defined LAN architecture as well as generalize the Ethernet specification to promote interoperability between the numerous implementations. The generalized LAN protocol architecture developed by the IEEE standardization bodies (IEEE 802.n) is shown in Fig. 12-5, together with its relationship to the ISO protocol reference model outlined earlier.

In the IEEE 802.2 LAN model [12.4] the data link layer may be sublayered into the so-called logical link control (LLC) sublayer and the basic LAN options described by standardized terminology as shown in Fig. 12-5. For example, the term "10 BASE-T" refers to the commonly used 10 Mbit/sec baseband Ethernet over twisted pair (unshielded) cable, whereas "10 BROAD 36" refers to 10 Mbit/sec line rate over broadband coaxial cable Type 36, and so on. Similarly, "100 BASE-T" refers to 100 Mbit/sec baseband over twisted pair copper cable. This terminology therefore enables a convenient categorization of the various LAN types and is widely used. The logical link control (LLC) layer was added to enable more flexible extension of the Ethertype function as well as provide multiplexing capability when required. The structure of the LLC layer is described later when considering the IP layer capabilities, but it may be noted here that the LLC concept has also been extended for the case of IP over ATM protocol architecture, as will be described later.

The generalized Ethernet protocol specified by the IEEE 802.3 [12.5] Committee is shown in Fig. 12-6, based on the IEEE 802.2 architecture concepts outlined above. By comparison with the "basic" Ethernet frame structure depicted in Fig. 12-4, it may be seen that the IEEE 802.3 protocol is an extension of conventional Ethernet frame structure to include the IEEE 802.2 LLC field and explicitly define a one-octet "start flag" in conjunction with the seven-octet "preamble" (or synchronization) field, corresponding to the earlier eight-octet preamble field. In addition, the earlier two-octet Ethertype field is renamed as a length field in the IEEE 802.3 format to indicate the number of octets in the payload field, including the LLC field, which may be up to eight octets. More significantly, it should also be noted that the structure of the six-octet MAC address fields (both destination and source) are also changed by utilizing one bit in the three-

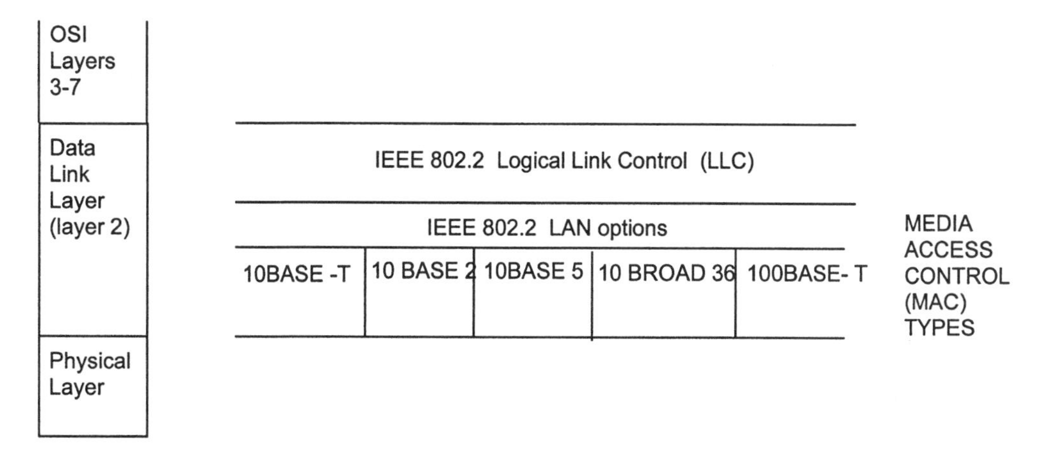

Figure 12-5. IEEE LAN protocol architecture.

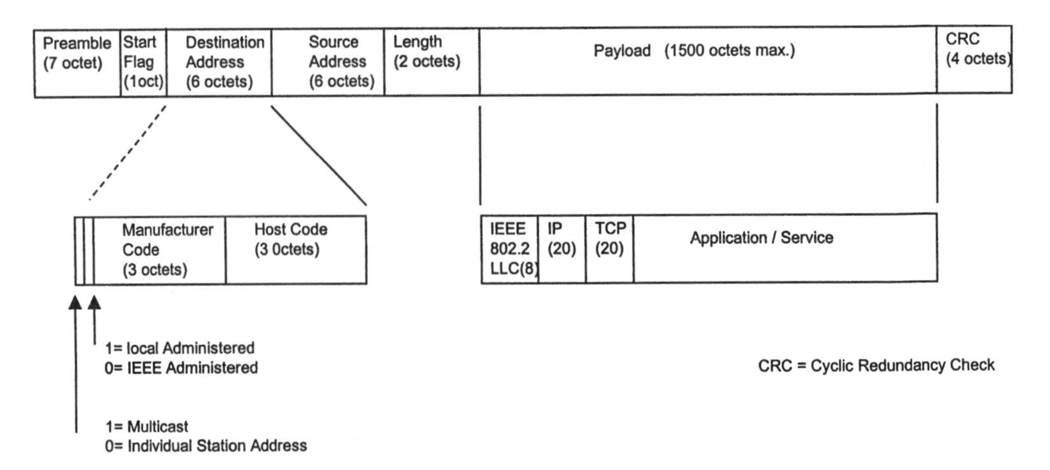

Figure 12-6. IEEE 802.3 protocol.

octet manufacturers code field to indicate whether the address is administered locally (i.e., not necessarily unique), or by the IEEE Committee. Apart from these extensions to the initial Ethernet frame structure, it is important to recognize that the IEEE 802.3 Ethernet protocol uses the CSMA/CD MAC mechanisms and MTU sizes as described earlier. Therefore, by maintaining essentially the same general frame structure and underlying MAC mechanisms in this way, interoperability between the initial basic Ethernet LANs and IEEE 802.3-based LANs may be achieved.

Although the relative simplicity and low costs of the CSMA/CD-based Ethernet and IEEE 802.3 LANs has enabled them to become the most widely used LAN technology, an alternative MAC procedure based on "token passing" is also widely used and is capable of providing relatively high performance as well. The LAN configurations based on the token passing concept have been specified by the IEEE as the so-called Token Ring (IEEE 802.5), [12.7] and the so-called Token Bus (IEEE 802.4), [12.6]. Of these, the Token Ring LAN configuration is more widely used, and hence may be considered in this overview.

The basic frame structure and MAC mechanism is shown in Fig. 12-7, which illustrates the basic principle underlying media access control using token passing. The hosts, H1–H4, attached to the shared physical media may only transmit data frames to the other hosts when they have received a "token" frame, which is circulated around the LAN ring and passed from host to host as shown. If a host receives the token but does not have any data frames to send at that particular instant, it simply retransmits the token to the next host (station) on the LAN, and so on. The protocol clearly avoids the possibility of collisions and hence corruption of data due to simultaneous transmission of data frames from more than one station on the LAN. It may be intuitively recognized that the protocol implies constraints on the time that any given station may hold the token to maintain fair sharing of bandwidth on the LAN.

The frame format of the IEEE 802.5 Token Ring protocol shown in Fig. 12-7 includes the inevitable destination and source MAC address fields, which identify the (physical) port for which the data frame is intended or is sourced from. The Token Ring protocol allows for MAC addresses ranging from two octets up to six octets in length, although the address structure and provision of multicast capability is similar to that of the IEEE 802.3 Ethernet-based protocol outlined previously. Despite the differences in details, it is important to note that here the addressing mechanism is also connectionless, implying the need for a unique MAC address for every station attached to any given LAN. In other respects such as frame delineation and

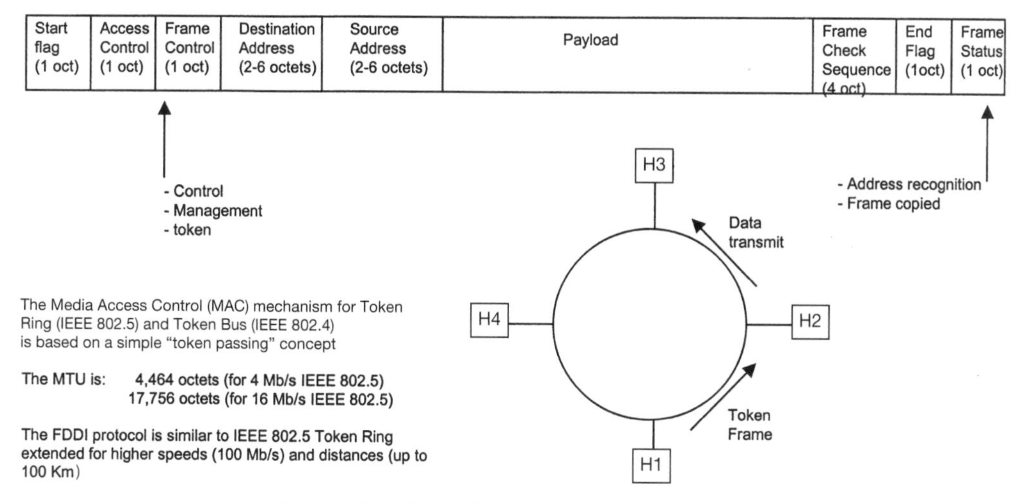

Figure 12-7. IEEE 802.5 protocol (Token Ring).

control, it may be seen that there are differences between Token Ring and Ethernet protocols, some of which are the result of the different MAC contention resolution mechanisms utilized. The IEEE 802.5 Token Ring frame delineation is achieved by use of the one-octet "start" and "end" flag sequences, analogous to the commonly used HDLC framing mechanism outlined earlier in the case of FRBS. The one-octet "frame control" field is used to identify the token frame that is passed from host to host to enable the MAC procedure, indicating that the "token" is simply an empty frame. In addition, codepoints in the frame control field may be used to identify frames used for management functions or other control purposes. The four-octet frame check Sequence field is used for error detection purposes and the one-octet frame status field is used to indicate whether the data frame has been copied and the MAC address has been recognized (validated) by the station. Details of these and other possible control capabilities are outside the scope of this overview, but it may be recognized that the IEEE 802.5 Token Ring is capable of providing richer functional capabilities than the basic Ethernet protocol outlined previously.

Of more interest here is the fact that the MTU sizes for IEEE 802.5 Token Ring LANs are significantly larger than those specified for IEEE 802.3 Ethernet protocols. As shown in Fig. 12-7, the MTU specified for the 4 Mbit/sec Token Ring is 4,464 octets whereas that for the 16 Mbit/sec line speed is up to 17,756 octets. The combination of high transmission (line) speeds and large MTU enables potentially high data throughput performance to be achieved on the LAN, and is often cited as a key advantage of Token Ring LANs. On the other hand, the performance advantages need to be balanced against the additional complexity and hence cost of Token Ring LANs, which may have inhibited their more widespread use in competition with the relatively low-cost Ethernet-based LANs. Nonetheless, the evolution and continued development of robust high-performance Token-Ring-based LANs needs to be recognized. In this context, it may also be noted that the widely used fiber distributed data interface (FDDI) protocol intended primarily for LAN backbones is essentially similar in concept to the IEEE 802.5 Token Ring, although extended for operation at higher speeds (100 Mbit/sec or more line rates) and distances of up to 100 km, exploiting the benefit of optical fiber transmission capabilities. The extended range and robustness of the FDDI protocols led to their consideration for use in metropolitan area network (MAN) applications or for networking of large computers.

12.4 INTERNETWORKING PROTOCOL (IP) LAYER FUNCTIONS

This brief overview of two of the most commonly used LAN protocols serves to underline the fact that from the outset, data networking connectivity between hosts (computers or servers, etc.) has been based on connectionless packet transfer. Networking connectivity in LANs originated by specifying a global physical (i.e., MAC) address embedded in the individual hosts, which enabled them to extract the data packets with matching MAC addresses that were detected on the LAN. Although this simple mechanism works well in the LAN environment, it was soon recognized that for data internetworking between geographically remote LANs, the use of only physical MAC addresses at Layer 2 would soon prove somewhat unworkable. The solution was to develop a globally unique "logical" address structure for all hosts in the network, which could operate in the layer above the LLC + MAC layers, namely at Layer 3 in the corresponding OSI protocol model. The use of such a Layer 3 logical address (also sometimes loosely referred to as a "Protocol" or "Network" address) to provide more universal connectivity between remote LANs or hosts clearly results in advantages, since it provides independence from the various physical (LAN) addressing schemes to the application requiring the connectivity. In other words, the physical connectivity is essentially "hidden" from the higher-layer application or service. Since the LAN connectivity mode is connectionless, this mode is also extended to the logical (Layer 3) address mechanism, thereby providing a connectionless (or "datagram") service to the layer above.

In addition to the primary function of providing a connectionless logical addressing mechanism at the IP layer, this layer also includes the capability to isolate the higher layers from the MTU limitations of the commonly used LANs outlined earlier.

As shown below (see Section 12-5, Fig. 12-14 part a), this isolation is achieved by providing for the so-called "fragmentation" and subsequent "reassembly" of arbitrarily long higher-layer messages within the IP layer, thereby accommodating the MTU limitations at the lower (LAN) layer. The mechanism underlying the fragmentation/reassembly function within the IP layer will be described below when considering the IP packet structure, but this should not be confused with the ATM-related segmentation and reassembly (SAR) procedures in the AAL, which were described previously. The IP layer also includes other capabilities important for internetworking, as will be summarized below, but it is useful to recognize that the primary functions incorporated in the IP layer protocol control information (PCI) relate to

1. Provision of the connectionless mode logical addressing of hosts to enable the routing of the data packets from source to any geographically remote destination
2. Enabling the higher-layer application to be independent from the lower-layer capabilities or limitations such as MTU size

From the perspective of IP and ATM internetworking, it is evident that a fundamental difference that will need to be interworked is the connectionless mode inherent in IP, with the connection oriented mode underlying ATM connectivity. As was described earlier in discussing ATM addressing and routing protocols, there is an intimate relationship between routing protocols and addressing structures for any transfer technology, be it IP or ATM. It is therefore useful to summarize the basic structure and mechanisms underlying IP addressing and its related routing to enable a clearer understanding of the requirements for IP and ATM interworking. The most widely used IP layer addressing structure is defined by the IETF for the ubiquitous IP Version 4 protocols deployed in the Global Internet. The IP Version 4 specifications utilize a 32-bit (four-octet) IP address whose structure is shown in Fig. 12-8.

It may be seen that the 32-bit IP Version 4 address is divided into the so-called "network ID"

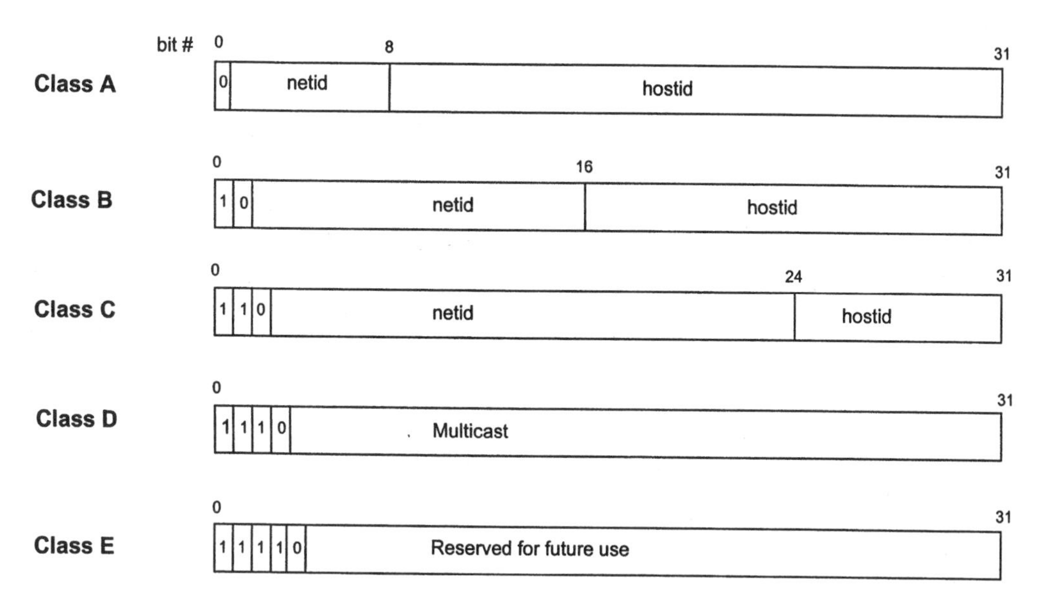

Figure 12-8. IP address classes. The 32-bit IP address (for IP Version 4) is divided into "netid" and "hostid" parts and five address "classes" to cover a wide range of network configurations. Note: IP Version 6 extends address space to 16 octets.

(netid) and "host ID" (hostid) parts and consists of five "IP address classes" termed Class A to Class E. The intent of the IP address classes is to provide for a wide range of possible IP network configurations, depending on the size of the network and number of attached hosts. The IP address classes are identified by the codepoints of the "leading" bits of the address field as shown in Fig. 12-8. Thus, Class A is identified by the binary "0" in bit number 0 signifying a netid field of one octet, whereas Class B has a netid field of two octets and a hostid field of two octets, as identified by the leading "1 0" codepoint, and so on. It should be noted that the IP address Class D is intended for IP multicast address applications in general, whereas the IP address Class E is reserved for future applications, and hence not generally in use at present. Thus, the most widely used IP address classes are A, B, and C.

The relative sizes of the netid and hostid fields for the various classes indicates the potential IP network applications, as summarized in Table 12-1. It may be seen that IP Version 4 (IP v.4) address Class A may be applicable to very large IP networks, since a potentially large number (up to 16,777,214, corresponding to a three-octet hostid field) of hosts per netid may be accommodated. The Class A IP address format provides unique netid codepoints for up to 127 IP networks. For many medium-sized networks, such as those deployed in enterprises or large cam-

TABLE 12-1. IP address class applications

Address class	Number of networks (netid)	Number of hosts (hostid)	Typical network application
Class A	127	16,777,214	Very large networks
Class B	16,383	65,534	Medium sized networks, e.g., LANs, enterprise
Class C	2,097,151	254	Large number of small networks
Class D			Multicast use in all networks

puses, the IP v.4 address Class B may be applicable, allowing up to 16,383 unique netid values (14-bit netid codepoints) with up to 65,534 unique hostid values per netid, corresponding to the 16-bit hostid field. Finally, the IP v.4 address Class C may be applicable for configurations where there are a very large number (i.e., up to 2,097,151 netids) of relatively small (i.e., up to 254 hostids) networks. The IP v.4 address Class D is intended for multicast use in all IP networks, with defined multicast address types for designated multicast groups. It may be noted that the wide range of flexibility inherent in the IP v.4 address class structure implies that IP network designers need to carefully consider which address class and range provides the best fit for the evolution of the IP network. It should also be pointed out that as a consequence of the enormous growth of IP-based internetworking applications recently, with the corresponding demand for IP address space, concerns were raised that the 32-bit v.4 IP address would be rapidly exhausted. To overcome this potential problem, an enhanced version of IP called "IP Version 6" has been developed by the IETF; it extends the IP address field from the four octets in IP Version 4 to 16 octets (128 bits) in the newer IP Version 6 (IP v.6). Although IP v.6 implementations have been deployed recently, other techniques to conserve IP v.4 address space have also been developed, which may delay the general introduction of IP v.6 in the Internet.

It should be recognized that the IP v.4 address structure depicted in Fig. 12-8 makes no assumptions regarding hierarchy, other than identifying a given IP network (netid) and the individual host (hostid) attached to the network. It may be recalled from the earlier discussion on ATM addressing and routing that for efficient routing decisions, it is important that there is a direct relationship between the addressing and routing hierarchy, as is the case for the ubiquitous E.164-based addressing structure. For the case of the connectionless IP addressing scheme, routing hierarchies (often termed "subnetting") have been overlaid on the basic IP address class structures to enable more efficient and scalable routing of the IP packets across the Internet. These mechanisms are outside the scope of this brief overview, but it should be recognized that the same underlying principles apply. However, it is important to stress that, despite their obvious differences in structure and coding, both the ATM-related E.164 [6.14] (or, more generally, the AESA) and the IP address are essentially "logical" network-wide or Layer 3 addresses, and hence equivalent in function and intent. The differences in effect stem from the way in which the Layer 3 address is used in the connection-oriented ATM methodology and the connectionless IP-based hop-by-hop routing. Thus, in the connection-oriented ATM, the logical address (E.164 or AESA) is signaled to enable the network to set up an end-to-end (virtual) connection prior to data transfer, whereas in the connectionless IP case, the logical address (IP address) is included in each IP packet (datagram), enabling each IP router in the IP network to forward the packet by analyzing the IP address in the datagram on a packet-by-packet basis.

To obtain a more "user-friendly" mechanism for notating or memorizing an IP address, a decimal IP address format is generally used, as shown in Fig. 12-9 for a class B IP v.4 address. In this so-called "dotted decimal" notation, the 32-bit binary IP address space is divided into four octets as shown. Each octet is then converted into the equivalent decimal number (d1, d2, d3, d4) in the range 0–255, and the IP address is then notated in the well-known dotted decimal format as "d1.d2.d3.d4." An example of an IP v.4 address given in such a notation is 171.25.92.45. It is clear from this example that the dotted decimal notation for representing IP addresses is substantially easier for human manipulation than long strings of binary digits representing the netid and hostid parts of the IP address. It should be noted that the dotted decimal representation may be used for all the IP address classes described above. In fact, as shown in Fig. 12-9, the address class can be derived from the range of the first decimal number d1 in the dotted decimal notation, since d1 includes the IP address class identifiers as its leading bits. As listed in Fig. 12-9, each of the IP v.4 address classes will result in a value d1 in the range listed, for example for Class B the d1 will lie between 128 and 191 (inclusive), and so on. It may also

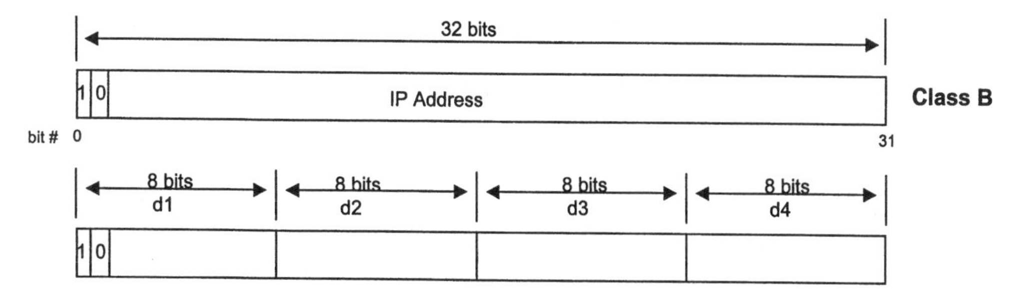

- 32 bit binary IP address is divided into 4 octets
- each octet is converted into equivalent decimal number, (d1, d2, d3, d4) in range 0 - 255
- IP address given in general dotted-decimal notation as d1.d2.d3.d4 (e.g. 171.25.92.45)
- all 0's and all 1's not valid

Address Class can be derived from range of first decimal number, d1

Address Class	Minimum d1	Maximum d1
Class A	1	126
Class B	128	191
Class C	192	223
Class D	224	239
Class E	240	247

Figure 12-9. Decimal IP address format.

be noted that the binary values "all zeros" and "all ones" are not considered as valid addresses, as they may be used for preassigned capabilities, as described below.

In terms of the dotted decimal notation, the general IP v.4 address ranges for each of the classes can be represented as listed in Table 12-2 in terms of the maximum and minimum netid and maximum and minimum hostid for any given class. In addition, some general rules and special cases are followed as indicated there as well. In particular, it should be noted that the value 127.x.x.x is not assigned as an address. This value is only used for maintenance purposes as an internal loopback test within a host in order to verify internal continuity, for example between the transmitter and receiver buffers in a given host. Although the Class D address is intended for general multicast use, it may be noted that some implementations may use a limited "broadcast" address of 255.255.255.255 (i.e., all ones) as a broadcast address. More specifically, for any par-

TABLE 12-2. IP address ranges

Class	Minimum netid	Maximum netid	Minimum hostid	Maximum hostid
A	1	126	0.0.1	255.255.254
B	128.0	191.255	0.1	255.254
C	192.0.0	223.255.255	.1	254
D	224	239	—	—

Note: IP v.4 address rules.

1. The value 127.x.x.x is not assigned. May be used only for internal loopback test on host (transmit to receive buffers).
2. Limited broadcast address of 255.255.255.255 may be used in some implementations.
3. Broadcast addresses have all ones in hostid field (e.g., 171.25.255.255).
4. Hostid of 0 not assigned to an individual host (denotes network only).

ticular netid value, a hostid value corresponding to all ones signifies a broadcast address, i.e., to all hosts on a given network. For example "171.25.255.255" will be interpreted as a broadcast to all hosts attached to netid 171.25. An additional rule is that, in general, a value of 0 is not assigned to any individual host, so that all zeros in the hostid field signifies that the address identifies only the network (netid), and the IP packet is not intended for any individual host on that network.

The recent rapid commercialization of the IP-based internetworking concept, in particular the widespread use of e-mail, has necessitated further development and classification of IP addressing, as well as its overall governance and regulation. Although these broader aspects are outside the scope of this overview, the relationship between the commonly used "e-mail addresses," e.g., xyz@pqr.com, often erroneously assumed to be an IP address, and an actual IP address as described above needs to be clarified. In principle, an e-mail address of the type given above is not strictly speaking an IP address but is generally "translated" into the corresponding IP address by means of a so-called dynamic name server (DNS) protocol operating within the access part of the Internet. The DNS protocol operating between a name server database and an IP router enables the exchange of a corresponding <email address + IP address> pair to the router, which enables it to convert the "name" of the destination user into an equivalent IP address, which may then be used to route the data packets across the Internet. The concept of address translation is not unique to IP-based data networking, as it may be recalled that analogous notions are used in telecommunications in relation to the so-called "intelligent network" services based on the SS7 network, although in that case the messaging protocols are quite different. In addition to the higher-layer protocols applicable for address translations between strict IP addresses and names such as e-mail addresses as provided by the DNS protocol, IP data networking also utilizes another address translation protocol that is more relevant to IP and ATM internetworking, and is known as the "address resolution protocol" (ARP).

The address resolution protocol (ARP) concept originated in the need to determine physical (i.e., MAC) addresses in LANs when only the logical (i.e., IP) address of an attached host was known. The basic ARP concept is illustrated in Fig. 12-10, which depicts a number of hosts (stations) attached to a typical LAN, e.g., Ethernet, which are in general identified by the MAC, or physical, address. Any given host, say host A, need not be aware of the MAC address of any "target" host, say host C, to which it intends to send data. However, if the source host A "knows" the IP address (i.e., the global logical address) of the destination host C or of an ARP

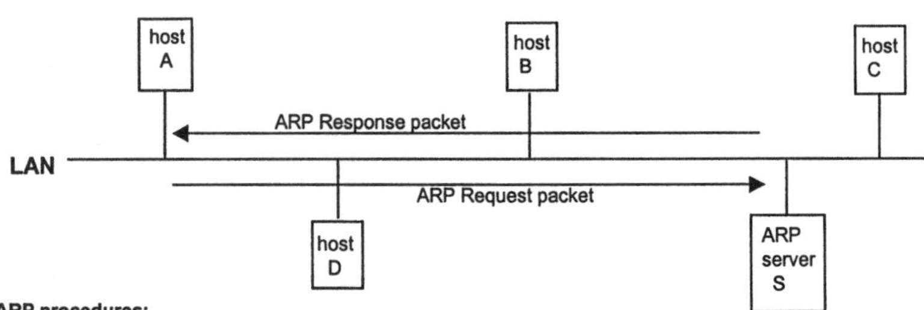

1. Hosts broadcast ARP request packets with the IP address of ARP server to request physical (and/or logical) addresses of "target" hosts
2. ARP server responds to address resolution request with physical (and/or logical) address of target host
3. Hosts may use ARP responses to "store" or cache the <IP : Physical> addresses of other hosts on network
4. Hosts may broadcast their <IP: Physical >address pairs when initially turned on (Boot up procedures)

Reverse ARP (RARP) procedures may be used by host to "learn" its own IP address from a RARP server (also known as Inverse ARP)
- useful when local host has limited memory storage capability

Figure 12-10. Address resolution protocol (ARP) concept.

server, it may obtain the (physical) MAC address by using the ARP technique, as summarized in Fig. 12-10. The ARP messages may be used in several ways. In the simplest case, the host broadcasts an "ARP request" packet that contains the IP address of the target host, as well as its own (MAC address + IP address> pair. Upon detecting its IP address, the target host C responds with an "ARP response" packet, which includes its own <MAC address + IP address> pair. The source host A may then use this information to transmit data frames to the target host C, as well as possibly store or "cache" the <MAC address : IP address> pair for future use. Alternatively, the source host A may broadcast the "ARP request" packet to a so-called "ARP server" attached to the LAN, which is intended to perform the function of an address database for the network configuration in question. The ARP server is designed to respond to ARP Request messages with the <MAC address + IP address> pair of the target host C, which in turn enables the source host A to transmit data to the (MAC) address of the target host C.

The ARP messages may also be used by the hosts to broadcast their <Physical : IP> address pairs (also termed an address duple), when initially attached to a given network, or when initially switched on, as part of the so-called "boot-up" procedures (initialization). The address pair information in the ARP response messages may be cached by the individual hosts, or by an ARP Server, for subsequent use in transmitting data when required. In an alternative mode of operation referred to as reverse ARP (RARP) or inverse ARP, a modification of the basic ARP procedures may be used by any given host to "learn" its own IP address from a so-called "RARP server," which provides the requested <MAC : IP> address duple to the host. The RARP or inverse ARP mechanism is useful in instances when a given host has limited memory storage capability, for example. The general ARP message (request and response) structure is shown in Fig. 12-11 and the same packet format is also used for RARP or inverse ARP.

Although it is not necessary to describe the ARP messaging semantics and procedures in detail here, it is useful to bear in mind the basic structure, in particular the address pairing included in the "sender address pair" and "target address pair" as described above. In addition, it may be noted that both hardware (i.e., MAC or physical) and protocol (i.e., IP or logical) address field length may be varied based on "H Len" or "P Len" field coding, and the "operation" field coding is used to indicate whether the message is an ARP request or response, or RARP message

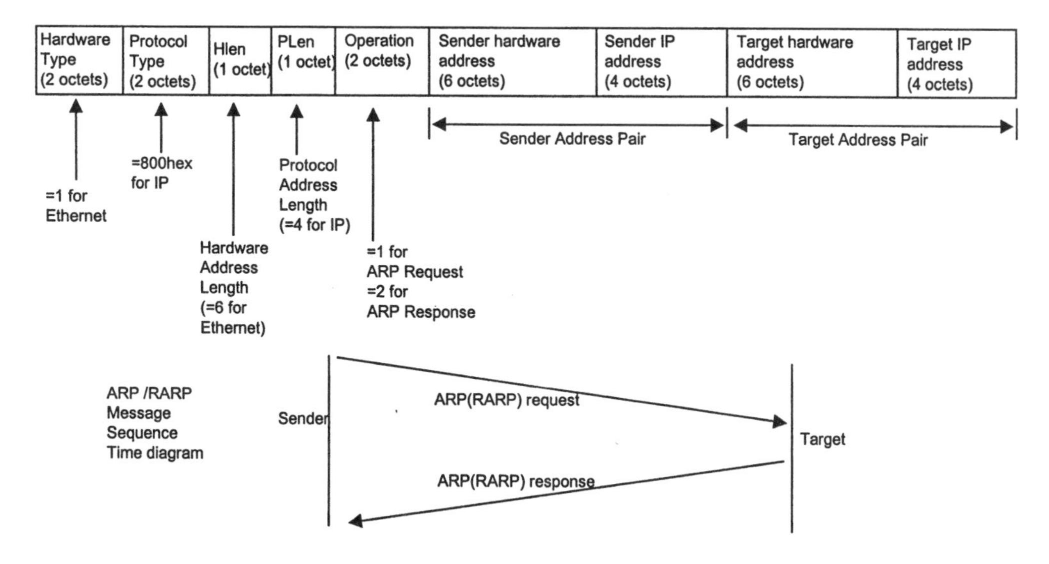

Figure 12-11. ARP packet format.

type. The ARP messages are encapsulated in normal LAN packets for (broadcast) transmission, and the procedures may be generalized to operate across multiple IP subnetworks if required.

The ARP mechanism provides a relatively simple methodology for the administration and maintenance of LAN addressing databases, but more importantly for the purposes here, the underlying concepts may be generalized to enable address interworking (or address resolution, in the data networking terminology) for networks other than LANs. In particular, as described in more detail below, the basic ARP concept has been generalized for the case of IP and ATM internetworking, where it enables essentially the same function in "pairing" the very different IP and ATM addressing schemes. In general terms, the ARP methodology may be viewed as a kind of automated means to provide a "directory inquiry" function from a database (i.e., an ARP server), by which the telephone number (analogous to the MAC address) of a subscriber may be obtained if the "name" (analogous to the logical or IP address) of the subscriber is known, for example. It is evident that a dynamic, automated protocol to perform a function analogous to the traditional manual directory inquiry, such as the ARP mechanism, may be readily generalized to any association of address (or name) pairs, and hence provides a useful tool for control plane interworking, as will be required for IP and ATM internetworking.

In summarizing this overview of IP addressing basics, it is useful to contrast the differences between the ATM addressing mechanisms and those that relate to IP addressing. In a sense, this constitutes one of the major challenges of IP to ATM interworking and lies at the heart of a number of the IP/ATM internetworking architectures considered subsequently. As described previously, the addressing (and hence related routing) methodologies pertaining to ATM are derived from the conventional connection oriented telecommunications switching infrastructure, in which numbering plans are generally hierarchically organized based on geographical distribution. This is typically exemplified by the ISDN-based E.164 numbering plan structured typically as [Country Code + City Code + Area Code + Local exchange + subscriber number]. Typically, common channel signaling is used to set up a connection (or virtual connection) along a path determined by either a static or dynamic routing algorithm performing digit analysis on the strictly hierarchical destination number. The E.164-based geographical hierarchy enables very efficient routing of calls globally, with robust connection management based on SS7 network protocols, based on the one-to-one correspondence between addressing and routing hierarchies.

In contrast, the IP addressing methodology was derived from the need to route connectionless data packets between hosts (i.e., typically computers) attached to remote and possibly diverse LANs, by providing a globally unique logical address independent of the physical (MAC) addresses used for connectionless data packet routing on LANs. Since LANs generally provide connectionless networking, the same paradigm was naturally extended to the networking (i.e., IP) layer. The basic IP (Version 4) address structure was intended to uniquely identify individual networks (netid) and hosts (hostid) for a wide range of possible IP network configurations by means of the IP address classes, but no explicit geographical routing hierarchy is implied in the structure. The IP routers within any given logical IP subnetwork (LIS) or administrative domain route the IP packets on a packet-by-packet basis by matching the IP address to routing tables within each IP router NE. The IP routing tables maintained within each router are dynamically built up by means of separate routing protocols that operate automatically between the routers to enable computation of "best" routes across the IP subnetwork and between subnetworks. To facilitate the routing process, a subnetworking hierarchy may be superimposed on the configuration, whereby the IP packets are forwarded by each router based on a longest match of the address space, analogous to the case of ATM PNNI routing described previously. A major part of IP routing complexity involves the automatic dynamic routing protocols themselves, which are used to update and maintain the routing tables efficiently. A number of sophisticated automatic routing protocols, such as the so-called "open shortest path first" (OSPF), also used as the basis of ATM PNNI routing, have been developed for this purpose. This continues to be a fertile area of study for IP-based networking.

At first glance, considering the widely different antecedents and evolution paths of both ATM and IP-based networking technologies it may appear difficult to reconcile them within a unified interworking scenario. However, the extensive study in both the ATM Forum and the IETF has resulted in the development of protocols and related network architectures that attempt to combine the strengths of both types of technology, thereby facilitating internetworking. These protocols are described in more detail subsequently, but it is useful first to examine the IP functions as well as packet structure in somewhat more detail to clarify the internetworking requirements underlying the various approaches adopted in practice.

12.5 IP PACKET STRUCTURE AND FUNCTIONS

The IP Version 4 (IP v.4) packet format is shown in Fig. 12-12 in its common representation in terms of a 32-bit (four-octet) word-based structure. The IP layer PDU is of variable length and only the IP v.4 header format is shown, which may also be of variable length, the actual IP header length being indicated by the four-bit "IP header length" field, encoded in 32-bit word. The total IP packet length (in octets) is encoded in the two-octet "total packet length" field, as shown. For convenience, the semantics of each of the fields is summarized in Table 12-3, and may be described in turn as follows.

1. IP Version Function (four bits). The first field of the IP header is used to specify which IP protocol version is used. As noted earlier, at present IP Version 4 is in common use

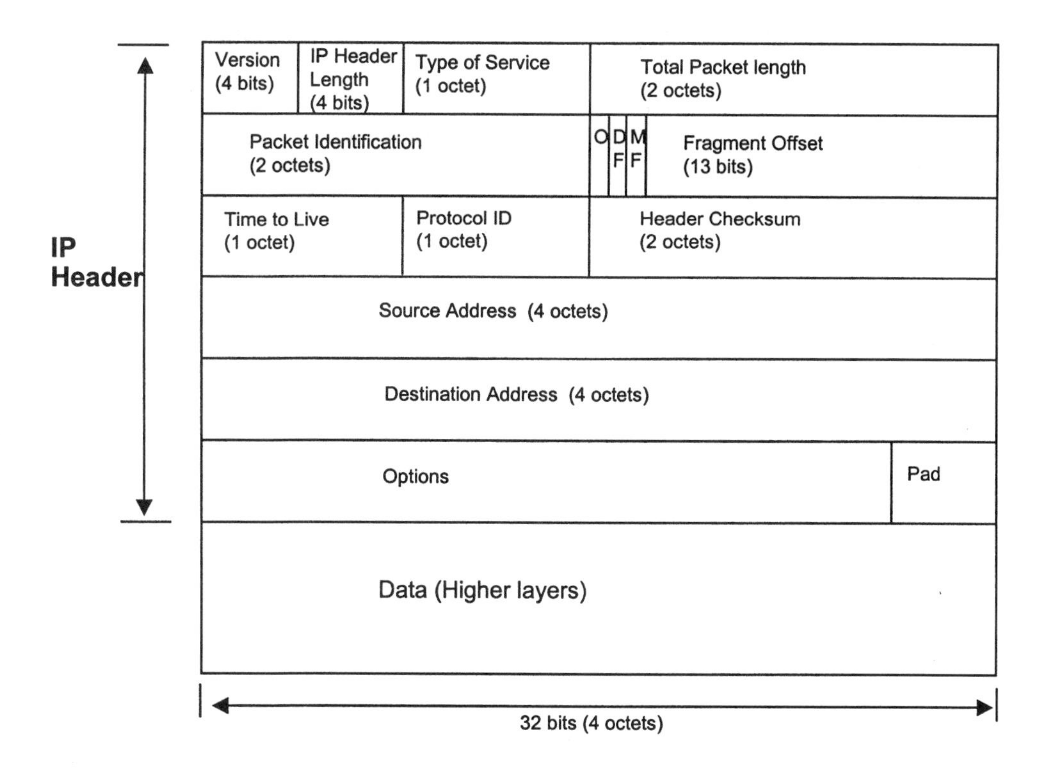

Figure 12-12. IP packet format.

(IP v.4), although IP routers may support several IP versions to maintain backwards compatibility if required. IP Version 6 (IP v.6) has also been developed with substantial enhanced capability, notably with increased IP address space of 16 octets (128 bits).

2. IP Header Length (IHL) (four bits). As mentioned above, the IP header length may vary as a result of the variable length options field described below. The IP header length is specified in 32-bit words in the range 5–15 with a default length of 5 (i.e., when there is no options field).
3. Type of Service (TOS) (one octet). Conventionally, this field was used to indicate the required Quality of Service (QoS) in terms of either delay, discard priority, throughput, or reliability, as encoded in Fig. 12-13. The so-called "precedence" bits are intended to perform a control capability to enable IP packets carrying control information (e.g., routing or management related) to take precedence over normal data traffic. However, the original TOS functions have not been widely used in keeping with the "best efforts" nature of earlier IP network applications such as email or general file transfer. More recently, with increased demand for improved QoS capabilities over the Internet, the TOS capability has received renewed attention. In fact, the semantics and uses of this field have been effectively redefined in accordance with the concept of the so-called "differentiated services" (Diffserv) initiative by the IETF. This concept, which essentially introduces multiple discard priority levels in the handling of the IP packet, is discussed in more detail later, in relation to the associated concept of "integrated services" (Intserv) in IP networks.
4. Options Field (variable length). The options field in the IP header is used to indicate additional optional capabilities if required. Typical optional capabilities include functions such as timestamps and source routing, which itself may be strictly according to a selected route or "loose" in the sense that it is a suggested route, but not mandatory. Other ca-

TABLE 12-3. Description of IP packet fields

IP field	Brief description of function
Version	Identifies IP specification version (presently Version 4).
IP header length (IHL)	Indicates length of IP packet header in 32-bit words. Header length may vary due to Options field.
Type of service (TOS)	Indicates the requested QoS in terms of delay, priority, throughput, reliability.
Total packet length	Indicates total length of IP packet in octets.
Packet identifier	Indicates datagram number for fragmentation/reassembly function.
Don't fragment (DF), More fragment (MF)	Flags used to indicate and control fragmentation function.
Fragment offset	Indicates number of fragment relative to initial fragment. Used to reassemble in sequence.
Time to live (TTL)	Indicates maximum time an IP packet may remain on network in seconds. Not including processing/queuing time.
Protocol identifier (PID)	Identifies the upper-layer protocol using IP.
Header checksum	Provides error detection for IP header.
Source and destination address	Indicates (unique) IP source and destination addresses for routing function.
Options	Used to indicate IP options if present (e.g., security, timestamp, source route, etc.)

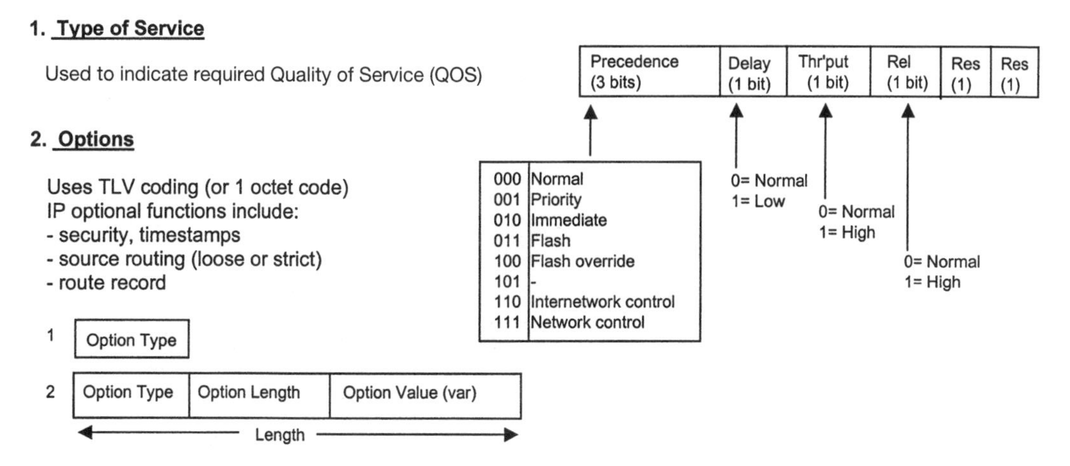

Figure 12-13. IP functions—TOS and options.

pabilities that may be encoded in the options field include security-related functions or a record of the route followed by an IP packet. As shown in Fig. 12-13, there are two methods to encode the options field: (1) by codepoints in a one-octet options field, or (2) by using a type–length–value (TLV) format when flexibility is required in the encoding of the required function. It should also be noted that since the IP header structure is aligned to a 32-bit word pattern, a PAD field is also included to make up the options field length to the requisite 32-bits. Other uses of the options field, such as value-enhancing proprietary features, may also be implemented in some cases.

5. Total Packet Length (two octets). As noted earlier, the coding of this field is used to indicate the total IP PDU length in octets, up to a maximum of 65,535 octets, including the IP header length. In practice, the maximum IP packet sizes are generally much smaller than this "theoretical" maximum length, and are typically implementation-dependent, although hosts should be able to handle at least a maximum packet size of 576 octets, consisting of 64-octet IP header plus 512 octets of payload.

6. Packet Identification (two octets). It was noted previously that in addition to providing the (connectionless) unique logical addressing capability, the other main function of the IP layer was to provide the "fragmentation and reassembly" function to accommodate physical layer MTU limitations. This capability is implemented by using the packet identification field in conjunction with the fragment offset field and fragment control flags described below. The packet identification field is used to assign to each IP packet (or datagram) a unique (message) identifier value at the sending end. At the receiving end, this message identifier value is used for the reassembly of the IP packet from the individual fragments, since each of the fragments will contain the same packet identifier value (modulo 65,535).

   The packet fragmentation and reassembly process is shown schematically in Fig. 12-14 part b, which depicts a long IP packet or datagram being fragmented into a number of (smaller) fragments. The same value of the packet identifier value is inserted into each of the individual fragments. Consequently, the receiving end may reassemble the fragments based on this packet identifier value, in conjunction with the fragment offset number inserted into the fragment offset field. It is of interest to note that the packet identifier mechanism described here is similar to the use of the message identifier (MID) function used in the AAL Type 3/4 SAR process described earlier, which is used for the

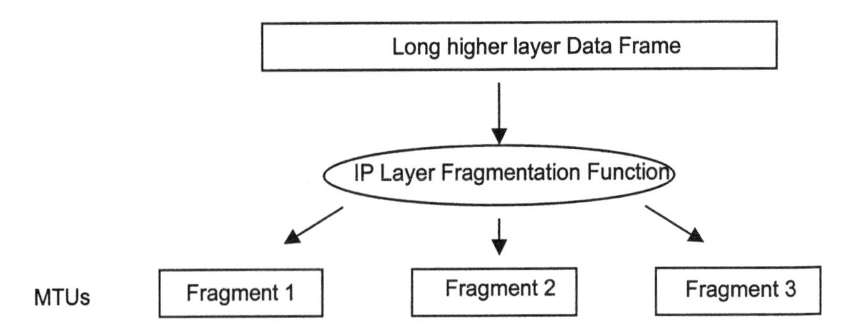

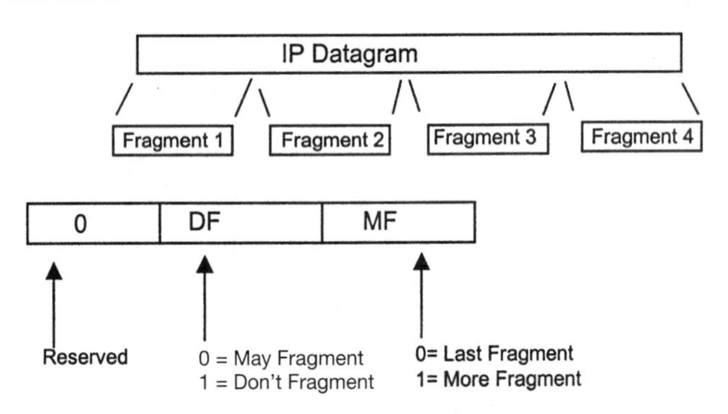

Figure 12-14. IP functions—fragmentation.

support of the connectionless services such as SMDS (or CBDS) over ATM VCCs or VPCs.

7. Fragment Offset (13 bits). As mentioned above, the fragment offset field is used in conjunction with the packet identifier value described above to enable the fragmentation and reassembly of the IP packet, as shown in Fig. 12-14 part b. The fragment offset indicates the number of the fragment relative to the "start" of the IP packet, in the range 0–8191. The sending end assigns a separate fragment offset number in sequence to each fragment, which is used by the receiver end to resequence and check the individual IP datagram fragments in the event of misordering during the routing process. It should be noted that each of the IP fragments includes the entire IP header containing the same packet identifier value, as noted above. It may be noted that the entire fragmentation/reassembly process is somewhat processor-intensive, and will slow the datagram service in the event of fragment misordering.
8. Fragment Control Flags (DF, MF). The one-bit-each fragment control flags are used in conjunction with the other functions described above in the control of the fragmentation/reassembly process, as shown in Fig. 12-14 part b. The "don't fragment" (DF) bit may be set to inhibit the fragmentation process, whereas the "more fragment" (MF) bit is used to indicate the presence of the last fragment of the sequence, thereby terminating the reassembly process.
9. Time to Live (TTL) (one octet). The "time to live" function is used to time out the IP packets on a network to avoid problems due to recirculating packets in the event of routing "loops" resulting from link failures. The TTL value indicates the maximum time the

IP packet may remain on the network (in seconds), in the range 0–255 sec (i.e., 4 min and 15 sec!). This value does not include the actual processing or queuing time in the individual IP router NEs, which are counted as 1 unit (sec). The IP packet is discarded if the TTL = 0.

10. Protocol Identifier (PID) (1 octet). The coding of the protocol identifier field is used to identify the (upper-layer) protocol using the IP datagram. The use of the PID enables the "multiplexing" of various protocols at the IP layer as shown in Fig. 12-15, which also lists typical PID values assigned to some widely used protocols. It may be recognized that some of the PID values relate to the routing protocols such as OSPF or EGP, etc., which may be effectively viewed as the "control plane" of the connectionless IP layer, whereas other values indicate actual higher-layer protocols such as TCP or UDP.
11. Header Checksum (two octets). As may be anticipated, the header checksum field is used to provide error detection over the entire IP header. IP packets with errors in the header are discarded to avoid misrouting. It is of interest to note that, analogous to the case of ATM header, IP also provides error detection only over the IP header and not over the payload field. In the case of IP, the payload error detection is performed at the MAC/LLC or the data link layers, whereas in the case of ATM it is generally within the AAL (e.g., AAL Type 5).
12. Source and Destination Address (four octets each). In addition to the fields described above, the IP header includes the four-octet destination and source IP addresses necessary for the hop-by-hop connectionless routing process outlined earlier. The IP v.4 address structure has already been described above and hence need not be considered further here. It may be noted that the minimum IP header length (with no option field) is 20 octets, of which 12 octets relate to addressing and fragmentation/reassembly processes.

Protocol Identifier

Used to identify the upper layer protocol using the IP datagram. Allows for "multiplexing" of various higher layer protocols at the IP Layer

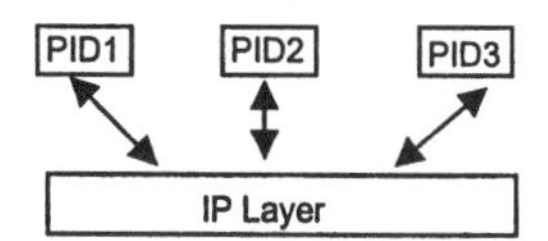

PID value	Description
0	Reserved
1	Internet Control Message protocol (ICMP)
6	Transmission Control Protocol (TCP)
8	Exterior Gateway Protocol (EGP)
9	Interior Gateway Protocol ((IGP)
11	Network Voice Protocol (NVP)
17	User Datagram Protocol (UDP)
22	Xerox Network System Internet Datagram protocol (XNS IDP)
29	ISO Transport Protocol Class 4 (ISO TP4)
89	Open Shortest Path First Protocol (OSPF)

Figure 12-15. IP functions—protocol identifier (PID).

The description of the IP header functions summarized above provides an overview of the extensive capabilities and flexibility of the IP-based networking layer. The use of the protocol identifier (PID) field enables a rich variety of upper-layer protocols to utilize the capabilities of the IP layer, as noted above. Earlier when outlining the underlying LAN protocol architecture (see Section 12.3) and its generalization to include a logical link control (LLC) sublayer by the IEEE 802.2 [12.4] Committee, its relationship to the encapsulation of IP packets was mentioned. It is useful to revisit this relationship to obtain a fuller view of the methodology used, since the LLC approach also essentially generalizes the concept of multiplexing based on a protocol identifier. The use of the logical link control/sub network access protocol (LLC/SNAP) protocol is shown in Fig. 12-16.

In the early implementations of IP packets carried on Ethernet LANs, the "Ethertype" field in the Ethernet frame header was used to indicate the encapsulation of an IP packet in the payload, with a value 800 hex pointing to an IP packet as shown. In this case, the Ethertype field was essentially performing the function of a protocol identifier. As shown in Fig. 12.16, the IEEE 802.2 based specifications utilize a more general procedure to identify the higher-layer protocols encapsulated in the MAC frames, based on the use of the logical link control (LLC) sublayer.

The three-octet LLC header includes three fields as follows:

1. Destination service access point (DSAP) 1 octet
2. Source service access point (SSAP) 1 octet
3. Control field 1 octet

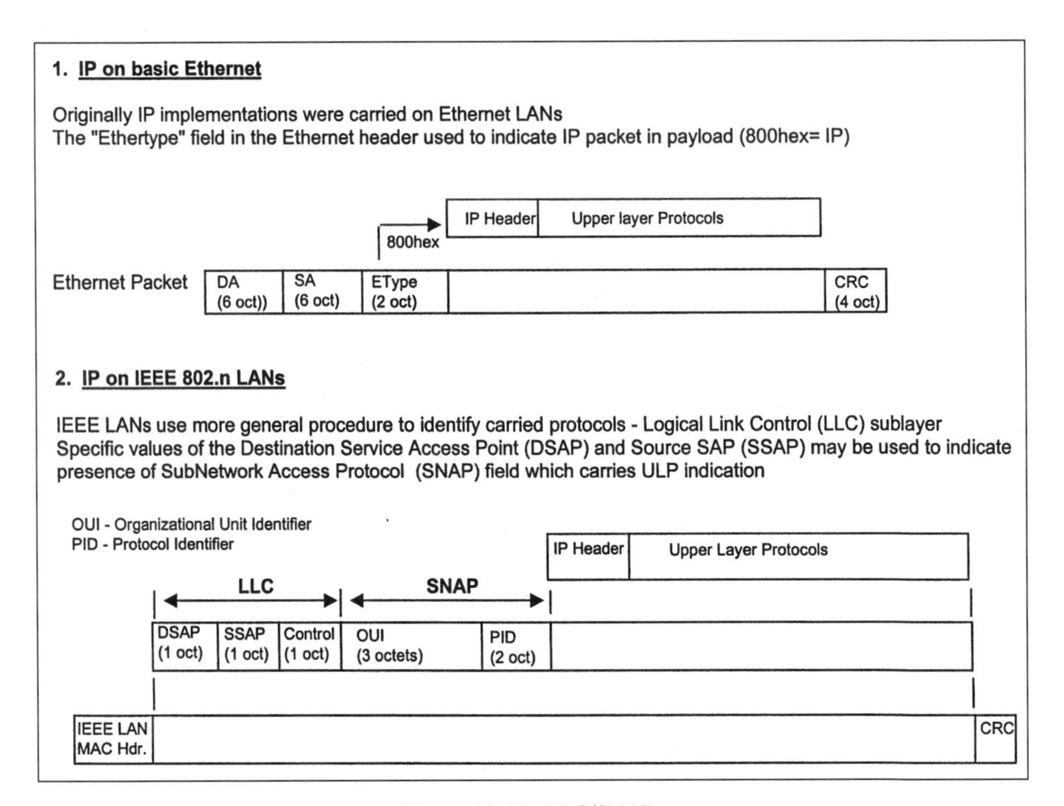

Figure 12-16. LLC/SNAP.

The destination service access point (DSAP) and source service access point (SSAP) fields may be used to indicate separate SAPs at the destination and source protocol entities, thereby enabling additional multiprotocol multiplexing if required by the network application. The control field is used for additional control or management functions at the subnetwork layer if required. Of particular interest here is that for a preassigned value of the DSAP and SSAP fields, an additional header called the SNAP is appended to the LLC as shown. The five-octet subnetwork access protocol (SNAP) header includes two fields as follows:

1. Organizationally unique identifier (OUI) 3 octets
2. Protocol identifier (PID) 2 octets

The protocol identifier (PID) field in the LLC/SNAP header is used to indicate the type of protocol (e.g., IP packet) encapsulated in the IEEE LAN frame. The function of the organizationally unique identifier (OUI) is to identify any given organization (or administration) that is assigned the specific protocol being carried, thereby serving to "customize" the protocol if required for proprietary purposes. The OUI values are administered by the IEEE and may be authorized for use by any organization for proprietary features when required.

It will be recognized from the LLC/SNAP structure shown in Fig. 12-16 that substantial multiplexing flexibility is possible with this eight-octet PCI, particularly for multiprotocol carriage, given the two-octet PID field. Although, as described above, the LLC/SNAP sublayer was developed for the encapsulation of multiple protocols in the IEEE LAN MAC frames, of which IP packets are just one example, it is interesting to note that the same encapsulation methodology has also been adopted for the transport of IP packets over ATM, as will be seen later. Thus, the LLC/SNAP-based encapsulation may be viewed as a general mechanism by which "multiprotocol" transport of any Layer 3 protocol (IP, IPX, SNA, DECNET, etc.) over a Layer 2 protocol (e.g., ATM) may be carried out. However, it should be recalled that an exception to this generalization is the case of IP over frame relay bearer service (FRBS), where an alternative multiprotocol encapsulation mechanism is used based on the network layer protocol identifer (NLPID) PCI structure, as described earlier. The use of two different multiprotocol encapsulation mechanisms for the transport of IP (Layer 3) over ATM and FRBS necessitates protocol translation for FR-to-ATM service interworking, as described previously in Chapter 11 in the section on translation mode for FR/ATM service interworking.

In summarizing the basic (connectionless) IP addressing structure earlier, it was pointed out that a substantial part of the technology underlying IP-based networking between remote hosts lay in the incorporation of suitable dynamic routing protocols, to enable more efficient routing of the IP packets to the destination host. To this end, numerous automatic routing protocols and algorithms have been developed and specified by the IETF for a variety of IP networking applications. The description of such automatic routing protocols itself constitutes a vast field of study, and is outside the scope of this overview. Nevertheless, it is useful to summarize some of the key underlying concepts, particularly in respect to their relationship to IP internetworking with ATM.

As depicted in Fig. 12-17, which illustrates the basic functional model of an IP router, the function of the routing protocols is to enable routers to automatically maintain and update their routing tables by exchanging routing-related information, such as topology or link status, between themselves. Since the network configuration or operational states of a given IP network may change with time, the exchange of routing messages between routers within an administrative domain (also referred to as a logical IP subnet or LIS) occurs periodically to enable efficient updates of the routing tables, analogous to the case of PNNI routing protocols in ATM.

As evident from Fig. 12-17, the term "router" is generally used to describe a device that is capable of processing the Layer 3 "networking" PCI, for example, the IP packet header, thereby enabling the interconnection of separate, physically remote and possibly different LAN net-

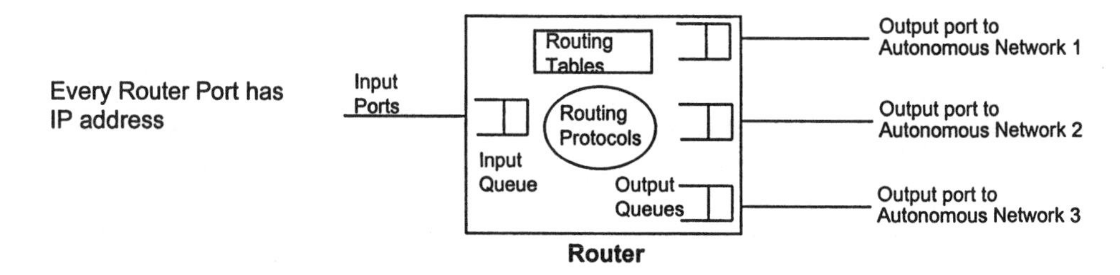

Figure 12-17. IP routing principles.

works. It is important to distinguish the routing function from the so-called (LAN) "bridge" device, which generally only implements Layer 2 (i.e., the data link layer) functions, including possibly the translation between different types of MAC (i.e., physical) addresses, such as Ethernet or Token Ring. The simplified functional breakdown of a typical IP router serves to indicate the basic routing process. IP packets arriving at a given input physical port are queued in an input queue while the destination IP address in the packet header is analyzed for a "best fit" to a destination port using the set of input-to-output mappings available in the "routing tables" stored in the router. These "routing tables" are built up and maintained by the automatic routing protocols that operate between the routers in the network. It should be noted that each of the router ports has an associated IP address to enable the router to transfer the IP packet from an input to the relevant output port as determined by the mapping in the routing tables. The process of transferring the IP packet from the input to the output queue is sometimes also referred to as packet "forwarding" to distinguish it from the purely routing aspects related to routing table look-ups and address processing. As shown in Fig. 12-17, the router ports are connected to so-called "autonomous" IP networks, in the sense that these may be administratively distinct IP networks or independent routing "domains."

The routing protocols exchange messages between the routers independently of the process described above, namely the packet store—IP header processing including table look-up—and packet forwarding operations. The periodic exchange of the routing information serves to update and maintain the routing tables within each router, so that the per-packet table look-up necessary for (connectionless) routing may be based on the most recent routing information on the network topology and state. Thus, for example, if any given (physical) link fails (e.g., due to a cable cut), the routing messages should be capable of relaying that information to all affected devices, so that the routing tables may be modified accordingly and an alternative path computed. For IP routing, two broad categories of routing protocols have evolved based either on (1) distance–vector algorithms (DVA) or (2) link state algorithms (LSA). Although the detailed message structures and procedures of the numerous routing protocols within either of these broad categories are complex and need not be entered into here, it is useful to outline the basic operation underlying these two routing methodologies to assist in the understanding of IP and ATM internetworking applications.

In the typical distance–vector routing protocol, the routing information is broadcast as "distance" (i.e., number of hops) and "vector" (typically a path identifier) pairs within the routing domain. A commonly used DVA-based mechanism is the conventional routing information protocol (RIP). The information in the RIP messages, or equivalently in "hello" messages exchanged between the routers, is then used to build up a topology "view" of the domain based on the number of hops for any given path (vector) to a destination (IP address). Although extensively used, distance–vector-based protocols such as RIP are sometimes viewed as poorly converging and hence inefficient and generally not useful for IP/ATM networking applications. In addition, their scalability to building very large IP networks has also been called into question, but it is likely that DVA oriented routing protocols such as RIP will continue to be used for many IP networking scenarios in practice.

In the case of typical link-state-based routing algorithms, the routers periodically exchange messages (which may also be termed "hello" messages) indicating whether a given link between routers is "up" or "down." The routing algorithm then uses this information to construct a topological "view" of the network depending on the availability state of the individual links making up the routing domain. A typical and widely used example of a link-state-based routing protocol is the so-called open shortest path first (OSPF) protocol, which, it may be recalled, also serves as the basis for the ATM PNNI routing protocol described in detail previously. The link-state-based protocols such as OSFP are generally considered more efficient, as well as scalable, than DVA-based approaches and hence are of more interest to IP/ATM networking applications. In addition to introuting domain protocols such as OSPF, the so-called border gateway protocol (BGP) used for interdomain routing is also categorized as a link-state-based protocol, somewhat related to OSPF.

It is important to recall that all the routing protocols summarized above (as well as others not mentioned here) use the services of the IP layer for the transfer of the routing information between the individual routers in the IP network. The routing messages containing either DV or LS information are encapsulated in the IP packet and routed in the same way as "user data" packets, but are distinguished from the user data packets by the coding of the protocol identifier (PID) field in the IP packet header, as described earlier. For example, the PID codepoint of 89 indicates that the IP packet payload contains OSPF messages, which are then passed to the routing protocol entity for further processing.

The IP PID field codepoints are also used to distinguish another category of messages that are typically used for "maintenance" or control functions in IP networks. Thus, the IP header PID value of 1 identifies the so-called Internet control message protocol (ICMP), whose structure is shown in Fig. 12-18 and whose typical functions include the provision of diagnostic capabilities for maintenance in IP networks.

Some typical examples of ICMP functions are also listed in Fig. 12-18 and are identified by the relevant codepoints in the "message type" field of the ICMP header. Commonly used ICMP functions include:

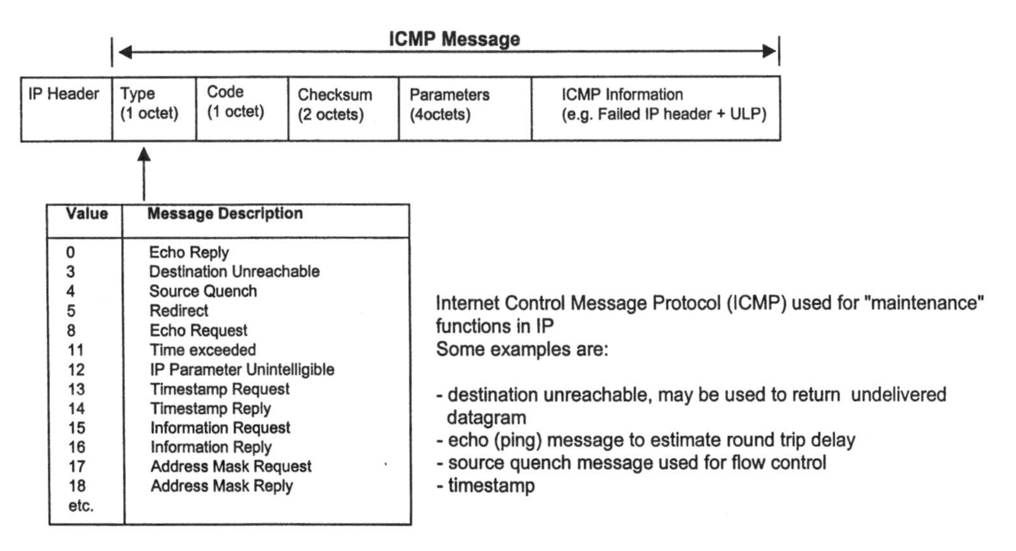

Figure 12-18. Internet control message protocol.

1. Use of the "echo" (or "ping") messages to estimate round trip delay, as well as verify connectivity.
2. Return of an undelivered datagram (IP packet) in the event that the destination is "unreachable."
3. Use of a "source quench" message to provide flow control of IP packets, e.g., in the event of failure or congestion.
4. Support of timestamp for more accurate delay measurements.

It is useful to recognize that the use of the ICMP "ping" (echo) messages "echo request" and "echo reply" are analogous to the ATM OAM "loopback" function, which is also used to verify connectivity along an ATM VPC or VCC, as described previously. However, apart from the differences of detail between the ICMP and ATM loopback OAM cells, it is important to bear in mind that the ICMP "ping" or "echo" function is based on connectionless IP packet routing and forwarding, whereby the path followed by the ICMP echo messages may not necessarily coincide with the IP packets carrying other (higher-layer) user data. For the case of ATM OAM loopback cells, the connection oriented nature of ATM implies that the loopback cells always follow the same path as the user data cells on the VPC or VCC, thereby enabling a more reliable measure of connectivity. It may also be questioned whether the notion of "connectivity" has the same meaning in a connectionless IP network as it does in a connection oriented network such as ATM (or FRBS). In such networks, it may be preferable to consider more generally the "reachability" of any given destination, whereby the use of ICMP echo messages may be viewed as providing a means to verify its reachability.

Although the use of the ICMP functions do not in themselves improve the reliability of IP networks, it is clear that the ICMP messages provide a valuable, as well as flexible, diagnostic tool for the operational maintenance of IP networks, somewhat analogous to the in-band OAM cell flows in ATM described in detail earlier. However, it is of interest to note that the transport of OAM-related information is split between SNMP and ICMP functions in the case of IP network management, in contrast to the case of ATM, where a clear distinction is made between the in-band OAM flows and the resulting management information passed across the management interfaces (e.g., Q3 or X interfaces) as described previously. In the case of IP, the SNMP messages may provide the capability to convey fault management information directly to the network management system across the management interface by use of the TRAP message. However this approach may not be viewed as analogous to in-band OAM information such as the AIS/RDI functions in ATM, but as more akin to information passed across an out of band OS/NE management interface. In addition, for the case of performance management functions, it is clear that no direct analogy exists between the IP and ATM methodologies, although ICMP provides the possibility for transfer delay estimation based on timestamp or echo messages. In particular, it may be noted that mechanisms for the estimation of packet loss or errored packets are not available in either ICMP or SNMP messages, so that these capabilities may need to be provided by the underlying layers if required. These observations indicate that significant differences of approach exist between IP and ATM network management methodologies; these require reconciliation when considering IP and ATM interworking.

12.6 TRANSMISSION CONTROL PROTOCOL (TCP) STRUCTURE AND FUNCTIONS

From the description of the IP Layer PCI in the previous section, it will be evident that the IP layer does not provide functionality for the reliable transfer of the information in its payload (whether it be higher-layer user data or routing or control information). Strictly speaking, the service provided by the IP layer constitutes a basic "unassured" datagram service, implying that any lost information cannot be retrieved. The functionality for providing for reliable transfer of the user data resides in the transport layer (Layer 4 of the equivalent ISO OSI protocol reference model), where the most commonly used protocol to provide for assured delivery of the data is called the transmission control protocol (TCP). It may be noted that TCP is not the only protocol capable of assured mode transport, and not all applications require the use of TCP capabilities. However, its widespread applicability may be gauged from the fact that reference is often made to the "TCP/IP" protocol in many popular descriptions of IP-based networking, even though not all applications may require the use of TCP for assured mode transport. Common examples of these include most real-time applications such as voice over IP (VoIP), or management information (e.g., SNMP or ICMP).

The primary functions provided by TCP are listed in Fig. 12-19, which also shows the structure of the TCP packet header format. As noted earlier, these functions include:

1. The reliable (i.e., assured mode) delivery of data
2. Window-based flow control
3. Mutiplexing of the upper-layer protocols using "port" numbers (also known as TCP "sockets")
4. Full duplex "virtual circuit" connectivity

The mechanisms required to provide these functions may be understood by examining the purpose of each of the TCP header fields, as listed in Fig. 12-20. The multiplexing of the (higher-layer) applications may be achieved by use of the two-octet "source port" or "destination port" fields, which essentially function as protocol identifiers to indicate the specific application using

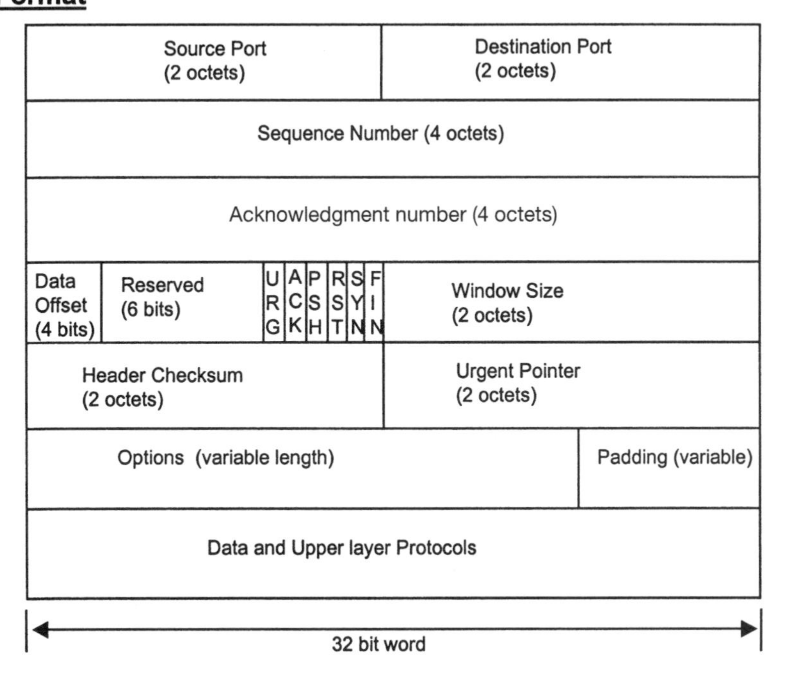

Figure 12-19. Transmission control protocol (TCP).

TCP Field	Description
Source / Destination Port	Used to identify the source / destination ULP
Sequence Number	32-bit sequence number to identify every octet transmitted
Acknowledgment Number	32-bit number identifying successfully received octet
Data Offset	TCP header length in 32-bit words
Flags: URG,ACK,PSH,RST,SYN,FIN	Control flags
Window Size	Indicates size of flow control window (in octets)
Checksum	Error control for TCP header only
Urgent Pointer	Indicates location of "Urgent" message.Used for "interrupt" function
Options	Specifies TCP options. Max. Segment Size only used

Flag	Description
URG	Urgent Indicator
ACK	Acknowledgment (used with SYN)
PSH	Push. Deliver data now. (eq. to end of message)
RST	Reset. Used to reset connection
SYN	Synchronous. Used with ACK for connection control
FIN	Finish. Used to terminate connection

Port Number	Description
20	FTP data
23	TELNET
25	SMTP
53	DNS
80	WWW
104	X.400
etc.	

Figure 12-20. TCP functions.

the services of the TCP layer. Some examples of commonly used application port numbers (also sometimes called TCP sockets) are given in Fig. 12-20. The similarity of this function to the use of the PID field in the IP packet header may be evident, although it may be noted that in TCP both the source and destination port numbers are identified.

The four-octet "sequence number" field is used to assign a 32-bit sequence number to identify every octet transmitted by the sending entity. Similarly, the "acknowledgment number" field contains the 32-bit number identifying the successfully received octet number at the receiver entity. The sequence number, acknowledgment number, and the (two-octet) "window size" fields are used by TCP to ensure reliable transmission of the data from end-to-end, as illustrated schematically in Fig. 12-21.

The TCP "window size" corresponds to the number of octets of data that may be transmitted and is determined by the TCP receiver entity, based on the available buffer space to avoid loss of

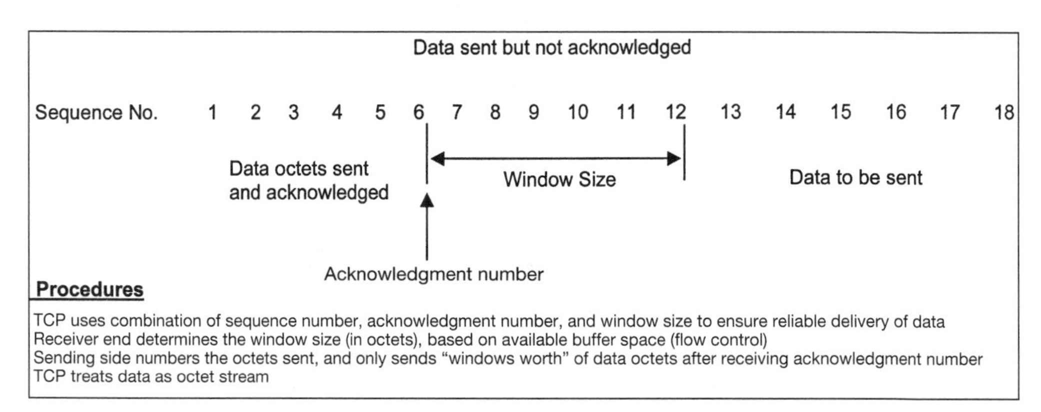

Figure 12-21. TCP flow control and acknowledg1ment.

any of the received octets. It may be noted that the use of such a "window" of adjustable size essentially provides a flow control mechanism determined by the available buffer space at the TCP receiver, thereby reducing the probability of data loss due to buffer overflow. Using the sequence number field, the sending entity sequentially numbers each of the octets specified by the window size (i.e., sends only a "window's worth" of data after it receives an acknowledgment number from the receiver. The acknowledge number indicates the number of octets successfully received by the TCP receiver entity, which then expects to receive a further window's worth of octets from the TCP transmitter. In the event of a discrepancy between the acknowledgment number and the sequence number at the TCP transmitter, it is assumed that the difference has been lost and retransmission of the octets will occur. However, if the acknowledgment number corresponds to the last sequence number sent, it is assumed by the TCP sending entity that all the octets have been received, and an additional window's worth of octets is transmitted, with the sequence number incremented by the window size.

The above mechanisms combining individual octet sequence numbering, positive acknowledgment of each octet received with retransmission (PAR), and variable window size (in octets) serves to provide TCP with the capability of reliable data transfer (assured mode transfer) and flow control to ensure low data loss with finite buffer sizes. In effect, TCP treats the (higher-layer) data as a "stream" of octets, each of which is individually numbered and accounted for in any given TCP session. The transmission of a finite number of octets determined by the window size allows for flow control of the octet stream end-to-end, whereas unacknowledged octets are retransmitted by the sending TCP entity. It may also be recognized that the entire process, although ensuring reliable data transfer end-to-end, is relatively processing-intensive and may also limit performance under certain conditions. In addition to the basic window flow control mechanism outlined above, TCP also generally incorporates procedures for the rates at which window sizes may be incremented or decremented in the event of octet losses, in order to mitigate the effects of congestion or other transient phenomena. Such procedures need not be considered here, but may play a part in the overall traffic design of IP over ATM networks.

It was noted that TCP also provides the capability of a full duplex (i.e., bidirectional) "virtual circuit" for the duration of a TCP session. These procedures, enabled by the use of the TCP "control flags" listed in Fig. 12-20, allow for the initiation and termination (as well as reset) of the TCP data transmission session, in effect emulating a "virtual circuit" between the source and destination ports for the transport of the octet stream. Although this capability is sometimes described as implying that TCP provides a "connection oriented" service to the higher layers, it is evident that the initiation/termination of the TCP session using the control flags is not strictly connection oriented in the same way as an ATM VCC. However, the association of the source and destination ports, coupled with the initiation of the TCP "connection" using the SYN and ACK flags, and its termination with the FIN flag setting may be viewed as effectively emulating a virtual connection, with the control flags serving as a rudimentary "signaling" protocol for the TCP session. It may also be noted that the TCP header includes a two-octet "urgent pointer" field, whose value indicates the location (in octets) of a so-called "urgent" message, which may be interleaved with an octet stream for the priority transfer of an urgent message in the event of a long octet stream. The use of the urgent pointer field in conjunction with the URG flag provides an interrupt function for higher-priority messages.

The four-bit "data offset" field is used to indicate the length of the TCP header in 32-bit words, since the header length may vary if the "options" field is used, as for the case of the IP packet header. The TCP options field is generally used to specify a maximum segment size of the TCP packet when used. The TCP header also includes a two-octet "header checksum" field for error detection over the TCP header, as for the case of the IP packet header checksum, also utilizing a similar error check algorithm.

For applications that do not require the complex reliable data transfer and session control capabilities provided by TCP, an alternative simpler protocol may be used called "user datagram protocol" (UDP). UDP does not include the sequence numbering and flow control functions as well as the control flags provided by TCP, thereby reducing overhead and processing requirements. However, UDP provides for source and destination port numbers (UDP sockets) and simple message length and header checksum functions to provide a basic datagram capability to the application. In addition, it may be noted that other Layer 4 protocols have also been developed for specific IP networking applications such as the so-called "real time protocol" (RTP) for the transport of voice or video signals over IP. The description of these and other such special-purpose transport protocols are outside the scope of this survey.

12.7 ENCAPSULATION OF IP OVER ATM

In summarizing the primary functions and evolution of the IP layer, it was pointed out that IP is intended to isolate the higher layers from variations in the underlying data links and transport mechanisms. This essentially implies that, by design, IP datagrams may be mapped and transported by any Layer 2 protocol deemed suitable. Consequently, it is not surprising that in considering the carriage of IP packets over ATM, the IETF specified a generic encapsulation based on the IEEE 802.2 LLC/SNAP structure, also used for encapsulation of IP packets in IEEE LAN frames, as described earlier. It was pointed out in that description that the LLC/SNAP methodology may be generalized to provide for encapsulation of multiple network layer (Layer 3) protocols over a Layer 2 protocol.

The general encapsulation of IP packets over ATM is shown in Fig. 12-22, as described in IETF RFC 2684 [11.14]. It is important to note that two options for IP packet transport over ATM are specified, both based on the use of the AAL Type 5 common part convergence sublayer (CPCS) described earlier. The most commonly used option utilizes the IEEE 802.2 LLC/SNAP-based encapsulation, which enables the "multiprotocol" multiplexing of IP or other (e.g., IPX, SNA, etc.) network layer protocols onto a single ATM VCC (or VPC) by using the "protocol identifier" (PID) field in the LLC/SNAP header to distinguish the various protocols, as described earlier. In the other option shown, termed "VCC-based multiplexing" or "direct" encapsulation, it will be noted that no intervening LLC/SNAP sublayer is present and hence each network layer protocol is mapped directly to a separate ATM VCC after encapsulation in the AAL Type 5 CPCS. In this case, there is a one-to-one relationship between the Layer 3 (network layer) protocol and the ATM VCC.

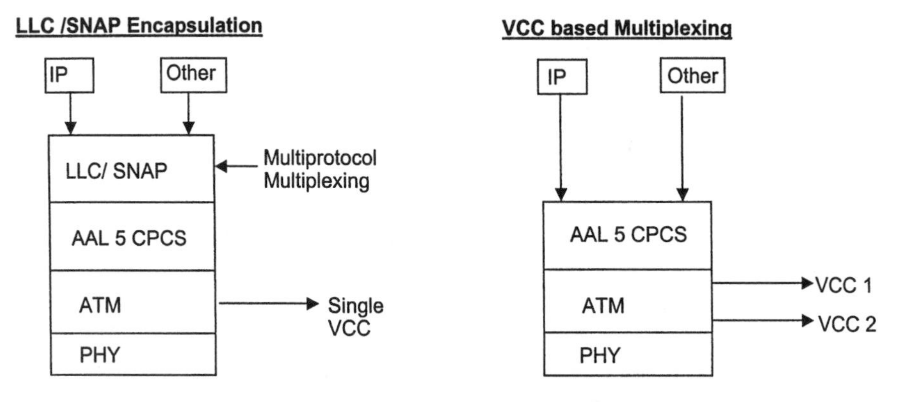

Figure 12-22. Encapsulation of IP in ATM.

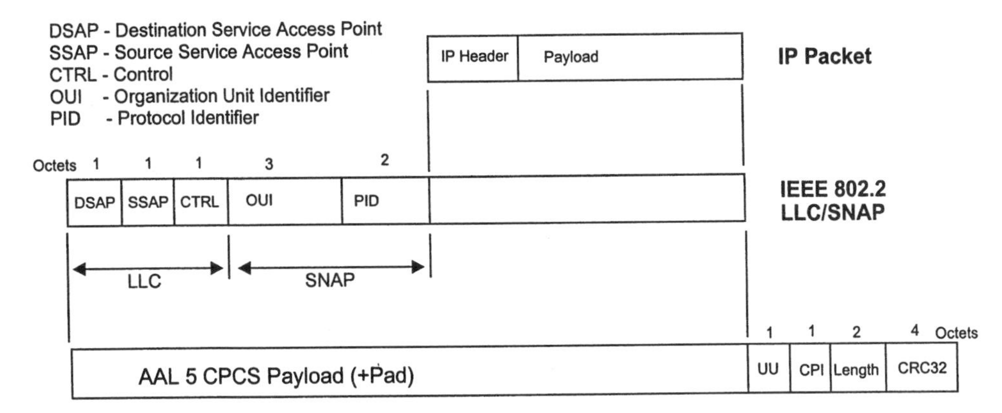

Figure 12-23. LLC/SNAP format.

For the case of the IEEE 802.2 LLC/SNAP encapsulation option, the mapping of IP packets into the AAL Type 5 CPCS payload is shown in Fig. 12-23, where the detailed structure of the LLC/SNAP header and the AAL Type 5 trailer may be recalled from the description given earlier. From the perspective of the AAL protocol architecture, it is of interest to note that the IEEE 802.2 LLC/SNAP sublayer may be viewed as analogous to a "service-specific convergence sublayer" (SSCS) for the interworking of IP [or other network (Layer 3) protocols] with ATM, in much the same way as SSCS are specified for other interworking scenarios such as the FR/ATM SSCS described previously. It is also important to recognize that the primary role of the LLC/SNAP sublayer here is to enable the transport of "multiprotocols" over an ATM VCC, since in this context the use of the other LLC/SNAP functions, such as OUI, and DSAP/SSAP, although not precluded, are not commonly utilized. The use of the two-octet PID codepoint to identify the encapsulated network layer protocol potentially allows for multiplexing of a very large number of such protocols within a single ATM VCC and would appear to be excessive, given the already large VPI/VCI-based multiplexing capability available within the ATM layer itself. In addition, the provision of a further six octets assigned to functions such as OUI, DSAP, SSAP, and controls with no clear network application, at least for public carrier networking, would also appear to be unnecessary. However, the benefits of using a common multiprotocol encapsulation technique for (almost) all Layer 2 protocols would appear to outweigh the drawbacks noted above, since the IEEE 802.2 LLC/SNAP encapsulation is widely accepted as the default methodology for carriage of IP packets over ATM in accordance with the sublayering depicted in Fig. 12-23. The VCC-based multiplexing option, although generally not precluded, is not widely used. The selection of which of the RFC 2684 [11.14] based encapsulation options to use may be implemented either by provisioning, or by signaling.

12.8 THE "CLASSICAL" IP OVER ATM NETWORK ARCHITECTURE MODEL

The so-called "classical IP over ATM" (termed CIPOA for convenience here) network architecture essentially describes a "network interworking" scenario, in the sense defined previously, for IP and ATM networks, as shown in Fig. 12-24. In this model, originally defined by the IETF in RFC 1577, now superseded by RFC 2225 [12.8], the IP packets originating in the IP network (or router) are encapsulated according to the methodology described above, utilizing one of the options specified in [11.14] and AAL Type 5 CPCS. The IP packets (segmented into the ATM

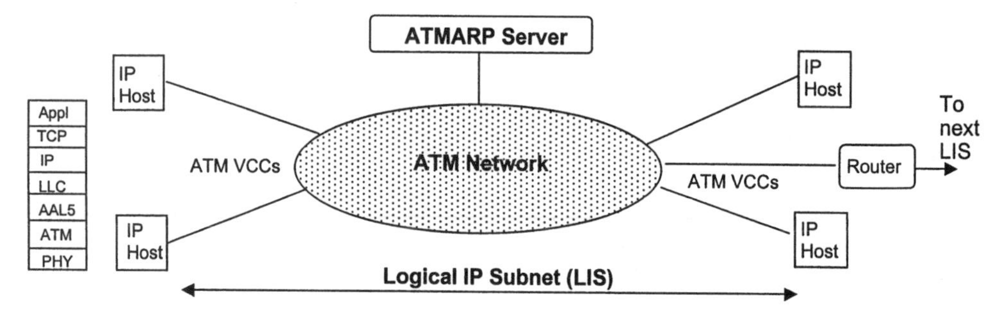

Figure 12-24. Classical IP on ATM.

cells) may then be switched through the ATM VCC to the remote destination IP network (or router) where the IP packets are reconstituted and forwarded to the destination host. As shown in Fig. 12-24, the "interworking function" between the IP and ATM networks in the transfer plane essentially consists of the encapsulation of the IP packets into an ATM VCC using the RFC 2684 based protocol stack as shown. If the IEEE 802.2 LLC/SNAP encapsulation option is used, multiprotocol encapsulation may be multiplexed within any given ATM VCC, as described above. It will be noted that the classical IP over ATM (CIPOA) model assumes that an ATM VCC (or VPC) is set up to the required destination router, but since this destination is identified by the IP address in the packet header, it is clear that the underlying ATM network will require a "translation" from the IP address to the relevant ATM destination address, which may use an E.164-based or AESA (NSAP-based) ATM address, which, as noted earlier are different from IP addresses.

This need for address resolution between the (Layer 3) IP addresses and the very different ATM addresses results in the main architectural functional element of the CIPOA model, termed the ATM address resolution protocol (ATMARP) server. The function of the ATMARP server is to provide the "translation" or mapping of the destination IP address to the corresponding destination ATM address in order that the ATM VCC can be set up either by on-demand signaling protocols or by the management interface for the case of semipermanent virtual connections. The CIPOA model essentially defines extensions of the conventional IP ARP capability to provide the <IP address : MAC address> pair as described earlier for the case of IP over LAN protocols. In the CIPOA model, the ATMARP server essentially treats the ATM address as "equivalent" to a LAN MAC address, thereby enabling the underlying "physical" connectivity to be established for the transport of the IP packets from source to destination router or host. In the ATM case, of course, the "physical" connectivity is actually by means of a VCC or VPC, within an actual (physical) transmission path. It is also of interest to note that the ATMARP server function essentially corresponds to a "control plane" interworking capability, which enables mapping between a (connectionless) protocol address such as an IP address, and a (connection oriented) "signaled" address such as an E.164 or an AESA, required for connection set-up.

The CIPOA model depicted in Fig. 12-24 does not specify a physical location for the ATMARP server function since this is essentially a logical entity, which clearly would need to be accessed by potentially any host/router comprising the logical IP subnetwork (LIS), to enable ubiquitous connectivity. The concept of the logical IP subnetwork (LIS) in the CIPOA network model also needs to be clarified. In general terms, the LIS corresponds to a given routing or administrative domain, or IP subnetwork, which may be served by a single ATMARP server that comprises a set of unique IP address prefixes to ATM addresses. Thus, in the basic CIPOA model, it is assumed that the ATMARP server is capable of providing the <IP address : ATM ad-

dress> pairing required to enable connectivity between each and every host or router device comprising the LIS. As depicted in Fig. 12-24, the basic model also assumes that connectivity to an adjacent LIS is achieved by means of a router interconnecting the two LIS domains. The intervening router terminates the ATM VCCs, thereby reassembling the IP packet, which must then be forwarded onto the next LIS in the same way as for the preceding LIS, and so on, until the destination host is reached.

The messaging sequence required within each LIS domain of the basic CIPOA approach is summarized in Fig. 12-25. For initiation purposes, it will be clear that each host or router comprising the LIS must at first establish an ATM VCC between itself and the ATMARP server function in order to access the ATMARP server capability. Typically, this ATM VCC (or VPC) may be provisioned as a PVC, since access to the ATMARP server may be required on a permanent basis for address translation, although the model does not specify whether the procedures for setting up the VCC (or VPC) utilize on-demand signaling or provisioning via a management interface. Once the connectivity is established between the ATMARP server and the host or router device, the ATMARP server may send the inverse ATMARP_Request message (InATMARP_REQUEST) to the host or router to identify (or initialize) the host or router. As shown in step 3 of Fig. 12-25, the host/router may then respond by sending an InATMARP_REPLY message to the ATMARP Server to "register" itself, thereby enabling the ATMARP Server to add the hosts/routers <IP address : ATM address> to its database for subsequent use. This exchange of the InATMARP_REQUEST and InATMARP_REPLY messages between the host/routers and the ATMARP server completes the initialization procedures for the LIS domain. However, it may be recognized that such an initializing procedure may also be carried out by provisioning of the <IP address : ATM address> pairs corresponding to each attached host/router by means of management procedures directly to the ATMARP server.

On completion of the ATMARP server initialization procedures (steps 1 to 3) by whichever means for a given LIS domain, the basic CIPOA model assumes that ATM VCC (or VPC) connectivity may then be established between any of the attached hosts or routers by extension

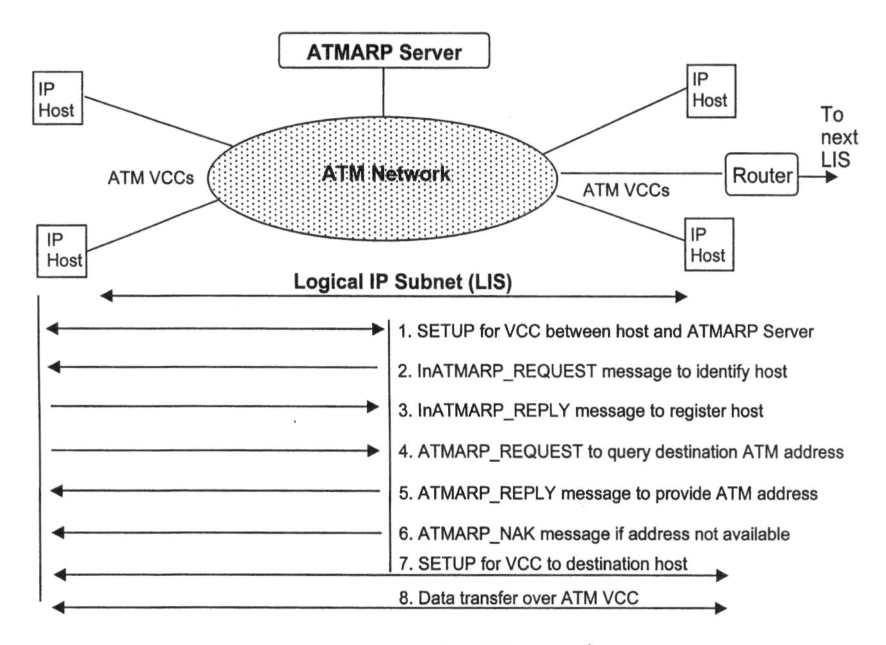

Figure 12-25. ATMARP messaging.

of the ARP methodology represented by steps 4 to 7 in Fig. 12-25. Thus, a host sends an ATMARP_REQUEST message to the ATMARP server to query the destination ATM address (E.164- or AESA-based) of the host or router for which it knows the destination IP address (step 4). The ATMARP server responds with an ATMARP_REPLY message that contains the <IP address : ATM address> pair (duple) of the destination host/router, as represented in step 5 of Fig. 12-25. In the event that the destination is unknown or not available, the ATMARP server may respond with an ATMARP_NAK message to indicate a negative acknowledgment to the request message. Finally, as indicated in steps 7 and 8 in Fig. 12-25, once the sending host or router has obtained the ATM address of the destination host or router, it can set up the ATM VCC (or VPC) using normal on-demand ATM signaling via the ATM network to the destination host and then proceed to transfer data as required. The requisite ATM VCC or VPC may either be released on completion of the data transfer session or left connected in the event that subsequent sessions are anticipated, for some programmable interval. Clearly, the duration of such an interval would depend strongly on the type of session as well as the specifics of the network applications. These aspects are not generally addressed as part of the basic CIPOA model. It may also be recognized that the ATM connectivity between the attached hosts/routers comprising the LIS domain may also be set up using management procedures as for a full mesh PVC network, although this approach may only be practicable for relatively small CIPOA networks.

It is of interest to note that the ATMARP message set that forms the basis of the CIPOA network architecture essentially uses the same basic ARP procedures and packet formats as described earlier for conventional (LAN-based) IP, with extensions needed for the additional ATM specific parameters. The ATMARP message set, together with the designated "OP code" (or message type) values and the ATMARP parameters, are summarized in Fig. 12-26, based on the RFC 2225 naming conventions. It will be noted from the ATMARP parameter list that for the case of the ATMARP messages, the MAC (or "hardware") address in the conventional ARP is simply replaced by the relevant ATM address (plus any subaddress if present). In other words,

ATMARP Message Set	OP Code
1. InATMARP_REQUEST	8
2. INATMARP_REPLY	9
3. ATMARP_REQUEST	1
4. ATMARP_REPLY	2
5. ATMARP_NAK	10

ATMARP Parameter	Name	Length (bits)
Hardware Type	ar$hrd	16
Protocol Type	ar$pro	16
Source ATM number type/length	ar$shtl	8
Source ATM subaddress type/length	ar$sstl	8
Operation Code	ar$op	16
Source protocol address length	ar$spln	8
Target ATM number type/length	ar$thtl	8
Target ATM subaddress type/length	ar$tstl	8
Target protocol address length	ar$tpln	8
Source ATM number	ar$sha	q octets
Source ATM subaddress	ar$ssa	r octets
Source protocol address	ar$spa	s octets
Target ATM number	ar$tha	x octets
Target ATM subaddress	ar$tsa	y octets
Target protocol address	ar$tpa	z octets

1. RFC 2225 (formerly RFC 1577) ATMARP message set uses same basic ARP procedures and packet formats as in conventional IP, with some additional ATM specific parameters.
2. The MAC (hardware) address in conventional ARP is replaced by the ATM address + subaddress.
3. The maximum default MTU size is 9180 octets (+ LLC/SNAP makes AAL5 PDU =9188 octets).
4. The <IP address: ATM address> pairs in host or server cache are "aged" and revalidated in (suggested) timescales of 15 min. (client) and 20 min. (server), or as updated.
5. Hosts have responsibility to establish ATM VCCs (PVC or SVC) to ATMARP Server.

Figure 12-26. ATMARP parameters and procedures.

from the perspective of the address resolution procedures, the underlying "ATM network" simply behaves as if it replaced the LAN for the transfer of the IP packets. In this context, it is immaterial that the underlying ATM network provides considerably more extensive capabilities than any conventional LAN. By interworking using the ATMARP procedures for address translations, the IP-based applications can exploit any (or all) of the QoS, signaling, advanced intelligent network (AIN), or OAM, etc. capabilities of the ATM backbone network required for the delivery of any desired service.

The <IP address : ATM address> pairs stored on the ATMARP server or "cached" in any given host may be updated or revalidated at regular intervals in order to ensure that the reachability information required for routing purposes is fully up-to-date. This "aging" process is customary for the typical IP-based routing protocols to allow for dynamic changes in the network configuration and state. The intervals may be programmable depending on network stability, with suggested time scales in the order of 15 to 20 minutes or more as required. In the CIPOA model, the individual hosts (terminals or routers) attaching themselves to the LIS domains are assumed to have the responsibility to establish (and maintain) the ATM VCC (which may either be PVCs or SVCs) to the ATMARP server to initially register and subsequently update their IP and ATM addresses with the server for the LIS. It may also be evident that if the "downloaded" <IP address : ATM address> pairs can be cached in the hosts/routers, no subsequent ATMARP messaging is required (for that particular destination) prior to the data transfer phase. For frequently used destinations, this consideration may be used to significantly reduce the latency before data transfer in the CIPOA architecture, although at the expense of more cache memory requirements in the host. In addition, to enable more efficient data transfer, the CIPOA procedures allow for a maximum (default) MTU size of 9180 octets. Thus, for the case where the LLC/SNAP encapsulation is used, the maximum length of the AAL Type 5 payload is 9188 octets.

The brief overview of the basic CIPOA approach to IP/ATM internetworking summarized above clearly serves to illustrate the conceptual simplicity of this model. This can result in a relatively straightforward implementation and network configuration procedures for IP-based applications over ATM backbone networks. More importantly, the simple CIPOA architecture allows any IP-based application to take full advantage of the extensive capabilities provided by ATM, particularly in terms of the QoS classes supported by the ATM traffic control mechanisms, as well as the OAM functions described earlier. In addition, this substantial networking advantage may be achieved with no major modifications required by any of the IP-based or ATM protocols for signaling or routing. Only a relatively minor extension to the well-established ARP procedures are needed in order to provide for the required IP to ATM address translation. As indicated earlier, the basic CIPOA internetworking scheme simply constitutes a "network interworking" scenario in the general sense described earlier, with a form of "control plane" interworking relating to the IP to ATM address translation by the ATMARP server function. The inherent simplicity and ease of implementation (assuming that the attached hosts or routers can be equipped with the appropriate ATM hardware and signaling software interfaces) may be viewed as the main advantages of the CIPOA network architecture for IP-to-ATM internetworking, and accounts for its widespread deployment in existing ATM networks at present.

However, it should also be recognized that, despite these significant advantages, the basic CIPOA network architecture also possesses some inherent limitations. In fact, it would be fair to observe that much of the subsequent work on IP/ATM internetworking has focused on various ways to overcome the main disadvantages of the initial CIPOA approach, often with limited success. The main disadvantage of the basic CIPOA approach is that the provision of a single ATMARP server per LIS limits its scalability for evolution to large networks, particularly since for optimum performance (i.e., low latency) it is generally necessary to maintain PVC connectivity between each attached host or router and the ATMARP server function for a given LIS. In addition, it is clear that the ATMARP server may also be viewed as a single "point of failure" for the entire network, raising questions about the consequent robustness of the CIPOA architecture.

It may be noted that these disadvantages are to some extent inherent in the essentially "client–server"-based architecture adopted for the CIPOA approach, as evident from Fig. 12-25, and, in principle, would be valid for any client–server-based scenario. Another limitation of the basic CIPOA model, somewhat related to its scalability disadvantage, is the consideration that for inter-LIS communication, the data transfer path would still need to traverse the routers located between each LIS, thereby limiting performance. From a conceptual viewpoint, the fact that the CIPOA architecture essentially maintains a clear separation between the IP and ATM routing domains is seen by some as an inherent disadvantage of this approach, although it must be stressed that this view is not universally shared by all workers in this field. Thus, the question of whether it is desirable to "integrate" IP and ATM routing and the extent to which this should be done is still an open issue, although the current work in the IETF and elsewhere on the so-called multiprotocol label switching (MPLS) technology (as outlined later) to some extent circumvents this debate by introducing an alternative methodology for integration.

Irrespective of the somewhat theoretical consideration as to whether or not it is desirable to integrate IP and ATM routing for purposes of internetworking, several enhancements have been made to the basic CIPOA architecture to address the limitations concerning scalability and performance aspects mentioned earlier. In the basic CIPOA model, the ATMARP server role is restricted to providing IP to ATM address resolution to a given LIS domain, but it is evident that this concept may be extended across several LIS domains, as illustrated in Fig. 12-27, by using the concept of the so-called "next hop resolution protocol" (NHRP) [13.12]. In this model, the basic CIPOA architecture is supplemented by the introduction of the next hop server (NHS) between each LIS, to provide end-to-end reachability across multiple LIS domains. As shown in Fig. 12-27, the next hop resolution protocol message set was specified by the IETFs Routing Over Large Clouds (ROLC) working group as an extension of the ATMARP message set to enable inter-LIS address resolution.

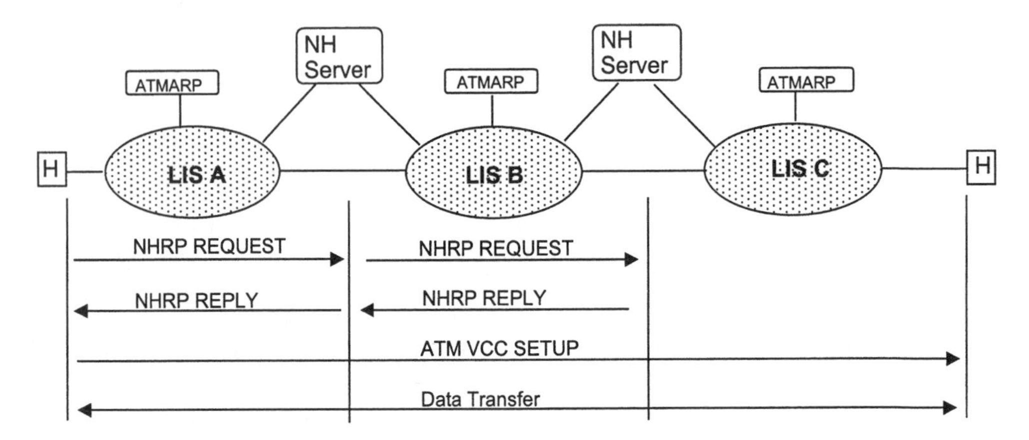

1. IETF Next Hop Resolution Protocol (NHRP) is an extension of ATMARP protocols to enable inter-subnet address resolution
2. Uses concept of Next Hop Server (NHS) to enable hosts / routers to resolve destination IP address with ATM address
3. Hosts or routers may then establish direct ATM VCC (PVC or SVC) to destination for data transfer
4. Removes limitation of router processing between the subnetworks by use of ATM VCC bypass (cut-through)
5. May be used in conjunction with conventional IP Internal Gateway or External Gateway Protocols (IGP/EGP)
6. NHS function may be physically co-located with ATMARP Server

Figure 12-27. Next hop resolution protocol (NHRP).

The NHRP message set specifies two additional messages that operate between a host/router and the NHS, or between NH servers as shown. If a given host intends to send data to a destination host or router attached to a remote LIS, or if the "local" ATMARP server does not recognize the destination address, the host sends a NHRP REQUEST message to the NHS. The NHRP REQUEST message may be relayed onwards from NHS to NHS until the destination address pair can be identified, whereupon it is returned to the source host/router in an NHRP REPLY message as shown, exactly analogous to the ATMARP messaging procedures described earlier. The source host now knows the <IP address : ATM address> translation for the destination host or router, and may thus set up the ATM VCC (or VPC) using normal ATM signaling protocols (or management procedures for the case of PVCs). It is important to note in this case that the ATM VCC can be an end-to-end VCC across multiple LIS domains, thereby eliminating the need for processing of the IP packets by routers located between the LIS domains, as was required in the basic CIPOA model as mentioned earlier. This process, often termed ATM VCC "bypass" or "cut-through," significantly enhances the end-to-end performance of IP packet transfer by enabling the full benefits of ATM QoS and speed without incurring the processing delay caused by intermediate routers between the LIS domains. Thus, the NHRP network architecture is intended to overcome both the scalability and end-to-end performance limitations of the initial CIPOA approach by simply extending the ARP concept across multiple LIS domains.

In Fig. 12-27, the partitioning between the ATMARP server and the next hop server may be viewed as functional, so that, in practice, the NHS function may be physically colocated with the ATMARP server. Administratively, the NHS function may in fact be partitioned on a per-LIS basis, whereby each LIS domain is viewed as responsible for the maintenance and validity of its NHS function. It is important to recognize that the NHRP is not an extension of a routing protocol, which may be used essentially unchanged, but simply provides for intersubnet IP-to-ATM address resolution capability. As for the case of the initial CIPOA model, for which the NHRP concept may be viewed as a logical extension, the NHRP approach also has advantages and disadvantages, which have been the subject of extensive debate. Its main advantages are that it embodies a relatively simple conceptual extension of the ATMARP server concept as well as implementation to enable direct intersubnet (or LIS) connectivity, thereby enabling the benefits of end-to-end ATM VCCs or VPCs. It may also be pointed out that the NHRP model may be extended to apply for the case of any underlying (Layer 2) protocol (e.g., such as IP over frame relay bearer service), and not just to the case of IP over ATM, as described here. On the other hand, the NHRP approach also possesses some of the disadvantages noted for the basic CIPOA architecture, in that the NHS may be viewed as a single point of failure. In addition, it has also been pointed out that the detailed coordination of the many NHS functions required to build large networks may be subject to some administrative uncertainty, involving negotiations between multiple LIS domains for the regular updating and maintenance of the NHS databases. Despite these general caveats, it should be noted that the NHRP approach does remove some of the limitations of the initial per-LIS-based CIPOA model and has been widely accepted, notably by IETF and ATM Forum specifications relating to IP and ATM internetworking, as described later for the so-called multiprotocol over ATM (MPOA) specification developed by the ATM Forum.

It will be recognized that the initial CIPOA network model as well as its enhancement using the NHRP extensions were not intended to allow for multicast capabilities between the IP hosts or routers attached to the ATM network. Since the capability to support multicast is viewed as important for some network services (e.g., video conferencing or distance learning applications), the IETF developed the concept of the "multicast address resolution server" (MARS) as a further enhancement to the basic CIPOA network architecture.

As shown in Fig. 12-28, the MARS model, together with its associated message set, can be

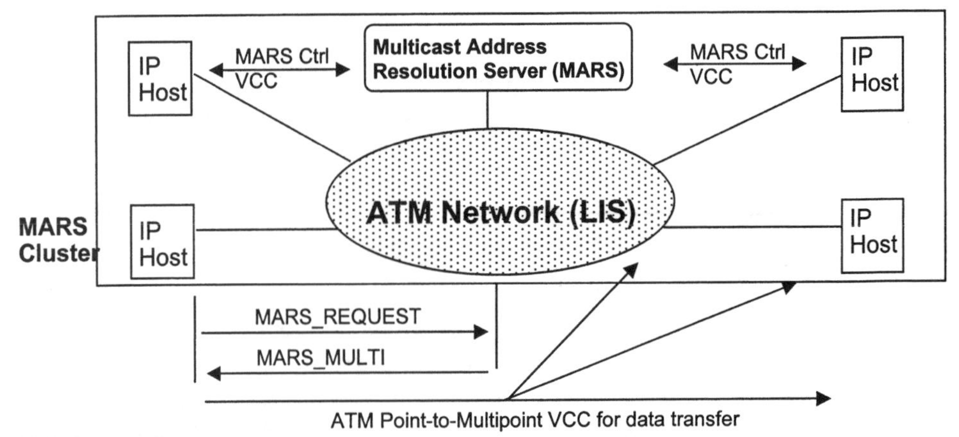

MARS Message Set

1. MARS_REQUEST: used by hosts to request destination ATM addresses of IP multicast group
2. MARS_MULTICAST: used by server to provide hosts with ATM addresses of IP multicast group
3. MARS_JOIN: used by hosts to join MARS multicast group (called a MARS Cluster)
4. MARS_LEAVE: used by host to leave MARS cluster

MARS Control:

MARS establishes point-to-multipoint ATM VCC to update group addresses

Figure 12-28. Multicast in IP over ATM.

viewed basically as an extension to the ATMARP server model to provide the <IP group address : ATM group address> pairs to a given host or router to enable it to set up the point-to-multipoint ATM VCCs to the destination hosts/routers. The MARS concept may also be used for the purposes of mapping (Class D) IP multicast addresses to ATM addresses if required. The concept may also be extended to be used in conjunction with an ATM multicast server function if this capability is provided within the underlying ATM network. In general, the MARS-related procedures only define unidirectional point-to-multipoint connections, since the more general case of multipoint-to-multipoint connectivity, though feasible in principle, will clearly be more difficult to implement. Note that as for the case of the NHS, the MARS function is shown as a separate entity, but in practice may be physically collocated with the ATMARP server.

The MARS message set with associated procedures are summarized in Fig. 12-28. It should be noted there that an ATM VCC termed the "MARS control VCC" needs to be established between each attached host/router and the MARS entity (either as an SVC or a PVC). The "MARS control VCC" is initially used for address registration purposes, as for the case of the ATMARP server described earlier, using the InATMARP_REQUEST and InATMARP_REPLY messages. However, for the MARS case, the IP and ATM group address pairs are registered for subsequent use by the hosts or routers attached to the LIS. To do this, the host sends a "MARS_JOIN" message to the MARS entity to register itself with the MARS multicast group, also called a "MARS cluster." The host may leave the MARS cluster by sending a "MARS_LEAVE" message to the MARS entity at any stage if it chooses to do so. When the host wishes to transfer data to a given multicast group, it sends a "MARS_REQUEST" message to the MARS entity over the MARS control VCC, to request the destination ATM addresses corresponding to the IP multicast group. The server responds by sending a "MARS_MULTICAST" message back to the host, which includes the destination ATM addresses of the desired IP multicast group. The host may then establish an ATM point-to-multipoint VCC to the destination hosts, as shown in Fig. 12-28, using normal ATM signaling protocols as described earlier. The multicast data transfer may commence once the point-to-multipoint ATM VCC has been set up.

It will be evident that the MARS messaging procedures outlined above simply extends the basic ATMARP paradigm to provide multicast capabilities for IP over ATM. It is therefore not surprising that some of the same advantages and disadvantages inherent in the basic CIPOA (client–server) architecture may also be identified for the MARS enhancement to basic CIPOA network architecture. Thus, the MARS approach may be subject to the same overall scalability and performance (i.e., latency due to the underlying ARP messaging delay) considerations described earlier for the initial CIPOA as well as NHRP enhancements. As before, robustness may also be an issue since the MARS entity may be viewed as a single "point of failure" for the network, unless more complex redundancy or back-up capabilities are provided. It may be noted that MARS maintains the separation between the ATM and IP-related multicast mechanisms, which may be viewed as desirable by some, but inelegant by others seeking a more "integrated" solution to IP/ATM internetworking. Although it is clear that the debate regarding the relative merits and demerits of the MARS approach will continue, it should be recognized that provision of multicast capabilities remains a challenge in general for both IP and ATM technologies. Despite this, there has been general acceptance of the MARS approach by the ATM Forum as part of the MPOA specification.

12.9 QUALITY OF SERVICE IN IP NETWORKS

In discussing ATM and IP internetworking so far, it will be recognized that the primary emphasis was placed on the question of address interworking (or translation) in order to establish end-to-end connectivity for the transfer of the IP packets. The IP-to-ATM address translation function constitutes part of the overall "control plane" interworking, necessitated by the fundamentally different nature of (connectionless) IP routing and (connection oriented) ATM addressing paradigms. However, even if connectivity may be established using the ARP-based procedures, coupled with conventional ATM signaling procedures, the question of interworking between the different "Quality of Service" (QoS) requirements inherent in IP and ATM networking remains. It will be recalled that, in essence, conventional IP internetworking was not primarily intended for providing QoS, in the sense of achieving bounded (or quantifiable) delay and packet loss parameters. Such a QoS is often characterized as a best efforts (BE) service, in the sense that the statistical packet delay and loss performance parameters are unspecified, and may largely depend on the overall traffic load and dimensioning of the IP network resources such as buffer sizing and available link bandwidths. For many IP-based applications such as e-mail, the best efforts service is perfectly adequate, as long as any packet loss can be recovered by TCP-based retransmission procedures.

However, with growing interest in utilizing IP-based networks for the carriage of network applications requiring "real time" (i.e., low, bounded delay) performance, such as voice or video signals, substantial effort has gone towards enhancing IP functionality to provide for support of QoS in IP. As may be evident from the earlier descriptions, ATM is inherently capable of providing multiple QoS classes as a result of well-developed traffic management and network engineering functions as well as the associated signaling procedures for resource allocation in network elements. In contrast, conventional IP-based networks were not designed for the support of delay-sensitive traffic such as voice or video signals and the traffic engineering was based primarily on the concept of "overprovisioning" of network resources (e.g., bandwidths and buffer capacities) to meet the requirements of the best efforts packet delivery. With the recent growth in internet traffic, it is clear that the overprovisioning of bandwidth alone is likely to prove a somewhat inefficient (and possibly costly) solution to the problem of QoS support. It is evident that more systematic solutions to the problem of providing QoS in IP-based internets are required if delay sensitive performance requirements are to be met.

To develop a systematic framework for the provision of QoS capability in IP-based networks, the IETF initially adopted the so-called "integrated services" (IntServ) architecture, together with its associated protocol termed "resource reservation protocol" (RSVP) to provide a form of "signaling" for the negotiation of network resources. More recently, an alternative framework, based on assigning IP packet relative priority, has also been developed by the IETF and is generally termed "differentiated services" (DiffServ). To more clearly understand the mechanisms for QoS interworking between IP and ATM both the IntServ and DiffServ procedures are briefly summarized below.

12.10 INTEGRATED SERVICES ARCHITECTURE (INTSERV) AND RSVP

The main components of the IP integrated services framework architecture are shown in Fig. 12-29 part a as they relate to both a host and an IP router. Since the primary purpose of IntServ is to enhance IP-based networks to support defined QoS capabilities, the components necessary broadly include:

1. A set of new "service classes" to provide the defined QoS capabilities for the IP traffic streams. Initially, four distinct IntServ QoS classes were defined, as will be described in more detail below. However, two of these classes are essentially in more common use in practice, as they appear to encompass the majority of envisaged applications.

a) Integrated Services Functions

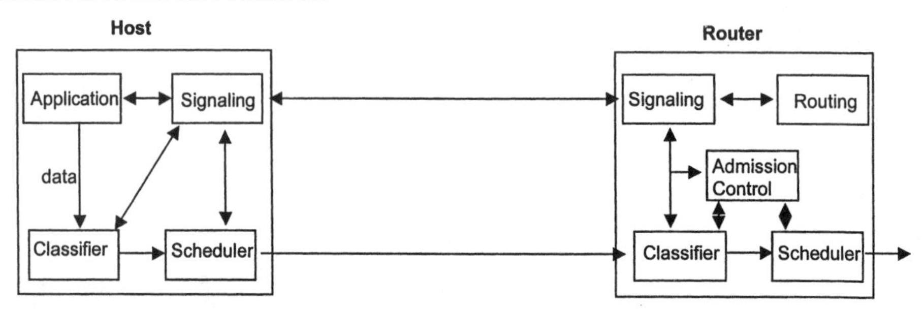

1. Classifier - Associates IP packets with a specified "service class"
2. Scheduler - Queue management and packet scheduling according to the service class
3. Signaling - Protocols to enable resource reservation in hosts / routers along the path (e.g. RSVP)
4. Admission Control - Determines if the QOS request can be met and allocates resources accordingly

b) Resource Reservation Protocol (RSVP)

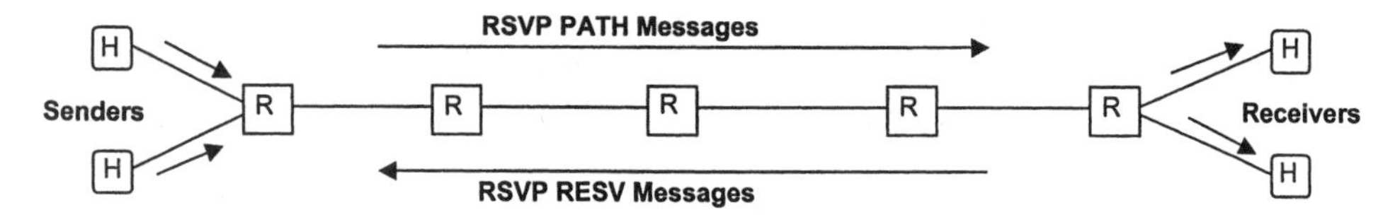

Figure 12-29. Integrated services architecture and RSVP.

2. "Signaling" messages and the associated procedures to enable the hosts and routers to request and negotiate the necessary network resources to support the required QoS. Although a number of different "signaling" approaches were considered initially, the most commonly used protocol is termed the "resource reservation protocol" (RSVP), and is described in more detail below.
3. IP traffic management mechanisms and the associated "traffic parameters" to enable the IP hosts and routers to maintain (and protect) the QoS classes negotiated by means of signaling across the IP network. These IP traffic parameters are essentially analogous to the ATM traffic descriptors defined earlier.
4. The concept of "admission control" to enable the routers to determine whether the requested QoS can be provided and maintained for the IP packet flow. This concept is analogous to the ATM "Connection Admission Control" (CAC) function described earlier, although it may be recalled that the concept of a "connection" as such does not apply in the case of IP packet delivery.

In order to support the main concepts of IntServ outlined above, the four main functional elements required in the host or router are identified in Fig. 12-29 part a. These are:

1. Classifier. The classifier function associates each IP packet with one of the specified IntServ "service classes" (or QoS classes).
2. Scheduler. The scheduler functional block is responsible for scheduling of the IP packets and the associated queue management within the buffers and links of the IP routers and hosts.
3. Signaling. The signaling-related functional block is responsible for the processing of the signaling messages (protocols) used to enable resource reservation in the hosts and routers along the path. Typically, this constitutes the handling of the RSVP functions for IntServ.
4. Admission Control. As noted earlier, the admission control function is responsible for ensuring that sufficient resources (buffer capacity and link bandwidth) are available in any given IP router to enable the requested QoS to be maintained, by allocating capacity within each router along the path for congestion avoidance.

Despite the use of different terminology and the fact that the IntServ functions and concepts relate to the connectionless (IP) transfer mode, the close parallels between the IntServ approach to QoS capability and that defined for the ATM case should be evident. Apart from the obvious close analogies between the IntServ admission control function and the ATM CAC function, as well as the similarity of RSVP signaling procedures to conventional ATM signaling protocols, the use of IP traffic "parameters" and traffic control mechanisms such as the scheduler are similar to the ATM traffic control mechanisms described earlier. Moreover, although the definitions and procedures underlying the IntServ- and RSVP-based approach are clearly not the same as those underlying ATM traffic control and signaling, the broad similarities evidently help in achieving QoS interworking between IP and ATM networks. Although a number of approximate mappings between the IntServ QoS classes and the ATM QoS classes have been proposed, it is of interest to note that for the case of the IntServ QoS classes, no quantitative parameter values have so far been generally adopted. This is in contrast to the case of ATM QoS classes, for which objective targets have been specified for both end-to-end network performance as well as NE performance, as described earlier. However, as IP-based services requiring QoS become better defined and more widely used, it is evident that quantitative QoS parameter values will need to be defined in the interests of interoperability.

Prior to listing the IntServ classes, it is of interest to summarize the main aspects of the RSVP signaling mechanisms, since these are essentially central to the support of the IntServ QoS model and comprise a key element in QoS interworking between IP and ATM. However, it must be pointed out that RSVP in itself constitutes a complex and vast area of study that lies outside the scope of this text. Here we focus on the salient features necessary for the understanding of IP-to-ATM internetworking, but the interested reader may refer to the extensive RSVP-related material developed by the IETF (e.g., RFC 2205 [12.10]).

The resource reservation protocol (RSVP) is essentially a "signaling" protocol developed for both IP v.4 and IP v.6, and it is important to recognize that it is not a routing protocol. In effect, the RSVP signaling messages defined below use the path that is assumed to be already computed by any conventional IP routing protocol (e.g., RIP, OSPF, BGP, etc.). The RSVP message set is defined to be encapsulated in and carried as IP packets in much the same way as ICMP or routing messages, or they may be transported over UDP if required. A primary feature of the RSVP is that it is based on a so-called "soft state" methodology. The soft state inherent in RSVP means that the reservation state (e.g., for bandwidth or buffer) in any given node (i.e., RSVP router) lapses unless it is periodically refreshed or updated. Thus, any given reservation initiated by the RSVP signaling will "time-out" unless periodically updated by refresh messages. Although the soft state aspect of RSVP is intended to allow for dynamic changes of QoS as well as rerouting inherent in connectionless IP networks, it may be noted that the need for periodic refresh of state adds complexity and overhead that may adversely affect scalability to large networks. On the other hand, it has also been argued that RSVP is not intended as a signaling protocol for connection oriented networks such as PNNI (or BISUP) for ATM, so that the soft sate capability is an underlying requirement for the case of connectionless networks such as IP-based internets. This aspect of RSVP will be further considered subsequently in discussing other IP-related QoS control protocols.

In addition to its soft state orientation, a key feature of RSVP is that it is "receiver oriented," which means that the actual (resource) reservation request is initiated by the receiver, based on the original request for a given QoS from the source (or sender). The protocol is also "simplex" in the sense that the resource in each node traversed is allocated for unidirectional data transfer from a sender to a receiver.

As shown in Fig. 12-29 part b, there are two key RSVP messages that are of interest here for its basic operation. These are:

1. The RSVP PATH Message. The PATH message is the basic set-up (or initiation) message sent from the sender to the receiver and must follow the same path as the subsequent data packets. The PATH message includes the sender traffic specification (TSpec) and the session identifier (i.e., the source and destination addresses and port numbers). As an option, the PATH message may also include other information, such as the "advertisement specification" (AdSpec), whose primary function is to enable the receiver to access resource availability along the path, as described further below.
2. The RSVP RESV Message. The Reserve (RESV) message is sent from the receiver to the sender to reserve resources (or update a reservation) in each RSVP router along the path, hop-by-hop. Note that the RESV message must follow the same path as the PATH message in the reverse direction. The RESV message includes path state information (to enable it to follow the same path) as well as QoS and resource reservation information in the form of the so-called:

   a. Flow specification (FlowSpec), specifying the QoS class requested and the resource parameters derived from the traffic specification (T Spec) in the PATH message.
   b. Filter specification (FilterSpec), which specifies the parameters required to update the classifier function.

RSVP also specifies other messages for path tear-down, confirmation of reservations, and indication of error states. However, these messages, together with details of associated procedures and message formats, need not be considered here in order to understand the basic reservation mechanisms. However, it is of interest to summarize some underlying concepts related to IP packet flow as used by RSVP prior to describing the reservation procedures related to the PATH and RESV messages. These concepts include:

1. IP Flow. RSVP introduces the concept of an "IP packet flow"—a stream of unicast or multicast IP packets from a sender having the *same* QoS requirements.
2. IP Session. In the context of RSVP, the notion of an "IP session" is defined as a stream of unicast or multicast IP packets that have the *same* IP destination address and port number. It may be recognized that an IP session may include one or more IP flows.
3. Traffic Specification (TSpec). In RSVP, the TSpec is a description of the IP traffic characteristics from the source. The TSpec is included in the PATH message from each sender host, and may be modified by the receiver in the FlowSpec, as defined below. It is important to note that the TSpec is generally defined in terms of "token bucket" (or "leaky bucket") parameters, analogous to the case of the ATM traffic descriptors when related to the GCRA model described earlier.
4. Flow Specification (FlowSpec). The FlowSpec refers to the information in the RESV message indicating the specific service class requested by the receiver, together with the allowable traffic parameters as derived by modifying the TSpec from the sender. This procedure enables the receiver to essentially determine the reservation request based on the source request, as noted above. The FlowSpec also includes other QoS-related reservation information (termed RSpec) that may be used by the routers. The FlowSpec information is used to update the RSVP scheduler function.
5. Filter Specification (FilterSpec). The FilterSpec is the information in the RESV message that indicates the set of IP packets that receive the specified QoS, namely the IP packet flow. The FilterSpec also includes the IP session identifiers (i.e., the destination address plus port number), as well as the "reservation style" as described below. The FilterSpec information is used to update the classifier function, as described earlier.
6. Reservation Styles. To allow for different levels of statistical resource sharing of the available or requested resources, RSVP introduces the concept of "reservation styles." In essence, a reservation style is a means for specifying a desired level of statistical multiplexing for the IP flows to a receiver. Three reservation styles are defined in RSVP at present. These are termed:

   a. Fixed Filter. The "Fixed Filter" style allows the receiver to make a *separate* and distinct resource reservation per sender. The requested QoS in the FlowSpec is then applied to the IP packets from the sender.
   b. Shared Explicit. In the "shared explicit" reservation style, the QoS requested in the FlowSepc is *shared* by the packets generated by all the named senders in the session. This implies sharing of the bandwidth resources, similar to case of statistical multiplexing in connection oriented networks such as ATM.
   c. Wildcard Filter. The "wildcard filter" reservation style indicates all the senders to the IP session without explicitly naming each one, thereby allowing for additional flexibility in the overall sender traffic. In this case, the QoS requested by the FlowSpec is also (statistically) *shared* by all IP packets generated by all the sources in the IP session.

In general, RSVP does not attempt to specify the level of resource sharing, viewing this aspect as largely a matter of network engineering policy and dimensioning. The notion of reservation styles is intended to allow substantial flexibility in the way network resources are utilized for a range of applications requiring different QoS and traffic profiles. In broad terms, it may be recognized that the fixed filter sytle roughly corresponds to a "deterministic" resource allocation in the ATM traffic control sense. In the same sense, the shared explicit and wildcard filter styles roughly correspond to statistical bit rate approaches. However, it must be admitted that such comparisons are somewhat approximate at best, and the differences between the RSVP reservation styles and ATM transfer capabilities will make direct interworking more complex.

Putting aside these interworking considerations for the moment, the basic RSVP procedures may now be summarized with reference to the RSVP router functional model shown in Fig. 12-30. To initiate the RSVP session, each sender host in an IP session sends a PATH message to the unicast (or multicast) destination address (and port number). The PATH message includes information relating to the session identifier and source IP address, as well as the TSpec specifying requested source traffic parameters. As an option, the PATH message may also include the AdSpec information, whose intent is to convey to the receiver the available resources in each RSVP node along the path. Since the receiver has no other (in-band) means to "know" the available resource state along the path, the AdSpec information inserted by each node along the path provides a mechanism by which the receiver may assess whether a reservation request may be met by the network. This concept is termed "one path with advertising" (OPWA) in the RSVP literature. To enable this mechanism, it is assumed the AdSpec is updated by each router along the path. The RSVP processor function in each router processes the PATH messages to determine the per-sender state for each IP session. Each node also maintains state information for the address of the previous hop (P-HOP) from the PATH message, to enable the RESV message from downstream nodes to follow the same path traversed by the PATH messages from the upstream nodes.

The receiver (host) processes the PATH messages to determine the required QoS for any given IP flow in the session. Depending on its available resources, the receiver may modify the TSpec and QoS requests and send a RESV message along the reverse path, hop-by-hop, using the "previous hop" address stored in each node from the PATH message to follow the same path in reverse. In addition to the session and next hop information, the RESV message includes the

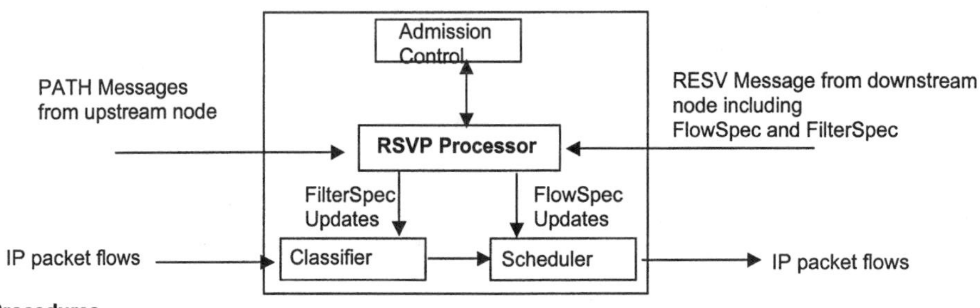

Procedures

1. Each sender in a session sends PATH message to unicast/multicast destination address.
2. PATH message includes: Session ID and source IP address; TSpec indicating traffic parameters; AdSpec (optional) which enables receiver to determine resource availability along data path. The AdSpec is written/updated by routers along path based on concept of "One Path With Advertising" (OPWA).
3. Each router processes PATH message to determine per sender state for each session. Receiver determines required QOS and available resources to update TSpec.
4. Receiver sends RESV message along reverse path hop-by-hop. RESV message includes session and next hop information; FlowSpec which includes service class and modified TSpec; FilterSpec which includes QOS required and reservation style information.
5. Each router performs Admission Control procedures on RESV message to determine if resources are available to meet requested QOS and allocates bandwidth/buffers accordingly (towards receiver). If not router may reject message and generate reject message to receiver.
6. Node updates Classifier with FilterSpec parameters.
7. Node updates Scheduler with FlowSpec parameters.
8. Reservation state must be updated periodically (Soft State) or resource will be released.

Figure 12-30. RSVP functional model and procedures.

FlowSpec containing the (modified) TSpec and service class required. The RESV message also carries the FilterSpec, which contains the QoS required and the reservation style information as described above.

Each RSVP router along the path performs admission control (as well as so-called "policy control," which may determine whether a user is entitled to reserve resources, for example), based on the information contained in the RESV message. This enables each node to determine if sufficient resources (e.g., bandwidth/buffers) are available to meet the requested QoS, and to allocate resources accordingly. Each node reserves resources in the direction toward the receiver. If the admission control (or policy control) function determines that insufficient resources are available, it will reject the reservation request and may generate a reject message toward the receiver. If the reservation request is accepted, the node updates the classifier function with the FilterSpec parameters and the scheduler function with the FlowSpec parameters, as indicated in Fig. 12-30. In keeping with the "soft state" nature of RSVP, the reservation state in each RSVP node along the path must be periodically updated by use of the PATH and RESV messages or the reservation will "time out" and be released. The sender or receiver may also initiate release of a given reservation by the use of dedicated "tear down" messages to facilitate more dynamic control over the network resources. The use of these and other RSVP procedures is outside the scope of this brief outline and may be found in RFC 2205 [12.10] and related RSVP literature.

It was mentioned above that the TSpec information in RSVP was based on a "token bucket" representation of IP traffic characteristics. As modeled in Fig. 12-31, this representation is equivalent to the familiar "leaky bucket" model widely used in ATM traffic management for conformance definitions based on the generic cell rate algorithm (GCRA) test. As depicted in Fig. 12-31, TSpec defines the IP traffic parameters in terms of a leaky bucket model in which b = "bucket size" (in bytes) and r = bucket fill rate (in byte/sec), and the token rate corresponds to the rate at which the bucket is allowed to drain. In essence, the algorithm (as implemented in the classifier and scheduler functions) requires that the IP packets are transmitted if the number of tokens is greater than the packet size. If not, the packets may be discarded or marked (i.e., tagged) for potential discard. The traffic parameters specified in TSpec from the integrated services QoS framework include:

p = peak rate, where $p > r$

m = minimum policed unit, where packets of size $< m$ are counted as packets of size $= m$ bytes

M = maximum packet size; the network rejects the flow if the packet size is $> M$ bytes

R = a rate allocated to a reservation. This may be a link rate, for example

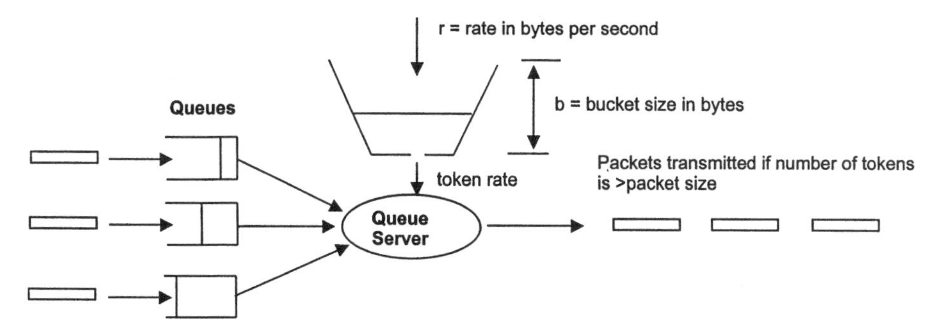

Figure 12-31. Traffic specification in RSVP.

The TSpec parameters above are used in accordance with the guidelines summarized in Fig. 12-31. These rules stipulate that the IP traffic sources (or routers) should not generate traffic at rates in excess of $M + pT$ byte/sec over any interval of time T (sec). The policer algorithm (e.g., in the scheduler function) enforces the data rate to be less than $rT + b$ to maintain the QoS requirements; the delay is bounded by the ratio b/R as long as $R > r$.

The traffic parameters and the policer algorithms needed to achieve the required QoS were developed initially as part of the IETF's "integrated services" (IntServ) framework and it should be noted that RSVP essentially provides a messaging mechanism to transfer this information between the nodes. As summarized in Table 12-4, IntServ initially identified four "service classes" for envisaged IP- based network applications requiring distinct types of QoS capability. These IntServ service classes are:

1. Guaranteed Delay. This class was intended for "real-time" applications requiring strict delay sensitivity, and hence provides for a bounded maximum delay. The traffic parameters and enforcement criteria to achieve this QoS requirement are listed in Table 12-4 and carried by RSVP contained in the TSpec and RSpec information objects, as shown.
2. Controlled Delay. The controlled delay service class was intended for applications that were adaptable to different levels of delay. To accommodate these requirements, three levels of delay control were envisaged that could be indicated in the RSpec carried by the RESV messages. The controlled delay IntServ class is not generally used.

TABLE 12-4. Integrated service classes

Service class	Main characteristics	Main applications	Network element processing	Required parameters	Enforcement criteria
Guaranteed delay	Bounded maximum delay	Real-time application with strict delay sensitivity	Resource allocation with token bucket. R = dedicated rate. B = buffer size. C, D = deviation parameters	TSpec (r, b, M, p) RSpec (R, S)	$M + \min <pT, rT + b - M>$
Controlled delay	Three levels of delay control	Applications adaptable to different service delay levels	Admission control accepts flows within delay levels. Needs three measurable delay levels	TSpec (r, b, m, M) RSpec (service level)	$rT + b$
Predictive service	Three levels of delay control with bounded delay	Applications can accept some lost or late packets	Admission control accepts flows within delay levels. Needs three measurable delay levels plus delay bound	TSpec (r, b, m, M) RSpec (service level)	$rT + b$
Controlled load	Emulates best-effort over noncongested network	Applications requiring bounded loss due to network congestion	Admission control	TSpec (r, b, m, M)	$rT + b$

3. Predictive Service. The predictive service IntServ class is similar to the controlled delay class outlined above in that three levels of delay control are envisaged but, in this case, delay bounds are added in order to obtain a more "quantitative" or predictive level of QoS. However, as for the case of the controlled delay IntServ class, the predictive service class has not found widespread use in RSVP deployments.
4. Controlled Load. The controlled load IntServ class is intended for IP-based network applications requiring bounded packet loss due to network congestion. In effect, the controlled load QoS class may be seen as emulating a "best efforts" packet transfer over a noncongested network. In the IntServ model, this may be achieved by using admission control and enforcing a data rate less than $rT + b$, as described earlier based on the TSpec parameters. As might be anticipated, the controlled load IntServ class (CLS) is widely used in typical RSVP-based network applications.

In summary, although the initial IntServ QoS framework broadly identified four distinct IntServ QoS classes in an attempt to cater for all foreseeable IP-based network applications, primarily only two of these are in common use in typical (RSVP-based) IntServ oriented networks. These are the controlled load service (CLS) and the guaranteed delay (GD) IntServ classes. In order to understand more clearly how these IntServ-based QoS classes may be interworked with the analogous ATM-based QoS classes, it is useful to summarize the main attributes of the IntServ classes as follows:

1. Controlled Load Service (CLS). As noted earlier, the CLS effectively emulates the QoS an application (or IP-based service) would experience under "best effort" transport in an *uncongested* network, by ensuing that each CLS IP flow is allocated sufficient bandwidth/buffer capacity to avoid overload. Although it is not assumed that packet loss and delay are guaranteed, a statistically bounded packet loss rate may be achieved by traffic control based on admission control and policing of flows at each node for conformance to TSpec (or FlowSpec) parameters, such as r, b, m, and M. Here "conformance" implies that the number of bytes over a period T sec is less than $rT + b$, and the maximum packet size is less than M. If packets are "nonconforming," they may be treated as normal "best efforts" datagrams or discarded. In CLS the applications may assume a low packet loss, limited by the bit error ratio (BER) of the transmission medium, as well as low delay, which may be of the order of the minimum transfer delay through the network. It may be noted that CLS does not specify quantitative targets for these packet loss and delay objectives, as do the ATM QoS classes defined in ITU-T Recommendation I.356 [5.9]. CLS does not preclude statistical multiplexing of the IP flows and assumes that the details of the resource allocation algorithms and dimensioning within nodes are specific to any given implementation.
2. Guaranteed Delay Service (GDS). The guaranteed delay service effectively emulates the QoS an application (or IP-based service) would experience assuming a "dedicated" virtual circuit with a bandwidth of R byte/sec. The end-to-end transfer delay experienced by the IP packets in a session is bounded, although no limits are specified for packet delay variation. Based on a so-called "fluid model" for a token bucket representation of IP packet flows, the flow delay is bounded by the ratio b/R if $R > r$ as noted earlier. Moreover, to allow for processing variations in specific GDS implementations, the delay bounds may include additional "deviation" parameters (termed C, D, and S) if required. More importantly, the GD service assumes there is no statistical multiplexing of the IP flows, so that resources cannot be "shared" by the flows, as for the case of CL service.

From the perspective of QoS interworking between the IP IntServ classes outlined above and ATM, it is useful at this stage to recall the basic attributes of the ATM service categories as identified by the ATM Forum (which were described earlier in more detail). The ATM service categories (corresponding also to the ATM transfer capabilities or ATCs in the ITU-T based terminology) are summarized in Table 12-5 for convenience of comparison with the IntServ model. The traffic and QoS parameters requirements applicable to the six ATM service categories, CBR, rt-VBR, nrt-VBR, ABR, GFR, and UBR are shown, together with the potential intended network applications. Ignoring for the moment the detailed differences between the way the ATM service category parameters and IntServ parameters are defined, broad comparison between the two common IntServ classes and ATM service categories indicates that:

1. Best efforts (BE) IP service may be viewed as roughly corresponding to the ATM UBR service category, with no specified bounds on packet loss or delay. It may be noted, however, that UBR does require indication of the peak cell rate (PCR) and cell delay variation tolerance (CDVT) parameters to protect the QoS of other ATM VCCs within a transmission path.
2. The guaranteed delay (GD) IntServ class may be viewed as broadly analogous to the ATM CBR and rt-VBR service categories, where cell transfer delay (CTD) and cell delay variation (CDV) are bounded. As indicated earlier, GD service does not explicitly limit packet delay variation, but this may be generally constrained by strictly limiting the transfer delay.
3. The controlled load (CL) IntServ class may be viewed as analogous to the ATM nrt-VBR, ABR, GFR or even CBR service categories (it may be recalled that ATM does not in any way restrict applications to any given service categories, although some may clearly be more suitable for a specific application).

TABLE 12-5. ATM service categories

	ATM service category					
	CBR[a]	rt-VBR[b]	nrt-VBR[c]	ABR[d]	GFR[e]	UBR[f]
Traffic parameters						
PCR and CDVT	Yes	Yes	Yes	Yes	Yes	Yes
SCR and MBS	N/A	Yes	Yes	N/A	Yes	N/A
MCR	N/A	N/A	N/A	Yes	Yes	N/A
QoS parameters						
CDV	Yes	Yes	No	No	No	No
Maximum CTD	Yes	Yes	No	No	No	No
CLR	Yes	Yes	Yes	No	No	No

[a]Constant bit rate (CBR). Intended to support circuit emulation and delay-sensitive applications with strict bounds on delay, cell loss (CLR), and cell delay variation (CDV).

[b]Real-time variable bit rate (rt-VBR). Intended for bursty real-time applications (e.g., data, coded video). QoS specified by CLR, CDV, and CTD.

[c]Nonreal-time VBR (nrt-VBR). Intended for bursty nonreal-time applications. As in *b* but no bounds on CTD and CDV.

[d]Available bit rate (ABR). Intended for bursty (e.g., data) applications with no requirements for delay or CDV; can adapt to changing ATM transfer characteristics with flow control.

[e]Guaranteed frame rate (GFR). Intended for bursty (e.g., data) applications with no requirements for delay or CDV and high probability of frame delivery.

[f]Unspecified bit rate (UBR). Intended for bursty applications on a best-efforts basis, with no bounds on CLR, CTD, or CDV. An enhancement to this category called UBR+ includes a minimum cell rate (MCR).

For QoS interworking between IntServ-based IP and ATM networks, the broad correspondences listed above may be considered as a starting point or first-order approximation upon which more detailed parameter mappings may be defined. However, it must be admitted that in attempting to obtain more precise mappings between the parameters (either QoS or traffic related) in any given IntServ class and ATM service category, questions remain, and it may not be possible to derive unique or precise mapping. This ambiguity does not imply that QoS interworking between ATM and IntServ, as mediated by RSVP, is not possible, but that simple one-to-one mapping between IntServ and ATM QoS and traffic control parameters is difficult, primarily as a result of the underlying differences in approach taken in the two diverse technologies. The main differences and similarities between IntServ/RSVP and ATM attributes are summarized in Table 12-6

As noted earlier, a number of these differences in approach result from the need to preserve some of the "connectionless" attributes of RSVP, such as soft state of resource allocation. While this aspect in itself does not affect the mapping of QoS or traffic parameters in principle, the possibility that the RSVP QoS state may change arbitrarily during sessions (in extreme cases, with each refresh cycle) clearly introduces uncertainty in selecting suitable one-to-one mapping with the more typically "hard" state ATM VCCs. More problematic is the use of the "reservation styles" in RSVP to enable various levels of statistical multiplexing of the IP flows, which has no direct counterpart in ATM, which simply stipulates the use of an appropriate service category (or ATC) capable of supporting statistical multiplexing per VCC.

On the other hand, similarities between the IntServ and ATM QoS frameworks should also be noted, such as the adoption of similar parameters for specification of the traffic characteristics based on a token bucket representation. Here, for example, the mapping of PCR to the peak rate p, bucket size b to CDVT, and MBS to M may be made to obtain comparable traffic performance characteristics. In general terms, it may be recognized that the considerable flexibility inherent in the IntServ architecture, as well as the associated RSVP signaling, enables easier interworking with the more precisely defined ATM QoS and traffic parameters, specifically for the

TABLE 12-6. ATM versus IP QoS

Attributes	ATM	IP integrated services/RSVP
Connection type	Bidirectional point-to-point (duplex). Undirectional point-to-multipoint.	Many-to-many flows possible. Simplex. Reservation from sender to receiver.
Connection state	Hard. State maintained during connection.	Soft. State deleted if not refreshed within timeout period.
Resource reservation	Source-intiated. QoS may be different in either direction but typically constant per VCC (assuming no call renegotiation).	Receiver-initiated. Multicast flows may have different QoS and may be changed at any time.
Routing	Connection setup individually routed based on QoS requirements. Rerouting only by connection reestablishment.	Path established before flow starts. Soft state allows for reestablishment if route changes dynamically.
Reservation styles	Not defined	Three reservation styles to allow various levels of resource sharing in a session.
QoS classes	By traffic and QoS parameter mechanisms. ATM service categories are general. Provides for cell tagging.	By TSpec and RSpec parameters. Enforcement similar to ATM, but parameter mapping difficult.
Dynamic/static reservation	Static. Resource allocated by CAC for VCC lifetime. Deterministic or statistical bandwidth sharing possible by ATCs.	Dynamic. Resource reservation may change during session.

more common case where IP traffic is being carried over an ATM (backbone) network, as described earlier. In such cases, the ATM service category, QoS class, and traffic parameters may be selected by the network operator in accordance with a service level agreement (SLA). The parameters corresponding to the given IntServ class (CL or GD) may then be engineered for "best fit" to the underlying ATM-based transport to enable a desired end-to-end QoS level. Since IntServ does not specify precise quantitative bounds to the QoS parameters, the mappings need not be precise and may largely be dependent on the underlying (Layer 2) network performance.

It is important to recognize that the IntServ QoS framework together with the supporting RSVP mechanisms imply major changes to conventional "best efforts" IP-based service. Thus, even the brief outline above should serve to show that these enhancements will have significant impact on the design of hosts and routers in terms of significant added complexity. In particular, the flexibility incorporated in RSVP requires the need for substantial state information in the network, as well as the need for periodic refresh (update) of this state to maintain the soft state characteristics deemed useful in a connectionless network. These aspects, coupled with the enormous anticipated increases in IP traffic, have led to considerable discussion in the IETF and elsewhere as to the scalability and even suitability of RSVP for large networks. It is argued that if RSVP is used in such networks to support QoS-based IP services, the potentially enormous number of IP sessions together with associated IP flows would require unmanageable amounts of state information in the RSVP-enabled nodes, as well as a significant amount of overhead required for the periodic refresh of this state information. The overall robustness of such an RSVP-based network has also been questioned, particularly since the protocol in itself does not require reliable transfer of the PATH and RESV messages, relying instead on the refresh messages to maintain state in all nodes along a given path.

On the other hand, proponents of RSVP have argued that the protocol has operated successfully in numerous trials and network deployments, albeit for relatively small enterprise as well as backbone networks, in comparison with most large-scale public carrier networks. Consequently, it is argued that extension of RSVP-based services to larger or more ubiquitous use should not present undue problems. Despite these assurances, concern regarding the suitability of RSVP and its complexity for widespread deployment has led many workers in IETF to seek alternative mechanisms for the provision of QoS-based IP services. As a result, a significantly different QoS framework, termed "differentiated services" (DiffServ), has been developed by the IETF and is considered by many to be more suitable for widespread deployment, as described below. It is also of interest to note that, in parallel with the development of the DiffServ approach, the RSVP procedures have also been modified and enhanced in an attempt to improve its scalability and robustness for use in large networks. However, given its relative complexity in comparison with other more recent protocols for enhancing IP-based services, it remains questionable whether RSVP will find wider usage.

12.11 THE DIFFERENTIATED SERVICES (DIFFSERV) MODEL

The potential scalability and robustness limitations of RSVP-based networks has resulted in the development of an alternative framework for providing QoS differentiation in the IP layer, as mentioned above. This approach, termed differentiated services (DS or DiffServ) is essentially based on the provision of "relative priority" for IP packets in the DS-enabled nodes by "marking" each IP packet header with a (unique) codepoint that identifies the packet to receive a given priority level within the DiffServ domain. The DiffServ architecture and general requirements are described in RFC 2475 [12.11]. The model assumes that IP traffic is "classified" on entry to the network into one of a set of "DiffServ classes" or behavior aggregates identified by a single DS codepoint (DSCP) in the IP packet header. The ingress traffic may also be "conditioned" or modified in order to assign it to a given behavior aggregate if required. Within the DiffServ network or domain, the IP packets are then forwarded in accordance with the per-hop behavior associated with each DSCP. The DiffServ model introduces the notion of a per-hop behavior (PHB) within each DS node, which is essentially a description of the (externally) observable forwarding behavior the node applies to the set of IP packets with a given DSCP value, in terms of different levels of relative priority for packet loss or delay QoS.

The concept of the DS-associated PHBs or, more generally, the PHB groups, is central to the DiffServ architecture and hence needs to be clearly understood. In essence, the PHB (or PHB group) is simply the characterization of the way in which the node allocates resources (i.e., buffers and/or bandwidth) to any given DS traffic aggregate to distinguish it from other aggregates in the network. The PHB includes the related buffer management and packet scheduling mechanisms implemented within the network element to support the different priority levels for both packet delay and drop (i.e., discard) behavior. An example of a PHB is to specify that a given DS node allocates a certain fraction (e.g., 20%) of the total (transmission path) bandwidth to an aggregate of IP packets identified by a specific DS codepoint (DSCP). Thus, the PHB provides a convenient way of describing how the DS node handles IP traffic aggregates belonging to different QoS classes, without necessitating detailed specification of the mechanisms for packet scheduling, queue processing, and bandwidth allocation, etc., which may vary depending on implementation. Although, in principle, a number of PHBs may be described based on the general guidelines given in RFC 2475, initial attention has focused on two broad PHB groups—(1) expedited forwarding (EF) PHB and (2) assured forwarding (AF) PHB—which are described further below.

In addition to introducing the notion of PHBs to describe the way in which a DS node processes IP packets to "differentiate" its QoS (i.e., its relative packet delay or loss probability performance) from others, the DiffServ model also provides a schema for "marking" IP packets for different QoS levels. For the case of IPv4, DiffServ essentially redefines the original type of service (TOS) field in the IPv4 header, described earlier in relation to Figs. 12-12 and 12-13, which the DiffServ model renames as the "DS field." For IPv6, DiffServ utilizes codepoints in the "traffic class" octet (also called the priority field) in a similar way. Only the commonly used IPv4 case need be considered here. It may be recalled that originally the one-octet TOS field was intended to indicate the requested QoS in respect to relative priority of the packet delay, precedence, throughput, and reliability (see Fig. 12-13). In principle, the original TOS functions may be viewed as roughly similar to DiffServ in intent, although less well defined for the support of IP-based services. Conventionally, the main use of the TOS function has been in utilizing the precedence field comprising the three bits numbered 0 to 2 in the TOS field to support priority handling of Internet "control" traffic such as routing packets, which may obviously need to be forwarded preferentially to ensure timely routing-table updates. Specifically, the precedence field codepoints "110" and "111" are widely used for preferentially forwarding routing and other control information in existing IP-based internets. Consequently, in redefining the TOS field to constitute a new "DS field" for more general support of IP QoS, it was felt desirable to subsume these uses of precedence within the new DS field coding structure.

However, it should be stressed that the structure and coding of the DS field as specified in RFC 2474 [12.12] is generally incompatible with the original definition of the IPv4 TOS field (as specified in RFC 791 [12.3] and further developed in RFC 1122 [12.13] and RFC 1812 [12.14]). Nevertheless, an attempt to maintain a limited level of backward compatibility with current usage of the precedence codepoints as well as the ubiquitous "best efforts" service was made in DiffServ, essentially by "reserving" currently used codepoints in the overall DSCP field. These reserved codepoint values are generally referred to as "class selector codepoints" in the DiffServ coding structure, as outlined below. The one-octet DS field is structured as shown in Fig. 12-32 using the conventional IP field bit numbering scheme.

The six-bit differentiated services code point (DSCP) field value is used to select the PHB for any given IP packet in the behavior aggregate as noted above. Bits 6 and 7 are currently unused (CU) and are reserved for future enhancements. Note that these bits were also reserved in the TOS field structure as well. The specific DSCP value "000 000" is reserved to indicate conventional "best efforts" IP traffic and therefore maps into a "default = PHB" that provides this forwarding behavior. IP packets containing an unrecognized DSCP value may also be mapped into this default (i.e., best efforts) PHB. It may also be noted that the default all zeros coding also enables backward compatibility to be maintained with the existing TOS field coding for normal IP traffic.

In principle, the six-bit DSCP field may provide up to 64 independent values capable of being mapped to any given set of PHBs supported within a DS network. However, in order to maintain a minimum level of backwards compatibility with some existing uses of the TOS precedence field for distinguishing packets carrying routing and other network control information, a subset of the DSCP values is reserved for consistency with this usage. This set of DSCP values are known as the "class selector codepoints" and are generically designated by the values "xxx 000," where x may take the values 0 or 1. In particular, the values xxx = 110 and xxx = 111 are reserved for use by the network control traffic such as routing packets, thereby maintaining a limited backward compatibility with the TOS precedence codepoints typically used for the same purpose. The class selector codepoints are intended to be mapped into the associated class selector PHBs when these are implemented within a router for the preferential forwarding of packets carrying routing information. It is of interest to note that other values of the precedence field are not similarly constrained, since these are generally considered as having limited usage in existing networks.

In addition to the specification of the class selector codepoints xxx 000 to provide limited compatibility with commonly used precedence functions, the 64 DSCP values are also grouped into three DiffServ pools to simplify the management and assignment of such a large codespace. A pool of 32 codepoints, which includes the class selector codepoints as well as the all zeros default value for best efforts service, is designated for general assignment as standardized codepoints. A second pool of 16 codepoints is set aside for so-called "experimental or local use" to facilitate proprietary designations within local (e.g., private) network usage. Finally, a third pool of 16 codepoint values is available initially for experimental/proprietary assignment, but may be eventually transformed to standardized values if required, in the event that Pool 1 is exhausted, for example. The codepoint space representation of the three pools can be summarized as shown in Table 12-7 (where x = 0 or 1):

As noted earlier, the all zeros "default" value and the class selector subset xxx 000 fall within Pool 1 since these are intended for standardized allocation for backward compatibility. It should

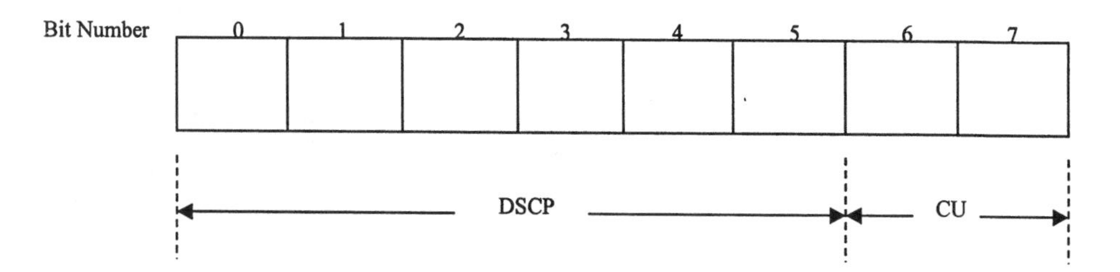

Figure 12-32. Differentiated services (DiffServ) coding structure.

TABLE 12-7. Differentiated services assignment pools

Pool	DS codepoint space	Assignment policy
1	xxx xx0	For standardized designation
2	xxx x11	For experimental or local use (EXP/LU)
3	xxx x01	For initial experimental or local use (EXP/LU) (but may be used for future standard allocation)

also be recognized that the DiffServ model allows for the association of more than one DSCP value to any given PHB (so-called N to 1 mapping), and also permits the modification of DSCP values within the DS nodes, assuming this is within the constraints of any given service agreement. Typically, the DSCP values may be changed at the edges (boundaries) of DS domains, within which different DSCP → PHB mappings may apply.

Assuming that a DSCP value has been selected from any one of the codepoint pools in accordance with the DS field guidelines summarized above and assigned to an aggregate of IP packets, each node along the path maps this value to a specific PHB capable of providing the required level of QoS (i.e., delay and/or loss probability) to the traffic aggregate. As noted earlier, two main groups of PHB have been specified to date and these are intended to be used for the support of currently envisaged IP-based services requiring improved QoS levels in comparison with typical best efforts service. It is of interest to summarize the main characteristics of these PHB groups with a view to exploring the interworking of DiffServ-based QoS capabilities with ATM networks. In this context, it is important to recognize that interworking between any given PHB and ATM service category (or ATC) is not appropriate, since ATM does not specify the notion of per-hop behavior in describing any given ATC or service category. However, an end-to-end IP-based service constructed on a given PHB may be compared, and possibly interworked, with an end-to-end ATM-based service using a given ATC and QoS class as defined earlier. Consequently, from the perspective of QoS interworking, it is important to bear in mind that the PHB is essentially one component in the construction of an IP service. Other important components may include the classification and conditioning (i.e., shaping and policing) of traffic aggregates at the edge of the DS domain, as well as the network resource allocation policies and QoS levels targeted by the network administrators in question.

1. Expedited Forwarding (EF) PHB. The EF PHB has been described in detail in RFC 2598 [12.15] and is intended to be used in the construction of IP services that require low delay, including delay variation or jitter, and low loss performance characteristics. Such a service would be that provided by a "virtual leased line" or dedicated circuit from an end-to-end perspective. This kind of IP-based service has also been referred to as a "premium service" in the sense that it implies very low packet delay, loss, and jitter and possibly a level of assured bandwidth as its main characteristics. In order to achieve this, the aggregate of packets must experience very low queuing delay within each DS node along the path. Considering that packet queues arise when the instantaneous traffic arrival rate exceeds the departure rate at the output link for the EF packet aggregate, the EF PHB specifies that the minimum departure rate should equal or exceed the maximum ingress rate of the aggregate within the DS node. The EF PHB assumes that other mechanisms such as traffic aggregate conditioning, including policing and shaping, are used to limit the maximum rate at which a given aggregate enters a node, so that the EF PHB simply configures the departure rate to equal (or exceed) this value. It will also be recognized that configuring the exiting packet aggregate rate to equal or exceed the maximum (i.e., peak) packet

aggregate rate entering the node is essentially equivalent to a deterministic bandwidth allocation to a traffic stream, somewhat analogous to a DBR ATC in the ATM case. Clearly, in such cases the packet queue lengths in any given node can be minimized and hence delay (as well as delay variation) maintained at low levels.

The EF PHB may also be implemented by means of various priority queuing disciplines, examples of which may be found in RFC 2598 together with comparisons of their relative delay performance. However, description of such implementation-related details is outside the scope of this summary. It may be noted that the EF PHB, however implemented, requires the enforcement (i.e., policing) of the aggregate packet rate at the ingress to the network and possibly within the network. Packets exceeding the maximum permissible aggregate rate are discarded (dropped) as a result of this traffic conditioning process. Shaping of the aggregate traffic streams at network edges may also be considered, but since shaping essentially involves buffering of packets, the resultant impact on delay and jitter needs to be factored into the end-to-end delay targets. The EF PHB uses the DSCP value "101 110" for the selection of this PHB within each node. Note that this DSCP value falls within Pool 1 as outlined above.

2. Assured Forwarding (AF) PHB Group. The AF PHB group is described in RFC 2597 [12.16] and comprises four independent forwarding classes (in terms of emission priority of packets), within each of which packets may be assigned to three levels of drop precedence (i.e., discard priority). The DSCP values, taken from Pool 1 of the DSCP code space, may be tabulated as shown in Table 12-8 for these twelve categories of relative delay and drop priority.

   It may be noted that the DSCP values to be associated with the AF classes are consistent with the general coding rules for the DS field outlined earlier and do not overlap with the class selector or other default values noted previously. It is also of interest to compare these values with the earlier use of TOS precedence field (see Fig. 12-13), which indicates that a limited level of backward compatibility has been attempted. In the AF PHB classification, it should be noted that the class with higher numerical value has higher forwarding priority. Similarly, the "high drop precedence" corresponds to relatively higher packet discard probability within each class. The AF PHB specification also stipulates that additional classes and drop levels may be defined for "local" use in the future if necessary. Quantitative values for packet loss or delay are not specified.

   Although the AF PHB does not specify detailed algorithms or buffer management mechanisms for the AF classes, the implication is that each AF class is allocated a certain configurable amount of "forwarding resources" in terms of buffer space and bandwidth within each AF-capable DS node. Although the intent is to avoid or minimize congestion within the DS network in the event that traffic congestion does occur, the AF PHB requires that nodes selectively discard packets marked for high drop precedence relative to the medium or low drop precedence, respectively, within each AF class. As may be recognized, this requirement implies the provisioning of congestion monitoring thresholds as

TABLE 12-8. DSCP values for assured forwarding classes

AF class	1	2	3	4
Low drop precedence	001 010	010 010	011 010	100 010
Medium drop precedence	001 100	010 100	011 100	100 100
High drop precedence	001 110	010 110	011 110	100 110

> part of the buffer (queue) management processes implemented within a given node. The AF PHB also requires that there should be no reordering of the packets within any given AF class, irrespective of the discard procedures employed. This requirement essentially applies to reordering of individual end-to-end application-to-application packets (generally termed a "microflow") within the aggregate assigned to any given AF class. In general, this requirement implies the use of a first in, first out (FIFO) queuing discipline as part of the buffer resource management procedures used to implement the AF PHB in any given class. Although RFC 2597 [12.16] provides some examples of mechanisms that may be used to implement the behavior required for AF, as well as possible "IP services" that may be constructed using the AF PHB classes, these are described in somewhat general terms, leaving significant scope for service providers and vendors to configure more specific instances.

As for the case of the EF PHB outlined earlier, the AF PHB may be viewed as a building block for the construction of a variety of end-to-end IP-based services when combined with the appropriate traffic conditioning and classification actions at the edges of a given DS domain, as well as the network policy underlying the service agreements negotiated with a traffic source. Simply taken by themselves, the PHB requirements outlined above cannot be viewed as constituting a "service" in any meaningful sense, since they simply indicate that a given packet aggregate may be subjected to three levels of relative discard priority in any one of four levels of relative forwarding priority within a node. In somewhat general terms, it has been suggested that the AF PHB group may be used for the construction of the so-called "Olympic" IP-based services, since it comprise three service classes designated as "gold," "silver," and "bronze" after the Olympic medal classifications. The main characteristics of such Olympic IP-based services is that the gold service would exhibit the highest forwarding (i.e., emission) priority and hence would be associated with AF Class 4 (or 3, depending on network operator policy and configuration). Similarly, the silver service would utilize AF Class 3 (or 2), and the bronze service would be built on AF Class 2 (or 1). In keeping with the characteristics of the AF PHB group, each of these resulting IP-based services may exhibit up to three levels of packet discard probability, depending on network dimensioning and congestion conditions within any given DS node.

From the perspective of QoS interworking between DiffServ-based IP and ATM networks, it may be conjectured that although certain underlying functional similarities may assist interworking, other differences of approach will make direct mapping of DiffServ QoS to ATM QoS somewhat arbitrary, as was the case for the IntServ framework. It will be recognized that the basic underlying principles of the DiffServ model, whereby specifically marked packet aggregates are accorded differential forwarding (or discard) priority by configuring sufficient resources and queue management procedures, are also inherent in ATM. As described earlier, ATM traffic management procedures may use either explicit or implicit priority control for congestion control and provision of QoS. For VBR traffic, explicit priority is indicated by using the cell loss priority (CLP) field in the ATM cell header, which allows for two levels of loss priority within a given VCC. This may be compared with the three levels of (packet) drop precedence available in the DiffServ approach, assuming that the AF PHB is selected in the definition of any given IP-based service. In ATM, implicit priority may be used for QoS differentiation by simply selecting groups (or ranges) of VPI/VCI values to correspond to certain QoS classes and allocating resources accordingly. This process may be viewed as analogous to the creation of the so-called Olympic services based on the three (out of four) AF classes outlined above.

It is important to recognize that DiffServ, unlike both ATM and IntServ/RSVP, does not require signaling procedures in order to allocate resources within nodes to maintain a given QoS level. The DiffServ model assumes that resource allocation, coupled with the necessary queue management and packet scheduling mechanisms, are configured within each DS node in accordance with the service level policies offered by the network operator, as described by the PHBs implemented.

In the DiffServ framework, the "signaling" is effectively "in-band," in the sense that each packet within the specific behavior aggregate is "marked" with the appropriate DSCP value that signals that it is to be mapped to the relevant forwarding behavior defined by a PHB. The DiffServ model also makes a distinction between the so-called "boundary nodes," which may mark packets as well as "condition" the traffic entering a DS domain, and so-called "internal nodes" which simply forward packet aggregates in accordance with the selected PHB. However, the distinction between such boundary and internal DS nodes is not strict, since it is clear that any node that allows the ingress of IP traffic may need to incorporate traffic conditioning functions whether within or at the edge of a DS domain. The DS traffic conditioning functions include the classification of packets into the DS aggregates, policing, and shaping of aggregates to ensure that any maximum aggregate rates are not exceeded.

In the DiffServ framework, a QoS-enabled IP-based service, such as the proposed "premium" service or the so-called Olympic set of services (gold, silver, and bronze), requires not only the selection of an appropriate PHB within each node, but also the specification of the traffic conditioning rules at the edge of the DS domains and the service agreement policies followed by the involved network operators. This aspect has resulted in some confusion for the case of DiffServ interworking with ATM because it was initially assumed that simply mapping the PHBs to an appropriate ATM service category (or ATC) would be sufficient to provide the service interworking guidelines. However, it is now recognized that PHBs by themselves do not constitute a "service" in the same sense as an ATM service category such as CBR, VBR, ABR, etc., and consequently simply associating a given PHB to an ATM service category is incomplete, as well as misleading. Subsequently, work in the IETF and ATM Forum on DiffServ–ATM service interworking has attempted to represent DiffServ-based IP services as:

$$\begin{aligned}\text{DS Service} = {} & \text{Traffic conditioning rules at network edge} + \\ & \text{PHB} + \text{network policy on service agreements}\end{aligned}$$

Assuming that this general relationship holds for the construction of any given DS-based end-to-end service, some general guidelines for the interworking of DS-based IP QoS to ATM may be envisaged. However, it must be stressed that these are at best somewhat approximate, as it will be clear that no one-to-one correspondence exists (or was intended) between the two approaches, despite some underlying similarities.

The simplest case is the interworking of DS-based IP services utilizing the expedited forwarding (EF) PHB within a node. It was noted that such services (e.g., the so-called "premium" service) are intended to emulate the properties of a "virtual leased line" by providing the very low transfer delay, jitter, and packet loss required by "real-time" services such as voice or video signals over IP (VOIP). Since the same low delay and loss characteristics are expected to ensue from the use of the ATM CBR service category (equivalently the DBR ATC), and to a somewhat lesser extent from the "real-time VBR" service category, it is clear that IP services based on the EF PHB may be interworked with ATM services based on CBR or rt-VBR service categories (or DBR ATC). As described earlier for the ATM case, such services are characterized by a fixed peak cell rate (PCR), and cell delay variation tolerance (CDVT) coupled with deterministic QoS parameters that are well defined. Since the EF PHB operates essentially by constraining the maximum ingress packet aggregate rate to be less than or equal to the minimum egress aggregate rate to eliminate packet queues (and thereby eliminate delay), suitable mapping of aggregate rates to peak cell rates may be derived to maintain the required delay and loss performance in an end-to-end IP-to-ATM interworked service. Consequently, it may be assumed that in interworking between a CBR (or rt-VBR) ATM VCC and the corresponding DS-based IP service, the most appropriate PHB to select is the EF PHB, with the related DSCP value of each IP packet in the aggregate set to "101 110" as outlined in RFC 2598 [12.15].

For the case of DS-enabled IP services constructed from the assured forwarding (AF) PHB, interworking with ATM is more problematical, since a wider range of possibilities may be envisaged. In the first place, as noted above, the AF PHB may be configured to allow up to three levels of packet discard priority within each one of the four AF classes, whereas ATM allows for two levels of selective cell discard priority within a VCC (or VPC) for only the VBR service category (or SBR 3 ATC). Consequently, for either network or service interworking scenarios, correspondence between AF-based packet discard precedence and ATM selective cell discard may only be configured by coalescing the AF discard behavior to support two levels of loss for the interworked connection, assuming VBR VCCs are used for IP packet transport. In addition, it may be recognized that the four AF PHB classes, which are intended to provide four distinct levels of packet emission delay bounds within any given node, do not directly correspond to any well-defined ATM service categories or QoS classes. Although this does not imply that interworking between AF based services and ATM is not feasible, it is clear that unique one-to-one mapping of parameters may not be possible in general. As indicated earlier, the AF PHB classes essentially require the allocation and partitioning of resources (buffers) coupled with suitable packet scheduling mechanisms to distinguish the packet forwarding performance within any aggregate. Such behavior may be emulated in ATM by logical partitioning of the buffer space for implicit priority based on "grouping" VPI/VCI values to correspond to different cell emission priorities. If such a mechanism may be configured within the nodes, it is evident that aggregates belonging to a given AF class may be mapped into the corresponding ATM VCC (groups) exhibiting the cell emission delay bounds required by the AF class. Thus, DS QoS services based on AF PHB classes may be interworked by utilizing the implicit VCI-value-based priority capability inherent in ATM (though not always available in all cases). Since DiffServ does not specify quantitative delay bounds for the AF classes, the actual mapping of the AF classes to any given VCC group remains largely a matter of configuration and dimensioning.

It may be noted that the QoS interworking mechanism considered above is largely independent of the ATM service category, although it is generally accepted that the CBR (or DBR) service category is not primarily intended for AF-based services (although this is not precluded). Initial interest has focused on the use of UBR service category to support AF-based DS services since this category is typically most widely used for the transport of IP-based services. Although the configuration of AF class related relative cell emission priority behavior is feasible using UBR VCCs for AF-based aggregates, the fact that the UBR service category inherently implies a form of "best efforts" service class betrays some confusion of intent in such attempts. However, since in many cases UBR may be the only "general purpose" ATM service category available in a given network implementation, the question of interworking with DS-based IP services remains of interest. In such cases, the introduction of various "subclasses" of cell delay performance within the UBR service category is required, typically by implementing some form of queue scheduling mechanism, such as a weighted round-robin (WRR), coupled with buffer partitioning to accommodate the interworked IP-to-ATM VCCs. Alternatively, if the ATM network is capable of providing other service categories such as rt-VBR, nrt-VBR, GFR, or ABR providing various levels of delay performance, mapping of the AF classes into VCCs based on these service categories may also be possible, although in this case the discard priority levels may not always be available (e.g., in GFR or ABR).

The DiffServ QoS framework is evidently capable of supporting a very large range of configurable IP services with different (relative) QoS types, to provide flexibility in the provision of future envisaged IP networked services. In particular, services based on the AF PHB may use up to twelve distinct priority levels, with each of the four forwarding (delay) classes capable of up to three levels of discard priority. Whether such a range of flexibility is useful or even manageable in practice remains an open question, and it may be premature to hazard predictions given that these concepts are relatively recent and largely untried in general use. It may also be recognized that since DiffServ requires not only the implementation of the various per-hop behaviors in individual nodes coupled with provisioning of resources and also the IP traffic classification and conditioning functions such as policing, shaping, congestion thresholding, and marking, it requires a substantial enhancement of the capabilities of IP-based networks, as does RSVP/IntServ. It is not within the scope of this outline to compare which of these two quite different approaches is preferable or simpler, but it may be safe to assume at this stage that both DiffServ and RSVP may coexist in different parts of the QoS-capable IP network. It is sometimes conjectured that RSVP may be applicable to smaller (enterprise) networks, whereas DiffServ may be used in the WAN or public network domain. Whether or not such distinctions can be maintained may be arguable but, clearly, QoS interworking between both IntServ- and DiffServ-based services will be required in such configurations, in addition to the need for interworking with any ATM networks used for transport of IP-based services.

If past experience with other (albeit connection oriented) packet networks such as X.25, frame relay, or ATM may be relied upon for the potential usage of priority-based IP QoS, it would appear unlikely that the large flexibility designed into DiffServ will be useful in practice. The complexity inherent in managing and provisioning multiple classes of service with a large number of QoS parameters typically restricts practical offerings to two or three main QoS classes. In practical terms, the majority of envisaged network applications may be adequately served by three broad QoS categories. These may be characterized by analogy to ATM as:

1. A general purpose "best efforts" service
2. A limited (engineerable) packet loss level and delay-tolerant service
3. A strictly delay bounded and relatively low-loss service (for real-time applications)

These considerations do not imply that other categories, or finer granularity within these broad categories, may not be useful as well in future applications. However, it seems questionable whether the 16 possibly distinct (comprising 12 AF based + two default precedence + one EF based + one default best efforts) classes initially identified in the DiffServ framework are useful, not to mention the possibility that the DSCP codespace allows up to 64 possible gradations within the three identified pools.

It may also be noted in passing that the somewhat unqualified use of the (often overused) term "service" in the context of the IntServ or DiffServ framework may result in confusion, particularly if the term is also used to describe conventional telecommunications services such as voice or data services. To add to the confusion, the term "service" is also used in the context of a "layer" service as in the ISO OSI layer model, wherein a lower layer provides a "service" to a higher layer characterized by "primitives" passed across a service access point (SAP) between the layers. It is important to distinguish such a purely protocol layer service (i.e., the exchange of PDUs between protocol layers in the transfer plane) from a broader network application service such as voice telephony or data file transfer service, which may also involve components from the management and control planes in its characterization, for example, as part of a general service level agreement between end user and service provider. From this perspective, it may be postulated that the IntServ or DiffServ frameworks do not specify "services" but rather IP "transfer capabilities" that may be used as components in the construction of IP-based network application services such as VOIP. In this sense, DiffServ and IntServ/RSVP essentially define a range of "transfer capabilities for IP QoS" analogous to the ATM transfer capability (ATC) described earlier, which are used to support the various ATM QoS classes intended for ATM-based network application services. If these distinctions in the use of the term "service" are kept in mind, particularly for the case of network and service interworking between ATM and IntServ- or DiffServ-based IP networks, confusion between support of conventional telecom services and the required capabilities may be avoided.

CHAPTER 13

Internetworking Framework Architectures in IP and ATM

13.1 INTRODUCTION

The previous chapter pointed to some of the vast amount of work undertaken to enhance IP-based networking to accommodate anticipated demands in growth as well as multiple classes of service or QoS capabilities. In developing protocols and QoS frameworks such as IntServ and DiffServ, it was seen that a number of concepts inherent to ATM traffic management, such as admission control, traffic policing, and/or shaping, were also adopted by IP, although in somewhat modified form to allow for the fundamental connectionless nature of IP networking. From the perspective of internetworking between IP- and ATM-based services, such functional commonality may be viewed as largely desirable, despite the differences of detail that make simple one-to-one mapping of parameters somewhat difficult. The use of RSVP "signaling" for on-demand resource reservation introduces connection-oriented aspects into IP networking, regardless of its complexity in attempting to maintain soft state capability. By coupling either DiffServ or RSVP (or both) based QoS with the underlying ATM traffic management in interworked scenarios, it is anticipated that the desired end-to-end (service/application level) QoS can be delivered for any foreseeable network application.

In parallel with these QoS-oriented enhancements, substantial effort has also been expended in developing various approaches to increase the throughput of IP-based networks. These activities have been motivated by the exponential growth in Internet traffic in recent years, primarily as a result of widespread use of the Worldwide Web (www) and e-mail applications. The resulting enormous increases in IP traffic throughput have been largely accommodated by deploying routers capable of very high packet throughputs (such devices are often known as gigabit or even terabit routers, depending on their total packet throughput capacity), coupled with very high speed (e.g., SDH- or SONET-based) transmission systems between the routers. Alternatively, IP over ATM backbone networks utilizing very high speed (e.g., 10 to 40 or more gibabit/sec) ATM switches are also widely used to provide high-capacity switching of the IP packets over PVC or on-demand switched VCCs between routers, in accordance with the "classical IP over ATM" (CIPOA) architecture and its enhancements described earlier. In addition to these conventional architectures, there have also been attempts to "combine" or integrate various aspects of ATM and IP functions into a unified architectural scheme in attempts to utilize the perceived "best" features of either technology. In this chapter, the basic concepts underlying a number of these approaches are summarized, with a view to examine potential trends in IP and ATM internetworking.

As may be expected, the evolution of IP–ATM internetworking architecture will depend largely on the initial network or technology premises used as its basis. Studies in the ATM Forum and, more recently (from around 1998), in the ITU-T, have been based on enhancing an "ATM network-centric" approach, resulting in an overall architectural framework termed "Multiprotocol over ATM" (MPOA), which also incorporates the ATM Forum's "LAN emulation" (LANE) protocols, as outlined below. Although the MPOA framework architecture was primarily developed within the ATM Forum, it is of interest to note that it is based on enhancing the initial classical IP over ATM model that originated from the earlier work in the IETF. Nonetheless, it will be recognized that the MPOA framework constitutes an ATM network-centric path to IP–ATM internetworking evolution, thereby providing all the benefits of both the IP and underlying ATM technologies. On the other hand, the IETF, which originally developed the CIPOA model and its enhancements (i.e., NHRP, MARS) as described earlier, has more recently adopted an IP network-centric model termed "multiprotocol label switching" (MPLS) for its studies on an evolution path for IP networking. Interestingly, MPLS, as its name implies, utilizes essentially the same label multiplexing/switching principles as does ATM (or FR for that matter). Moreover, in requiring some form of "signaling" protocol for the exchange of labels between nodes, this approach moves IP further along the path to becoming a connection-oriented networking technology, more functionally akin to ATM and FR.

Ironically, the architectural approach that originally gave impetus (albeit in somewhat different ways) to both the MPOA and MPLS approaches, and which embody its essential concept, is now generally not regarded as a viable evolution path for IP–ATM internetworking. This approach, termed "IP switching" was introduced by Ipsilon Networks Inc. as a model for "combining" ATM switching and IP routing mechanisms in order to improve IP forwarding throughputs. In this sense, the IP switching model, described briefly below, may be viewed as the forerunner for the various "integrated" IP–ATM architectures being currently studied for ATM and IP internetworking evolution. In the brief survey below of all these approaches, the LAN emulation, IP switching, and MPOA models are considered first, followed by the MPLS architecture, but this should not be taken as implying any chronological ordering, since work on these architectures has generally proceeded in parallel, often with close interaction between the various groups.

13.2 LAN EMULATION OVER ATM (LANE)

The LAN emulation service and underlying architecture was developed by the ATM Forum [13.1] primarily to enable ATM networks to use the large amount of existing (i.e., legacy) LAN application software directly, by making the underlying ATM network "appear" (i.e., emulate) as a LAN to the given application. In addition, and just as importantly, the LANE protocols would enable ATM networks to interwork with legacy LANs such as Ethernet and IEEE 802.3 [12.5] or 802.5 [12.7] by means of conventional bridging. Since it was assumed that users are unlikely to discard using their existing LANs in the event that new ATM networks are installed, the need for interworking will be evident. From the brief outline of LAN protocols earlier, it may be recalled that the key LAN characteristics that would need to be emulated by an ATM network in order to "look like" a LAN to the higher layers include:

1. Provide a connectionless service. It will be recalled that commonly used LAN protocols are connectionless and do not need to set up connections to send data between stations. As described earlier, LANs use a unique (hard-wired) media access control (MAC) address in the LAN frame header to identify individual stations. In contrast, ATM is connection oriented.
2. Support Multicast or Broadcast Operation. Traditionally, LANs are based on a shared media physical transport, so that broadcast or multicast of LAN frames is inherent in the mechanism and hence readily achieved. This is generally not the case for point-to-point switched technologies such as ATM, which generally do not use shared media transport.

3. Support MAC Driver Interfaces. Existing LAN application protocols generally utilize standard MAC driver interfaces and, therefore, LANE should be assumed to provide the same "service" to the higher layers. Consequently, ATM-based LANE is also expected to use a typical MAC driver interface such as "network driver interface specification (NDIS)," "open data-link interface" (ODI), or "data link provider interface" (DLPI). Details of such driver interfaces are outside the scope of this account.

It is important to distinguish between LAN "emulation" and the more typical use of ATM for direct LAN "interconnection" between remote sites. This difference may be described as shown in Fig. 13-1. For the case of direct LAN interconnection applications, the ATM network transparently carries the legacy LAN (e.g., Ethernet or Token Ring) frames between remotely located routers, which may be assumed to be equipped with ATM interface cards. The intervening ATM network essentially behaves as a "leased line replacement" (sometimes called a virtual leased line, since an ATM VCC is used instead of a conventional DS-0, DS-1, or E1 leased line between a customer's remote sites). In this application, no additional protocols are required within the ATM network or the legacy LANs, assuming the router devices can encapsulate the IP packets as described previously in the CIPOA architecture. The ATM VCCs interconnecting remote routers may be set up as a mesh, either as semipermanent virtual channels (via the management interface) or by normal ATM signaling procedures such as PNNI or B-ISUP.

In contrast, for the case of LAN emulation, the ATM network provides a "LANE service" by means of additional functionality incorporated in a LANE server, which allows the ATM network to provide "LAN-like" behavior to the attached end stations across the so-called LANE UNI (or LUNI). This approach, which must also clearly be based on a mesh of underlying ATM VCCs set up between the end stations as well as the LAN server functions as described in further detail below, requires the use of additional protocols operating across the LUNI. However, use of these additional LANE protocols enables service interworking between the resulting emulated LAN (ELAN) and legacy LANs such as Ethernet by conventional bridging mechanisms.

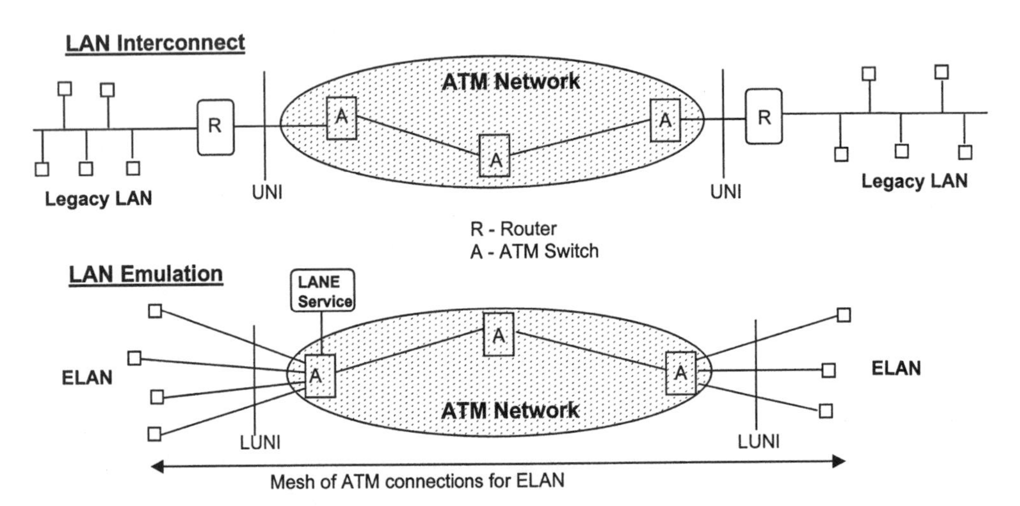

Figure 13-1. LAN interconnect versus LAN emulation. In a LAN interconnect application, the ATM network transparently carries legacy LAN frames between remote routers; no additional protocol is required. The ATM network essentially acts as "leased line" replacement. In a LAN emulation, the ATM network provides a "LANE service," enabling the ATM network to provide LAN-like behavior across a LANE UNI (LUNI). This requires additional protocols, but enables service interworking with legacy LANs.

From the perspective of the (higher-layer) applications within any given end station attached to such an emulated LAN, comprising the "mesh" of ATM VCCs and LANE server functions, the fact that the LAN frames are transported and switched over conventional ATM VCCs, as opposed to shared media (or switched) connectionless LAN transport, should be immaterial. However, these applications may now benefit from the enhanced performance and QoS capabilities made possible by using ATM VCCs for the underlying transport, without requiring any changes within the application itself.

The main functional components of the basic LANE network architecture are shown in Fig. 13-2, which also summarizes the essential role of each functional component in the model. It may be seen that the LAN emulation service (as provided by the LANE server) is made up of three broad components:

1. LAN emulation server (LES) entity
2. LAN emulation configuration server (LECS)
3. Broadcast and unknown server (BUS)

In addition, each end station attached to the ELAN controlled by the LANE service contains the LAN emulation client (LEC) functionality to enable it to communicate with the LANE service and other LECs attached to the ELAN across the LUNI by means of one or more ATM VCCs. The role of each one of these LANE entities may be described as follows:

1. LAN Emulation Client (LEC). The LEC entity in the end systems comprises the functions that perform forwarding of data as well as the control plane functions required to set up the ATM VCCs across the LUNI to other entities in the ELAN.
2. LAN Emulation Server (LES). The LES entity provides some of the basic services in the ELAN. It provides overall coordination and control functions as well as the basic address registration and address resolution functions required to maintain the ELAN. To enable this, each LEC joining the ELAN must register its <ATM:MAC> address pair information with the LES entity, as well as query it for the address pair information prior to setting up VCCs for data transfer to other LECs in the ELAN. Thus, the LES may be viewed as providing the basic LANE address resolution protocol (ARP) capability as described earlier for conventional (i.e., legacy) LANs as well as IP-based networks.

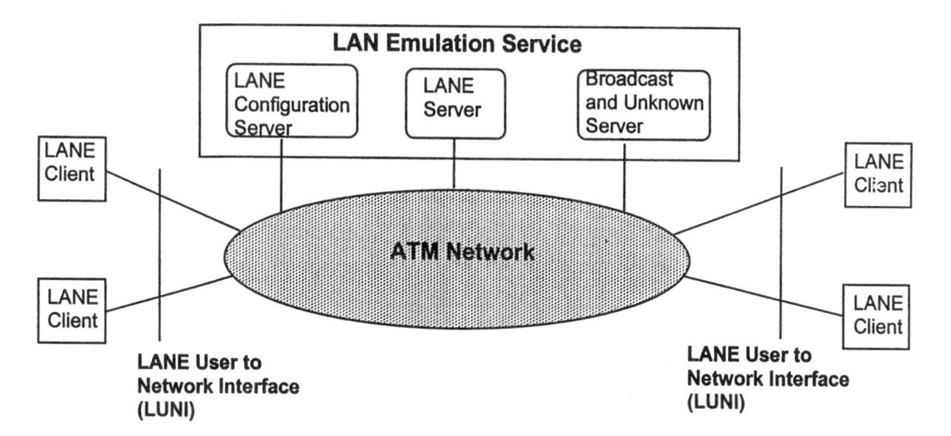

Figure 13-2. LAN emulation components.

3. LAN Emulation Configuration Server (LECS). The LECS entity comprises those functions that maintain the configuration database of the various ELANs and assigns individual LECs to different ELANs that may be configured within an ATM network. The LECS entity may thus be viewed as providing an overall configuration management capability to the LANE service, analogous to the configuration management of conventional ATM (or other) networks. The LECS function also provides information in response to configuration requests from the other LANE entities.
4. Broadcast and Unknown Server (BUS). As its name implies, the BUS entity provides for the broadcast or multicast of data frames by enabling the registration of multicast address groups. It also includes the handling of unknown addresses or routing information in the event that connectivity with unknown (i.e., unregistered) addresses is requested by any given LEC.

It will be recognized from the functional partitioning above that the LANE service utilizes the "client–server" architecture model commonly used in data networking. In addition, LANE also employs an extension of the address resolution protocol (ARP) approach coupled with normal ATM signaling procedures to convey the <ATM:MAC> address pairs necessary to support LAN interworking and emulation. The LANE components may also be colocated within any network element in the ATM network, or be distributed in different elements, provided the required connectivity is maintained. Thus, the LANE architecture shown in Fig. 13-2 should be viewed as a logical (and not physical) grouping of the LANE functionality.

The LANE protocol architecture is shown in Fig. 13-3, where it is seen that for the transfer plane the LANE layer is effectively interposed between the AAL Type 5 CPCS (with a null SSCS) and the LAN logical link control (LLC) layers (or bridging functions) of the particular LAN being emulated over ATM. From the perspective of the control plane, the LANE layer functions lie above the call control layer since, as noted below, LANE introduces two additional messages to the existing ATM UNI signaling message set for LANE-specific information transfer. It is also of interest to compare the LANE protocol architecture with the reference ISO layered model out-

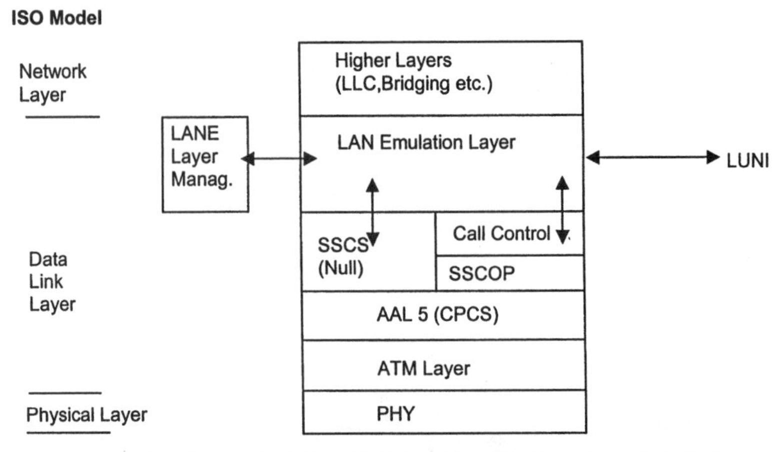

1. Communications between the LECs and LES, LECS and BUS is by ATM VCCs (PVC or SVC) established in the normal way. Separate VCCs are used for control functions and data frames. The VCCs form a mesh of ATM connections between the LECs and the LECS, LES and BUS entities in each ELAN. Consequently each LUNI may have multiple VCCs across it.

2. The LANE protocols may be implemented in ATM end systems, intermediate systems or ATM switches. In ATM end stations the LANE functions may be part of software driver or ATM adapter card hardware.

Figure 13-3. LANE protocol architecture.

lined earlier. This comparison indicates that the LANE layer functions lie within the data link layer (DLL) of the ISO model together with the ATM plus AAL, as indicated in Fig. 13-3.

Although communication between the LECs and the LES, LECS, and BUS entities is by means of ATM VCCs (either established as PVCs or on-demand SVCs using conventional procedures), it is important to note that LANE uses separate VCCs for the transfer of control information and for user data frames. These ATM VCCs form a mesh of ATM-based connectivity between the LECs and the LECS, LES, and BUS entities in each ELAN. A consequence of this architecture is that each LUNI, for any given LEC, may have multiple VCCs (for either data or control information transfer) across it. The LANE protocols may be implemented in ATM end systems (i.e., terminal devices or wherever ATM VCCs are terminated), intermediate systems or ATM switches. The LANE functions may be implemented as part of the software driver or in hardware as part of the ATM adapter card within the ATM terminal devices.

The basic connectivity required across the LUNI for the LANE capabilities outlined above is shown in Fig. 13-4 for the case of two LECs (or bridges). As described in more detail below, control VCCs are set up between the LECs and the LECS, LES, and BUS entities in the initialization phase of the formation of an ELAN. As noted above, the control VCCs are used to carry control information such as LAN emulation ARP (LE_ARP) frames but are not used to carry user data frames between the LECs (or bridges). For the transport of such user data frames, data VCCs are set up directly between the LECs and also between the LEC and BUS entity. The data VCCs carry the Ethernet (or IEEE 802.3) or IEEE 802.5 data frames. In addition, as described later, the data VCCs are also used to carry the so-called "flush" messages. It may also be noted that although LANE utilizes conventional ATM UNI signaling (and call control) for the establishment (or release) for the required mesh of VCCs, it introduces two "new" LANE-specific messages in addition to the normal ATM UNI message set described earlier. These LANE-specific messages, termed (1) READY_INDICATION and (2) READY_QUERY are described in more detail below; they introduce some additional complexity in the case of LANE signaling procedures across the LUNI.

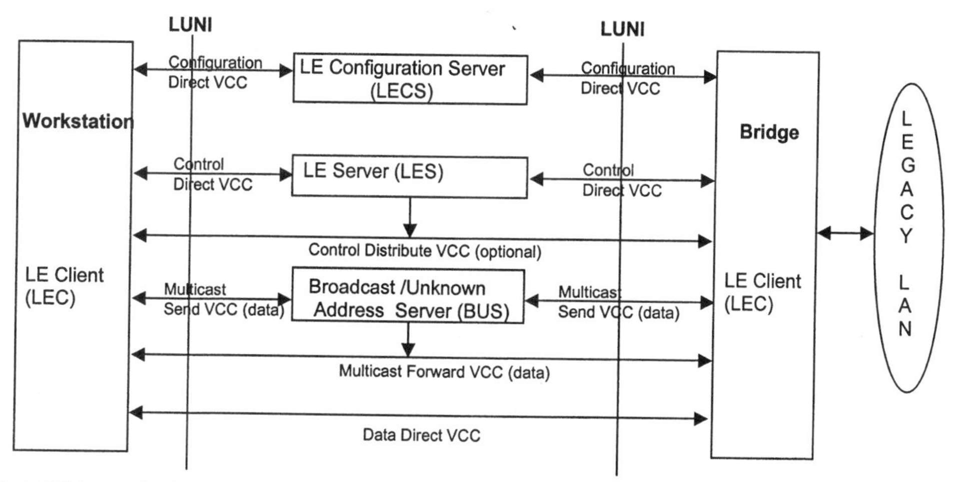

1. Control VCCs are set up between the LECs and the LECS, the LES and BUS in Initialization phase. The control VCCs never carry data frames. The control VCCs may carry LE_ARP and Control frames.
2. Data VCCs are set up between the LECs and also between the LEC and BUS. Data VCCs carry Ethernet/ IEEE 802.3 or 802.5 data frames and Flush messages.
3. LANE adds two "new" messages to ATM call control (UNI Signaling): 1) READY_INDICATION and 2) READY_QUERY.

Figure 13-4. LANE connections.

The sequence of operations required in setting up or joining an emulated LAN may be conveniently grouped in a set of phases as summarized in the flow diagram of Fig 13-5. Referring to both Figs. 13-4 and 13-5, the LANE procedures may be summarized as follows:

1. Initial State. In the initial state it is assumed that the LECs and LES are aware of parameters such as their addresses (ATM or MAC), the type of ELAN, the maximum frame size permitted (e.g., MTU) and any other information that may be required for the ELAN. This information may be provided by configuration or provisioning by means of the management interfaces to individual devices.
2. LECS Connect Phase. In this phase, each LEC establishes a "configuration direct VCC" to the LAN emulation configuration server (LECS) entity, as indicated in Fig. 13-4. The individual LECs may obtain the ATM address of the LECS entity in a number of ways, including (a) use of the ILMI "GET" or "GETNEXT" commands to obtain the LECS address from an ILMI Server, or (b) use a "default LECS address," as specified by the ATM Forum LANE specifications (also known as a so-called "well-known address" in LANE). Alternatively, the configuration direct VCC may be provisioned as a PVC, using a preassigned VPI/VCI value of VPI = 0, VCI = 17.
3. Configuration Phase. In the configuration phase, the LEC is assigned to a given ELAN by the LECS entity, by providing it the address (or ELAN name) of the LES responsible for that particular ELAN. The LEC also obtains the specific LANE service parameters for the ELAN to which it is assigned from the LECs, which may depend on whether the ELAN to which it is assigned is Ethernet or Token Ring based.
4. Join Phase. During the join phase, the individual LECs establish the "control direct VCC" to the LES to which it has been assigned. (i.e., the LES entity responsible for the ELAN of

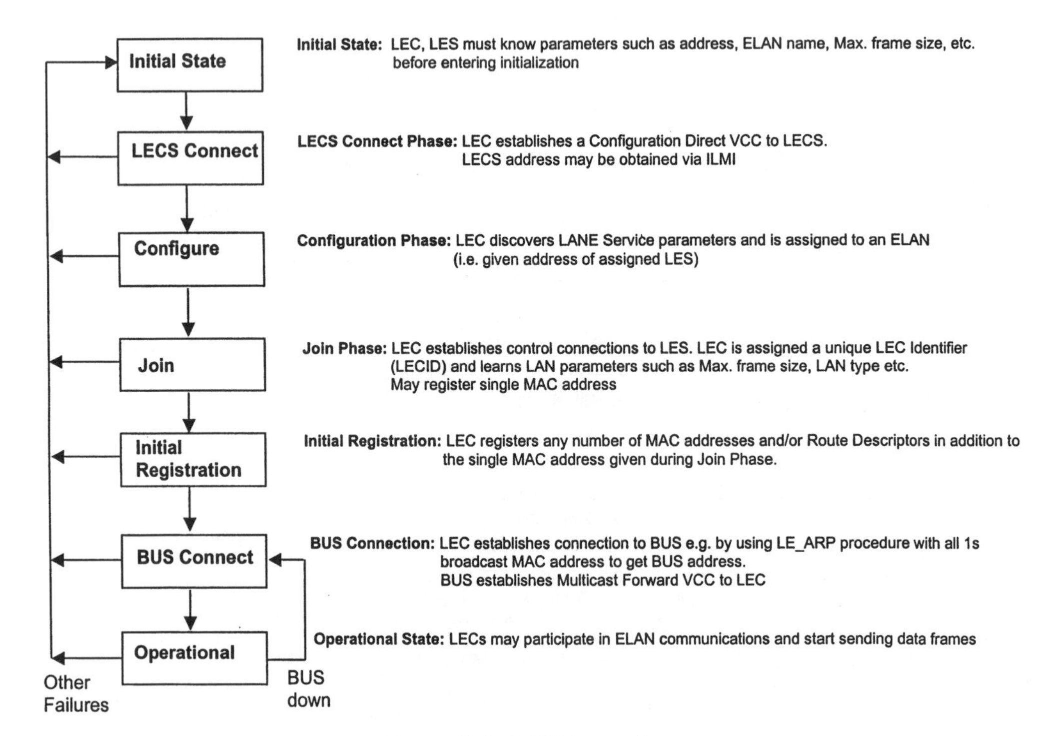

Figure 13-5. LANE operations.

which the LEC is a member, as outlined above). The control direct VCC is used for the control messages flowing between the LECs and the LES entities as indicated in further detail below. The control direct VCC must not be used for transport of user data frames, as noted earlier. Upon establishing the control direct VCC, each individual LEC is assigned a unique identifier called a LEC identifier (LECID) by the LES, which may also provide any relevant ELAN parameters such as the maximum permissible frame size, or LAN type. In addition, the LEC may register a single MAC address with the LES entity at this stage. The LECID constitutes an important parameter in LANE operation and is utilized in both control and data transfer protocols to uniquely identify each LEC within a particular ELAN, as evident below in describing the LANE frame structures.

5. Initial Registration. During this phase of the LANE establishment, any given LEC joining the ELAN may register additional MAC addresses or route descriptors for which it is responsible. In this sense, the term "initial registration" is misleading, since true initial registration actually occurs during the join phase described above in which only a single MAC address can be registered with the LES. It may be recognized that if no additional MAC addresses or route descriptors are applicable in the creation of the given ELAN, this initial registration phase need not occur.
6. BUS Connection. During the BUS connection phase, the individual LEC establishes a "multicast send VCC" to the BUS entity, followed by the BUS establishing a point-to-multipoint "multicast forward VCC" to each LEC in the multicast group. It should be noted that both the point-to-point multicast send VCC and point-to-multipoint multicast forward VCC are intended for transfer of user data frames and not control frames, since the BUS is primarily intended to provide the multicast (or broadcast) capability inherent in LAN operation. The individual LECs obtain the address of the BUS (responsible for their particular ELAN) by using the LAN Emulation ARP (LE_ARP) procedures with an "all ones" broadcast MAC address to the LES or LECs. On receiving such a LE_ARP request message, the LECS (or LES) responds with the appropriate BUS address pair.
7. Operational State. Assuming the procedures listed above are all completed successfully, resulting in the LANE connectivity indicated in Fig. 13-4 for each LEC, these may now participate fully in the given ELAN and commence the exchange of user data frames between individual LECs (or bridges), by establishing the so-called "data direct VCCs" between pairs of LECs for point-to-point transfer of user data. For multicast (or broadcast) of user data frames, only the multicast send and multicast forward VCCs must be used via the BUS in the LANE architecture.

In Fig. 13-5 it may also be noted that apart from the BUS connect phase, failure of any of the other phases effectively involves the return to the initial state, whereupon the step-by-step establishment of the ELAN (or joining of an individual LEC entity to an existing ELAN) may commence again. For the failure of the BUS connect phase, the LANE procedures simply require an iteration of the BUS connect process outlined above, since it is not necessary to return to an initial state for set up of any control related VCCs in this case. Assuming that all the initial configuration parameters, addresses, and other required LAN-associated information is correctly installed in the LANE servers and clients during initialization, it may be noted that the step-by-step establishment of the ELANs may be highly automated in accordance with the operational phases listed above. Each individual ELAN, as administered by its own LES and BUS entities, may be built up step-by-step as each LEC "joins" the ELAN, or may be provisioned between a number of LECs by configuration through the relevant management interface procedures if required. Such details depend largely on the specific network application or the implementation of the LANE equipment. It may also be pointed out that there is a significant amount of flexibility inherent in the LANE procedures and architecture to allow for a range of applications. For example, as shown in Fig. 13-4, it is possible to set up and utilize the point-to-multipoint "control distribute VCC" between the LES and individual LECs as an option to transfer control information between LES and LECs if required. This may save on the number of control direct VCCs between LEC and LES in applications where unidirectional control transfer is sufficient. It will also be recognized that the "configuration direct" and "control direct" VCCs may be set up as PVCs since they will in general be required for extended periods for the exchange of control information (e.g., LE_ARP procedures) between the LES and LECs.

As indicated earlier, the client–server architecture underlying LANE is based on the use of LE_ARP messaging, which also presupposes registration of the relevant address pairs within the LANE server entities. The address registration mechanism enables individual LECs to provide the LES with the <ATM address:MAC address> pair information during the "initialization" or "join" phases as noted above, using dedicated LANE messages. These control messages are listed below, although it is not the intention here to enter into a detailed account of the associated procedures and message structure, which may readily be found in the LANE specifications issued by the ATM Forum. The address registration procedures also require that the LEC must register all the address pairs for which it is responsible, in the event that a given LEC entity is performing the role of a bridge or gateway to a number of MAC end stations. In addition to address registration, the ARP procedures used in LANE are simply an extension of conventional LAN ARP mechanisms, adapted for the purpose of transferring <ATM address:MAC address> pairs information to enable LECs to set up ATM VCCs for user data transport.

Although the ARP mechanism in LANE is directly analogous to the ARP used in conventional LANs and in IP networks as described earlier, a key difference is that LE_ARP messages utilize (control) ATM VCCs between the client–server entities for messaging. These VCCs include the control direct VCC and, optionally, the control distribute VCC shown in Fig. 13-4. Once these are set up, the LE_ARP process is relatively straightforward as outlined below:

1. If any given LEC has user data frames it wishes to send to a destination LEC whose LAN (i.e., MAC) address is unknown, the LEC sends an LE_ARP REQUEST message over the control direct VCC to its assigned LES entity (i.e., the LES responsible for the particular ELAN of which the LEC is a part).
2. The LES may either (a) respond directly by sending an LE_ARP RESPONSE message containing the <ATM address:MAC address> pair, or may (b) forward the LE_ARP REQUEST message to the appropriate (destination) LEC itself for it to return an LE_ARP RESPONSE message containing the destination address pair information.
3. In order to reduce such messaging latency (which is inherent in any ARP-based approach), the LECs may maintain a "cache" of <ATM address:MAC address> pairs for future (or frequent) use if required. The address pair caching mechanism allows for virtually immediate set-up of the data direct VCCs for user data transfer between the LECs in question. However, the cached address information needs to be updated periodically, based either on a "time-out" or "aging" procedure, or on requested topology information contained in a so-called LE_TOPOLOGY_REQUEST message, which can be used by the LEC to update cached information.

It will be evident from the above description that LANE requires extensive use of ATM signaling protocols and in fact specifies the use of the ATM Forum's UNI signaling and call control procedures for the establishment/release of both control and data VCCs between the LANE components. However, LANE also imposes some additional requirements to the basic UNI signaling capabilities, in terms of two LANE-specific messages as noted earlier, as well as the need for mandatory information elements (IEs), which may be optional for the UNI signaling case. As an example of the latter, LANE requires the use of the so-called "broadband lower-layer information" (B-LLI) as a mandatory IE, whereas in normal UNI signaling the B-LLI IE is optional. LANE also specifies two additional signaling messages together with their associated procedures. These are:

1. READY_INDICATION message. This message is used to let the "called" LEC party know that the "calling" LEC can receive user data frames.
2. READY_QUERY message. This message is sent by the called LEC if it does not receive the READY_INDICATION message within a certain time-out period.

The use of these messages between two arbitrary LECs is shown in Fig. 13-6. The intent of this message sequence is to ensure that the originating LEC entity may also be in a position to receive user data frames sent by the destination LEC, but it may be inferred that this information is also provided by the normal CONNECT message, assuming that the data direct VCCs are bidirectional between the communicating LECs. Consequently, the need for these additional LANE-specific messages has led to some controversy, since they add some complexity to the normal UNI signaling procedures by introducing two additional messages and associated timer, which may not be necessary in most LANE applications, assuming full bidirectional connectivity based on normal UNI signaling procedures is available. In more general terms, by introducing these and other differences between normal UNI-based signaling and the LUNI signaling procedures specific to LANE, it is arguable that LANE may have unnecessarily complicated its interoperability with respect to an implementation based on only UNI signaling procedures, which could be assumed to provide essentially the same capability. It was also noted above that the LANE architecture indicates the use of point-to-multipoint signaling capability, in particular for the use of the BUS function for multicast connectivity. Although this clearly simplifies emulation of the multicast function of LANs, it may be noted that the BUS may use either point-to-multipoint VCCs if available, or individual point-to-point VCCs to the LECs for sending multicast frames, in the event that the point-to-multipoint signaling is not implemented. However, this latter option may be somewhat unwieldy for large multicast trees.

It was noted earlier that the LANE architecture was intended for emulation of the different legacy LAN types as well as interworking with them by means of a bridging model whereby a LEC performed a bridging role, as shown in Fig. 13-4. However, it must be admitted that this need to emulate the different legacy LAN types effectively complicates ATM-based LANE. As shown in Fig. 13-7, not only are the data frame formats for the two main LAN types different, but also it may be recalled that Ethernet and Token Ring LANs differ in the transmission order of the bits and bytes in the frame structures. These somewhat basic differences in the LAN structure and mode of

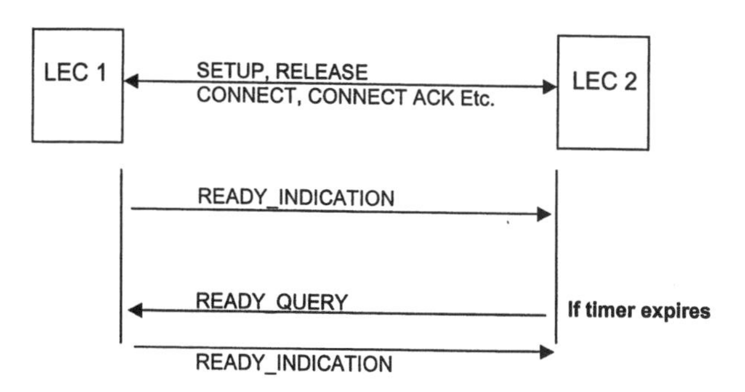

Figure 13-6. Signaling in LANE.

operation require multiple, or switchable options in the necessary hardware and software implementation of the LANE protocols. For the case of IEEE 802.3 [12.5] or Ethernet LANs, the data frame format shown in Fig. 13-7 appends a two-octet LANE header that consists of the LAN emulation client identifier (LECID) field to the Ethernet header. The LECID value is assigned to each LEC uniquely by the LES and may be used to identify the LEC entity within any given ELAN for either data or control information transfer purposes. It may be noted that apart from the LECID no other LANE-specific field is used for the user data frames for both the Ethernet/802.3 frames and Token Ring frames. If no specific value is assigned to the LECID by the LES, a default coding of (hexadecimal) X0000 is inserted in this field. The frame check sequence (FCS) field [and the access control (AC) field in the case of Token Ring] is not used in LANE, but the IEEE 802.2 LLC/SNAP capability [12.4] may be used as described previously (see Fig. 12-4 to Fig. 12-7 for the LAN frame structure and Fig. 12-16 for the LLC/SNAP structure). It will also be recalled that the maximum and minimum frame sizes indicated in Fig. 13-7 for the LANE data frames correspond to the MTU's noted previously for both Ethernet and Token Ring LANs.

The general structure of the LANE control frames is shown in Fig. 13-8, using conventional 32-bit word representation. It may be noted that all control frames include a common 16-octet "control frame header" which includes a two-octet so-called operation code (Op_Code) field that essentially serves as a message-type function to distinguish the various control messages used in LANE operations. Although it is outside the scope of this brief overview to describe each of the control messages and associated procedures in detail, it may be useful to list for reference the LANE control messages and Op_Code values as shown in Table 13-1. The Op_Code values (as for some other protocol control fields in LANE) are given in hexadecimal notation in LANE specifications, whereas the LANE message names typically reflect their primary function in the overall LANE operations outlined earlier. The similarity of the general LANE control frame structure in Fig. 13-8 to the ARP messages outlined previously should also be noted, although there are clearly differences in detail, mostly as a result of the larger number of functions required for LANE operations. In the common LANE control frame header, the "status" field is used to encode cause values for the various functions, whereas the four-octet "transaction identifier" is used to correlate the control messages exchanged between the sender and receiver pair in any given set of message exchanges, or transactions, analogous to a correlation tag or signaling identifier function. The control frame header also includes the LECID value of the requester

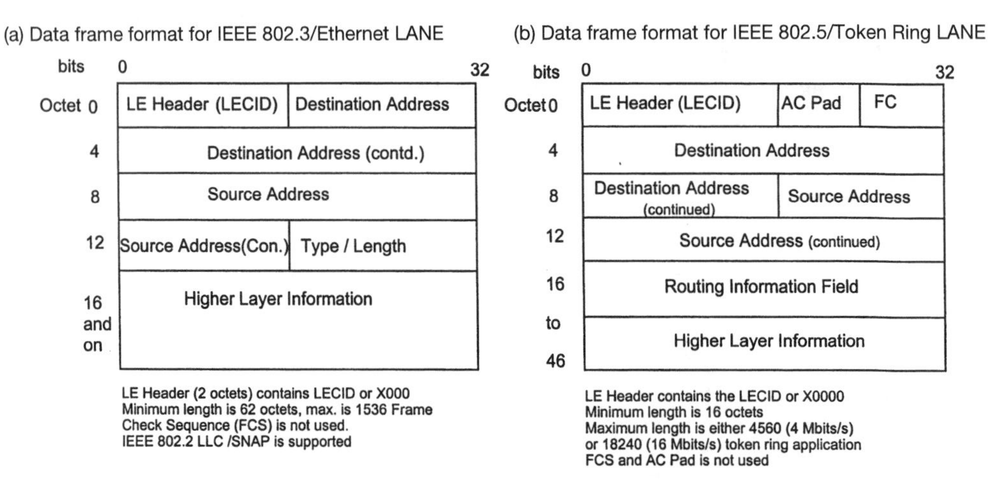

Figure 13-7. LANE data frames.

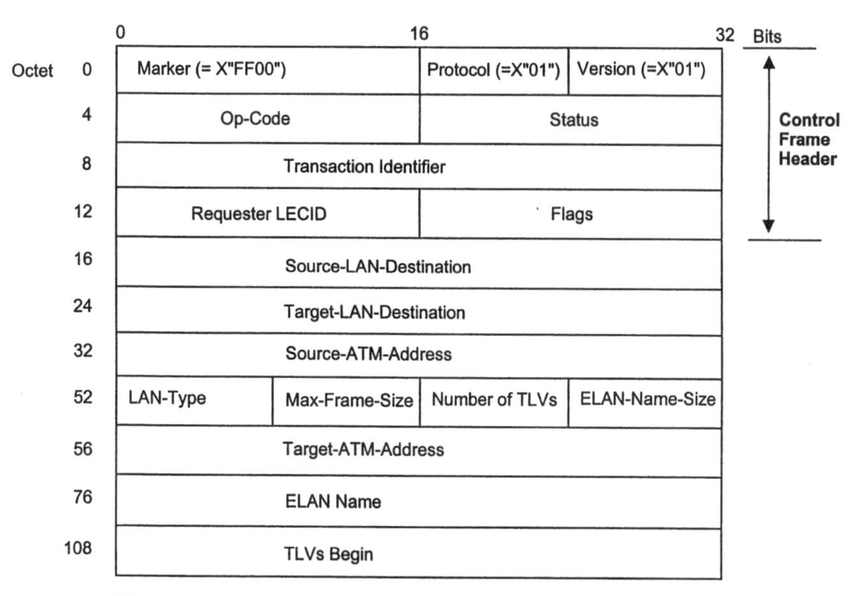

Figure 13-8. LANE control frame.

LEC, as well as fields to indicate the protocol version and type and various flags. It may also be noted that the source and target ATM address size may incorporate the complete 20-octet AESA structure, whereas the source and target LAN (or MAC) addresses include the six-octet MAC address structure outlined earlier.

The LANE control message types listed in Table 13-1 may be compared with the operational flow diagram outlined earlier with respect to Fig. 13-5 to obtain a general overview of how the

TABLE 13-1. Control message types (Op_Code values)

OP-CODE value	Function
X"0001"	LE_CONFIGURE_REQUEST
X"0101"	LE_CONFIGURE_RESPONSE
X"0002"	LE_JOIN_REQUEST
X"0102"	LE_JOIN_RESPONSE
X"0003"	READY_QUERY
X"0103"	READY_INDICATION
X"0004"	LE_REGISTER_REQUEST
X"0104"	LE_REGISTER_RESPONSE
X"0005"	LE_UNREGISTER_REQUEST
X"0105"	LE_UNREGISTER_RESPONSE
X"0006"	LE_ARP_REQUEST
X"0106"	LE_ARP_RESPONSE
X"0007"	LE_FLUSH_REQUEST
X"0107"	LE_FLUSH_RESPONSE
X"0008"	LE_NARP_REQUEST
X"0009"	LE_TOPOLOGY_REQUEST

various LANE control messages are used. Thus, in the configuration phase the LEC obtains the LES address from the LECS entity using the LE_CONFIGURE_REQUEST and LE_CONFIGURE_RESPONSE message set. Similarly, in the join phase the LEC establishes the control direct VCC with its assigned LES and obtains the information to join the ELAN using the LE_JOIN_REQUEST and LE_JOIN_RESPONSE message set. In the initial registration phase, a given LEC may register additional addresses after successful completion of the join phase using the message set (a) LE_REGISTER_REQUEST and LE_REGISTER_RESPONSE to register each additional address or (b) LE_UNREGISTER_REQUEST and LE_UNREGISTER_RESPONSE to remove address from the LES if required.

As noted earlier, the LECs may connect to the BUS by using the LE_ARP_REQUEST and LE_ARP_RESPONSE control messages to the LES with an "all ones" (signifying broadcast) MAC address to obtain the BUS address. The LEC then establishes the "multicast send VCC" to the BUS, which itself establishes a "multicast forward" VCC to a given LEC. These VCCs may be either point-to-point or point-to-multipoint, assuming this latter capability is available. In addition to the relatively straightforward use of the above messages, LANE also utilizes a number of other messages whose function may be considered more specific. For example, a LEC may use LE_NARP_REQUEST to advertise changes in remote address pairs, or use a LE_TOPOLOGY_REQUEST message to indicate address change information to all LECs attached to the ELAN. It was pointed out earlier that LANE introduces the use of two "new" signaling messages termed READY_QUERY and READY_INDICATION, whose utility may not always be necessary if correct and complete UNI signaling procedures are incorporated. More controversially, LANE also introduces the concept of "data flushing," using control messages as described below, to avoid the possibility of "out of sequence" data frames in the event that more than one VCC is set up to a given destination in the multicast case.

The so-called "flush procedures" may be described with reference to Fig. 13-9, derived from the general LANE architecture model considered in Fig. 13-4 for the case of the BUS entity and two LECs. It was recognized that in some LANE configurations there may be the possibility that a LEC may send user data frames via both the data direct VCC and the multicast forward VCC from the BUS to the same destination LEC. This may result in potentially "out of sequence" user data frames (or cells) in such a situation, which is difficult to avoid procedurally without imposing constraints on connectivity or architecture. Consequently, to avoid the possibility of out of sequence frames in such configurations, LANE introduces the concept of "flush" procedures, using two additional control messages. The flush protocol uses the following message set:

1. "LE_FLUSH_REQUEST" message. This (control) message is sent by the source LEC on the "data direct VCC" or the "multicast send VCC" to ensure that all the user data frames have reached the destination LEC. Thus, the LE_FLUSH_REQUEST message effectively clears (or flushes) the user data frames sent towards the destination LEC along the data VCCs.

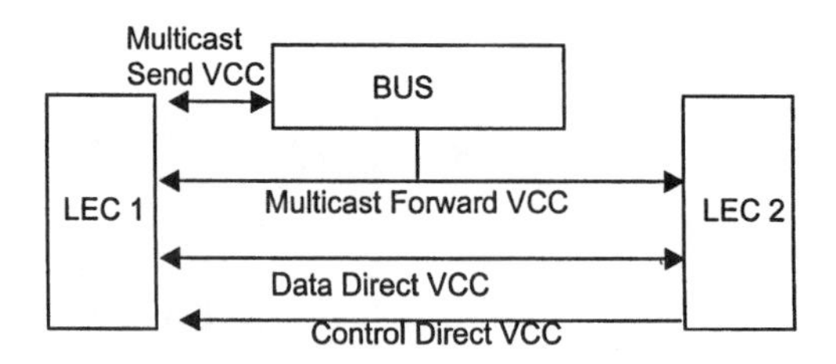

Figure 13-9. Flush procedures.

2. "LE_FLUSH_RESPONSE" message. This control message is sent by the destination LEC on the "control direct VCC" and "control distribute VCC" to confirm that the earlier user data frames have been "cleared," or flushed, from the LEC buffers in response to the LE_FLUSH_REQUEST message described above.

It is important to stress that the flush procedures above must be invoked *before* a given LEC can start sending user data frame or cells on a new path, to ensure that the previous path is "cleared" of all the user data frames or cells sent previously. It should also be recognized that the LE_FLUSH_REQUEST messages are sent on data path VCCs, even though they are essentially control messages, whereas the LE_FLUSH_RESPONSE messages are sent on the control VCCs as would normally be required for control messages. The flush procedures thus introduce an undesirable asymmetry in the protocol, as well as a potential additional latency before sending the user data frames to wait for arrival of the LE_FLUSH_RESPONSE acknowledgment that earlier user data frames are cleared. This may adversely affect performance for some applications in which latency needs to be minimized. In such cases, it may be preferable to configure the LANE network connectivity to avoid the use of the flush procedures if possible. It should also be pointed out that the flush mechanism is not used for the case of bridging (or interworking) between ELANs and legacy LANs. More generally, the usefulness and even the need for the flushing procedures summarized above has been questioned, although with careful configuration of LANE connectivity it may be possible to avoid the use of this option.

It will be recognized from the brief account of LANE procedures above that LANE procedures are relatively complex when compared with the simple LAN interconnect model mentioned earlier. It is clear that extensive use of ATM UNI signaling capabilities must be made to efficiently emulate the "shared media" connectivity inherent in legacy LANs and, consequently, the ATM signaling processes implemented in the LANE components must be efficient and exhibit low latency, although this may be an issue for some critical applications. The fact that LANE requires "modified" signaling procedures across the LUNI compared to the basic UNI (or Q.2931-based [6.8]) signaling, in terms of both additional messages and mandatory IEs, may prove to be disadvantageous, since specific implementations are entailed and reuse of UNI signaling software may result in interoperability issues. It may also be noted that LANE also specifies some enhancements to the basic (ILMI) MIB to accommodate address registration database and ELAN naming. In addition to these aspects, as well as the need for the data flushing procedures noted earlier, the overall scalability of the LANE architecture has been questioned, particularly in the need for efficient call control and signaling processes for the establishment and release of large numbers of ATM VCCs. Although the summary here has addressed only the LUNI procedures, it should be pointed out that a substantial amount of work has also been done in the ATM Forum to develop protocols applicable to the so-called "LANE NNI" (LNNI), based on extensions of LUNI for large or wide area networks (WANs). These extensions include the use of different QoS classes for multiservice support, server redundancy mechanisms for enhanced robustness, and other features useful for WAN applications. These aspects of LANE extensions are outside the scope of this account but may be found in the ATM Forum's specifications.

The LANE approach may be viewed as a specific ATM (or B-ISDN) network application or service which, despite its disadvantages in terms of complexity or scalability, provides the key advantages of enabling ATM-based networks to interwork relatively seamlessly with the ubiquitous legacy LANs, as well as enabling them to utilize existing data applications (in the sense of the IP stack described earlier) without any modification. LANE may thus be viewed as providing the potential for an evolution path for ATM use in campus (or enterprise) data networks, currently dominated by legacy LANs. Although numerous LANE implementations have been tested and deployed in such network scenarios, it may also be noted that evolution of traditional shared media LANs to high-performance "switched" LANs, coupled with their relatively low cost compared to LANE in practical use, has generally inhibited more extensive use of LANE. The recent development of the so-called "gigabit Ethernets" and other LAN technologies no doubt provides strong competition to ATM-based LANE, particularly in terms of relative cost of deployment. But in comparing these alternatives, it should be noted that the LANE approach also enables the potential use of other ATM capabilities, not only in terms of QoS capability, but also with respect to management and signaling features that may be used to provide additional services.

13.3 IP SWITCHING

In conventional IP routing, it will be recalled that each router device makes an independent routing decision on each IP packet (datagram), based on a "best fit" matching of the IP address in the packet header to a routing table built up in the device from independently operating routing protocols as outlined earlier. Such a "store-and-forward" routing and forwarding process, though simple in principle and convenient in practice, may limit the potential throughput of the IP packet flows and hence the attainable overall performance of IP-based internetworking. A number of approaches to overcome such throughput limitations have been proposed and studied by the IETF, in addition to the somewhat direct one of utilizing very high speed processing technology to boost packet forwarding as well as routing algorithms. Apart from such purely IP-based methods to implement so-called "gigabit routers," which have resulted in very high throughput devices, it was recognized by some that the "label switching" hardware fabrics could be combined with IP routing to provide enhanced throughput. The initial attempt to "combine" IP routing with ATM switching to obtain the perceived advantages of both technologies is called "IP switching" and was proposed by Ipsilon Networks Inc. in 1996 [13.2, 13.3], both as a commercial product offering and as IP technology evolution in the IETF.

The fundamental concept underlying Ipsilon's radical approach to combining IP routing with ATM switching hardware in an IP switch is shown in Fig. 13-10. In this architecture, an IP router is modeled as the IP layer software required to provide conventional routing and forwarding of the IP packets, coupled with the data link and MAC layer functions implemented in the layers below the IP layer functions, as outlined earlier. Similarly, a generic ATM switch may be modeled as the combination of the ATM control software, which implements typical control plane (i.e., signaling and call control) functions, coupled with the ATM switch "hardware," which implements functions related to cell switching and buffer management. As indicated in Fig. 13-10, the IP switch concept as proposed by Ipsilon essentially combines the "IP layer software" with the ATM switch "hardware," coupled by means of a so-called "controller" functional block, which provides the internal control between the IP layer software functions and the ATM switch hardware functions as outlined below. The controller also implements the local (or internal) control of the IP flows for either conventional routing and forwarding by the IP layer software, or for high-speed switching through the ATM switch hardware function block.

This notion, whereby the IP switch may dynamically choose between conventional IP routing (and forwarding) and high-speed (cell-based) ATM switching, depending on the characteristics of the IP flow, was termed "cut-through switching." The concept of "cut-through switching" between IP and ATM is central to the IP switching model initially proposed by Ipsilon Inc. and, as will be seen later, also formed the basis of other approaches to interworking between IP and ATM, notably the so-called "multiprotocol over ATM" (MPOA) framework developed by the ATM Forum. In essence, the notion of cut-through switching enables the ATM switch hardware func-

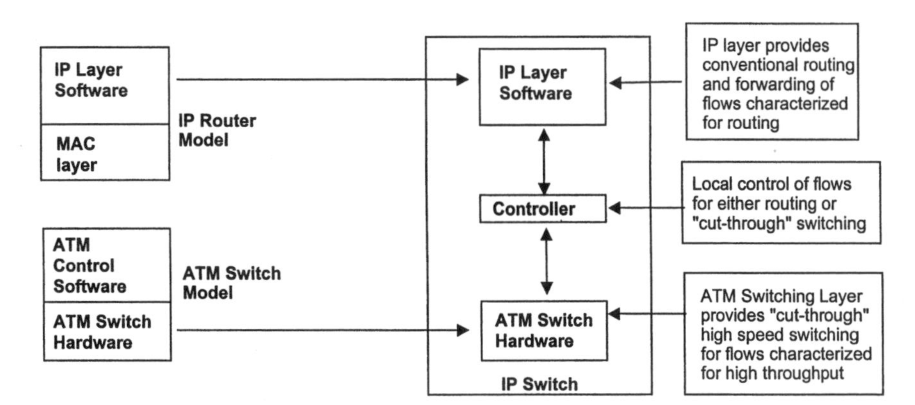

Figure 13-10. IP switching, an initiative by Ipsilon Networks Inc. to combine advantages of IP routing and ATM switching in one network element. IP switch dynamically chooses between IP routing and ATM switching depending on traffic characteristics. Enables significant improvement in packet throughput for low-cost overhead. Initially targeted at campus/enterprise networks evolving to switched Ethernet technology. Initial IP switch implementations use proprietary "signaling" protocol to establish ATM VCCs for cut-through switching. The ATM switch hardware acts as high-speed bypass of routing. Based on "soft-state" flow classification to enable dynamic routing and local control. Can incorporate RSVP capabilities for QoS support.

tional block to act as a local "high-speed by-pass" of the inherently slower routing/forwarding process provided by the IP layer software function block. In the IP switch architecture shown, the controller function block performs the role of determining which IP flow undergoes conventional routing/forwarding and which flow should be cut through to the ATM switching hardware, based on a "soft state" flow classification of the IP traffic flow characteristics. It will be recognized that the IP switch architecture does not constrain the mechanisms whereby such a dynamic flow classification can be performed for cut-through switching, although the detailed algorithms used will determine the overall packet throughput improvements achievable by the IP Switch implementation. In the initial simulation studies, based on actual IP traffic data obtained from Internet backbones, Ipsilon Networks Inc. showed that significant improvements in packet throughput, of the order of 2–5 times, could be obtained relative to basic IP forwarding.

The initial IP switch implementations assumed the use of a proprietary and simplified "signaling" protocol to establish/release the ATM VCCs for cut-through switching, since it is evident from Fig. 13-10 that the normal ATM "control" software (e.g., based on Q.2931 [6.8] or PNNI protocols) is not incorporated in the IP switch. However, it will be recognized that the basic IP switching model does not preclude the use of conventional ATM signaling protocols if required for the establishment of the cut-through VCCs. Since the initial IP switch designs were primarily intended for enterprise or campus network applicatons, it was assumed that not all the functionality (and hence complexity) provided by conventional ATM signaling (whether UNI or PNNI based) would be required and hence a low-cost, "light-weight," signaling procedure would be sufficient for IP switching.

The basic cut-through IP switch operation and associated protocols are summarized in Fig. 13-11 and may be described as a sequence of steps as follows.

1. Initially, each of the IP Switches in the network establishes ATM VCCs on the physical links (i.e., the transmission paths) connecting or meshing the IP switches, using a proprietary (or conventional) signaling protocol.

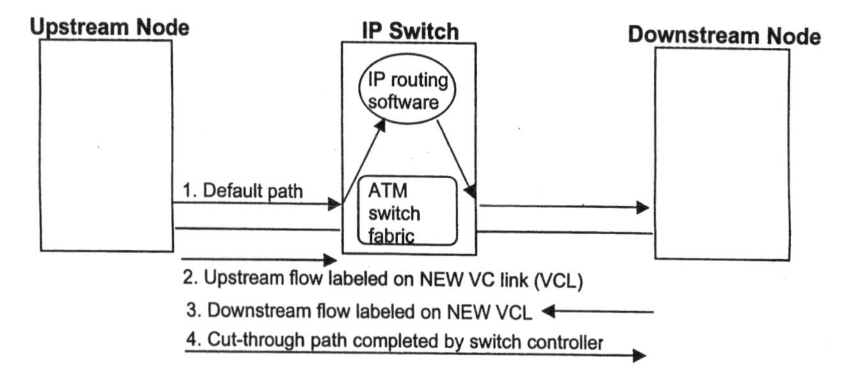

Figure 13-11. IP switch operation and protocols.

2. The IP packets are routed on a so-called "default" path using conventional store-and-forward IP routing, by-passing the ATM switching hardware (fabric) in the IP switch NEs in this phase. It will be recognized that this implies the IP switches incorporate a conventional IP routing protocol (e.g., OSPF, etc.) to compute a path through the network in question as for a normal IP router.
3. The IP switch controller functional block performs "flow classification" on the IP packets traversing the node to identify which of the IP flows would require, or benefit from, cut-through switching by the ATM switching fabric in the given IP switch NE. The criteria for the flow classification, coupled with the detailed algorithms used for this purpose, constitute a broad field of discussion in themselves, but it may be intuitively recognized that IP flows that are relatively "long lived" would be more suitable for the cut-through switching process. This aspect of the IP switch operation is discussed further below, although detailed algorithms for flow classification are clearly outside the scope of this summary.
4. Once an IP flow requiring cut-through switching is identified, by whatever criteria considered suitable and implemented in the IP switch controller, the IP switch requests the upstream node to "label" the IP traffic in question *using a new VCL* (i.e., a new VPI/VCI value on the upstream link). In the initial proposals (and implementations) the node-to-node signaling protocol used for communicating this information between the IP switch nodes was called Ipsilon flow management protocol (IFMP), and was used primarily to associate IP flows with ATM VCCs. However, it may be noted that the process of setting up a new VCL link-by-link may be performed by conventional signaling protocols, with the difference that in IP switching the process is initiated by any node requesting cut-through switching, whereas in conventional ATM signaling a source (terminal or node) initiates VCC set up on a link-by-link basis.
5. Assuming for now that the Ipsilon flow management protocol (IFMP) is used between each of the IP switch nodes to associate IP flows with ATM VCLs, the downstream node may independently request the IP switch controller (in the center node in question, as shown in Fig. 13-11) to label the selected IP flow on *a new VCL* (i.e., other than the default VCL), as identified by a new VPI/VCI value. At this stage in the cut-through procedure it is evident that, from the perspective of the center IP switch node in Fig. 13-11, the IP flow selected for cut-through is associated with new VCLs from both the upstream and downstream IP switch nodes.

6. The IP switch controller (in the given node shown as the center in Fig. 13-11) is now able to identify the selected IP flow to the new input and output port VPI/VCI values and update the context tables used for the ATM switching process accordingly.
7. The IP switch controller completes the "cut-through" switching process via the ATM switch hardware (or fabric) by establishing the VCI mapping between the input and output VCLs on the appropriate ports to the upstream and downstream IP switches. At this stage, it is evident that the IP routing/forwarding functions are by-passed by the IP flow in question, which is now entirely switched (as cells) through the ATM switch hardware or fabric.

The IFMP procedures may also be used to release the ATM VCCs on a link-by-link basis once the IP flow (or session) has been terminated. Various mechanisms for indicating the termination of a flow or session may be used, ranging from the use of a simple "time-out" procedure to sending a dedicated message downstream to signal the termination state. In this, as in other respects, it is evident that the IFMP procedures act as a rudimentary signaling mechanism between IP switch nodes, somewhat blurring the conventional distinction between the connectionless IP domain and connection-oriented ATM protocols. It is also important to recognize that, although the basic IP switching model outlined above makes no assumptions regarding QoS capabilities, both the IP- or ATM-related QoS mechanisms may be incorporated within the model to provide a full range of QoS classes for both the default or cut-through paths. In addition, although not explicitly indicated in the initial IP switching model, it may be assumed that use of the conventional ATM traffic control mechanisms or in-band OAM functions are not precluded for the cut-through ATM VCCs once they are established.

The IP switching model also introduces a new protocol for the "internal" control of the cut-through switching process between the IP routing layer and the ATM switch hardware function block as indicated in Fig. 13-10. This protocol, termed the generic switch management protocol (GSMP) [13.14] is responsible for the internal management and control of the IP switching nodes, as represented by the controller function block in Fig. 13-10. It is of interest to note that the GSMP concept, although initially introduced for the specific purpose of IP switching, has been extended and generalized as a potential "open" ATM control/management protocol for other network element architectures, notably by recent work in the Multi-services Switching Forum (MSF). The primary motivation underlying this extension of the GSMP procedures is that it allows for the specification of a standardized interface between any switching or routing software, which may in principle be remotely located, and the actual switching fabric (or hardware) implementing the packet transfer functions. Thus, in principle, such an interface enables a "controller" element to manage and/or control one or more remote "switching" elements using GSMP messages to set up/tear down connections and perform other management functions while maintaining the call or routing intelligence on a common platform. It has been suggested that such an architecture may result in cost reductions by enabling a unified and common platform to manage and control several diverse switching fabrics, analogous to a "master–slave" control model. Although the use of such an extended GSMP-based approach is outside the scope of this description, it may be noted in passing that the overall performance and reliability implications of such an architecture, as well as its inherent complexity, will need to be considered as well as any potential cost savings, particularly for large-scale networks.

Returning to the basic operating principles for the IP switch (see Fig. 13-11), it is evident that the effectiveness of the cut-through switching in enhancing IP packet throughputs depends largely on the soft state flow classification mechanisms inherent in the IP switch concept. As mentioned earlier, as well as in the context of RSVP-based QoS procedures, the use of IP flow classification mechanisms has been extensively debated by internetworking experts, many of whom argue that the need for IP flow classification limits the scalability of the network. It is argued that growth of IP based internets will result in exponential increase in the number of IP flows and consequently flow-based triggers will be unable to cope, and hence limit scalability. On the other hand, proponents of the IP switch concept have claimed that the initial IP switch implementations, based on simple application-oriented flow classification mechanisms, can significantly enhance IP packet throughput rates up to 5 million packets per second (PPS) using actual Internet backbone traffic profiles. In addition, studies and simulations of IP traffic profiles have been undertaken on either side of the debate in an effort to prove or disprove the feasibility of IP flow characterization as a useful mechanism for IP enhancements. For example, based on studies of Internet traffic traces by Ipsilon Networks Inc. it has been claimed that typically more than 80% of IP packets (or 90% of bytes) may qualify for cut-through switching, since they comprise the so-called "flow-oriented" (or relatively long-term) traffic, as opposed to short-lived IP traffic.

The applications that typically generated such flow-oriented or relatively long-term IP traffic include:

1. File transfer protocol (FTP)
2. TelNet data
3. Multimedia audio or video
4. Web image transfers (www)
5. HyperText transmission protocol (HTTP) data

as well as others exhibiting characteristics similar to these. On the other hand, applications resulting in relatively "short-lived" traffic, which therefore may not benefit from the cut-through switching process, include:

1. Domain name service (DNS) queries or responses
2. Network timing protocol (NTP)
3. Point of presence (POP) exchanges
4. Simple mail transfer protocol (SMTP) data
5. Simple network management protocol (SNMP) queries

Since such categorizations remain at best somewhat approximate, and IP switch throughput performance depends strongly on the ability to identify and rapidly set up the cut-through flows, it will be recognized that use of application-oriented IP flow classification may not result in optimum throughput. Although more complex and efficient mechanisms may result in enhanced throughput, the application-oriented flow classification provides a relatively simple mechanism for triggering the cut-through switching procedures as described earlier. It may be recalled that, in principle at least, an IP switch node may infer the application in question by monitoring the relevant protocol identifier and TCP or UDP port (or socket) codepoints as was outlined earlier when summarizing the TCP and IP coding structures.

Apart from the complexities and even viability of mechanisms based on IP flow classification, the proponents of the IP switching model have also claimed that implementing (in software) the IP switching procedures in IFMP and GSMP is relatively simple when compared to other ATM-based alternatives such as LANE/MPOA, as proposed by the ATM Forum and described below. Underlying such arguments is the assumption that the ATM signaling and routing protocols used in the conventional IPOA network architectures are much more complex than the rather rudimentary signaling and control capabilities in IFMP and GSMP as required for IP switching. At first glance this may well be the case, since the complete ATM signaling and routing (UNI, NNI, or PNNI) protocols defined by the ITU-T and ATM Forum are functionally rich as outlined earlier, whereas the initial specification of IFMP and GSMP were intended simply to support the cut-through switching process in a an IP switch node. On the other hand, if a function-by-function comparison between the two protocol structures is carried out, it remains questionable whether the relative complexity of one or the other is of practical significance. More often than not, such arguments are the result of commercial posturing or wishful thinking, rather than sound engineering practice, and hence may be safely disregarded in a more objective assessment of competing technologies.

Irrespective of any detailed protocol-related comparisons, it is important to recognize that the concept of IP switching introduced a radically new architectural element into the evolution of IP and ATM internetworking in the notion of "cut-through switching" using an ATM VCC, however the latter may be set up. In this respect, it will be noted that the IP switching architectural model is significantly different from the conventional or classical IP over ATM (CIPOA) architecture described earlier. In the CIPOA network architecture (and its extensions), the IP-based routing is clearly separated from the underlying ATM connectivity, which is established using ATM-specific routing and signaling (whether SVC- or PVC-based). Internetworking between these separate networking paradigms is mediated by the ATMARP-based mechanism to provide the binding between the IP and ATM destination addresses for routing purposes, while maintaining the separation between the IP and ATM domains. However, in the IP switching model the notion of cut-through switching effectively blurs the separation between the IP routing domain and ATM switching over connection-oriented virtual channels as seen above. In the IP switching model there are no address resolution mechanisms required, since it is assumed that the end-to-end path is computed using conventional IP routing protocols (e.g., OSFP, BGP, RIP, etc.) based *only* on the IP address. There is no ATM routing or signaling required in this model, since the node-by-node cut-through process simply provides a high-performance, QoS-enabled, bypass virtual channel when required for enhanced throughput.

The IP switch architecture also incorporated the concept of the generic switch management protocol (GSMP), initially for internal control of the cut-through switching process, but subsequently extended substantially for somewhat wider use as a standardized switch interface for the control of remote switching elements. Such extensions of the GSMP concept, although of interest for switching element architecture evolution, are not of primary concern to the evolution of IP and ATM internetworking addressed here.

In effect, the IP switching concept, in attempting to combine the perceived advantages of both IP routing and ATM switching within a network element, triggered a wider interest in the development of so-called "integrated" networking architectures, which utilize both IP- and ATM-related functions in various ways. The two main examples of such approaches to be considered here are the so-called multiprotocol over ATM (MPOA), as developed by the ATM Forum, and the multiprotocol label switching (MPLS) approach adopted by the IETF. As will be evident below, both these internetworking architectural frameworks incorporate concepts introduced by IP switching, even though the underlying protocols used are quite different. The IP switching model itself has been deployed and tested, typically in campus or enterprise environments, but its reliance on nonstandard protocols such as IFMP and GSMP, as well as the uncertainties relating to scalability aspects, has hampered its wider acceptance. The technical work on IP switching was incorporated in the IETF, although the subsequent interest in alternative, more general architectural approaches such as MPLS has overshadowed the IP switching model to a large extent. At this stage, it remains questionable whether the initial IP switch approach will make significant headway in competition with alternatives such as MPLS or MPOA, despite the fact that key underlying concepts in all these architectural frameworks are similar in many respects.

Needless to say, commercial interests and not just purely technical aspects play a significant part in the development of any given architectural approach, particularly in the rapidly changing field of data internetworking, where numerous parallel technological options compete ostensibly for the same ends. In this volatile milieu, the introduction of the IP switching model by a relatively small player in the internetworking marketplace led to other competitive proposals, as would be expected, resulting in a range of potential architectural solutions with varied levels of integration between IP and ATM as well as different starting assumptions. This has resulted in a healthy debate as to the potential evolution of internetworking, and perhaps even a possible convergence between traditional connection-oriented or connectionless networking approaches. Whether or not this may be achievable remains for now an open question. It is also worth recalling that the initial intent underlying the IP switching approach was to substantially enhance the packet throughput of internetworking in comparison with conventional IP-based store-and-forward routing. This objective has also been addressed by the development of the so-called "gigabit routers" capable of very high speed packet forwarding throughputs, in some cases approaching "line" or transmission speeds along the data path. These advances, particularly when coupled with the introduction of very high bit rate SDH- or WDM-based optical transmission systems between the gigabit routers, have also overshadowed the IP switching approach to a large extent.

Although it may reasonably be argued that the same technological advances in optical fiber transmission systems, processor speeds, and routing software may be used to enhance packet forwarding throughputs somewhat irrespective of the network architectural model selected, it must be admitted that the architectural discussion proceeds independently of the technological building blocks. To obtain a more objective assessment of the actual throughput (or QoS) performance of any given approach, it seems clear that an "apples-to-apples" comparison needs to be undertaken, including the hardware and software building blocks used in the network element design. However, this is rarely possible given the varied commercial interests involved, as well as the inherent costs in undertaking such tests. Consequently, the network designer must often resort to a careful function-by-function comparison of the various internetworking architectural approaches on offer, coupled with a critical evaluation of the relative protocol complexity involved for the control (e.g., signaling) as well as management functions. In this respect, it may be of value to construct a simple functional model of the network or protocol architectures being evaluated, assuming the specification of a common set of requirements or capabilities. For the case of ATM and SDH, such an undertaking may be relatively straightforward, given the availability of detailed functional models as outlined earlier. However, this is generally not the case for IP-oriented network elements, particularly for the more recent enhancements for QoS support or increased throughput. Although such functional models will be developed as the architectures are consolidated, at this stage detailed functional comparison may be limited to the conventional network element architectures.

13.4 MULTIPROTOCOL OVER ATM (MPOA)

In the so-called classical IP over ATM (CIPOA) model outlined earlier, it will be recalled that the IP layer (including the applications in the layers above) is simply treated as any other application to be transported and switched by the underlying ATM network. The "differences" between the IP routing and ATM switching paradigms are taken care of in this simple model by the ATMARP functionality and its extensions such as NHRP, MARS, etc., as indicated earlier. Although it is clear that this straightforward approach to IP–ATM internetworking is architecturally complete as well as relatively simple to implement, the development of the IP switching concept resulted in the desire to achieve more "integrated" internetworking architectures, which could be viewed as extensions of CIPOA and IP switching based concepts. In the ATM Forum, this work was viewed as an extension of IP over ATM and was termed "multiprotocol over ATM" (MPOA) [13.9], although it incorporated a variation of the IP switching concept in an attempt to build a more integrated architectural model. In addition, a primary objective of the MPOA approach to integration was to utilize existing protocols such as LANE, NHRP, MARS, etc. as well as ATM signaling and routing (i.e., UNI and PNNI) wherever possible, to minimize the need for new developments, as well as achieve interoperability with existing networks.

The use of the term "multiprotocol" in MPOA (as well as other IP-related architectural approaches) needs to be clear in this context. The multiple protocols referred to here include typical Layer 3 data networking protocols such as IP, IPX, AppleTalk, SNA, etc., which may be identified by means of the protocol identifier (PID) codepoints in the LLC/SNAP encapsulation procedure outlined earlier. In this sense, the "multiprotocol" capability implied in MPOA is identical to that available in CIPOA, assuming that the LLC/SNAP encapsulation described in RFC 2684 [11.14] is used. As will be evident below, the MPOA specification mandates the use of the LLC/SNAP encapsulation for IP and other Layer 3 packets to provide the multiprotocol multiplexing capability. It is also of interest to note in passing that the MPOA architecture may also be extended for use with other Layer 2 technologies than ATM, notably frame relay. For this reason, MPOA is sometimes referred to as a "framework architecture" that may be applicable to other underlying transfer technologies. However, the focus here is on ATM, since it clearly provides the most comprehensive capabilities from either transport, control, or management perspectives at this stage of MPOA definition.

The basic MPOA architecture is shown in Fig. 13-12, which also introduces some of the terminology used in this framework, which may be seen to based on the typical "client–server" model inherent to data networking. The two main functional components in MPOA are termed the edge device or MPOA host, and the "MPOA server" or router (MPS). It should be noted that these components are connected by means of an emulated LAN (ELAN) utilizing the LANE protocols outlined earlier. The incorporation of LANE functionality in the overall MPOA framework marks a key point of difference between the MPOA approach and the CIPOA architecture,

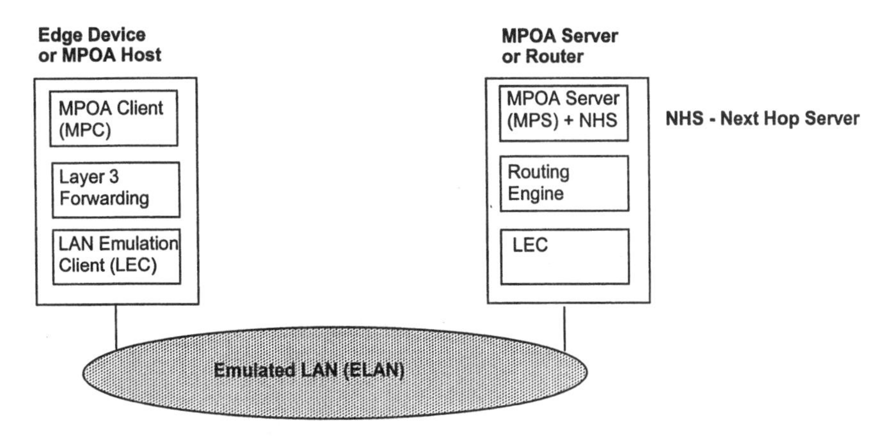

Figure 13-12. Basic MPOA architecture.

which does not imply the use of LANE, as will be recalled. Since connectivity between the edge devices and the MPOA server is enabled over an ELAN, it is evident that each of these components needs to incorporate the LAN emulation client (LEC) functionality as part of its structure. The edge device or MPOA host comprises two main functional blocks as shown, apart from the LEC. These are:

1. MPOA Client (MPC). This block includes the functionality to provide IP flow characterization and the control of the so-called shortcut paths via ATM VCCs. It is also responsible for the establishment and release of the ATM VCCs used for the transport of the IP (or other Layer 3) data packets, using conventional ATM signaling protocols. The MPC also includes the capability to initiate the NHRP procedures (query/response messaging) to identify the destination ATM address when required. These procedures are described in further detail below.
2. Layer 3 Forwarding. The edge device (MPOA Host) includes the capability to "forward" the IP packets by incorporating a forwarding engine functionality analogous to that provided in a conventional "store-and-forward" router. It is important to recognize that the related "routing" functionality, whereby the optimum path is computed based on some routing protocols and algorithms, is *not* part of the edge device or MPOA host, but is logically "separated" to be part of the MPOA server as outlined below. This concept, whereby the packet forwarding capability may be logically "separated" from the routing functionality, is termed the virtual routing concept in MPOA, although it may be applicable to other networking architectures as well.

Similarly, the MPOA server (or router) comprises two main functional elements in addition to the LEC functional block as shown in Fig. 13-12. These include:

1. MPOA Server (MPS). This block includes the full next hop resolution protocol (NHRP) server functionality, as outlined earlier, to provide the next hop or internetwork address resolution information to the MPCs required for forwarding of IP packets. It may also provide data link layer (DLL) related information if required for egress MPCs as described in more detail below.
2. Routing Engine. As indicated earlier in relation to the concept of virtual routing, the routing engine comprises the "routing" part of a conventional Layer 3 router device, which processes the dynamic routing protocols and related algorithms for computation of the optimum path through the (connectionless) network for transport of the IP packets.

In effect, the MPOA server component may be viewed as a conventional IP router that has been enhanced by incorporating the next hop server (NHS) capability as well as a LEC interface (or LUNI) to enable it to operate as part of an emulated LAN. The packet forwarding functions are "removed" from this component and are logically considered as part of the edge device or MPOA host in keeping with the notion of virtual routing in MPOA framework.

Although for simplicity Fig. 13-12 shows an MPOA edge device comprising a single MPOA client with a single LEC interface, in principle an MPOA edge device may consist of one or more MPCs; each MPC may be associated with one or more LECs as shown in Fig. 13-13. This hierarchical relationship between the edge device, the MPCs, and LECs shows that any given LEC is associated with one and only one MPC. More generally, any given MPC in the hierarchy may communicate with one or more MPS if required. In this way, the MPOA architecture is intended to provide maximum flexibility, within the constraints of the virtual routing concept, without constraining the way in which the functional blocks are implemented in any given phys-

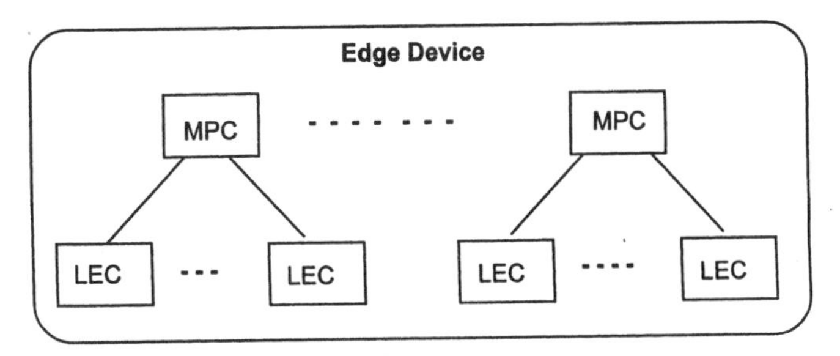

Figure 13-13. Relation between edge device, MPC, and LEC.

ical device. The MPOA architecture is also intended to improve the overall scalability and manageability of the network by reducing the number of nodes required for the routing function within any given subnetwork and thus also resulting in fewer configurable (routing) nodes. However, it has also been argued that these potential advantages of MPOA may be offset by the significant overall complexity of the MPOA approach, which essentially requires both LANE and IP router functionality together with the underlying ATM related protocols. In practice, there are always trade-offs between the desired flexibility and resulting complexity within any given network design, so that any detailed description based on a rigorous function-by-function comparison with alternative approaches remains outside the scope of this general overview.

It is of more interest here to examine the underlying principles for data transfer in the MPOA architectural model (or framework). This can be readily understood by analyzing the information flows in the MPOA model, as shown for a simple two-element MPOA network in Fig. 13-14 and listed in Table 13-2. In the first place, it is important to note that MPOA distinguishes between two categories of information flow. These are (1) the MPOA control flows and (2) the MPOA data flows. In the basic MPOA network configuration shown, comprising ingress and egress MPOA Hosts together with their corresponding MPOA servers or routers connected over ELANs, all the information flows between the MPOA functional components are identified. For simplicity, only the LAN emulation configuration server (LECs) functional block is shown separately from the other LANE components described earlier, in order to highlight the configuration flows for MPOA. However, it may be assumed that the ELANs represented in Fig. 13-14 include all the required functionality for conventional LANE operation over the underlying ATM connectivity as described earlier.

The MPOA Control Flows represented in Fig. 13-14 include the following:

1. Configuration Flows. As the name implies, the MPOA configuration flows are used to exchange configuration-related information between the LECs and the associated MPCs and MPSs elements. In addition to the normal LANE related configuration information, this may also include MPOA- (i.e., router-) specific information such as node identifiers or IP addresses as well as topology information necessary for initialization of the nodes comprising the MPOA network. It is also of interest to note that the MPOA specification does not necessarily require the use of ATM VCCs between the MPOA devices for the configuration flows, in contrast to the other information flows outlined below. The configuration information used for network initialization may be exchanged over any available interface, including the management interfaces if required.

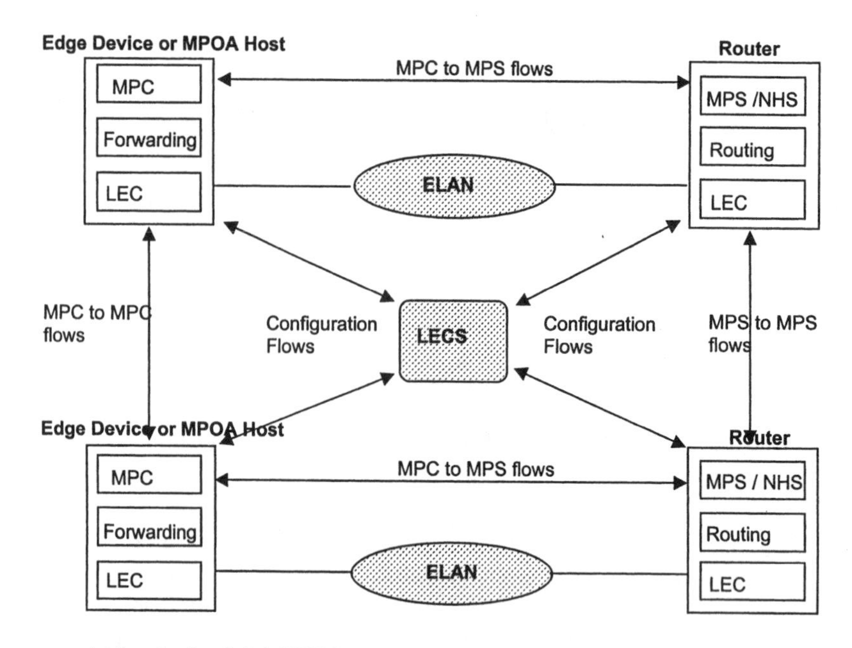

Figure 13-14. Information flows in MPOA.

2. MPC to MPS Control Flows. These messages are used by the MPC to obtain and update the forwarding information from the MPS routing function. In addition, the MPS may also initiate the request for forwarding information in the event that the MPC has not initiated the procedure. In essence, these flows are analogous to conventional address resolution information exchanges, as further described below.
3. MPS to MPS Control Flows. These flows refer to the conventional routing protocols between IP routers as well as the NHRP message exchanges for internetwork address ex-

TABLE 13-2. Information flow descriptions

MPOA control flows	Description
Configuration flow	Used for configuration information between LECS and MPCs and MPS.
MPC to MPS	Used by MPC to obtain (and clear) forwarding information. MPS may also trigger request for forwarding information.
MPS to MPS	Uses conventional internetwork routing protocols (as between routers) and NHRP. MPOA does *not* define any new control procedures between routers.
MPC to MPC	Used for data "purge" messages if misdirected packets are received.
MPOA data flows	**Description**
MPC to MPC	Used to transfer data over "shortcut" ATM VCCs. The MPC to MPC flow is the *only* data flow defined by MPOA protocols. Other data flows use conventional routed path via routers.
MPC to NHC	Used for unicast data to and from next hop client (NHC)

changes. It is important to recognize that MPOA does not define or require any new control procedures (or messaging) between the routers and relies entirely on any existing routing protocols coupled with the NHRP capabilities.

4. MPC to MPC Control Flow. This control flow is used only for the "purging" of misdirected packets, analogous to the flushing procedures used in LANE, as mentioned earlier. As in that case the MPOA, purge messages are sent along the data path but should be distinguished from the normal user data flows from MPC to MPC as described below.

The MPOA data flows represented in Fig. 13-14 include:

1. MPC to MPC Data Flow. The MPC to MPC data flow is the only data flow specific to the MPOA protocols, and constitutes the user data packets transferred over the so-called "shortcut" ATM VCCs established between the source (ingress) and destination (egress) MPCs in the MPOA network. The criteria and procedures used for the "shortcut" ATM VCCs are described in further detail below, but it may be noted here that this aspect signifies the key underlying mechanism in the MPOA architecture. User data flows that are *not* earmarked for transfer over the MPOA shortcut ATM VCCs use the conventional hop-by-hop routed path as established by the routing capability in the MPOA routers. These paths are termed default paths in the MPOA framework.
2. MPC to NHC Data Flow. This flow is used to refer to the (unicast) data transferred between the MPC and the next hop client (NHC) in a different logical IP subnet (LIS) for the case where an MPOA network communicates with a conventional IP over NBMA (ION) network. This flow may be viewed as a special case of the general MPC to MPC data flow listed above, where the destination "MPC" is in fact a client node attached to an ION network which includes the NHRP capability. This type of MPOA data flow is not further addressed here as it relates to a particular interworking scenario.

13.5 MPOA FUNCTIONAL OVERVIEW AND DATA TRANSFER PROCEDURES

The general MPOA architectural model and associated information flows outlined above demonstrate how MPOA attempts to extend the concepts underlying classical IP over ATM, NHRP, IP switching, as well as LANE in an integrated networking framework. This meets the primary objective of utilizing existing internetworking protocols and network architectures wherever possible while allowing for a flexible evolution path from presently deployed ATM- and IP-based networks with minimum "new" development costs. It may be recognized from the MPOA information flows listed above that in essence the only new information flow introduced corresponds to the so-called "shortcut path" between ingress and egress MPCs to transfer IP packets over ATM VCCs when required, in accordance with a set of criteria. The notion of such a shortcut path for IP data transfer is clearly a "network-wide" extension of the basic concept underlying IP switching, namely, that of a cut-through path within a node using ATM-based switching to enhance throughput. This similarity between IP switching and MPOA extends to the use of some form of IP flow characterization (which may well be simply application-based) to select between packets destined for the shortcut path and those that may be conventionally routed using hop-by-hop "store and forward" IP routing between source and destination hosts. In describing the MPOA functionality in further detail below it is useful to bear in mind this underlying similarity between the IP switching concept and MPOA, while recognizing that whereas IP switching relied on new specific protocols such as GSMP and IFMP to establish cut-through paths, MPOA utilizes existing ATM signaling, LANE- and CIPOA-related protocols in an integrated framework.

Somewhat analogous to LANE operation, on which it clearly builds, MPOA functions may be categorized into five broad groups as listed below:

1. Configuration. The individual MPCs and MPS functional elements comprising the MPOA network clearly need to exchange and process network-configuration-related information during initialization using the MPOA configuration flows mentioned above. The "default" procedures rely on the LAN emulation configuration server (LECS) using the LE_CONFIGURE request and response message set listed earlier and enhanced to include the specific MPOA device type, namely MPC or MPS. However, configuration may also be performed using other mechanisms such as information exchange over the external management interfaces, drawing on provisioned data contained in a MIB.
2. Discovery. In the discovery phase, the MPCs and MPSs automatically "discover" each other's addresses and any related topological information using an extension to the LANE protocols to include the MPOA device type (i.e., whether MPC or MPS). The messaging is based on the LE_ARP procedures described earlier. The topological information may be updated periodically, or as required, depending on whether such information is subject to aging.
3. Target Resolution. In the MPOA framework, the "target resolution" procedures relate to dynamic address resolution based on extensions to the next hop resolution protocol (NHRP) described earlier and used in much the same way as for extending the CIPOA network architecture as outlined earlier. Thus, in the MPOA target resolution phase, the ATM address of the destination (or egress) MPC is provided by the NHRP address resolution procedures and is then used to establish the "shortcut" ATM VCC for subsequent MPC to MPC data transfer, as described in further detail below.
4. Connection Management. In MPOA, the connection management functions include the set-up, release, and maintenance of the ATM VCCs used for the short-cut data transfer paths. It is important to recognize that the MPCs and MPSs utilize conventional ATM signaling (and routing) procedures for ATM connection management, whether it is UNI- or NNI-based. Consequently, the shortcut path may potentially provide the complete range of QoS and OAM capabilities available to the underlying ATM network for the data transfer.
5. Data Transfer. The data transfer functions in MPOA relate to the forwarding of the internetworking layer (e.g., IP) data packets. As described below, MPOA essentially defines two modes of data transfer. The so-called "default" mode uses conventional (Layer 3 or IP) routing procedures and therefore does not require any address resolution (i.e., ATMARP) procedures. In principle, the default mode may utilize any Layer 2 transfer mode including ATM or may use conventional IP networks. Alternatively, in the shortcut mode, as indicated earlier, the data packets are transferred over a "shortcut" ATM VCC established directly between the ingress and egress MPCs using the target resolution mechanism alluded to above.

In describing the MPOA target resolution and data transfer procedures, it is useful to consider the simplified MPOA network configuration shown in Fig. 13-15, which consists of an "ingress" and "egress" MPC and MPS pair reachable over one or more ELANs and/or LISs. Although not explicitly shown, it is also assumed that the ingress and egress MPCs may be linked over an underlying ATM network, which may or may not support any intermediate ELAN connectivity. For both conventional IP packet transfer and ATM-based transport, it may be assumed that the intermediate (or wide area) networks are not constrained by the MPOA procedures outlined below. Consequently, the general MPOA procedures may conveniently be described at the

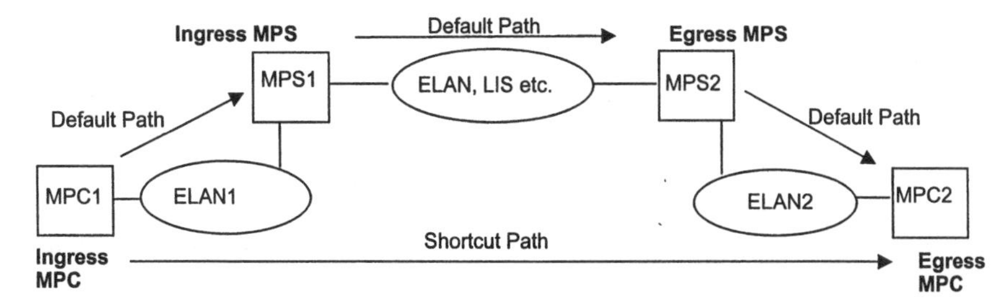

Ingress MPC: Obtains MAC address of attached MPS1 using LE_ARP and device type. Performs flow detection based on Layer 3 destination addresses. Decides if a given flow should be via default or shortcut path based on (programmable) threshold.Default forwarding of packets is by normal router-based forwarding. If shortcut flow required Ingress MPC performs MPOA target (address) resolution to ingress MPS to obtain ATM address of egress MPC.Sets up shortcut ATM VCC to egress MPC and forwards data packets directly.

Ingress MPS Processes target (address) resolution requests from MPCs. If local may respond directly. Otherwise uses NHRP to obtain from remote NHS.Ingress MPS needs to re-originate NHRP resolution requests by using its internetwork address as source. All other MPC information is sent to NHS. After receiving destination ATM address ingress MPS sends to ingress MPC.

Egress MPS: When NHRP resolution request for attached MPC2 is received the egress MPS generates MPOA Cache Imposition Request message to egress MPC. The MPOA Cache Imposition message provides state maintenance and encapsulation information to egress MPC to check if egress MPC can accept ATM VCC request. If Cache Imposition Reply from egress MPC indicates ATM VCC may be accepted, egress MPS sends NHRP Resolution Reply towards ingress MPS.

Egress MPC: Replies to MPOA Cache Imposition Request from egress MPS after determining if resources available for new VCC. If not error status is returned. If yes, sends ATM address and relevant state / tag information in MPOA Cache Imposition Reply message to egress MPS. Tags may be used to improve egress cache look-up or resolve conflicting forwarding information.

Figure 13-15. Target resolution/data transfer procedures—ingress and egress.

ingress and egress nodes in turn, while assuming that the intervening data transfer depends on the specifics of the intermediate wide area networks employed. It should also be noted that although the description here refers to the "ingress" (i.e., source) and "egress" (i.e., destination) MPC/MPS pair, this simply pertains to unidirectional data transfer. It should be clear that for bidirectional data transfer, both the ingress and egress functions will be required at each end of the MPOA network shown in Fig. 13-15.

The ingress MPC (MPC1) obtains the MAC address of the associated MPS (MPS1) during the discovery phase, as indicated earlier by using the LE_ARP message set, which includes the MPOA device type, assuming that the initial configuration of the MPOA network has been completed using either LANE procedures or the management interfaces (MIBs) as indicated earlier. The ingress MPC performs the "flow characterization" process on incoming data packets based on the destination (Layer 3) address and any other criteria selected by the MPOA network. Although MPOA does not constrain the mechanisms whereby the flow characterization may be implemented in the MPC, a simple default flow detection mechanism based on packet rate is provided in the MPOA specification, as described further below. The flow characterization of the incoming data packets is used by the ingress MPC to determine whether a given packet flow should be transferred via the default path shown, or over a direct shortcut path to the destination (or egress) MPC. As noted earlier, the default path involves the normal router-based forwarding of the IP packets hop-by-hop from the ingress MPC to the egress MPC (or more strictly the MPOA host or edge device). It will be recalled that MPOA assumes that the forwarding function is logically associated with the MPOA host, whereas the Layer 3 routing functions reside in the MPOA server, in accordance with the notion of virtual routing. If the MPC determines that a given IP flow requires a shortcut path, it must perform a target (address) resolution process to obtain the ATM address of the egress (or destination) MPC (MPC2). Target resolution in MPOA simply utilizes the NHRP procedures over multiple LISs to obtain the destination <IP address:ATM address> pair as described earlier. Once the destination ATM address is known, the ingress MPC can establish the shortcut ATM VCC directly to the egress MPC using conventional (UNI and NNI) ATM signaling procedures and transfer the IP packets directly.

The ingress MPS (MPS 1) functions include the processing of the target (address) resolution requests from the associated MPCs, which can be used to set up the shortcut ATM VCCs. If the required <ATM address:IP address> pair is available locally, the MPS may respond directly, otherwise it uses the NHRP messaging procedures to obtain the required address pair from the remote next hop server (NHS). An important function performed by the ingress MPS is to "reoriginate" the NHRP resolution requests by using its IP address as the source address in the NHRP messages. The concept of IP address reorigination in MPOA is used to ensure that the NHRP address resolution response is only returned to the originating MPS and not directly to the associated MPC or edge device. This is consistent with the notion that the MPS is performing the role of a (virtual) router to the edge devices and, therefore, the IP address of the MPCs need not appear as a source address in the NHRP message set. Once the ingress MPS receives the ATM address of the egress MPC (MPC2), it sends this to the ingress MPC to initiate the shortcut ATM VCC establishment. Alternatively, for the default path, the MPOA server simply functions as a conventional IP router device to provide hop-by-hop routing of the selected IP packet over the default path. It may be noted that the MPOA framework does not constrain the underlying transfer mode for the IP packets over the default path in any way, so that ATM (or FBRS) virtual circuits may be used for these if available.

When the egress MPS (MPS2, as shown in Fig. 13-15) receives the NHRP resolution request for the associated MPC2 attached to it via ELAN2, it sends a control message to the relevant MPC (called a MPOA cache imposition message as described below) to ascertain whether the egress MPC can accept the anticipated ATM VCC set-up request in terms of available resources and other related compatibility parameters. This procedure allows the shortcut data transfer to be limited by the resources available at the receiver (i.e., MPC2) in much the same sense as for the case of RSVP-based receives resource limitation. Assuming that the egress MPC can accept the anticipated ATM VCC, it will indicate this in the corresponding cache imposition reply message to the egress MPS. The egress MPS may then return the NHRP resolution request message to the ingress MPS with the required ATM address information for the egress MPC. It is important to recall that as a result of the IP address reorigination at the ingress MPS side, the NHRP address resolution request and response messaging will only occur between the ingress and egress MPS functional elements and not directly to the attached MPCs. In the MPOA architecture, separate messaging is used between the MPS and associated MPCs to convey the relevant addressing (and other) information.

The egress MPC (MPC2) responds to the MPOA cache imposition request messaging from the egress MPS as indicated above, after determining that sufficient resources are available at the receiver to accommodate the new ATM VCC. If the receiver cannot accommodate the addition of a new VCC, an error status is returned to the egress MPS. If the receiver can accept the anticipated shortcut VCC, it responds with the relevant state information and ATM address in the MPOA cache imposition reply message to the egress MPS. MPOA also introduces the notion of "tags" in the address resolution messages, which are intended to speed up the cache look-up process and resolve conflicting forwarding information resulting from address look-ups. The use of such tags in MPOA cache management procedures is optional and may be viewed as a refinement for performance enhancement and hence not of primary interest here.

The MPOA message flows required for the shortcut path procedures outlined above are summarized in Fig. 13-16. It should be noted that no such messaging is required for selection of data transfer over the default path in MPOA, since this implies conventional IP store and forward routing of the data packets as indicated above. However, if the ingress MPC elects to send the IP flow over the short-cut path, the control messaging sequence shown may be described as follows:

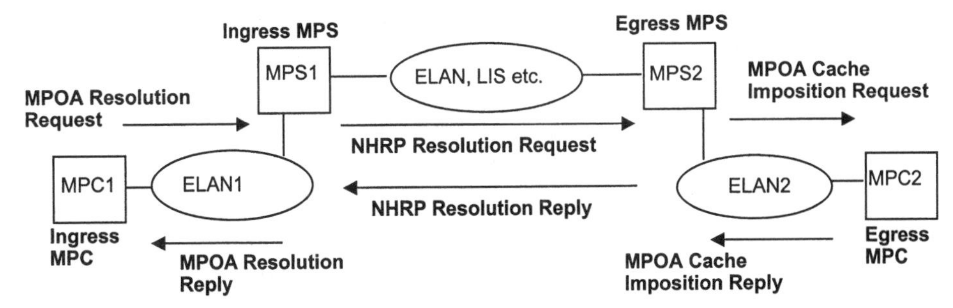

MPOA Resolution / Reply Messages:
Exchanged between ingress MPC and MPS to obtain destination ATM address corresponding to destination internetwork layer address: <ATM address: Internetwork address> tuple+other parameters. This exchange only occurs if MPC determines the monitored flow needs shortcut VCC.

NHRP Resolution Request / Request Messages
Exchanged between NHS function in ingress and egress MPSs to obtain remote <ATM address: Internetwork address> tuple. MPS "translates" MPOA Resolution Request/Reply to NHRP resolution/Reply, based on concept on "Re-Origination". Required since MPC may not have internetwork address.

MPOA Cache Imposition Request / Reply Messages
Exchanged between egress MPS and MPC to provide state + address information from target.
Receiver confirms if sufficient resource available to accept VCC and protocol compatibility.

Figure 13-16. MPOA resolution messages.

1. The ingress MPC sends the MPOA resolution request message to the ingress MPS to obtain the destination <ATM address:IP address> pair together with any other required parameters for the shortcut data transfer. The MPOA resolution request message is directly analogous to a conventional ATM–ARP request and is transferred over the ELAN1 by using LLC/SNAP-based encapsulation into AAL Type 5 CPCS as described earlier.
2. If the required ATM address information is not already cached (or otherwise available) by the ingress MPS it will generate the NHRP resolution request message over the intervening IP subnets (NHS) capabilities to locate the relevant egress MPS. It will also be recalled that the ingress MPS translates the MPOA resolution request to NHRP resolution request by replacing its internetworking (or Layer 3 or IP) address in place of that of the ingress MPC, in accordance with the concept of reorigination as described earlier. An added advantage of the reorigination process is that it eliminates the need for MPCs to have a Layer 3 (or IP) address, which may simplify the MPOA host.
3. When the egress MPS receives the NHRP resolution request message it sends a "MPOA cache imposition request" message to the relevant egress MPC over the egress ELAN to obtain the required state and address information from the egress MPC. This message enables the receiver to confirm if sufficient resources are available to accommodate the requirements of the shortcut ATM VCC as well as check protocol compatibility at the data link layer (DLL) and MAC layer if required. Although the MPOA cache imposition request message is somewhat analogous to conventional ATM-ARP messaging (enhanced by the other MPOA-related requirements indicated earlier), the use of the term "cache imposition" in this message set highlights the role of the MPOA cache management procedures and the state information requirements in the MPOA framework. It will be evident that the MPOA cache imposition request message may initiate update in the MPC state in its local cache "memory."
4. If the egress MPC can accept the new ATM VCC for shortcut data transfer, it generates the "MPOA cache imposition reply" message to the egress MPS over the ELAN2 as shown, which contains the required state and address information.

5. After receiving and processing the MPOA cache imposition reply, including the Layer 3 address reorigination procedure when applicable, the egress MPS returns the "NHRP resolution reply" message to the ingress MPS over the intervening LISs (or other networks).
6. Finally, the ingress MPS sends the required ATM address and other related state information to the ingress MPC contained in the "MPOA resolution reply" message, to enable the ingress MPC to initiate the shortcut ATM VCC set-up directly to the egress MPC using normal ATM signaling procedures.

It will be evident from the above description of the MPOA messaging sequence that the procedures are essentially an extension of the earlier CIPOA-related ATMARP/NHRP capabilities to accommodate virtual routing as well as LANE concepts. As mentioned earlier, the use of a shortcut ATM VCC directly between the ingress and egress MPCs may also be viewed as an extension or generalization of the IP switching concept from individual nodes to a network-wide approach. However one chooses to view the MPOA framework and associated messaging sequence, it is clear that its basis is essentially ATM-oriented in the sense that it utilizes, as far as possible, ATM signaling protocols coupled with the relevant ATM transfer and layer management capabilities for QoS and traffic engineering requirements. Even from this somewhat cursory description of the MPOA messaging procedures, it will be evident that the inherent latency for the shortcut path establishment may be significantly improved by the efficient "caching" of the addressing and routing information in the MPOA network elements. As mentioned earlier in relation to the CIPOA/NHRP procedures, the address resolution mechanisms may be regarded essentially as an automated "directory service," whereby distributed databases may be polled for the required routing (i.e., address) information. Consequently, if the means can be provided for the caching and update of frequently used destination addresses, the latency resulting from ARP messaging may be effectively reduced. In practice, this aspect will need to be traded off against the scalability of the databases included, particularly for very large networks.

The significance of efficient "cache management" in MPOA will be evident from the perspective of enhancing latency and scalability performance, given the ARP-based procedures underlying the MPOA network architecture. In the MPOA framework, the term "cache" refers to the temporary storage or memory functions that are used to maintain the address-binding information as well as any other state information in the individual MPC and MPS. The "temporary" nature of such information, coupled with the need for periodic "aging" of state information, depends on a number of factors including the specific deployment scenarios, operator policy, and size of the MPOA network. It may therefore be viewed as relative, with programmable "holding times" that may vary widely with network application and service provider requirements. Although it is not necessary here to describe in detail the MPOA cache management procedures, it is useful to briefly summarize the underlying principles used, since these may also be of interest to other ARP-related (client–server) network architectures. In general, it must be recognized that the MPOA cache management and information control procedures are likely to be complex, considering that the architecture attempts to synthesize both IP routing and ATM signaling functions as well as LANE capabilities within a unified architectural framework.

In MPOA, the ingress and egress MPC and MPS cache management procedures are considered separately, although, in general, all MPOA-related caches are subject to "aging" and updates, with settable "holding times" that may be determined largely by network operator policy and configuration.

As shown in Fig. 13-17, the cache information is managed in terms of "keys" and "content" for both ingress and egress nodes, although the information contained will clearly be different in either case. The ingress MPC creates new ingress cache entries when it detects IP packet flows for which there are no existing cache entries. The table cache content "keys" in this case include

Ingress Cache Management

Keys		Content		
MPS ATM Address	Internetwork Destination Address	Destination ATM address or VCC	Encapsulation Information	Other control information (e.g. flow count, hold time)

The ingress MPC creates new ingress cache entries when it monitors packet flow for which there is no existing entry. The ingress cache entry is completed when the target resolution reply is received. Cache entries may be removed by "Purge Request" in event of errors.

Egress Cache Management

Keys			Content		
Internetwork Destination address	Source/Destination ATM Address	Tag (optional)	LEC	DLL Header	Other control information (e.g. hold time etc.)

The egress MPC may create cache entries as a result of Cache Imposition Request from MPS using source ATM address and destination internetwork address as keys, with holding time given by the MPS. If cache entry is invalid egress MPC may initiate data Purge procedures.

Figure 13-17. Cache management in MPOA.

the ATM address of the ingress MPS and the internetworking (i.e., Layer 3 or IP) address of the destination host or MPS. The information content of the ingress MPOA cache can be completed when the target resolution reply message has been received, since this provides the destination ATM address required for the shortcut VCC. The ingress cache content may also include other control and encapsulation (protocol compatibility) information as shown. In the event of error or outdated cache information, MPOA specifies so-called "purge messages," which may be used to remove the errored information. In this case, a given ingress cache entry may be removed by using a "purge request" message.

For the case of the egress MPC, the cache table entries are created as a result of receiving the cache imposition request message from the egress MPS as described earlier. The cache context table "keys" include the destination internetworking (i.e., Layer 3 or IP) address and source as well as destination ATM address. As mentioned earlier, MPOA introduces the notion of an optional "tag," which may be used in egress cache management to identify the destination <IP:ATM> address pairs, which allows for more efficient table look-up. The egress cache content may include the LEC identifier and related data link layer (DLL) header as well as other control information to maintain state relating to the MPOA operations. Some examples of the MPC cache control information are listed in Table 13-3, and although it is not the intent here to describe in detail the individual functions, their use can readily be inferred from the list in most cases. The cache control functions listed, however, point to the importance and relative cache management complexity involved in MPOA operation. It may also be noted that the cache management procedures require the use of "identifiers" for the correlation of the various request and response messages between the MPS/MPC pairs. The use of purge procedures for the deletion of errored cache entries should also be noted. The purge messages are sent along the shortcut data path, even though they constitute control messages as indicated earlier. The detailed procedures for the MPC cache management functions are given in the ATM Forum MPOA Version 1 specifications, together with the state diagrams.

13.6 FLOW CHARACTERIZATION AND CONNECTION MANAGEMENT IN MPOA

It was noted earlier that the monitoring and characterization of the incoming IP packet flows constitute an important function of the ingress MPC. Monitoring of the packet flows to a given destination internetworking (i.e., IP or other Layer 3) address is a requirement in MPOA in order

TABLE 13-3. MPC cache control information

Function	Ingress MPC	Egress MPC
State	Shortcut entry state Shortcut VCC state Ingress MPS ATM address	Shortcut entry state Shortcut VCC state Egress MPS ATM address Cache ID
Connection	ATM address of egress MPC Service category VPI/VCI for packet forwarding	Ingress MPC ATM address Service category VPI/VCI
Aging	Holding time from ingress MPS	Holding time from egress MPS
Retry	Request ID Retry timer Hold down timer	Not applicable
Usage	Number of forwarded packets Recycle information (optional)	Number of packets received Recycle information (optional)
Purge	Ingress MPS protocol address Egress cache tag	Egress MPS protocol address Ingress MPC ATM address

for the MPC to decide if the (IP) packets are to be sent over the so-called "default" path, whereby the IP packets are routed conventionally over the intervening IP subnetworks, or whether they are to be sent over a directly established "shortcut" ATM VCC to the egress MPC device. Whereas the mechanisms and criteria for selecting the threshold for default or shortcut paths is largely left to the ingenuity of the implementor, a simple rate-based flow thresholding procedure is specified in MPOA as a default mechanism in the event that more sophisticated approaches are not warranted.

As shown in Fig. 13-18, this simple rate-based flow thresholding procedure relies on two parameters that may be settable over a wide range. These parameters are defined as:

Parameter name	Variable	Description
1. Shortcut set-up frame count	MPC-p1	Count of packets
2. Shortcut set-up frame time	MPC-p2	Time interval

The suggested range and default values of these parameters are:

1. MPC-p1 Packet count (minimum = 1; default = 10; maximum = 65535)
2. MPC-p2 Time interval (minimum = 1 sec; default = 1 sec.; maximum = 60 secs)

Based on these parameters, the MPOA default flow monitoring threshold procedures simply require that if there are more than (or equal to) MPC-p1 frames (packets) in an interval of MPC-p2 seconds, then the ingress MPC may initiate the MPOA target resolution messaging procedures and subsequently establish the shortcut ATM VCC for the IP packet flow. For example, for the suggested default parameter values, a flow rate of 10 frames or more per second would qualify for packet transfer over a shortcut ATM VCC. As may be evident from the wide range of the settable parameters, this threshold may be varied over a wide range of flow rates. While conceptually straightforward, it may be evident that such a simple rate-based threshold may not be sufficiently efficient for many network applications in which it may be more effective to select the

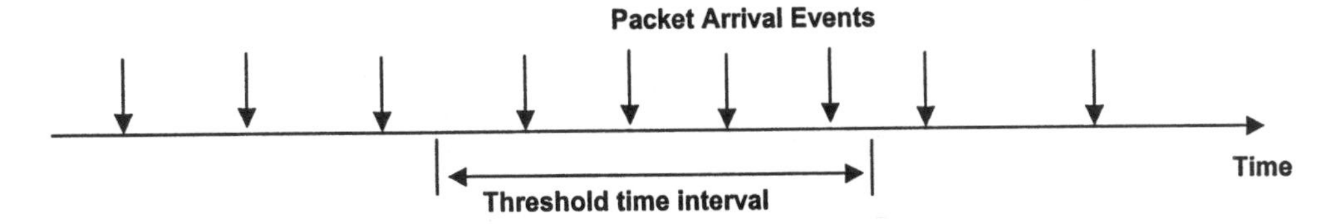

Flow thresholding is based on two parameters, and may be settable over a specified range:

Parameter Name	Variable	Description and Values
Shortcut-Setup Frame Count	MPC-p1	Count of packets (Min=1,Default=10,Max=65535)
Shortcut-Setup Frame Time	MPC-p2	Time interval (Min=1sec.,Default=1, Max=60 seconds)

Threshold Procedure

If there are more than or equal to MPC-p1 frames in an interval of MPC-p2 seconds than the MPC may initiate MPOA Target Resolution procedures and shortcut ATM VCC setup for the packet flow. Other flow monitoring / thresholding mechanisms (e.g. RSVP) may also be used for flow characterization in MPOA.

Figure 13-18. Flow thresholding in MPOA.

shortcut path based on the characteristics of the service or application itself, or other relevant parameters. These alternatives are clearly not precluded within the MPOA framework, but their wider discussion has been somewhat overshadowed by the more general controversy over flow characterization as a scalable mechanism for internetworking. The same arguments that have been voiced regarding the use of IP flow characterization in relation to RSVP as well as IP switching may also be raised for the case of MPOA, irrespective of the actual mechanisms used for the flow thresholding. While the need for some form of IP flow characterization is viewed by some as a shortcoming of the MPOA architecture, in practice this requirement needs to be assessed in relation to the intended network applications and deployment scenarios, particularly in comparison with the alternative internetworking mechanisms available.

It was noted earlier that connection management in MPOA required the use of ATM VCCs to be established and released for the transfer of both user data and control information, using conventional ATM signaling protocols described earlier. As may be expected for an architecture defined by the ATM Forum, MPOA typically assumes the use of the ATM Forum's UNI Version 3.1 or Version 4 [6.2] signaling protocols. The MPOA Version 1 specification typically does not specify the use of the PNNI signaling (or routing) protocols, although it is evident that, in general, MPOA is essentially independent of the type of signaling protocols used, so that these may be generalized to both PNNI or those based on the ITU-T's Q-series of Recommendations (i.e., Q.2931 and Q.2761 to Q.2764; see references to Chapter 6). The short-cut ATM VCCs are assumed to be point-to-point VCCs, although the architecture does not preclude the extension to point-to-multipoint connectivity in principle. MPOA requires the support of the NSAP-formatted ATM end system address (AESA) in the called and calling party number information elements (IEs), although the use of the native E.164-based ATM addresses and subaddress IEs may also be used as an option.

It is important to note that MPOA requires the use of the LLC/SNAP encapsulation procedure (RFC 2684 [11.14]) as described earlier for the carriage of the IP (or other Layer 3) packets in the classical IP over ATM (CIPOA) network architecture. The use of the LLC/SNAP-based multiplexing mechanism allows for the transport of multiple Layer 3 protocols over a single ATM VCC as was described previously. It is this capability that gives rise to the term "multiprotocol" in MPOA, in common with the approach adopted in the CIPOA architecture as well. The LLC/SNAP PDUs containing the IP (or other) packets are then encapsulated using AAL Type 5 common part convergence sublayer (CPCS) protocols described previously in the conventional way. Depending on the maximum transmission unit (MTU) frame sizes of the attached LANs, MPOA enables the negotiation of the AAL Type 5 CPCS_SDU sizes to accommodate different LAN protocols. A default size of 1536 octets is specified, corresponding to the more common Ethernet-based LANs. However, for direct IP over ATM encapsulation, such as may be used between remote MPS, the default size is 9188 octets. It is possible in MPOA that multiple VCCs may exist between any two MPOA components at a given time, since independent IP flows may trigger separate shortcut ATM VCCs. In this event MPOA specifies a simple rule that requires the user data to be sent over the VCC set-up by the MPOA component having the numerically lower ATM address value. It should be noted that the ATM VCCs may "time-out" in the event of inactivity or end of an IP session, with a settable threshold that may be varied over a wide range. A default time-out threshold of 20 min has been proposed, but this may be somewhat high for typical applications and hence may require some fine tuning for actual network deployments. It will be recognized that the MPOA VCC time-out threshold is a critical parameter in optimizing network performance, since it will determine the number of call set-up attempts and latency.

Since the MPOA shortcut path is an ATM VCC, it may be assumed that the QoS capabilities available to any of the ATM service categories (or ATCs) may be selected for the IP flows, thereby ensuring high performance and throughput over the underlying ATM networks when required. However, to simplify the initial deployment of the MPOA network, it was assumed that the basic unspecified bit rate (UBR) service category would suffice as a suitable "default" service category in the event that more stringent traffic control is not warranted. In this case, the default peak cell rate (PCR) may be taken as the "line rate" of the transmission path. Similar considerations would apply to other ATM network capabilities such as the in-band OAM functions described earlier. Whether or not these would serve a useful function in any given MPOA deployment depends largely on the operator's choice and intended network applications. Traditional data networking essentially relied on the OAM capabilities of the underlying transmission network to provide a suitable level of reliability. For IP transport over switched ATM VCCs, capabilities such as soft PVC (SPVC) based restoration or ATM layer protection switching and other OAM tools may provide added value, particularly for large carrier networks, but this choice is essentially outside the scope of the basic MPOA architectural framework outlined above.

Whereas the MPC functional entities provide the connection management (or control) and the flow thresholding at the MPOA network edges, the associated MPS functional architecture as shown in Fig. 13-19 is more related to the IP routing aspects of MPOA. The MPS may be viewed as a component of a router, which must also include both the NHS- and LANE-related functionality since it has to communicate with both the MPOA edge devices and the WAN. As indicated earlier, the MPOA router implements conventional Layer 3 routing protocols such as RIP, OSPF, IS-IS, or BGP to enable the computation of the default path for the selected IP packet flows. The MPS configuration and forwarding information is exchanged between the functional components via the LEC interface (LUNI) as well as conventional routing protocols. The MPS advertises its ATM address (AESA) to the attached LECs using the LE_ARP messaging described previously. Such configuration exchanges with the LECS must identify the LEC as an "MPS" (or "MPC") by means of a device identifier TLV as indicated earlier. An important function of the MPS is the processing of the MPOA resolution request messages from the attached MPCs, as described further below, and the subsequent "reorigination" of the NHRP request messages. Correspondingly, on the egress side, the MPS is responsible for the generation and processing of the cache imposition requests and replies to the associated MPCs. In general terms the MPS-related procedures may therefore be summarized into the four main categories as follows:

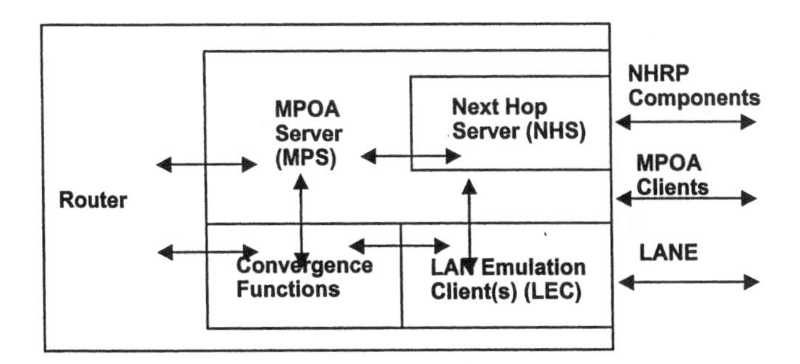

Figure 13-19. MPS functional architecture. MPS is a component of the router, which must include NHS and LANE functionality. The router implements conventional routing protocols and provides a default path for packet flows. The MPS interacts with the router, LEC, and NHS, exchanging configuration and forwarding information. The MPS advertises its ATM address to attached LECs via LE_ARP messages, processes MPOA Resolution requests from MPCs, and "reoriginates" the NHRP requests. It generates/processes cache imposition Requests/Replies to MPCs and maintains state for all ingress and egress cache entries sent to MPCs. It processes conventional NHRP address resolution queries from peer NHSs. Configuration exchanges with LECS must identify LEC as a MPS via device ID TLV.

1. Translations. These sets of MPS procedures relate to the conversion or "translation" of the MPOA resolution requests/replies messages to the NHRP resolution requests/replies messages as described earlier. Although the basic message formats remain essentially the same in this process, selected fields (e.g., the message type codings) may be modified as required. It is important to recognize that this process forms the basis of the IP address re-origination concept described earlier, whereby the MPS acts as the "source" of the NHRP request and not the individual MPC edge device. The MPS component is required to maintain state information relating to the request events by means of a state identifier parameter, to maintain correlation between the control messages. For the egress side, the translation procedures involve the conversion of the NHRP imposition requests and reply messages. The translation procedures maintain essentially the same message formats while modifying the required information elements (IEs) to provide the egress cache entity and DLL information to the relevant MPC.
2. Keep-alive Protocol. In order to enhance reliability, MPOA incorporates a Keep-alive protocol that is used to periodically inform the MPCs that the MPS is functioning normally. The keep-alive messages may be sent over any VCC that is established between the MPS and MPC, and the period for these messages is settable at configuration. The keep-alive messages include the internetwork (IP) address of the MPS, keep-alive lifetime, and a sequence number count to indicate message loss. The interruption of the keep-alive messages or sequence number is interpreted by the MPC as an error condition and may lead to the purging of the cached information.
3. Cache Maintenance. The significance of the MPOA cache management functions was outlined earlier, since the accuracy and timeliness of the cached information is an important element in the performance and robustness of the MPOA approach to internetworking. To enhance cache management procedures, the MPS maintains state information for all resolution request/replies as well as cache imposition requests exchanged between the MPS and MPCs. The cached information is considered valid up to a (programmable) "holding time" specified by the MPS, and may be periodically updated before expiry of the holding time. In the event that invalid (or errored) address bindings or other informa-

tion changes are detected by the MPS, the MPOA "purge" procedures may be initiated to clear the errored entries, or update procedures may be used to enter the corrected information in the cache tables.

4. Trigger Protocol. Although typically it is the role of the ingress MPC to characterize the inbound IP flows to determine the selection of default or shortcut paths as described earlier, MPOA also incorporates procedures whereby the MPS may initiate the characterization of inbound flows by "triggering" the ingress MPC. This "trigger protocol" operates by sending a trigger message to the ingress MPC to generate an MPOA resolution request message as if it were initiating a shortcut path. The MPS trigger message includes information relating to the destination internetworking (Layer 3) address and the ingress MAC address, and effectively enables the MPS component to initiate flow thresholding if required. The MPS trigger protocol may be used for maintenance purposes in the event that the normal MPC flow thresholding mechanism is malfunctioning or is not suitable for a given network application scenario in certain circumstances.

Although it is not necessary to describe in detail all the MPOA procedures of both MPS and MPC components, it may be evident from the brief account above that significant additional functionality must be added to ensure robust operations, even though MPOA builds on existing ATM- and IP-related protocols whenever possible. Nonetheless, the fact that the MPOA framework attempts to synthetize both ATM signaling and IP routing capabilities is viewed by some as an unnecessary complication, whereas others regard it as a logical extension of existing IPOA internetworking that maintains the architecturally valid relationship between the Layer 3 networking protocols (e.g., IP routing) and the Layer 2 switching and transport functionalities. Irrespective of these more general discussions of the applicability of the MPOA framework, it is of interest to summarize the basic MPOA message types and general formats, which as noted above are based on the NHRP packet formats, to which selected MPOA specific information elements (TLVs) and codepoints have been added. The MPOA message set and corresponding message type codepoints are listed in Table 13-4. The use of each of these messages has been described earlier and here the commonality with the NHRP message set should be noted in particular, where three of the NHRP messages use the same message type codepoints.

The MPOA/NHRP message header formats, using the notation adopted in the NHRP specifi-

TABLE 13-4. MPOA Message Types and Formats

Message type code	MPOA control message
128	MPOA cache imposition request
129	MPOA cache imposition reply
130	MPOA egress cache purge request
131	MPOA egress cache purge reply
132	MPOA keep-alive
133	MPOA trigger
134	MPOA resolution request
135	MPOA resolution reply
5	MPOA data plane purge
Note 1	NHRP purge request
Note 1	NHRP purge reply
Note 1	NHRP error indication

Note. Codings are the same as for NHRP.

cations, is shown for reference in Fig. 13-20, which shows both the "fixed header format" and the "common header format" used for the MPOA messages. In most cases, the functions provided by the header fields will be self-evident and generally follow the terminology used by the IETF for the naming of the ARP message structure fields. It may be noted that the common header fields allow for the inclusion of variable length non broadcast multiple access (NBMA) addresses (in this case an ATM address such as a NSAP-formatted AESA), as well as variable length protocol addresses (i.e., a Layer 3 or internetworking address such as an IP address). It should be noted that each MPOA control message listed above uses the same "fixed header" as well as the same "common header" as the NHRP packet, although with different codepoints to distinguish MPOA-specific messages. This close commonality between the NHRP and MPOA messaging structures and procedures is an inherent aspect of the MPOA design, as may have been apparent in the earlier description. In this way, it is possible to use simple extensions to existing NHRP message structures to accommodate the MPOA capabilities and hence minimize additional software development. The detailed message structures, codepoints and procedures have been described in the MPOA Version 1 document developed by the ATM Forum and hence need not be repeated here. It is also of interest to note that the MPOA Version 1 document includes the complete NHRP specification for reference, thereby stressing the relationship between these approaches.

As indicated earlier, the overall MPOA framework architecture provides a rather comprehensive synthesis of a range of protocols that have been developed for the purposes of ATM and IP internetworking. It builds on the extensions of the CIPOA architecture as defined in the initial RFC 2684 [11.14], RFC 2225 [12.8], and NHRP schemes and encompasses the notion of IP switching by extending it from a nodal to a network process while using conventional ATM signaling procedures and LANE concepts. Although such an approach would appear as a rational evolutionary approach for IP and ATM internetworking in general terms, the disadvantages of

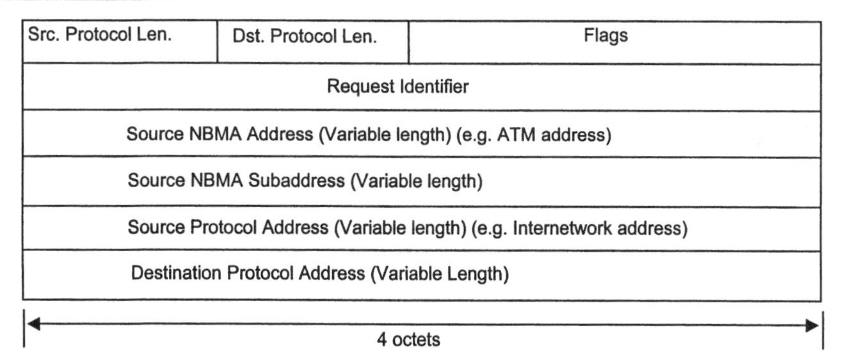

Figure 13-20. MPOA/NHRP header formats.

the MPOA approach have also been pointed out, resulting in concerns as to its scalability for large networks. The scalability issue has tended to focus on the incorporation of some form of IP flow characterization in MPOA, as indicated earlier. Since very large IP networks may involve very large numbers of instantaneous IP flows, it is often argued that these may make unmanageable demands on the processing requirements of the MPOA components. However, it may be recalled that the actual flow thresholding may be performed by the individual MPOA host or edge device in the MPC functional entity, where the number of IP flows may be relatively low, even though the number of flows in the backbone or core part of the networks are very large.

The use of ATM signaling procedures to establish the shortcut VCCs central to the MPOA framework also results in latency prior to the data transfer phase, as may be expected. This latency may be compounded by the latency inherent in LANE coupled with the user of the ARP-based procedures to obtain the destination address bindings. Although such inherent latencies may be minimized (or even reduced to zero) by suitable address cache management processes in the MPOA components, this aspect of MPOA performance limitation is essentially inherent to the client–server NHRP-based approach. It may be pointed out that there is sometimes confusion between the latency for signaling and ARP-based procedures and factors related to scalability in discussions of MPOA network performance. In evaluating MPOA for any given network application, it is important to distinguish performance limitations that may result from ARP-related or ATM VCC set-up latency from other factors such as flow thresholding efficiency, which may have a different bearing on the throughput performance of any given network application. In general terms, it may be pointed out that the term "scalability" is often used somewhat loosely in order to encompass all these aspects, whereas in fact the (spatial) scalability is more often than not related to the connectivity limitation of a fully meshed network, assuming PVC links. It may be noted that such spatial scalability considerations resulting from the well-known "n-squared" mesh connectivity has no direct bearing on the MPOA framework, since in this case it is assumed that connectivity is established and released based on on-demand ATM signaling for the shortcut VCCs.

The need for LANE within the overall MPOA framework has also been questioned, since it evidently adds complexity that may not be warranted for some network applications, which could otherwise benefit from the MPOA approach. Although the use of essential MPOA concepts without incorporating the LANE infrastructure between the MPC and MPS components is clearly feasible and in principle not very different in that case from the CIPOA architecture, such an approach has not been categorized independently as an alternative internetworking approach. From a purely protocol architecture perspective, the use of LANE within the MPOA framework is logical, by analogy with the use of the network layer (Layer 3) logical addressing capability for conventional networking between LANs. However, the use of normal Layer 2 connectivity as provided by ATM VCCs without the additional capabilities required by an emulated LAN, is also perfectly valid from the protocol architecture point of view, as is evidenced in the earlier CIPOA architectural approach with its extensions. Thus, it may be argued that the basic tenets underlying the MPOA approach may be employed with or without the LANE capabilities, if warranted by the networking applications. However, by disabling the LANE capability in any given MPOA network application, it should be recognized that interworking (or bridging) with any legacy LANs using the LEC interface will not be possible, which may prove a limitation in many cases.

While the MPOA/LANE framework model clearly provides a conceptually consistent evolutionary path for IP and ATM internetworking, incorporating as it does the key precepts of CIPOA and IP switching, and thereby providing the benefits of IP routing as well as ATM QoS capabilities, this approach has been perceived as overly complex and relatively unscalable by some. The overall complexity of the MPOA model clearly results from its synthesis of existing routing protocol suites (e.g., NHRP plus conventional IP routing protocols) with ATM signaling and LANE protocols, as well as flow characterization, which is often viewed as undesirable and inefficient, as mentioned previously. Thus, in attempting to be comprehensive in its scope, MPOA can appear as unnecessarily complex relative to other, although somewhat more constrained, internetworking architectural frameworks. As pointed out previously, the question of relative complexity of any given network or protocol architecture can only be objectively assessed when detailed analysis is undertaken on a function-by-function basis relative to any given networking application. Such assessments are often difficult to perform given the varied network requirements and applications that need to be factored in, as well as the often volatile nature of the commercial interests that may tend to be weighted in favor or against any given approach, depending on existing deployments that may already be in place.

As noted earlier, the question of MPOA "scalability" is sometimes confused with the issue of the inherent latency for both connection set-up and address resolution. In principle, it will be evident that the spatial scalability attainable within the MPOA network architecture will be of the same order as that provided by the CIPOA model, assuming that on-demand switched ATM VCCs are available in either case for the carriage of the user data (IP) packets. Spatial scalability limitations may be of relevance for the case when semipermanent virtual circuits (PVCs) are used in a partial or fully meshed configuration in either architectural model, but this so-called "n-squared" limitation is not specific to MPOA. On the other hand, scalability constraints resulting from the need for flow thresholding in MPOA are somewhat more difficult to pin down, and hence may need to be factored in when evaluating the use of MPOA for a given networking application, since this will clearly impact overall network performance. In addition, the performance impacts of ATM VCC set-up latency coupled with ARP-related delays clearly will need to be considered in relation to the MPOA model. These will largely depend on specific implementations to optimize connection establishment times and cache management efficiencies for address bindings. In the simplest of applications, considering that the inherent ARP mechanism is essentially an automated "directory service," the caching of frequently used destination ATM addresses may be used to minimize ARP-related delays in many cases. By also simplifying the ATM signaling procedures in MPOA, the shortcut set-up latency may also be reduced within reasonable bounds. It may be noted that such somewhat general considerations for IPOA network performance optimization are not specific to the MPOA model, but apply as well to CIPOA architectures.

13.7 MULTIPROTOCOL LABEL SWITCHING (MPLS)

In describing the notion of IP switching earlier, it was noted that this approach to internetworking attempted to "combine" the perceived advantages of ATMs "label switching" with IP routing to provide enhanced throughput and other benefits for IP internetworking. In the ATM forum's work on MPOA, the notion of "integrating" IP and ATM networking essentially resulted in the synthesis embodied in the MPOA architectural framework, which uses as its basis the CIPOA model and its extensions. Nonetheless, this approach may still be viewed as an overlay of IP over ATM, with no fundamental changes to either paradigm being required to obtain interworking between IP and ATM. In the IETF, however, the discussions regarding the IP switching model initiated a number of alternative and competing proposals for enhancing IP networking capabilities, some of which incorporated the concept of using "label" multiplexing for the switching of IP packets, analogous to the mechanism used in ATM or frame relay earlier. The primary competing proposal was initially termed "tag switching" since it incorporated a "tag" or "label" appended to an IP packet, which enabled switching of the packet by "tag or label swapping" at so-called "tag switching" nodes within the network. The tag switching proposal was initially introduced by Cisco Systems Inc., essentially as a counter proposal to IP switching and other related proposals whose primary intent was to enhance throughput and QoS performance of IP networking.

The initial "tag switching" concept was eventually consolidated and generalized into what became known as the so-called multiprotocol label switching (MPLS) framework architecture in the IETF. MPLS may be viewed as a significant evolutionary trend in IP-based internetworking intended to enhance the performance and throughput capabilities beyond conventional IP store and forward networking. It is of interest to note that the term "multiprotocol" in MPLS is used in the same sense as in the MPOA or CIPOA models, namely that any Layer 3 networking (connectionless) protocol may be multiplexed over the underlying transfer mode. While primarily intended for data networking, the carriage of other services in MPLS is not precluded. As for the case of MPOA, in its most general sense MPLS may also be viewed as a "framework architecture" or model, since a number of alternative mechanisms may be used for both the "label switching" aspect, as well as the distribution of labels between the MPLS-capable nodes, as will be described in further detail below. In this context, MPLS may use ATM VPI/VCI values as labels, or frame relay DLCI values, as an alternative to its own dedicated MPLS header label structure, which is sometimes referred to as an MPLS shim header in internet drafts. In this overview account, emphasis is placed on the underlying principles, using either MPLS header or ATM VPI/VCI values as the labels. The detailed procedures may be found in the related MPLS internet RFCs and drafts [13.4, 13.5].

It is important to recognize that in MPLS (or in tag switching as initially proposed) the basic mechanisms underlying "tag" or "label" switching are identical to those used in ATM (or frame relay) label multiplexing. Basically, an MPLS packet is switched between an input and an output link by translating (or swapping) the label values corresponding to the input and output links from a look-up table, in much the same way as would be done for an ATM cell or FR frame. The label values in the look-up or context tables relating the input to the output links or paths are derived by means of conventional routing procedures that compute a path through the network based on the destination IP address. Since each node along this path must obtain the "labels" corresponding to the path, MPLS requires a "signaling" procedure of some kind to enable the nodes to exchange and negotiate the label (or tag) values. For this purposes, MPLS has introduced a so-called "label distribution protocol" (LDP) for the exchange of labels between the MPLS nodes, although in general the MPLS framework model also allows for the use of other existing protocols such as RSVP or BGP to be extended for the purpose of exchanging labels, if required. However, the MPLS-specific label distribution protocol and its subsequent extensions evolved from the initial so-called "tag distribution protocol" (TDP), which was introduced by Cisco Systems Inc. specifically for the purpose of distributing labels in MPLS.

In its simplest form, the MPLS (or tag switching) networking model may be described as shown in Fig. 13-21, which represents a basic networking scenario with various LANs interconnected by means of "edge" routers to a backbone or core internet comprising so-called "label switching routers" (LSRs), which are capable of performing label (or tag) swapping as described earlier. The routers at the edge of the MPLS network are capable of attaching a "label" (or tag) to an incoming IP packet, whose value is associated or "bound" to a route or path through the network, known as a label switched path (LSP). The intermediate label switching routers (LSRs) forward the MPLS packets by translating the labels (or tags) in accordance with the look-up (or context) tables that associate the packets arriving at an input link with the appropriate output link, exactly as is done for the case of ATM or frame relay. The intermediate LSRs consequently do not need to route the packets based on the IP (or other Layer 3) address contained within the MPLS packets, as do conventional "store and forward" IP routers. The forwarding of packets

MPLS Network Architecture

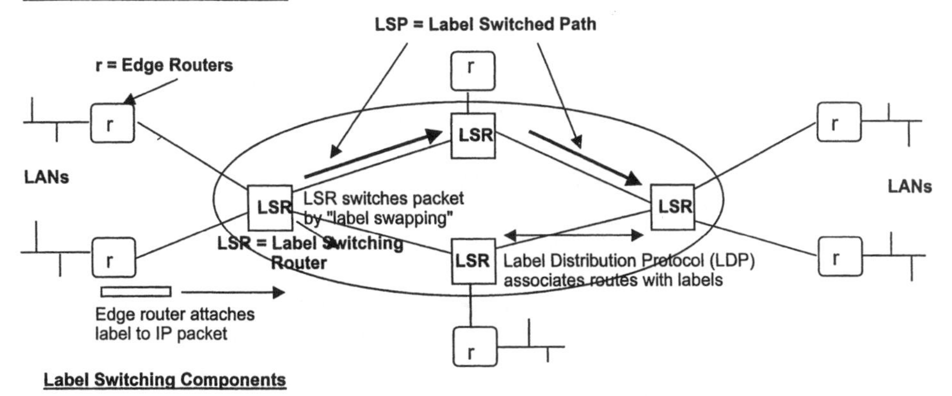

1. **Forwarding Component:** Based on concept of "label swapping" analogous to ATM , FR etc. LSR switches IP packet by associating input to output label (or tag) value.

2. **Control Component:** Provides "binding" (or association) between routes and labels in LSRs "Label Information Base" (LIB), based on normal routing protocols.

Figure 13-21. Multiprotocol label switching (MPLS).

based only on the "translation" of a label from input to output link results in significantly higher packet throughputs when compared with the conventional process of obtaining a "best fit" match with the entire IP address prefix before selecting a forwarding path. The actual label or tag used for this purpose may be an ATM header VPI/VCI value or FR header DLCI appended to the IP packet, or the alternatively specified MPLS "shim" header, as described in further detail below. It is of interest to note that some initial LSR implementations essentially adapted ATM switching hardware to provide the MPLS label swapping functions, which serves as an indication of the underlying similarity of concept between ATM and MPLS.

In addition to the similarities in the basic label multiplexing/switching mechanisms between the MPLS model and ATM, it may also be recognized from Fig. 13-21 that another basic similarity is the need for a "signaling" procedure in MPLS for the negotiation and exchange of the label information between the LSRs along any given LSP. Although the specific details of the signaling protocols used in MPLS are very different from those used in the conventional ATM signaling procedures described earlier, there are broad parallels in a number of aspects, as will become evident below. In particular, the need for signaling as well as the concept of label switching paths in MPLS appears to suggest that the MPLS model is more "connection-oriented" in nature than connectionless in its operating mode. From a protocol modeling perspective, it may also be recognized that the MPLS framework attempts to "integrate," or at least blur, the conventional distinction between, the Layer 2 (data link layer) and Layer 3 (networking layer) functionalities by more closely binding routing and label switching functions. Whichever way one chooses to describe such an approach, it will be evident that the MPLS model constitutes a significant change in the evolution of IP-based internetworking where the close synergy with the ATM model is clearly not coincidental. In some sense, it may even be argued that MPLS is essentially ATM "reinvented" for the internetworking environment.

It should be stressed that the MPLS framework architecture does not in principle require any changes to the conventional IP-related routing protocols, any one or more of which may be used by the label switching routers to dynamically compute an optimum path from source to destination based on IP address. In principle, the fact that routing is based on the IP address, and there is no requirement for an ATM address in determining a path, constitutes the fundamental difference between the MPLS approach and other internetworking schemes such as MPOA or CIPOA. The question of which addressing scheme forms the basis of the internetworking architecture is much more fundamental than the details of the signaling protocols, or even whether the transfer mode is connection-oriented or connectionless. This aspect will be dealt with later, following a more detailed description of the MPLS-related protocol structures.

Since the MPLS networking architecture is primarily an evolution of IP-based protocols, its description may at first glance appear superfluous in a text dealing essentially with ATM-based network applications. However, in noting the underlying functional similarities between ATM and MPLS above, and recognizing the potential need for interworking between ATM- and MPLS-based networks, it is of interest to outline the salient features of the MPLS architecture and associated control protocols. The actual detail of these protocols, some of which are still in a state of "work in progress," may be obtained from the extensive literature generated by the MPLS group actively working on these issues in the IETF. In addition, since the MPLS framework architecture may in principle embrace a vast range of protocol extensions within Layers 2 and 3, it is only of interest here to consider the MPLS aspects related to ATM interworking as well as the basic functions provided by the label distribution protocol (LDP) [13.6–13.8] for the control of the label switched paths (LSP).

13.8 MPLS ARCHITECTURE AND FUNCTIONS

In rationalizing the use of label multiplexing (or switching) mechanisms with the forwarding of (conventionally) connectionless IP packets, MPLS introduces the notion of the so-called forwarding equivalence classes (FEC), which is defined as the group of IP packets that are forwarded in the same manner over the same path between the LSRs. Essentially, all packets associated with a given FEC receive the same forwarding treatment by the node, and hence may be assigned a given label (or tag). These labels themselves may (or may not) have local significance between adjacent MPLS nodes (or LSRs), entirely analogous to the use of a VPI/VCI "label" in the ATM cell header or a DLCI "label" in the FR frame header. In fact, as noted earlier, the analogy is so close that a number of initial MPLS implementations simply used conventional ATM or FR headers appended to the IP packets assigned to any given FEC as the labels for MPLS, thereby making possible the adaptation of existing ATM hardware and software for switching of MPLS packets.

The assignment of any given IP packet to a given FEC is typically done by obtaining a "best match" of the address prefix in the nodes routing tables to the destination IP address in the packet header. As outlined earlier for the case of conventional IP forwarding, this process is carried out independently by each router as the packet traverses the connectionless network from source to destination node. In contrast, in MPLS the assignment of an IP packet to a FEC is performed only once at the ingress to the MPLS network, whereupon the label corresponding to the FEC is appended to the IP packet. At subsequent LSRs within the MPLS network, forwarding of the "labeled" IP packet is only carried out by translation (or swapping) of the label values based on a look-up table (also called a label information base) that associates the FEC between input and output links. The intermediate LSRs do not need to analyze the IP header any further for routing purposes, as forwarding is entirely based on the label exchange between input and output links. At the egress node (LSR) of the MPLS network, the current label may be removed (termed "label popping" in the IETF literature) to recover the original IP packet. Independently of this label multiplexing and forwarding process, it must be borne in mind that each LSR may be operating the conventional IP routing protocols (such as OSFP, RIP, BGP, etc.) or their extensions to enable it to compute the optimum path through the network. The results of the routing algorithms may then be used to update the label values corresponding to any given FEC. These concepts are indicated in the general functional model shown in Fig. 13-22, which relates LDP control to the LSR functions.

Although it is clear that the concept of a "connection" does not apply in the case of the (conventional) inherently connectionless IP-based networks, it will be recognized that MPLS may be viewed as "quasiconnection-oriented" since it introduces the need for a signaling protocol coupled with the concept of a label switched path (LSP). In this context, the notion of the "forwarding equivalence class," which essentially describes the group of IP packets that flow along a given LSP, may be viewed as being analogous to a "virtual connection" in the sense used in ATM or FRBS. In this sense, the set of MPLS "labels" that identify a given FEC and its corresponding LSP are somewhat analogous to the set of VPI/VCI values that identify the VPLs/VCLs that are concatenated to make up the conventional ATM VPCs or VCCs. However, in contrast to the ATM case, where the VPI/VCI labels directly identify the VPL/VCL between nodes, in MPLS the labels identify a given FEC only, reflecting the difference between strictly connection-oriented and the quasiconnectionless nature of MPLS. It is also of interest to recognize that MPLS essentially "decouples" the forwarding process (which depends only on label switching) from the routing process, which may operate independently for efficient computation of paths based on the destination IP (or other Layer 3) address. This separation of forwarding and routing procedures in MPLS, while conceptually analogous to the virtual routing concept outlined earlier as used in MPOA, is different from the MPOA mechanism, which relies on ATM cell-based forwarding of the encapsulated IP packets along the MPOA shortcut path. Nonetheless, the separation of IP routing from the actual forwarding mechanism does provide the advantage in MPLS (as for MPOA) that the routing protocols used may be enhanced and evolve independently of the packet forwarding mechanisms, which depend only on label multiplexing/switching.

Although for simplicity it was noted that the FEC is primarily derived from the analysis of the IP (or other Layer 3 protocol) address, it may be recognized that the packet's QoS or "prece-

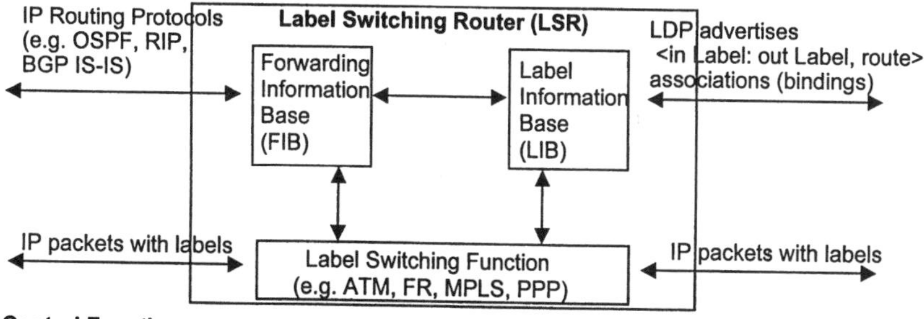

Control Functions

1. Primary function of control components is to provide "binding" between IP routes and labels.
2. Label switching router (LSR) uses conventional IP routing protocols such as OSPF, RIP, BGP, and IS-IS to construct/maintain its forwarding information base; e.g., for destination-based routing, the protocol will use the IP address.
3. LSR assigns label values to any given route (or group of routes) to build label information base (LIB).
4. Label Distribution Protocol (LPD) is used to "advertise" the binding between labels and routes to adjacent LSRs.

Label Distribution Procedures

1. **Downstream Unsolicited Distribution:** For each route in FIB, incoming label is allocated in LIB. LDP then advertises <in label, route> binding to adjacent LSRs.
2. **Downstream on Demand Distribution:** For each route in FIB, LSR sends request to next hop via LDP for label binding for that route. LSR allocates labels, updates LIB and sends binding information via LDP.

Figure 13-22. Label distribution protocol (LDP).

dence" may also be factored in the assignment to any given FEC. The use of the QoS class (as inferred from the IP packet header, for example, in DiffServ encodings) is an option in MPLS for the assignment of the label values. In MPLS terminology, the association between any given FEC, F, and the (typically locally significant) label value L is termed a "binding" between the FEC and the label, which must be locally unique in order to avoid misrouting of incoming packets from different upstream LSRs. The MPLS procedures require that the assignment of a particular label value, L, to a FEC, F, is performed by the downstream LSR, for any set of adjacent LSRs in the MPLS network. The downstream LSR then transfers the label binding to the upstream LSR by means of the appropriate label distribution protocol (LDP) messages (as outlined below), or other relevant signaling protocol if used instead of LDP. It will be recalled that the general MPLS framework architecture allows for the use of different signaling protocols by means of so-called "piggybacking" of the label binding information elements onto existing protocols such as RSVP, BGP4 [13.4], etc., even though the LDP messages were specifically developed for this purpose. It must be admitted that the possibility of multiple, potentially different, signaling mechanisms allowed for in MPLS, though enabling flexibility for different environments, also raises significant interoperability issues in practical implementations.

Irrespective of which specific signaling protocol (i.e., LDP or other) is used for the exchange of the label binding information, MPLS procedures identify a number of ways in which this information may be transferred and handled by the individual LSRs comprising the MPLS network or domain. These different signaling or handling modes enable considerable flexibility in the way the procedures may be implemented, depending on the various network scenarios envisaged. It may also be recognized that such flexibility may result in unnecessary complexity, or even interoperability problems. Since MPLS requires the "downstream" LSR to assign the label value for a given FEC, two modes have been identified for the way in which this may be done. In the so-called downstream label distribution mode, the LSR simply distributes the label bindings on an unsolicited basis, upon making an association between a FEC (i.e., a given address prefix match) and a suitable label. Alternatively, in the so-called downstream-on-demand (DOD) label distribution mode the procedures allow an upstream LSR to explicitly request a suitable label binding from its adjacent (downstream) LSR by generating a label request message for a given FEC, F. It may be noted that either of these label distribution modes may be used at the same time within a given MPLS network domain, but adjacent LSRs will need to negotiate which of the two modes of label distribution are in use to avoid superfluous messaging and processing of label bindings.

The MPLS procedures also identify two different mechanisms for the handling or "retention" of the label bindings received by any given LSR. It may be recognized that, in general, a given LSR may receive label binding information for a particular FEC, F, from other LSRs that are not necessarily the adjacent (or next hop) LSR for that FEC. In the so-called liberal label retention mode the MPLS node procedures allow for the retention (i.e., storage) of all such label bindings, whether or not the information is received from the next hop LSR. Alternatively, if the node adopts a so-called conservative label retention mode, the procedures allow for the discard of such label bindings, whereby the LSR in question only maintains the relevant next hop label binding in its label information base (LIB), hence requiring less memory for the forwarding process. The primary advantage of the Liberal label retention mode, although it requires more state information to be stored, is that in the event that rerouting is required or a topology change occurs, the relevant label bindings may be available already, thereby facilitating the forwarding process. The choice of whether to adopt either the liberal or conservative label retention mode (or both) is largely implementation- and network-application-dependent, but it is clear that no interoperability issues arise with either mechanism.

The MPLS architecture also allows for two different mechanisms for the set-up of LSPs, depending on whether so-called independent LSP control or ordered LSP control procedures are adopted within the MPLS domain. Independent LSP control procedures operate when any given LSP, on identifying a particular FEC, is able to assign a label corresponding to that FEC independently of any other LSPs assignment, and is able to distribute that binding to adjacent LSPs. In effect, the independent LSP control procedures are somewhat analogous to the way in which conventional IP routing operates, in that each IP router is able to make an independent (routing and forwarding) decision on the IP packet, based on the results of its own routing algorithms. Alternatively, for the LSP set-up based on the ordered LSP control procedures, the label to FEC binding is initiated only by the egress (or more generally edge) LSR nodes, which then propagate the label bindings upstream according to one of the label distribution modes described earlier in a step-by-step (or ordered) way, analogous to the way conventional signaling protocols may operate (e.g., in ATM or FR, except that these signaling messages are typically propagated downstream from calling to called party).

It is not necessary to enter here into a discussion of the relative advantages or disadvantages of independent versus ordered LSP set-up control procedures, since from the perspective of MPLS-ATM interworking it is evident that ordered LSP control is closer to the way in which ATM signaling operates. Although the independent LSP control mode may allow for less delay for label exchange than the ordered control mode, it may result in the creation of routing loops and hence may not be desirable for large networks. Although it is anticipated that both methods of LSP set-up control may interoperate, the independent control procedures may be less predictable, particularly for traffic engineering purposes, so that it is not surprising that the ordered control procedure has been selected for ATM network interworking by the ITU-T, as described in Recommendation Y.1310 [13.10].

In addition to the various operating modes outlined above, it is also important to note that the MPLS framework architecture allows for the stacking of multiple levels of labels appended to the IP packet, thereby allowing for the possibility of "tunneled" LSP or hierarchical networking. In MPLS, the label stack, if present, is structured as a "last-in, first-out" stack, where every forwarding decision by a given LSR is based on the label at the top of the ordered stack. This forwarding decision based only on switching (translation) of the topmost label in the stack is therefore independent of any underlying labels or whether any label existed above the one in question at some earlier time. Thus, the general concept of the MPLS label stack may be seen as providing for the possibility of a set of "nested" LSPs, with each of the labels in the stack corresponding to a given level in the nested hierarchy. It may also be noted that this concept is independent of any routing hierarchy that may also be set up to enhance the scalability of the MPLS network. Since each label in the MPLS label stack is processed independently (as it becomes the "top" label in the stack), this mechanism should not be confused with the "logical hierarchy" implied by the use of the VPI and VCI values in the ATM network to identify VPLs and VCLs, respectively, as described earlier. It may be recalled that for ATM the forwarding of cells belonging to a VCC involves the translation of both VPI and VCI values within the ATM NE in general. Consequently, in ATM the VPI and VCI labels are not "stacked" in the sense that labels may be stacked in the MPLS architecture, but simply form a "label multiplex hierarchy."

The possibility of providing for nested (or tunneled) LSPs by means of stacked labels in MPLS also implies that the MPLS nodes (or LSRs) are capable of adding and removing labels from the stack at specific points within the network. In the MPLS literature, the process of removing a label from the stack is termed "popping a label," whereas the process of appending a new label onto the label stack is termed "pushing a label." Thus in general, MPLS nodes require, in addition to the normal label switching (forwarding) functions, the capability to process label stacks if present, including the functionality to "pop" or "push" labels when required at specific points along a nested set of LSPs. It will be recognized that this capability of the MPLS architectural framework is different from the case of ATM, and while it may provide additional flexibility in the use of tunneled (or nested) LSPs for some network applications, it also adds complexity in the handling of the label stacks from both control and network management viewpoints. Although the potential utility of these capabilities in MPLS networks remains an open question, it is not intended here to describe in further detail the procedures related to the use of stacked labels since this does not directly impact the interworking aspects with ATM networks.

In introducing the concept of the FEC in MPLS, it was noted that this implied a grouping of different IP packets that will be subjected to the same forwarding behavior by an LSR, generally based on the address prefix match in the routing tables (coupled with other factors such as QoS requirements, etc., if applicable). In general terms, the partitioning of the IP flows destined for a given egress node into the set of FECs may be done in a number of ways, depending on the "granularity" with which the address prefixes may be mapped into FECs. In MPLS, this possibility of assigning a number of FEC-to-label bindings that end up corresponding to one common path and hence may be defined by a single common FEC to label binding is called "aggregation." The granularity of such an aggregation depends on the particular MPLS domain and configuration, but may vary from aggregating all packets to a single FEC or a set of FECs, up to assigning each address prefix match to a separate FEC (and hence labels). The aggregation process may be used to reduce the number of assigned labels (and associated control messaging), but in any case requires negotiation (or configuration) among the LSRs to determine a common level of granularity between adjacent LSRs in the network. This problem is clearly more relevant to the case of independent LSP control than to ordered control since in the latter case a single LSR may determine the level of granularity by initiating the label binding.

In effect, the notion of aggregation in MPLS is essentially analogous to the multiplexing of various IP packet flows along a path if they are designated to follow the same path or route and hence may be "bundled" into one or more LSPs for part or all of that route. The problem of assigning a suitable granularity arises essentially because there is no concept of a "virtual connection," as is the case for strictly connection-oriented packet transfer modes such as ATM or FRBS. Thus, whereas the notion of the MPLS label switched path (LSP) may be viewed as a sort of "quasivirtual circuit," it is evident that the LSP may be used to "aggregate" (or multiplex in conventional terminology) IP packet flows from various sources to various destinations over part of the overall route, where the address prefix allows for common forwarding behavior. It is also important to distinguish the aggregation process as outlined above from the possibility of merging of LSPs, which is also permissible in the MPLS framework. In this case, the merging of LSPs to a single LSP by label merging is essentially similar to the merging mechanism in ATM as described earlier in relation to defining multipoint-to-point (bidirectional) connectivity. However, in the MPLS case, the merging mechanism does not pose the same limitations as for the case of ATM, since the MPLS packets contain the destination IP address, which may be used to deinterleave the IP packets at the receiver end.

It will be recalled that in MPLS it is assumed that the individual LSRs are capable of running the conventional IP-based routing protocols in much the same way as do existing IP routers. This enables the possibility that each LSR may individually select a route "hop-by-hop" through the MPLS domain for any given FEC. This mode of operation, which is similar to the way normal IP routing occurs in existing IP-based networks, results in the so-called "hop-by-hop routed LSPs." On the other hand, the MPLS architecture also allows for route selection by the so-called "explicit routing" process. In using explicit route selection, a single LSR, which may typically be either the ingress or egress LSR, specifies which path is to be followed by the LSP by listing some or all of the LSRs along the path. Explicit routing (ER) in MPLS is analogous to the mechanism of "source routing" described in the PNNI routing procedures in ATM, whereby a "list" of the nodes to be traversed is specified by the routing protocol in the control messages that set up the VCC. In MPLS, the LSP is said to be "strictly explicitly routed" if a single LSR identifies the entire LSP. Alternatively, the LSP is said to be "loosely explicitly routed" if only a part of the LSP is specified by the LSR, thereby allowing the other part of the route to be selected hop-by-hop by individual LSRs.

The possibility of incorporating explicit routing (ER) in MPLS, typically by adopting a source routing methodology in the control messaging, is significant for the purposes of performing traffic engineering or policy-based routing of IP traffic and has hence received considerable interest. Explicit routing, whereby the path that a given IP flow follows is selected by the network operator policy according to some desirable characteristics such as load balancing or QoS provisioning, enables an IP-based network to behave more deterministically in terms of traffic performance, much as a connection-oriented network may be engineered to do so. In effect, an MPLS network using ER-LDP-based control protocols for setting up the LSPs may be viewed as essentially similar to a connection-oriented network such as PNNI-based ATM in terms of its potential traffic or QoS engineering capabilities. Although it is not necessary to enter into a detailed discussion of the pros and cons of these potential capabilities of MPLS here, it should be mentioned that the required enhancements to both the related routing and label distribution protocols are being pursued by the IETF with a view to meeting anticipated Internet growth demands for the future. From the perspective of interworking with ATM-based (or conventional IP over ATM) networks, it is of interest to point out that much of the functionality that is being accommodated within the MPLS framework is similar to that already defined and used within conventional ATM networks. Consequently, it may be anticipated that interworking between MPLS (and particularly MPLS based on ATM) and conventional ATM networks will be facilitated by this convergence in functionality. However, although this may be a valid assumption in somewhat general terms, it should be realized that detailed protocol interworking, and particularly service interworking in the sense defined earlier, may sometimes be difficult, often as a result of the somewhat different approaches adopted by IP-oriented and ATM oriented technologies. An example of this form of interworking divergence was evident in the case of DiffServ-based QoS, where direct parameter mapping to the more precisely defined ATM QoS capabilities is somewhat ambiguous.

Despite its somewhat quasiconnection-oriented attributes, the MPLS architecture also includes procedures for the avoidance of routing loops within an MPLS domain, mainly by including the possibility of a "time to live" (TTL) function in the MPLS packet header. In the summary of the IP packet header functions earlier, it will be recalled that the main function of the IP "time to live" protocol was to avoid uncontrolled circulation of IP packets in the event of routing loops. It is interesting that a TTL function is also included in the MPLS framework, although the probability of incurring routing loops, particularly for ordered LSP control with explicit routing, in all but pathological cases, is somewhat low. In the event that MPLS is used based on either ATM or FR label switching as indicated earlier, the avoidance of routing loops may be readily achieved by means of configuration or other management mechanisms, so that the TTL procedures are essentially superfluous in these cases. However, if MPLS is used over other protocols such as PPP or LANs together with independent hop-by-hop route selection and independent control, looping may occur in certain configurations. In such cases, the TTL procedures provide some level of loop protection in much the same way as for the case of conventional IP-based networks. It may also be noted that protection against looping may also be promoted by discarding MPLS packets whose label encodings are unknown (in the label information base) or not present, rather than attempting to route them by utilizing the (imbedded Layer 3) IP address.

As was pointed out briefly earlier, the MPLS protocols allow for the possibility of "label merging" whereby IP flows on incoming LSPs with different labels may be "merged" into a single outgoing LSP (identified by a single, possibly different label) by an LSR. The merging capability is viewed as an advantage in terms of enhancing the scalability of MPLS-based networking, since multiple IP flows bound for a given destination node may be merged within an MPLS network, thereby requiring fewer LSPs to be set up overall. As may be evident from the MPLS structure, the merging capability is available in principle because every MPLS packet encapsulates the IP packet, which includes a unique (IP) destination address that may be used to deinterleave the merged IP flows independently of the label merging. Although this is valid in general terms, it must also be recognized that for cases where ATM NEs are adapted for use as LSRs (and frame segmentation has been performed), the merge functionality may not be available in many earlier designs. This may also be the case for typical FR-based LSRs, since the merging of DLCIs for multipoint-to-point virtual connections was generally not supported by the hardware. It is of interest to note that in the ATM case, merging is feasible if the use of AAL Type 3/4 is incorporated instead of the more ubiquitous AAL Type 5, as described earlier. It may be recalled that AAL Type 3/4 was eventually designated for the support of connectionless data packets over ATM, and therefore included the so-called message identifier (MID) functionality in the SAR sublayer for the deinterleaving of the connectionless data frames switched over ATM VCCs. However, the use of AAL Type 3/4 is somewhat restricted in ATM-based networking, where it is only used in support of the SMDS (or CBDS in Europe) service, as mentioned previously.

It should be pointed out that although the analogy between the concept of the LSP and the ATM VCC (or more closely the VPC) is useful in understanding the basic mechanisms underlying MPLS, there are clearly differences in the two approaches that suggest that such an analogy may not be pressed too far. In particular, it will be recalled that the typical ATM VCC (or VPC) is essentially an "end-to-end" entity with precisely defined termination functions as set up either by on-demand signaling protocols or management-based provisioning. This essentially reflects the strictly connection-oriented nature of ATM VCCs and VPCs, however they may be used within any given network application. In contrast, an LSP is not necessarily an "end-to-end" entity in MPLS since any given LSP may only extend between LSRs that identify FECs that correspond to the longest match address prefix obtainable for the IP packets destination network layer address. At that termination point of the LSP, a different LSP may be initiated, provided that a best (or longest) match with the address prefix can be recomputed and assigned to a different FEC (and associated label). It may be evident that this procedure essentially reflects the underlying connectionless nature of MPLS, despite the fact that the label multiplexing paradigm implies a form of "quasiconnection-oriented" behavior for MPLS. Although other differences in detailed behavior between the two technologies may also be identified in specific networking applications, it is not necessary to elaborate on these further here.

Returning to the initial objective of interworking between ATM and MPLS networks, to a first approximation it may be assumed that MPLS may be treated much as any other Layer 2 (label multiplex) technology. Thus, in general terms it is possible to conceive of "network interworking" between MPLS and ATM, whereby the MPLS frames are encapsulated and transparently, transported by an ATM network in the case where ATM provides a backbone network scenario. The inverse scenario, whereby an MPLS network provides the backbone over which ATM cells are transported transparently has also been envisaged and studied within the IETF. While clearly technically feasible, this latter interworking architecture raises a number of issues, notably for control and routing as well as OAM interworking, that have yet to be clarified. More interestingly, the notion of direct "service interworking" between MPLS and ATM networks, analogous to that between FR and ATM networks described earlier, appears to be a more likely internetworking scenario for the future. The detailed protocol mappings to enable such types of interworking are receiving increasing attention in both the ATM Forum and the ITU-T as it becomes apparent to many workers that both ATM and MPLS are likely to be used as a basis for the deployment of backbone networks for data internetworking.

13.9 THE LABEL DISTRIBUTION PROTOCOL (LDP)

It will be evident from the earlier discussion that the use of LDP procedures are central to the operation of the MPLS framework, typically for the establishment, maintenance, and release of the LSPs, in direct analogy with conventional signaling protocols. As indicated earlier, the general MPLS framework architecture does not necessarily restrict label distribution procedures to use of a simple unique protocol suite, in contrast to the case of N-ISDN or B-ISDN (ATM) signaling. MPLS assumes that a number of existing IP-oriented routing and signaling protocols such as RSVP, BGP, OSFP, IS-IS, etc., may be enhanced by incorporating MPLS-related (label) information objects in their messaging structures. In specifying these enhancements to the existing protocols, the intent is that these protocols may be pressed into service as label distribution procedures, even though they clearly were not designed for that purpose in the first place. Although this may be considered a pragmatic approach by some, there are clearly disadvantages from the perspective of interoperability if different protocol structures are used for essentially the same functionality.

In parallel with the above essentially evolutionary approach, it is of interest that the IETF has also developed a dedicated LDP specification for MPLS, which provides the clear advantage that the protocol suite may be optimized for the inherent capabilities of the MPLS approach. The main features of this LDP protocol are outlined here, without entering into a description of the detailed message structures and associated procedures, some of which are still in the process of being defined in the ongoing work of the IETF MPLS-related Working Group. Such details may be obtained from a perusal of the copious MPLS-related documentation emanating from that body [13.6, 13.7]. The first point to be stressed is that all the LDP messages are based on the familiar, and highly flexible, nested type–length–value (TLV) structure, also used in ATM (and for that matter N-ISDN SS7) signaling, as seen earlier. In MPLS terminology, the "value" part of the LDP message, which may itself use a (nested) TLV structure, is often referred to as a "TLV-object" or simply a "TLV." Although the use of the ubiquitous TLV-based message structure in LDP (and its enhancements as described later) will serve to simplify control plane interworking between ATM signaling and MPLS, it should be noted that both the terminology and coding structures used in LDP messages are quite different from those familiar in ATM- (or SS7-) based signaling. Consequently, direct one-to-one correspondences or mappings may not always be found in developing a control interworking framework between LDP and ATM.

It may be evident that a point of difference between ATM-related signaling and LDP messaging is that ATM utilizes "common-channel" signaling, as described earlier, whereas no such concept exists in LDP messaging. In this sense, LDP does not require the use of a "dedicated" control network that may be subject to its own (often stringent) performance and maintenance requirements, as does, for example, the SS7-based network. The LDP messages are transported between LSRs over whichever Layer 1 and 2 protocols have been selected for the given MPLS deployment and, consequently, share the same transmission path bandwidth with the other (aggregated or multiplexed) user data IP packet flows traversing the physical links between the LSRs. However, as will be evident below, the LDP procedures mandate reliable and ordered delivery of the main label exchange messages between LSRs and hence require the use of TCP procedures for the assured and flow-controlled delivery of these messages. The use of a reliable, flow-controlled TCP based transport protocol for the important LDP messages is clearly a significant advantage of the LDP approach in comparison with the use of other (adapted) protocols such as RSVP, for example, which do not use reliable message transfer for the delivery of critical control information such as label bindings. It should be pointed out that the requirement to use TCP (and its extensions) as part of the LDP control architecture to ensure reliability constituted an important factor in the ITU-T's decision to select only LDP (with its explicitly routed extension known as constraint-based routed LDP, or CR-LDP) as the control protocol of choice for the implementation of ATM-based MPLS as specified in ITU-T Recommendation Y.1310 [13.10]. In this respect, LDP and its extension CR-LDP are functionally analogous to the SS7 protocol architecture outlined earlier, which ensures reliable message transport based on the message transfer part (MTP) functionality or, in the ATM case, on the SSCOP procedures for flow control and retransmission.

The LDP message set may be conveniently grouped into four main categories as outlined below:

1. Discovery Messages. These messages are intended to indicate and maintain the presence of an LSR in the network. These messages include the MPLS "hello" message that is sent periodically to identify the presence of the LSR within the MPLS domain, or the "keep-alive" message, which may be used to indicate the status of the links interconnecting LSRs.
2. Session Messages. These messages are used to initiate, maintain, and terminate (TCP) sessions between the LDP peers to initialize transport prior to the transfer of label binding information between the LDP peers.
3. Advertisement Messages. These are the messages used to distribute, update or delete the label/FEC bindings information between the LDP peers within the MPLS domain.
4. Notification Messages. These messages are used to distribute error information or advisory information between the LSRs for predominantly control maintenance purposes.

It is important to stress that all the above LDP messages with the exception of the periodic hello message utilize the TCP session set up by the session messages (used to set up the TCP session) for transfer of the control information. However, the hello messages are periodically transmitted as a UDP packet addressed to the so-called "all routers on this subnet" group multicast address configured on each LDP port. Since the MPLS hello messages are generated periodically (with an interval less than a programmable "hold timer" value), it is assumed that loss or corruption of hello messages need not utilize TCP procedures for retransmission and hence the UDP mode is sufficient. In any case, the LDP procedures also specify an additional "keep-alive" mechanism that is intended to monitor the integrity of the LDP sessions once they are initiated. The keep-alive mechanism operates by monitoring the receipt of a (settable) "keep-alive timer." In the event that no LDP message is received within the keep-alive interval, including a dedicated "keep-alive" message that may be sent if no other LDP message is available, the LSR assumes that the transport "connection" is defective or the peer LSR has failed and initiates closure of the session. These and other maintenance procedures, analogous to similar capabilities in conventional SS7-based control architecture, constitute an important strength of the LDP approach in comparison with other protocols (such as RSVP) which may be used for label distribution.

The LDP message set includes the following messages, whose primary function may be evident from their names, but may warrant a brief description as outlined below.

1. Hello message
2. Initialization message
3. Keep-alive message
4. Address message
5. Address withdraw message
6. Label mapping message
7. Label request message
8. Label abort request message

9. Label withdraw message
10. Label release message
11. Notification message

It is of interest to note that the LDP message set applies to any interface between the LSRs, so that MPLS makes no distinction between a UNI or an NNI, as is the case in ATM signaling protocols. Thus, in general, the LDP procedures are symmetric, although LDP does utilize the notion of a given LSR operating in an "active" or "passive" role in the message exchange process. The symmetric procedures are consistent with the MPLS architecture, which assumes that the "edge" LSRs generally append the label stack to the incoming IP packet, so that the intermediate LSRs operate as network node interfaces. In this scenario, the incoming IP packet entering the MPLS domain may be analogous to data transfer across a "UNI" link into the edge LSR.

The LDP procedures utilize the notion of an "LDP identifier," which is used to identify a given LSR as well as the label space to be used in the label exchange process. The LDP identifier, which is basically carried within the hello message, is a six octet entity comprising the (four octet) IP v.4 address assigned to the LSR in question, followed by a two-octet field that encodes a "label space identifier" whose value signifies the type of label to be used by the LSR, e.g., ATM-related or FR-related labels, etc. For the case where a so-called "platform-wide" label type is to be used, as opposed to an interface-specific label type, the label space identifier is simply coded as all zeros. In addition to the LDP identifiers that uniquely identify each LSR and label space to be used, LDP messages also include a four-octet "message identifier" field that encodes a 32-bit value used to identify the particular message in a sequence of message exchanges. The message ID is basically used to correlate message sequences, which may bear a relationship to each other, such as (error) notification messages relating to a given message sequence. The message ID function is somewhat analogous to the use of the "call identifier" procedures in conventional ATM signaling messages, whereas it may be recognized that the LDP identifier is analogous to the use of "node identifier" in PNNI procedures, where an AESA is used to uniquely identify the PNNI node.

An interesting aspect of the LDP procedures is that, unlike ATM signaling, the exchange of multiple LDP messages concatenated within a simple so-called "LDP protocol data unit" (PDU) can be transmitted. In effect, the LDP procedures assume that message exchanges are enacted by sending LDP PDUs, where each LDP PDU can be made up of one or more of the LDP messages listed above. This may not apply to the hello and keep-alive messages, since it is likely that these may be generated as periodic "single" messages on the UDP and TCP ports, respectively. As shown in Fig. 13-23, the LDP PDU structure comprises a "common" LDP header appended to a string of one or more LDP messages, each of which is structured in a (nested) TLV format with its own message ID value. The common LDP PDU header includes fields signifying the protocol version, PDU length in octets, and the LDP identifier entity described earlier. Although the maximum allowable PDU length is a negotiable parameter, a default maximum of 4096 octets is specified.

It should be noted that this capability to send multiple, possibly unrelated LDP messages "bundled" within a single LDP PDU structure between the LSRs is clearly different from the case of ATM signaling procedures, where it is assumed that single messages are transmitted over both UNI and NNI between the ATM NEs. It is evident that this somewhat basic difference in messaging procedures between LDP and conventional ATM signaling will need to be factored in when considering control plane interworking between LDP-based MPLS and ATM networks, particularly for the service interworking case, in which signaling protocols are supposedly terminated at the interworking function (IWF).

Bearing in mind the general comments above on LDP procedures and capabilities, it is of interest to briefly outline the key message functions and the TLVs used in the LDP process. Although the detailed message structures and codings are outside the scope of this account (but

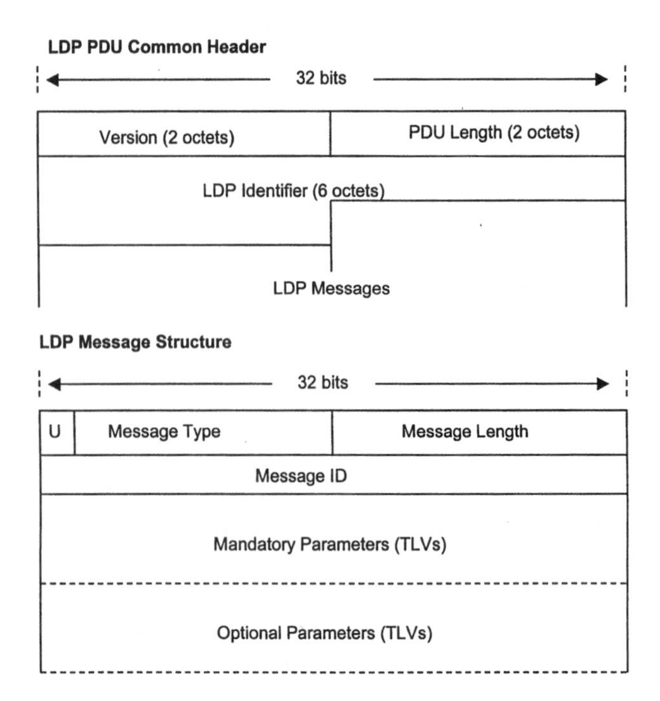

Figure 13-23. LDP message structures.

may be referred to in the LDP documentation), the features relevant to interworking with ATM are highlighted below.

1. Hello Message. As indicated earlier the LDP hello message is exchanged periodically between the LSRs in the MPLS domain to enable discovery of the network topology, analogous to its use in many conventional routing protocols, including PNNI. Two types of hello messages are feasible in LDP: those exchanged between adjacent LSRs, known as link hellos, and those that may be "targeted" to a specific LSR, known as targeted hellos. The targeted hello messages utilize the LDP identifier (i.e., the IP address of the target router) to establish the LDP peer relationship with the target. As noted earlier, the hello messages are the only ones sent over a UDP (unassured) transport, but contain a "hold timer" value specifying the maximum interval anticipated between hello messages on the links; default values of 15 secs for link hellos and 45 secs for targeted hellos are specified.
2. Initialization Message. The LDP Initialization message is used to set up the LDP TCP sessions as well as to initiate negotiation of a large number of parameters and operating modes for the subsequent LDP procedures. As such, the initialization messages as well as the associated procedures are relatively complex in comparison with some of the other LDP messages. Here it only needs to be noted that part of the negotiation procedures include the type and range of the label space, including whether it is ATM- or FR-related. The flexibility intended in the MPLS framework clearly implies that a significant number of options will need to be initialized in this message exchange, which may adversely impact LSP set-up latency for some network applications.
3. Keep-alive Message. As indicated earlier, the keep-alive message provides a mechanism to monitor the integrity of the LDP session transport (TCP) connection. In this process, receipt of any other LDP message (except a hello message, since this is a UDP "connec-

tion") or a keep-alive message within a settable keep-alive timer interval signifies that the TCP connection is operational. If no LDP PDU or keep-alive message is seen before the keep-alive interval expires, it is assumed the session is down.

4. Address Message. The address message, as its name implies, is sent by an LSR to its LDP peers to provide the set of interface addresses for update of the interface address database used in label exchange. If a given interface is activated or deactivated, the procedures require the use of the address or the "address withdraw" messages, respectively.
5. Address Withdraw Message. This message is used in conjunction with the address message above to signal the withdrawal or deletion of one or more interface addresses when these are deactivated or otherwise invalid.
6. Label Mapping Message. The label mapping messages are exchanged between the LDP peers to provide the FEC-to-label bindings necessary for MPLS operation. As may be expected, the procedures for the handling of the label mapping messages depends on the particular mode of operation of the MPLS domain. This includes whether the independent or ordered LSP control has been selected, and whether the "downstream-on-demand" or "downstream unsolicited" mode is to be used. The label mapping message may also be used in conjunction with the label request message described below, as will be evident.
7. Label Request Message. The label request message is used by the upstream LSR to request a label binding for any given FEC. As may be anticipated, use of this message is related to the corresponding label mapping message outlined above, specifically for the downstream-on-demand operating mode, which is of interest for the case of MPLS over ATM interfaces. However, the procedures recognize a number of cases where a label request message may not result in a corresponding label mapping, including (1) loop detection, (2) no label resources available, and (3) no route available. In these and other cases, the status may be communicated by means of the notification message as described below.
8. Label Abort Request Message. The label abort request message may be used to abort an earlier label request message in the event of an intervening change in the FEC or other situations in which the LSR may choose to "undo" the consequences of the label request message outlined above. A notification message may be used to acknowledge that the abort actions have been carried out. However, it should be pointed out that the abort procedures are not mandatory; the upstream LSR may choose to use or release the label binding information sent in a label mapping message after receipt of the label abort request message.
9. Label Withdraw Message. The label withdraw message is used to indicate that a specific FEC label binding should not continue to be used as valid. This may occur if the LSR has been reconfigured and hence no longer includes a FEC binding. It may be noted that unlike the above label abort request message, the label withdraw message requires the withdrawal of the label binding in question as mandatory, and the associated procedures require that the LSR in question must respond with a label release message, as described below, to confirm the deletion of the requested label binding.
10. Label Release Message. The label release message is used to signal that a specific FEC label binding is no longer needed or valid. As indicated above, this message is also sent in response to the label withdraw message to confirm the deletion of a requested FEC-to-label binding in the LSR database. If the "conservative operation" mode is configured, the label release message may be generated when labels not corresponding to the next hop are discarded as outlined earlier. Alternatively, if the "liberal operation" mode is adopted, no label release messages need be sent, since all received label bindings may be stored by the LSR.

11. Notification Message. The notification message is used to inform LDP peers of error events or provide advisory information such as the status or outcome of processing the LDP messages. As such, the notification message may be used for a large number of cases, a consequence of which is that the associated procedures are relatively complex. The LDP procedures specify a list of status codes and TLVs that may be used to convey the error or status information within the notification message structure. In the event that a "fatal" error condition is detected by the LSR sending the notification message, it is also assumed that the LSR will subsequently close the LDP session and discard the label bindings and FECs associated (or learned) from that session. In this sense, the notification message may also initiate fault diagnostic actions within the MPLS node.

As indicated earlier, the LDP messages outlined above contain the signaled information as TLV objects (often just called TLVs), which may be considered to be analogous to the "information elements" (IEs) or "parameters" familiar in conventional ATM (or SS7) signaling. The LDP procedures define a number of such TLVs, some of which may be carried as either mandatory or optional parameters in any given message. In addition, it may be recalled that LDP allows for the concatenation of more than one possibly unrelated message within one LDP PDU structure, which is transmitted between the LDP peers in any given LDP (TCP) session. It is not necessary to describe each one of these LDP TLVs here, although it may be of interest to list some frequently used TLVs, whose functions in general may often be readily inferred from their names. These TLVs include:

1. Forwarding Equivalence Class (FEC) TLV. The FEC TLV is made up of a list of one or more "FEC elements" of which three types have been currently defined. These FEC elements are: (1) address prefix, (2) Full host address, and (3) wildcard FEC element for which no specific value is assigned. The wildcard FEC element is only used in the label withdraw and label release messages to signify removal of all FECs associated with a specified label.
2. Label TLV. The Label TLVs are used to encode the label values and may appear in a number of the LDP messages outlined above. It should be recalled that the detailed structure of the Label TLVs depends on the type of label selected (during initialization or by configuration) in the MPLS domain. The Label TLV may be used to encode (a) a 20-bit generic MPLS label, as described later for the case of an MPLS "shim" header structure; (b) ATM labels, which may be either VPI or VCI or both; and (c) frame relay labels, which may be either the 10, 17, or 23 bit DLCI values.
3. Address List TLV. The address list TLV is used to carry the IP-address related information, typically in the address and address withdraw messages. The address is typically a full IP v.4 (four-octet) address.
4. Status TLV. The status TLVs are used to carry the error, advisory, or outcome-related state information, typically in the notification message, as outlined earlier. The status TLV may also be carried as an optional parameter in other LDP messages if required. The LDP procedures have identified a large (some 23) number of status codes and error indications, somewhat analogous to the "cause" codes used in conventional ATM- and SS7-based signaling protocols. The status TLV also includes the message type codepoint in order to identify the specific message to which the status data refers.

In addition to the above TLVs, it should be stressed that the LDP procedures also define a sizeable number of other TLVs. The Version 1 specification lists some 20 TLVs in all, including the "hop count" and "path vector" TLVs, which may be used for loop detection and avoidance as well as vendor-specific and experimental TLVs for proprietary use. It should also not be ruled out that as the functional requirements for MPLS-based networking increase, additional TLVs and status codes may be added in future versions of the protocol. The flexibility designed into the LDP procedures suggests that MPLS-based networks may be operated using the required level of functionality in the signaling mechanisms. However, as pointed out earlier, providing for a range of options may often result in interoperability issues arising when attempting to interwork networks configured with differing sets of optional capabilities. Admittedly, this caveat applies to other control protocols as well as LDP and may thus be raised here as a general observation on the often difficult trade-offs required between flexibility and complexity in protocol design.

13.10 EXPLICIT AND CONSTRAINED ROUTED LDP (CR-LDP)

In the brief outline of the LDP message set, TLVs, and associated procedures above, the focus was on describing the basic LDP signaling mechanisms and capabilities. However, as was pointed out earlier, an important potential network application of MPLS involves the capability to simplify IP traffic engineering or QoS-based routing by using explicitly routed LSPs. These additional capabilities have been developed by simply enhancing the basic LDP procedures with additional TLVs that allow for the possibility of using "source-routed" LSPs and traffic-related parameters to enable QoS support, in much the same way as was done in ATM using PNNI source-routed messaging. These enhancements to the basic LDP procedures are known as constrained routed LDP (CR-LDP) protocols [13.8], since they enable the use of explicit QoS policy or "constrained" routing as opposed to conventional hop-by-hop independent routing.

It may be recalled that in conventional IP routing protocols such as OSFP, the routing algorithm independently computes an "optimum" path through the network, which may then be used irrespective of the amount (or QoS requirements) of traffic flowing through that path. This may result in situations in which the "best' paths are overutilized, resulting in congestion conditions, whereas other possible paths are underutilized. This situation is also possible for the case of MPLS as long as independent (hop-by-hop) LSP control is used. In this case, the LSPs will still "follow" a best path with potential overutilization and congestion degradation, whereas other less optimum routes remain underutilized. However, if explicit routing is used, the network operator may select, or "constrain," the path based on policy, QoS considerations, or traffic load sharing to obtain more equitable traffic engineering capabilities not easily possible with conventional IP networking. It may also be pointed out in this context that although the option of using source routing is possible in principle in the conventional IP-based routing procedures, in practice these procedures have not been widely used. For the case of MPLS, the quasiconnection-oriented nature of the LDP procedures outlined earlier enable a more direct extension for deriving the benefits of explicit, source-routed LSP set-ups.

The basic CR-LDP procedures introduce four main additional TLVs (CR-LDP TLVs) to the LDP suite as outlined below. These include:

1. An Explicit (or Source) Route List. This list of typically IP v.4 addresses is included in the label request message from upstream to downstream LSRs, specifying the route to be followed by that message. It should be noted that the use of source routing in these procedures implies that ordered LSP control must be used in CR-LDP, essentially coupled with downstream-on-demand operation.
2. Traffic Parameters. In order to enable the LSRs to reserve resources and thereby provide the possibility of QoS categories along the explicit routes, CRP-LDP introduces MPLS

traffic parameters, somewhat analogous to the case of ATM and FRBS. These parameters are described in further detail below. The MPLS traffic parameter TLV is carried in the label request message from ingress to egress, but the CR-LDP procedures allow the individual LSRs to decrease specific traffic parameter values if they are flagged as negotiable. The final values of the downward negotiated traffic parameters are then inserted in the upstream label mapping message to enable each LSR to allocate resources accordingly for the CR LSP. Consequently, this traffic parameter negotiation procedure enables either the ingress, egress, or any intermediate LSR to determine the final traffic values used to allocate resources within each LSR along the path. It may be noted that this procedure is different from that used by ATM signaling renegotiation of bandwidth, where the source essentially initiates renegotiation of connection bandwidth.

3. Resource Class (also called "Color"). The Resource class parameter is used to provide additional degrees of constraint to the LSP set-up when required. Thus CR-LDP routes may be constrained only to links of a given resource class or "color." The resource class TLV essentially includes a 32-bit "mask" that may be used for constraint-based routing algorithms.
4. LSP Identifier (LSPID). This capability serves to uniquely identify LSP tunnels within an MPLS network for explicit routing in the case of "nested" LSPs.

It will be recognized that the source routing mechanism used for explicit routing in CR-LDP is functionally similar to that employed in PNNI signaling procedures. The essential difference is that PNNI nodes are identified in the designated transit list (DTL) by means of their ATM end system address (AESA), whereas the MPLS nodes or LSRs along the explicit route are identified by means of their IP v.4 addresses. Other differences in detail between CR-LDP and PNNI source routing procedures may also be identified, but these are not of primary interest here. More importantly, it is of interest to outline the differences between the CR-LDP MPLS traffic parameters carried in the TLV and the ATM traffic descriptors described in detail earlier.

The CR-LDP traffic parameters specified in the initial release of the specification include the following:

1. Peak Data Rate (PDR). The Peak Data Rate (PDR) is defined as the maximum rate at which traffic should be sent to the CR LSP ingress node, i.e., the LSR that demarcates the ingress to the explicitly routed LSP domain. The PDR is described in terms of a "token bucket" (i.e., leaky bucket) algorithm, as outlined earlier in relation to both RSVP and ATM traffic control, and is used in conjunction with the "peak burst size" parameter listed below.
2. Peak Burst Size (PBS). The peak burst size (PBS) traffic parameter is used in conjunction with the PDR outlined above as the second parameter in the token bucket algorithm to limit the traffic rate on the CR LSP. The PBS parameter signifies the maximum burst size at the peak data rate, assuming typical "on–off" traffic sources as may be anticipated for bursty data traffic, and may be viewed as somewhat analogous to the maximum burst size (MBS) descriptor identified in characterizing VBR ATM traffic.
3. Committed Data Rate (CDR). The committed data rate (CDR) parameter signifies the traffic rate actually committed to, or guaranteed, in the CR LSP; it may be less than the PDR described above. As for the PDR and PBS parameters, the CDR is used in conjunction with the "committed burst size," as described below, in a token bucket representation to indicate the bandwidth or resources that are reserved for the CR LSP.
4. Committed Burst Size (CBS). The committed burst size (CBS) is used in conjunction with the CDR parameter listed above in the token bucket representation to signify the resources

reserved for the CR LSP. As for the PBS parameter above, the CBS may be viewed as the maximum burst at the committed data rate for which buffer resources are reserved in any given LSR.

5. Excess Burst Size (EBS). The Excess Burst Size (EBS) traffic parameter also introduced in the CR-LDP procedures may be viewed as a measure of the traffic exceeding the CBS that may be sent on a given CR LSP, but for which resource may not have been committed. It may be used as an optional parameter in the committed rate token bucket representation, but its use as an additional traffic parameter, which may require enforcement, appears somewhat questionable.
6. Frequency. In addition to the "rate" and "burst size" related parameters listed above, CR-LDP also introduces a one-octet field that may be encoded with a parameter termed "frequency," which may be used to indicate how frequently the CDR traffic parameter should be assigned to a given CR LSP. The values assigned to the frequency parameter are relative and network-operator-specific, ranging from "very frequently" (no more than a single MPLS packet may be buffered), to "frequently" (several packets may be buffered). If the frequency parameter is unspecified it is assumed any number of packets may be buffered. In this way, it is assumed that the frequency parameter may be used to control the amount of variable packet delay introduced by the CR-LDP-based MPLS network. It may be evident that the frequency parameter has no direct counterpart in either ATM or FR traffic control procedures, and hence may be viewed as providing an additional "degree of freedom" in MPLS traffic control.
7. Weight. A one-octet parameter termed "weight" is also introduced in the CR-LDP traffic parameter TLV to provide a relative measure of the "share" of the given CR LSP in the available excess bandwidth. The larger the "weight" value, the larger the relative share the CR LSP receives of the available excess bandwidth. As for the case of the frequency parameter, the assignment of the weight parameter is network policy-specific, and its use may be viewed as providing an additional degree of freedom in allocation of bandwidth between the CR LSRs on a given transmission path. It may also be pointed out that as for the frequency traffic parameter, the weight parameter and related procedures do not have a direct counterpart in ATM or FR traffic management procedures, and will consequently impact service interworking between CR-LDP-based MPLS and ATM networks.
8. Negotiation Flags. Although not strictly a traffic parameter, the CR-LDP procedures include a one-octet field that encode a set of "flags" to indicate whether any or all of the above parameters are negotiable, with the exception of the frequency parameter. The parameter negotiation procedure was outlined earlier, and it should be stressed that each parameter may only be modified downward by any given LSR along the explicit (source) route from ingress to egress. The label mapping message returned upstream hence contains the lowest negotiated values of the parameters that may then be used to allocate resources in each LSR traversed by the CR LSP.

The CR-LDP traffic parameters and their associated procedures listed above may be compared with those defined earlier for ATM (or FRBS) from the perspective of QoS interworking and resource management. It is of interest to note that the use of traffic parameters such as peak data rate and committed data rate (coupled with their associated burst sizes in the ubiquitous token bucket algorithm) is somewhat closer to the approach adopted by the frame relay bearer service (FRBS) than by ATM. However, it may be argued that the CDR parameter may be interpreted in a similar way, as implied by the sustainable cell rate (SCR) traffic descriptor used in the description of the VBR (or SBR) service category (or ATC) in ATM traffic control. Moreover, it cannot be denied that the PDR parameter is analogous to the use of the PCR traffic descriptor central to the ATM traffic control mechanisms underlying the definitions of the ATM service categories (or ATCs in the ITU-T terminology). While these basic similarities will clearly assist in the interworking between CR-LDP-based MPLS and ATM, it is important to recognize some essential differences of approach that may complicate the overall internetworking scenarios. As already noted, the inclusion of the independent frequency and weight parameters clearly has no direct analog in either ATM or frame relay traffic management, and the somewhat relative nature of these mechanisms needs to be accounted for in allocated resources in an internetworking scenario between MPLS and ATM.

It should also be stressed that the CR-LDP traffic parameters apply to the specific CR LSP set up by the messaging procedures and *not* the individual (possibly host-to-host) IP packet flows that may be aggregated within each of the explicitly routed CR LSPs set up from ingress to egress LSR in the MPLS domain. Consequently, the CR-LDP approach assumes that the IP traffic flows aggregated in a given CR LSP are somehow controlled before the traffic enters the CR LSP, in accordance with the above parameters. Since the CR-LDP procedures are essentially a (signaling) protocol enhancement to the basic LDP procedures, they do not concern themselves with the detailed mechanisms or functions required to achieve this "network edge" traffic enforcement capability. In effect, analogous to the case of the DiffServ paradigm, CR-LDP based traffic control assumes that these network edge traffic enforcement capabilities may be separated from the resource allocation procedures signaled within the LSPs forming the core of the MPLS domain. The traffic control functions employed at the network edge may form part of an overall network traffic engineering or resource management framework for MPLS, which is outside the scope of the CR LSP signaling procedures. In contrast, ATM-based networks assume the specification of end-to-end VCC- (or VPC-) specific traffic conformance definitions based on the ATM traffic descriptors that may be enforced at the UNI or NNI in accordance with precisely defined conformance criteria for any selected ATM service category (or ATC).

It is also of interest to mention in passing that the CR-LDP procedures include a mechanism to prioritize the CR LSPs with respect to their importance in the allocation of the overall network resources. This procedure allows for the ranking of the CR LSPs so that higher-priority CR LSPs obtain network resources (i.e., bandwidth) before those ranked with a lower priority, enabling these to be effectively preempted or "bumped" in the event that sufficient bandwidth is not available. It should be noted that this CR LSP preemption mechanism operates independently of the traffic control procedures outlined above. The CR LSP priority or preemption procedures are based on introducing two additional parameters termed (1) "set-up priority" and (2) holding priority," and eight levels of priority are specified. In essence, when a CR LSP is set up, the set-up priority parameter is compared to the holding priority parameter of existing CR LSPs, whereby any with lower holding priority may be effectively preempted for their bandwidth. This process may be repeated until the lowest holding priority CR LSPs are either released or are relegated to the "worst" routes. This mechanism has no counterpart in ATM VCC set-ups, but its value remains questionable in public carrier networking.

13.11 THE GENERIC MPLS ENCAPSULATION STRUCTURE

It was mentioned earlier in several contexts that MPLS was intended to be used either over conventional Layer 2 (or link layer) protocols such as ATM or FRBS, or directly on its own as encapsulated in conventional LAN data links or the so-called "point-to-point protocol" (PPP) commonly used in IP-based networking [13.11]. To encompass transport over all these different link layer mechanisms, a "generic" MPLS header structure and encapsulation procedure has been defined. It may be noted that this mechanism is also sometimes referred to as the MPLS shim header encapsulation, since this header is inserted as a sort of "shim" between the conventional Layer 2 (LAN or PPP) header structure and the Layer 3 (i.e., the IP packet) header structure. It will also be recalled that since, in general, the MPLS architecture enables the "nesting" of multiple levels of LSPs, the generic or shim MPLS header is capable of being "stacked" according to a "last in–first out" discipline. Within any level *n* of the nested LSP hierarchy it was noted earlier that the label switching mechanism only operates on the top (or *n*th) label.

The structure of the four-octet generic MPLS header and the label stacking encapsulation protocol is illustrated in Figure 13-24. The MPLS header comprises four fields as shown and the function of each of these is summarized below as follows:

1. Label. This 20-bit field encodes the MPLS label value used for the label switching/multiplex mechanism within each LSR, as described previously. The label values from 0–16 inclusive are reserved for future functions and hence may not be used for general LSP assignment. It may be noted that the 20-bit MPLS label space is structured as a flat space, unlike the ATM label space, which is subdivided into the VPI and VCI logical hierarchy. In effect, the MPLS label is analogous to the DLCI label used in FRBS.
2. Experimental Bits. The 3-bit "experimental" field is initially unspecified, and may be used for future capabilities such as class of service (CoS) or OAM functions. This field has been used in some initial MPLS deployments to map a "class of service" indicator, which encodes values analogous to the DiffServ (DS) approach, to enable MPLS to support different levels of packet discard and/or delay priority. Whether this practice will continue or the field will be redefined remain open questions at this stage, since alternative proposals for in-band OAM functionality analogous to that used in ATM are also being examined.
3. The Stacking or S bit. This one-bit field is used to signify the bottom of the MPLS label stack, i.e., the first label in the stack. The value S = 1 signifies the bottom entry of

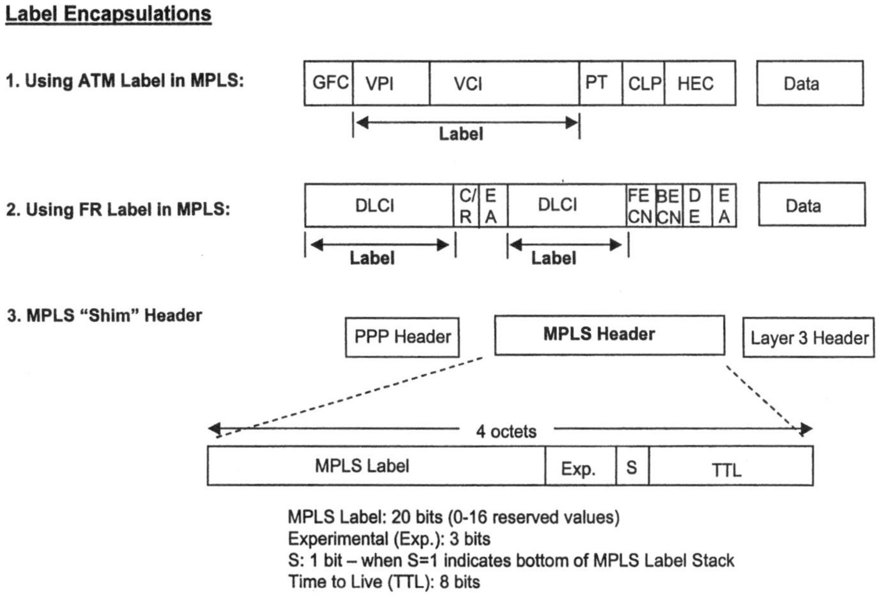

Figure 13-24. MPLS header structure.

the stack of MPLS headers. The value S = 0 signifies that additional labels follow in the stack.

4. Time to Live (TTL). This one-octet field encodes a time-to-live value, analogous to that used in the conventional IP header, to avoid endless circulation of packets in the event of routing loops. The value in the TTL field is set to the same value as in the IP header at the ingress LSR when the MPLS header is appended to the IP packet. As for the IP case, the MPLS TTL value is decremented by one each time an LSR is traversed. At the egress LSR, the MPLS TTL value is copied into the IP header TTL field, after which the IP TTL procedure may continue. As pointed out earlier, when MPLS is used in the ordered LSP control mode and for the case of CR-LDP, the formation of routing loops can generally be avoided so that in these as well as other scenarios, the use of the TTL function is largely superfluous. Typically, for MPLS utilizing underlying ATM or FR label switching protocols, routing loops are generally very unlikely due to the connection-oriented nature of these protocols. Consequently, the TTL function in the MPLS header is primarily of value in scenarios where independent LSP control procedures are used in conjunction with conventional hop-by-hop routing.

It may also be noted that the general MPLS header structure does not include an error check function over the header fields. Thus, it is assumed that error protection of the MPLS header, which may result in misrouting in the event the label value is corrupted, is provided by the error check (e.g., a CRC algorithm) in the underlying Layer 2 protocol. It is also assumed that this underlying Layer 2 protocol, whether it be PPP, FR, or ATM, also provides the capability for MPLS packet delineation, since no such functionality is included in the MPLS protocol control information (PCI). In the case of ATM, this delineation capability may typically utilize AAL Type 5 encapsulation of the MPLS packet, whereas with FR or PPP the conventional HDLC (flag-based) frame (PDU) delineation will be employed. From this perspective, it would appear that the MPLS PCI only provides a part of the function set typically associated with the conventional Layer 2 protocols, namely, that of label multiplexing. The other essential functions such as frame delineation and error detection/correction are not part of the MPLS "sublayer," but require the use of conventional HDLC or ATM protocol sublayers. Although the notion of a "transfer plane" does not play a part in the MPLS architecture, if this useful functional concept may be utilized here in a general sense, it would be evident that MPLS constitutes a "sublayering" of the conventional Layer 2 functions between label multiplexing and the other functions associated with the data link layer. Viewed in this light, the relationship of the MPLS protocol architecture to the other Layer 2 technologies described in some detail earlier may be clarified in a functional comparison for internetworking purposes.

In conclusion, the reader may be understandably forgiven if overwhelmed by a sense of déjà vu in reviewing the above outline of the MPLS framework and control procedures. The underlying functional similarities with ATM and probably more so with frame relay networking technologies, particularly for the case of CR-LDP-based MPLS, almost indicate a "reinvention" of earlier label multiplexing concepts. Nonetheless, terminology aside, basic differences between MPLS and ATM (or FR) also should be noted. These derive fundamentally from the IP v.4-address-oriented, connectionless roots of MPLS, despite the use of LDP (or other) signaling protocols to establish the quasiconnection-oriented entities termed LSPs to transport the IP flows. In contrast, both ATM and FR are strictly connection-oriented, with their traditional roots in ISDN-oriented signaling and routing based on the E.164/E.191 [6.15, 6.16] or AESA addressing system. The marked differences between the IP v.4 addressing (and associated connectionless routing) and the telephony-derived E.164/E.191 addressing/routing systems need hardly be stressed at this stage, but it is clear that the coexistence of both systems typify the central point of departure in any interworking network scenario.

CHAPTER 14

Perspectives in Networking Technologies

In the preceding survey of the main network and protocol architectural models developed for interworking between ATM and other networking technologies such as FRBS and IP, it is easy to lose sight of the simple, original intent of the ATM (or perhaps more broadly, the B-ISDN) vision. As described here, the many varied aspects of interworking, whether transfer-, control-, or management-plane related, between ATM and other networking approaches may be viewed as specific network applications of ATM technology. In this, the underlying implication is that ATM forms the core of the overall network, thereby serving as a unifying component for the transport and switching of a range of other technologies and services. Although such an "ATM-centric" perspective is consistent with the initial B-ISDN concept, it clearly constitutes only a part of the overall picture. As was stressed at the outset, ATM was conceived as both an overall packet switching and a transport (i.e., transmission) oriented technology, although this latter aspect has not generally been developed due to the overwhelming preponderance of conventional (SDH- or PDH-based) transmission systems in use, as well as in continued development.

It may also be observed that the intended potential of ATM even as a unifying packet switching technology has yet to be fully exploited, despite the many practical deployments of ATM-based backbone networks globally. In practice, initial deployments were primarily configured to provide meshes of semipermanent virtual circuits (PVCs) for the replacement of traditional leased line applications, analogous to the use of FRBS for the same purpose. Although these initial applications of ATM technology are clearly inevitable as a result of the significant cost savings made possible by the use of statistical multiplexing over virtual circuits, it is evident that they utilize only a fraction of the potential capability of ATM networking. Many of these potential capabilities are made possible by the use of ATM signaling/routing procedures, as described in detail in earlier chapters. In effect, the potential power of utilizing ATM signaling and routing procedures, not only for the provision of end-to-end QoS categories, but also for the creation of new services for generation of revenue, has yet to be generally exploited. The inherent power and flexibility of the common channel signaling capabilities underlying ATM, whether using the B-ISUP or PNNI-related protocol architectures, will only be realized in practice as more ATM network deployments utilize on-demand switched VCC or VPC connectivity in addition to the PVC-based applications in current use. This perspective would be analogous to the use of conventional SS7-based signaling and routing capabilities for new service creation in the existing voice networks, thereby harnessing the full potential of the advanced intelligent network (AIN) capabilities.

It will be recalled that the primary design intent underlying the development of the ATM networking approach initially was to provide *integrated* end-to-end communications capable of very high transfer rates, deterministic QoS, and reliability. With its roots in traditional ISDN-based telephony, many of these highly desirable network attributes are simply assumed to be a given with ATM, particularly since it was perceived that all the telephony applications as well as data-related traffic would eventually evolve to ATM for integrated (or unified) switching and transport. It was for this reason that the ATM layer protocol structure was deliberately designed to be "service-independent," oblivious of whether the cell payload contained digitized voice, video, data, or any other type of information. In this sense, ATM is inherently and fundamentally "multiprotocol" in a much more general sense than intended by subsequent network applications such as MPOA or MPLS. In addition, the use of a small set of AAL types, further sublayered into a "common part" and "service-specific parts" enables the accommodation of a wide range of higher-layer services or application protocol requirements. This results in a rationally layered protocol structure that may be readily extended to incorporate additional functionality when required for the accommodation of enhanced services. In addition to their multiprotocol properties, it will be recalled that the AAL protocol architecture also provides a relatively simple means for interworking between ATM and networks or services based on other transfer technologies, whether they be TDM, FR, or IP-based. The need for a straightforward interworking mechanism can hardly be overemphasized, considering the variety of transfer technologies already in existence and likely to continue in use.

Quite apart from its intended role as a unifying, high-performance switching and multiplexing technology for core network applications, the provision of comprehensive, detailed signaling procedures, closely coupled with traffic control mechanisms to ensure end-to-end QoS tailored to the selected service categories or applications, implies that ATM was designed for all internetworking purposes. However, unlike the traditional IP-based data internetworking approach, the ATM internetworking perspective is staunchly connection-oriented, with all the resultant benefits and drawbacks for the transfer of data frames. Although these issues have been extensively debated in the continuing and long rivalry between traditional connection-oriented and connectionless data networking approaches, it is increasingly recognized that in order to provide the benefits of QoS and traffic engineering capabilities, connectionless networking must be enhanced with some form of signaling procedure. This realization underlies the development of protocols such as RSVP and MPLS, based on LDP (or its enhancements such as CR-LDP), to provide IP-based networks with the traffic control and QoS capabilities required. With these protocols added to basic IP networking, it is clear that the traditional distinction between connectionless and connection-oriented transfer modes becomes somewhat blurred, and it becomes possible to consider a range of quasiconnection-oriented operating modes between the two limiting cases. Nevertheless, whether approached from the IP-oriented or ATM-oriented perspectives, it is now generally accepted that the benefits of QoS capabilities and traffic management may most readily be obtained by the use of signaling procedures in one form or the other. Since ATM was inherently designed with these capabilities in mind from the outset, it is hardly surprising that the ATM signaling procedures and traffic/resource management mechanisms are the most highly developed and precisely characterized at this stage, from both the theoretical and pragmatic perspectives.

Although the potential power of on-demand signaling for a unified networking approach may be readily recognized, the benefits of the common-channel signaling architecture inherent to the B-ISDN approach should also be borne in mind in evaluating the architectural options. The use of the common-channel control architecture, enables the "control" network to be separately engineered to provide the required extremely stringent levels of reliability and performance (as evidenced by the ubiquitous and robust SS7 networks) independently of the traffic load and performance of the user data path availability. In this model, the dedicated, highly redundant control network resources operate independently of fluctuations or failures that may occur in the user data streams, thereby ensuring that critical control messaging continues to be exchanged in the event of failures affecting other traffic. In addition, in normal operation the common-channel control architecture enables the independent exchange of control information during the normal user data transfer phase, since the control (signaling) virtual channel connection is available until the connection release is completed. This aspect may be useful in some network applications or services that can utilize the independent exchange of control information during the data transfer phase. It may also be noted that the common-channel architecture, as typified by the SS7-related capabilities, also enable the many attributes of the so-called advanced intelligent network (AIN) to be utilized and extended to the ATM case as required.

In IP-based networking, including its extensions to the MPLS framework, it should be recognized that there is no clear distinction drawn between the control network and data transfer between the routers, as may be expected in a connectionless network. However, in this case the routing messages exchanged between the IP routers (or LSRs) operate independently of the user data flows in order to maintain and update the routing tables, but these routing messages typically share resources with the other IP flows on any given link. The routing messages may be given precedence if marked, but may be subject to discard in the event of severe congestion. Similarly, in MPLS the label exchange (LDP) messages share bandwidth with other IP flows on a given link and may be subject to the same fate unless dedicated resources are allocated. However, this is typically not required as part of the basic MPLS framework, which draws no distinction between control plane and user plane resources. In this initial stage of definition, no performance or QoS requirements have yet been determined for the MPLS-based control mechanisms or even for the user traffic flow, but it is evident that for the design of large-scale, highly reliable, MPLS-based networks the performance and reliability of the control functions remain a key consideration. In this respect, it should be recalled that the LDP (and CR-LDP) procedures are clearly significantly more reliable than the RSVP-based extensions for label exchange, since LDP utilizes TCP procedures incorporating both flow control and retransmission mechanisms for assured delivery of the control messages. In contrast, RSVP does not use TCP-based transport, but relies on continuous (periodic) refresh messaging to maintain state, which results in unnecessary overhead and scalability limitations in practical deployments.

The benefits accruing from the clear separation of control/signaling functions from the transfer plane functions inherent in the ATM/B-ISDN architecture also extend to the case of the management and operational capabilities essential for the deployment of large-scale, robust carrier networks. In this regard, there can be no doubt that the comprehensive and detailed OAM mechanisms and related network management interfaces and procedures developed for ATM-based networking far outstrip any comparable capabilities in any packet mode transfer technology in use at present. In fact, the emphasis on OAM and associated network management capabilities, as typified by the TMN architecture, has sometimes drawn the criticism that ATM is somewhat "overengineered" and hence excessively complex from the perspective of management plane capabilities, thereby resulting in higher development costs relative to other, supposedly simpler, networking technologies. In respect to the ATM management plane capabilities, as described earlier, it is evident that the underlying philosophy has been to build on the traditional network management approach adopted by telephony or ISDN-related networking, including the stringent performance and availability criteria laid down for transport networks, particularly SDH-based optical networks.

Whether or not this overall network operations and management approach, rooted as it is in decades of traditional operational/craft experience, constitutes engineering "overkill" or is a vital component in ensuring robust, high-performance networking capabilities need hardly be debated in this context. However, it may be recognized that, as for the case of many of the potential control plane capabilities developed for ATM, the full potential of many of the OAM and associated network management interface capabilities (whether based on CMIP or SNMP approaches) have yet to be realized in many actual ATM network deployments. Obvious examples include the use of the ATM layer OAM performance monitoring (PM) and continuity check (CC) functionality, or the use of fast ATM protection switching capability, all of which are essentially intended to ensure extremely stringent availability performance in the network. Here again, as for the case of the myriad signaling/routing features and capabilities available in ATM, it may be assumed that the network or service provider can best select which of the specific set of network management functions are relevant for the networking applications or scenarios envisaged. This, of course, assumes that networking equipment vendors implement the required functionality in the hardware or software as appropriate, with a view to maintaining end-to-end interoperability of the OAM/NM protocols.

From the perspective of conventional IP-based networking, it may be recognized that the underlying network "management" philosophy is somewhat different from that adopted in the ATM/TMN framework. In essence, IP-based networking attempts to minimize network management overhead by the inherent reliability obtained from the use of dynamic routing protocols, as well as reliance on the operational capabilities provided by the underlying transport (transmission) systems. The IP-based networks may also utilize the capabilities provided by SNMP interfaces for event-driven fault reporting, and the basic ICMP messages to provide a level of management functionality if these protocols are supported by the routers or servers attached to the IP network. However, these capabilities may be viewed more as diagnostic in nature rather than providing comprehensive fault and performance management functionality as envisaged in the TMN B-ISDN model. Since conventional IP-based networking was initially designed to essentially support a "best efforts" service, with generally unspecified levels of packet delay or loss performance, it may be argued that the need for additional network management overhead was unwarranted. However, with the recent explosive growth in demand for IP-based services, such as the Worldwide Web (www) and e-commerce, as evidenced by the exponential growth in Internet traffic, there is increasing recognition of the need for IP network management enhancements. It is of interest to note that these capabilities are viewed as part of the "network policy" discussions initiated by the IETF, partly to distinguish these overall network management requirements from the protocol aspects dealt with by SNMP, and partly in recognition of the view that IP networking is assumed to be "self-managing," unlike telephony-derived networks.

As pointed out above, it is useful to recognize that IP networking derives as much benefit as any other traffic from the extremely high performance and reliability made possible by the wide deployment of modern optical fiber transmission systems such as SDH-based transport. The high throughput and low error performance, coupled with the comprehensive operations and maintenance functionality embodied in SDH-based transport is increasingly used to enhance the performance and reliability of the Internet. These transmission systems also enable the deployment of fast protection switching in the event of equipment failures, further improving the availability of the underlying transport links carrying IP traffic. These overall improvements in optical fiber (SDH) transmission systems may be viewed as applicable to all types of transported traffic, whether IP packets, voice signals, or ATM cells. However, the network management capabilities provided by these systems, though typically integrated into the TMN model for ATM or other transfer modes, needs also to be accommodated within the IP "policy" framework now under study in the IETF, assuming that a unified management or policy framework is desirable for the overall IP-based network.

The increasing deployment of ever higher capacity optical fiber transport systems between very high throughput (gigabit) routers within internet backbones has clearly enabled the Global Internet to accommodate the enormous increase in traffic experienced in recent years. With the advent of dense wavelength division multiplexing (DWDM) optical transport systems, the transport capacity available for IP-based traffic will be substantially increased even further in future. These developments have given rise to a model for IP traffic management that essentially relies on substantial capacity "overdimensioning," or overengineering to ensure that acceptable network performance/throughput can be achieved without the need for additional traffic control functionality in the transfer plane. In essence, it will be recognized that this so-called "fast routers and fat pipes" approach to traffic engineering is simply an extrapolation of the initial best-efforts IP networking approach, which was limited by low-capacity transmission paths and relatively slow routers in the past. Although a degree of overdimensioning is always going to be necessary in packet mode networks, which rely on statistical multiplexing of unpredictable traffic loads, it remains an open question as to whether transmission capacity increases, coupled with enhanced router performance alone, can adequately meet the demands of future internet traffic growth and diversification of IP-based services.

It may be evident that the above perspective on overall traffic engineering is significantly different from that adopted by the ATM/B-ISDN model, which defines detailed traffic control functions capable of "fine tuning" the traffic performance behavior of the network to optimize use of available resources. An example of this approach is the use of the available bit rate (ABR) service category (or ATC) to achieve very low cell losses with high link utilization by incorporating closed-loop feedback mechanisms for rate control. As described earlier, the ABR mechanisms for ATM traffic control are relatively complex but enable the service provider to make the best use of available capacity. Needless to say, the ATM traffic control model does not preclude the benefits of any overall increases in network resources, whether they be bandwidth increases resulting from DWDM, or increases in ATM cell switching throughputs. However, the ATM traffic control model relies on a range of resource management functions such as connection admission control (CAC), usage parameter control (UPC), and others to attempt to maximize traffic performance while protecting the QoS of existing VCCs or VPCs, assuming this is part of the end-to-end service agreement (or traffic contract). It is also of interest to recall that a similar resource management philosophy underlies the use of IP-related protocols such as RSVP, or even explicitly routed MPLS as described earlier.

On the one hand, it may be argued that deployment of "brute force" increases of network resources, made possible by technologies such as DWDM transport coupled with future "terabit" routers, including the use of either RSVP or DiffServ-based procedures for QoS differentiation, will be sufficient to meet future internetworking traffic growth requirements. On the other hand, there is also increasing interest in evolving the backbone networks toward the MPLS framework architecture, which offers the potential of more finely tuned traffic (or resource) management capabilities, somewhat analogous to the ATM model, although for the MPLS case the traffic control functions are much less well characterized. In principle, both approaches may be adopted in any given deployment, since it can be argued that they operate essentially independently of one another. Nonetheless, in the interests of cost containment, it may be necessary for a network operator to select one approach over another, based on the specific growth requirements anticipated over a period of interest. It must be admitted that at this stage of the development of either approach, it may be premature to select any given traffic engineering model in terms of a stringent cost/performance evaluation. As pointed out earlier in outlining the key attributes of both the IntServ/RSVP and DiffServ frameworks, issues relating to the scalability and complexity of the mechanisms involved need to be taken into account in any such evaluation. In addition, the apparent flexibility incorporated in either framework, coupled with the lack of quantitative performance/QoS objectives and enforcement procedures may result in interoperability problems in the provision of end-to-end QoS capabilities.

Some of the above considerations may also be viewed as applicable to the deployment of backbone networks based on ATM, but it must be acknowledged that ATM resource management and traffic control mechanisms are relatively well characterized and based on quantitative QoS and performance objectives. Capabilities such as CAC and UPC based on strict conformance definitions for any selected ATM service category imply stringent traffic enforcement on a per-connection basis to ensure that QoS objectives are met. Although seemingly more complex in terms of initial deployment, ATM traffic control is essentially an extrapolation of conventional traffic engineering and well proven network dimensioning principles, taking into account the possibility of varied levels of statistical multiplexing and QoS made possible by the use of packet mode virtual circuit connectivity. Networking based on these principles has been extensively characterized in practical deployments as well as simulation studies, lending a degree of confidence in the scalability and flexibility attributes of ATM-based backbone networks from the perspective of overall traffic engineering and dimensioning for optimum use of available resources.

In the broad networking perspectives outlined above, the underlying assumption is that the growth in (IP) data networking demand and associated requirements will essentially drive the overall character of the backbone carrier networks. Although it is clear that the exponential growth rates experienced by Internet-driven data traffic far outstrip the growth of traditional voice-oriented services (at least in the highly industrialized regions of the global economy), the need for conventional voice-based services will always need to be accommodated in any overall network scenario. From this perspective, there is no doubt that ATM-based networking provides significant advantages over other packet-based technologies, essentially because the technology was designed from the outset to accommodate the delay-sensitive, deterministic performance requirements inherent in voice communications. In addition, the ATM control protocol architecture as well as network management methodology is clearly based on extensions of the ISDN architecture, enabling relatively simple interworking with existing voice-based networks utilizing either AAL Type 1 or Type 2, as described earlier. As pointed out in several previous contexts, the underlying commonality between the traditional ISDN-based architectures and the B-ISDN or ATM approach to unifying networking for all types of applications results in a consistent evolution perspective that accommodates traditional telecommunications services with emerging applications fueled by the IP-based networking applications.

On the other hand, there is also considerable momentum in the drive to use these IP-based networks for the transport (and switching) of voice telephony signals, resulting in a plethora of protocols and applications in support of the so-called voice over IP (VoIP) network perspective. A description of the various VoIP approaches and related control architectures would require a treatise in itself, but it is only necessary to note here that the technical feasibility of using IP-based networks for voice telephony has been amply demonstrated and is increasingly used, albeit with somewhat reduced quality in many cases. Although it was initially believed that VoIP might eventually supplant conventional TDM-based telephony, particularly for long-distance communications in which cost constitutes a significant factor, this now remains an open question, given the success of deregulation and abundant bandwidth in driving down toll costs. Aside from these economic considerations, it may also be recognized that from a technical perspective no definitive architectural approach for interworking between VoIP and traditional ISDN-based voice services has as yet gained widespread acceptance. Although this may yet occur as the various proposals converge in the future, from a purely technical perspective it remains questionable as to whether widespread VoIP deployment will result in any significant cost/performance advantages over existing TDM-based voice telephony or, for that matter, even voice telephony over ATM (VTOA) approaches. As was evident in the earlier descriptions of the various IP and ATM interworking mechanisms, the primary problem to be considered for the control interworking protocols is the need for some form of address resolution between the two very different addressing structures used by IP and ATM networking. The addressing issue is also central to the question of VoIP interworking with ISDN/SS7-based services, where it seems inevitable that the ubiquitous telephony numbering plans will need to coexist with IPv.4 addressing mechanisms for the foreseeable future.

In general terms, it may even be argued that since ISDN/B-ISDN-based addressing (i.e., telephony-oriented numbering plans with their associated hierarchical routing topologies) will always coexist in parallel with the IP-based addressing/routing paradigms, the need for effective, seamless address translation (or resolution) capabilities is an inevitable consequence of any unified networking approach. From this perspective, it would seem immaterial whether the core networking technology is based on connectionless IP routing (or its extensions in MPLS) or on ATM switching, so long as the relevant control interworking is provided at "gateways" to accommodate the existing range of telecommunications services, including mobile (cellular) telephony. Considering the enormous installed base and investment in existing SS7-based networks, together with the rapidly expanding IP-based networks, both of which typically operate over a common transmission system infrastructure, the concept of unified networking may very likely imply mechanisms to provide seamless address interworking in gateways located between IP routers and voice (or ATM) switches. It will be recognized that this perspective of providing signaling and routing (i.e., address) interworking functions between the IP-based and SS7 based networks has given rise to numerous interworking protocols and gateway architectures in various forums including the IETF, ITU-T, and ETSI. It should also be pointed out in passing that in this context, but adopting a somewhat different approach, proposals to enhance the SS7-based procedures to accommodate IP as well as ATM-related routing and addressing, have also been extensively studied and adopted, notably by the ITU-T. However, it remains an open question at this stage as to whether such an SS7-based evolution to VoIP interworking, or the use of gateway interworking protocols will predominate in future deployments.

In surveying the growing number of different networking protocols and architectural approaches developed ostensibly toward the same ends, it is evident that strong competition between the various technologies plays an important role. As will have been apparent on numerous occasions in this text, networking technologies, whether they be FRBS, ATM, IP switching, MPOA, or MPLS, not only compete, but also influence each other in terms of enhancement of capabilities, and may even be interdependent on each other in certain scenarios. The interaction between ATM and FR, for example, may be somewhat obvious, given that both so-called "fast packet" approaches essentially evolved from somewhat similar control architecture roots in ISDN. The functional similarities between ATM and FRBS traffic control mechanisms was pointed out earlier, although in other respects such as OAM somewhat different approaches were adopted, resulting in much more comprehensive OAM functionality for ATM, given its intended role as a unifying, multiprotocol network technology. Thus, whereas ATM and FR may appear to compete for market acceptance in some network applications, it may also be noted that the phenomenal growth in FRBS was also dependent on the statistical multiplexing capabilities provided by the higher-capacity ATM backbone networks used to interconnect the FRBS networks, resulting in an interdependence between the two technologies.

A somewhat similar symbiotic relationship may also be discerned between ATM and IP-based networking, despite the fact that in this case considerable differences exist between the origins of these two networking paradigms. In the past, these differences have sometimes resulted in fierce criticisms by proponents of each technology against the other, often based on protocol details that may not prove critical in evaluating an overall networking approach. However, with the introduction of concepts such as IP switching and MPLS, as well as IntServ- and DiffServ-related notions of QoS differentiation in the evolution of the basic IP-based networking paradigm, the interdependence between ATM and IP-related networking is resulting in a convergence between these originally disparate networking approaches. Thus, functions deemed necessary by one technology may be incorporated in the competing approach, such as, for example, the use of dynamic routing protocols in ATM PNNI routing that are based on similar (link state) routing procedures used in IP. In addition, as noted earlier, traffic control and related signaling functions necessary to ensure deterministic QoS network behavior in ATM are now finding their way into the fundamentally connectionless IP-based networks, albeit under different, often looser, architectural frameworks, as characterized by RSVP and MPLS-related LDP/CR-LDP approaches. By disregarding the often divergent terminologies employed and focusing on the underlying functionality, the considerable influence of each technology on the other will be apparent. Here again, the interdependence also extends to the widespread use of high-capacity ATM backbone networks for the transport of IP packets, thereby providing statistical multiplexing gains with a given level of engineered QoS performance.

From an overall network planning perspective, the emergence of such interworked, interdependent networks evolving from a milieu of different networking technologies presents a mixed blessing. From the pragmatic viewpoint, it may be argued that such an approach simply builds on the extensive installed base of existing networks, whether in the enterprise or corporate network environment, or in the public carrier or network provider domain. As long as the interworking functions or gateways installed between the diverse technologies are capable of ensuring "seamless" data transfer as well as control and management information exchange without perceptible loss of expected performance or cost penalty, the end-user or customer should be indifferent to underlying technological discontinuities. In simple terms, it may be viewed that the primary purpose of any networking technology is to transfer information, whether it be voice or video signals or data frames, between remote terminals located anywhere with the required levels of performance and throughput and minimum cost or complexity. The fact that numerous networking protocol architectures have been developed and deployed over the years to achieve this end, with varying degrees of performance and functional comprehensiveness, is essentially a result of strong commercial and technological competitiveness in a rapidly evolving industry. This situation is likely to continue, since it is clear that each component of the networking protocol architecture will strive to evolve and enhance its capabilities in order to increase its market acceptance and value, as evidenced by the development of "gigabit" LANs and ultrafast IP routers, etc.

On the other hand it will be evident that the goal of achieving seamless interworking between the varying networking technologies is often difficult and may result in performance and/or complexity implications. In addition, the operation of multiple types of networks, each with its own network management structure and OAM procedures, may be viewed as somewhat undesirable from a service provider viewpoint. These considerations, coupled with the desire for creation of new services for revenue generation, have driven the development of "unifying" or integration technologies such as ATM, which provide a consistent overall protocol and network architecture for switching of all foreseeable services, thereby minimizing the need for interworking, at least in the longer term. More recently, the same objective has been espoused by the IP-based suite of protocols, which, although originally intended solely for data networking applications, now lay claim to the support of all types of services in the expanded Internet framework. Whether or not it will be eventually possible to evolve to a truly unified or integrated overall network solution using one or another of these technological approaches remains an open question, more subject to complex commercial interests and regulatory factors than relatively straightforward technical considerations. Nonetheless, there is a broad recognition of the benefits of integrating the switching and transport of all services or network applications within a unified technological framework, and this vision continues to drive the development of ATM technology and its convergence with IP-based networking.

From the perspective of a network operator or service provider, the proliferation of numerous functionally similar but procedurally different protocols may be viewed as a challenge. In order to remain competitive and exploit new opportunities for revenue generation in the increasingly deregulated and volatile telecommunications environment, a network provider may need to make fundamental choices between a range of technological approaches, many of which may appear promising but inadequately characterized. The protocols themselves may continue to evolve rapidly, thereby precipitating premature implementations by networking equipment vendors until a suitable level of stability is achieved in specifications. In some cases, a clear evolution path may not be evident, resulting in concerns that a given approach may lead to obsolescence or wasted investment. Although it may be argued that these concerns are inevitable given the rapid evolution of networking technologies, the problem is magnified by the enormous scale of the investment required for network growth and evolution. Consequently, the need to minimize or spread risk is evident in any underlying technology choice made by the network or service provider. In this respect, the deployment of standardized protocols and interfaces that are strictly capable of interoperable behavior across multiple administrative domains or network equipment types is clearly a minimum requirement, since the use of proprietary protocols exacerbate risk and may limit scalability. However, even with the use of strictly standardized protocols and interfaces, interoperability problems may sometimes arise if different functional capabilities are selected within a given subnet or region. Although it may be envisaged that the specification of a minimum functional subset would eliminate this problem, it must be admitted that this is rarely available. It is generally assumed that the network provider will select the appropriate functional requirements based on the specific services or network applications envisaged.

In principle, conventional network planning activities take into account all of the above considerations, by factoring in the growth in demand and traffic as well as the introduction of new services made possible by the advances in the underlying technology. In practice, this process has been rendered much more difficult as a result of the accelerated growth in both demand and the proliferation of networking technologies, which makes selection of any given approach more risky for long-term investment. As noted earlier, the unprecedented growth of Internet-related traffic has been accommodated so far by the deployment of high-capacity ATM backbone (or core) networks, coupled with the large increases in transmission bandwidth made available by the development of very high capacity optical fiber systems and high-throughput IP routers capable of gigabit forwarding rates. In order to meet the anticipated growth in future internetworking traffic, it has been conjectured that this trend is likely to continue, with the installation of DWDM optical transmission systems and IP routers capable of packet forwarding rates in order of terabits per second in the core networks. The role of ATM in this "brute force" approach to IP networking seems questionable, since ATM would be used primarily as a basic Layer 2 protocol, with many of its capabilities remaining unused. A simpler Layer 2 protocol such as PPP may be more suited for this application.

Alternatively, it may be argued that ATM can be used to provide the very high throughput rates required in the core network, since in any case the use of underlying transmission capacity remains common to either approach. The use of ATM provides the added benefits of enabling easier overall traffic control and QoS capability as well as the use of comprehensive OAM and protection switching for operational and reliability purposes. The integration of voice switching is also greatly facilitated by use of AAL Type 1 or 2 and CBR service category selection. In this more integrated perspective, the IP traffic may utilize one of ATM's transfer capabilities such as ABR or GFR, described earlier, to provide more efficient use of the available network resources in comparison with the brute force overengineering approach alluded to previously. Even in this perspective, the use of ATM signaling capabilities may not be fully utilized, since many backbone networks may simply be configured in terms of semipermanent virtual channels between the edge routers, provisioned via the management interfaces. However, in order to exploit the full flexibility and power of on-demand switched VCCs (or VPCs), the question of address resolution needs to be taken into account. As was stressed earlier, the coexistence of both E.164-

based numbering and IP addressing appears inevitable for the foreseeable future and constitutes the fundamental interworking issue from the perspective of any control or routing protocol architecture. In the classical IPOA network architecture described previously, address resolution procedures were deemed to introduce latency and scalability limitations, but it is conceivable that advances in server technology and/or directory services will also be used to improve the necessary translations between E.164-based numbering and IP addresses. This possibility provides a more consistent evolution path for existing ATM-based backbone networks, thereby enabling the benefits listed above in a more integrated networking framework.

While recognizing the inevitability of interwoked, interdependent networks in the development of any unified voice and data networking architectures, it may be recalled that the original simplicity of the ATM-based B-ISDN approach, intended for just that purpose, may be lost. In that concept, the full integration of all types of voice, video, or data signals was assumed to extend all the way to the end-users terminal equipment (or host) over the service-independent ATM layer functionality, and not just in the backbone part of the network. All networking functionality, whether related to on-demand (i.e., "dial-up") signaling and associated routing, traffic management with associated QoS capabilities, and OAM for robust availability performance, was intended for end-to-end applicability. Apart from initial limited trials and demonstrations of feasibility by some network operators, this simple approach, sometimes referred to as "direct" or "pure" ATM, has as yet been largely unexplored, primarily as a result of the low cost and predominance of existing LAN-based technologies, as indicated above.

Although it remains somewhat speculative at this stage that such a conceptually simple, end-to-end direct ATM-based networking perspective will gain a wider foothold, given the aggressive advances underway in competing (largely IP-based) networking technologies, its inherent potential and comprehensiveness makes it difficult to rule it out entirely as a possible candidate. Thus, it has been suggested that whereas enterprise (corporate) networks will likely remain LAN-based, the advances in access network technologies such as those based on the asymmetric digital subscriber line (ADSL) and "fiber to the home" concepts, coupled with demand for high-performance residential applications, are more readily enabled by direct ATM end-user connectivity. In this context, the enormous potential of providing "dial-up" (switched) ATM VCCs capable of data rates on the order of 10 to 50 Mbit/sec or more directly to a home computer may be envisioned. This capability is possible even with existing ATM access technology, although at prohibitive cost, considering the need to upgrade existing (low-speed) access facilities, as well as the "prototype" nature of the equipment at present. However, these are barriers that could conceivably be overcome if the often anticipated residential or "home office" demand materialized. Whereas personal computing equipment generally includes the IP-based networking software as a matter of course, the incorporation of simple ATM UNI signaling protocol software to enable direct dial-up connectivity is in principle no more complex. Consequently, it may be assumed that such ATM signaling software could be incorporated at comparable cost, provided that sufficient demand materialized. The same assumptions may be expected to apply to existing cost barriers for provision of the ATM interface hardware or line cards for terminal equipment in the event of more widespread demand.

In summary, it would appear that three broad perspectives may be identified for networking architecture evolution, considering the competition between the underlying technologies as well as their mutual interdependence. These somewhat general perspectives are:

1. The IP over optical network approach. This perspective essentially embodies the notion of increasingly fast IP routers coupled with increasingly higher-capacity optical networks based on technologies such as DWDM, and may be viewed as an extrapolation of the conventional IP-based networking architecture to substantially higher speeds. The extrapola-

tion relies on massive overdimensioning to accommodate traffic growth and QoS demand and the evolution of conventional IP routing protocols to support additional capabilities in support of QoS or policy-based routing and traffic engineering. In this essentially "IP-centric" architecture, interworking with existing voice (telephony) networks is achieved by means of suitable "gateways," although the voice and data networks may be considered to operate independently of one another (apart from possible sharing of a common underlying optical transmission path). Although in principle ATM may be used to provide the necessary Layer 2 functionality (e.g., frame delineation and error detection), in this protocol architecture, its extensive capabilities are largely unused. Consequently, a simpler HDLC-based Layer 2 protocol such as PPP may be more appropriate, so that, in general, the role of ATM in this overall networking perspective may be considered to be minimal, if used at all.

2. A unified or partially integrated networking approach. In this general category of networking architectures, possibly the broadest of the three listed here, varying levels of integration and control interworking between the ATM- and IP-based networking approaches are attempted. Examples of this perspective to internetworking evolution include the existing CIPOA or MPOA networks and their extensions or variants, as well as MPLS, although in the latter case the routing is assumed to be based entirely on the IP address. Although in these mixed ATM, IP, and voice networking protocol architectures different levels of integration or interworking may be achieved as seen earlier, it is clear that an attempt is made to utilize the advantages of both the ATM- and IP-based capabilities where possible. Thus, ATM may be used to provide traffic management (engineering), QoS, and possibly OAM functionality in the core network, whereas IP enables the use of dynamic routing, IP-based applications or services, and enterprise networking. Interworking with voice telephony may be achieved either via the ATM/AAL Type 1 or 2 protocol architecture or by means of interworking gateways to the IP-based part of the network. Although not all the potential ATM capabilities (whether control, management, or transfer plane related) may be utilized in any given type of network architecture in this perspective, it is evident that it is assumed to play a role as part of the partially integrated protocol architecture.
3. The B-ISDN or fully integrated network approach. In this perspective, which approximates to the initial ISDN to B-ISDN evolutionary view for the integration of all telecommunications services over a common ATM-based switching and transport infrastructure, the ATM capability is assumed to extend to the end-user terminals or hosts. In this perspective, the integration of IP-based services or applications and voice or video signals occurs by means of the appropriate AAL type at the ATM connection termination point. In this "ATM-centric" approach, it may be assumed that potentially all the extensive ATM functionality in the control, management, and transfer planes may be called upon as required for any given network application. As noted earlier, interworking with existing voice networks may be viewed simply as an extension of SS7-based networks and routing is based on E.164- (or AESA-) derived numbering. Although it may be regarded that this perspective (as well as the related extension mentioned earlier called "direct ATM") is somewhat speculative, considering its radical nature relative to the existing installed base, its potential advantages in terms of performance and services enhancement remains compelling.

The rapid pace of change in telecommunications in general and internetworking technologies in particular suggests that it would be difficult if not foolhardy to predict which approach or protocol architecture is likely to prevail among the competing technologies in this volatile field. Yet a relatively stable basis is desirable to encourage the large investments required for the development of new networks and services. This balance poses a challenge to both the network planner and equipment designer in considering which functionalities and protocol architecture need to be incorporated, and at what stage in the overall plan. Some general trends may be predicted with a high level of confidence, such as increased bandwidth capacity, use of optical transmission systems, and faster processing speeds and packet forwarding rates, while others may be more speculative, necessitating caution. It may also be recognized that in some cases the success or acceptance of a given approach may be more a matter of commercial interest and marketing "hype" than proven technological superiority. Nonetheless, a detailed functional and performance evaluation of each protocol architecture is clearly necessary in the network planning and design process to arrive at the "right" decision and facilitate interoperable solutions. In this process, the comprehensive functionality and detailed descriptions of the ATM protocols clearly help in ensuring interoperability. Moreover, the overall functional richness of ATM relative to any other existing packet mode technology, whether related to transfer, control, or management capabilities, ensures its continued importance in future internetworking applications.

References

Note. The ITU-T Recommendations listed here are published by the International Telecommunications Union (ITU), Geneva, Switzerland. IEEE specifications are published by the Institute of Electrical and Electronics Engineers, Inc., New York. RFC (Request for Comment) specifications are issued by the Internet Society and available on the IETF website.

General References on ATM, Frame Relay and IP

1. Martin de Prycker. *ATM: Solution for Broadband ISDN.* London, New York: Ellis Horwood, 1992.
2. Rainer Händel and Manfred Huber. *Integrated Broadband Networks. An Introduction to ATM Based Networks.* Reading, MA: Addison-Wesley, 1991.
3. William Stallings. *ISDN and Broadband ISDN with Frame Relay and ATM,* 3rd ed. Upper Saddle River, NJ: Prentice Hall, 1995.
4. David McDysan and Darren Spohn. *ATM: Theory and Application.* New York: McGraw Hill, 1994.
5. David Ginsburg. *ATM: Solutions for Enterprise Internetworking.* Reading, MA: Addison-Wesley Longman, 1996.
6. Uyless Black. *ATM: Foundation for Broadband Networks.* Upper Saddle River, NJ: Prentice Hall, 1995.
7. Radia Perlman. *Interconnections, Bridges, Routers, Switches and Internetworking Protocols.* Reading, MA: Addison-Wesley Longman, 2000.
8. W. Richard Stevens. *TCP/IP Illustrated,* Vols. 1–3, Reading, MA: Addison-Wesley Longman, 1998.

References for Chapter 2

[2.1] ITU-T Recommendation I.150, "B-ISDN Asynchronous Transfer Mode Functional Characteristics," 1999.

[2.2] ITU-T Recommendation I.121, "Broadband Aspects of ISDN," 1991.

[2.3] CCITT Blue Book Series.

[2.4] Roger L. Freeman. *Telecommunication System Engineering.* New York: Wiley, 1989.

[2.5] ITU-T Recommendation X.200. "Information Technology—Open Systems Interconnection—Basic Reference Model," 1994.

[2.6] Andrew S. Tannenbaum. *Computer Networks,* 3rd ed. Upper Saddle River, NJ: Prentice-Hall, 1996.

References for Chapter 3

[3.1] IEEE 802.6, "Local and Metropolitan Area Networks: Distributed Queue Dual Bus (DQDB) Subnetwork of a Metropolitan Area Network (MAN)," 1990.

[3.2] ITU-T Recommendation G.805, "Generic Functional Architecture of Transport Networks," 2000.

[3.3] ITU-T Recommendation I.326, "Functional Architecture of Transport Networks Based on ATM," 1995.

[3.4] ITU-T Recommendation I.732, "Functional Characteristics of ATM Equipment," 2000.

References for Chapter 4

[4.1] Bell Labs. *Transmission Systems for Communications,* 5th ed. Murray Hill, NJ: Bell Labs, 1983.

[4.2] Roger L. Freeman. *Telecommunication Transmission Handbook.* New York: Wiley, 1981.

[4.3] Walter J. Goralski. *Sonet,* 2nd ed. New York: McGraw-Hill, 2000.

[4.4] U. Black and S. Waters. *Sonet and T1.* Upper Saddle River, NJ: Prentice Hall, 1997.

[4.5] ITU-T Recommendation G.707, "Network Node Interface for the Synchronous Digital Hierarchy (SDH)," 2000.

[4.6] ITU-T Recommendation I.432 series, "B-ISDN User-Network Interface—Physical Layer Specification," 1995 to 1999.

[4.7] William Stallings. *ISDN and Broadband ISDN with Frame Relay and ATM,* 3rd ed. Upper Saddle River, NJ: Prentice Hall, 1995.

[4.8] ITU-T Recommendation G.782, "Types and General Characteristics of Synchronous Digital Hierarchy (SDH) Equipment," 1994 and merged with G.783 in 1997.

[4.9] ITU-T Recommendation G.783, "Characteristics of Synchronous Digital Hierarchy (SDH) Equipment Functional Blocks," 2000.

[4.10] ITU-T Recommendation G.702, "Digital Hierarchy Bit Rates" (*Blue Book*—Fascicle III.4), 1988.

[4.11] ITU-T Recommendation G.804, "ATM Cell Mapping into Plesiochronous Digital Hierarchy (PDH)," 1998.

[4.12] ITU-T Recommendation G.832, "Transport of SDH Elements on PDH Networks—Frame and Multiplexing Structures," 1998.

References for Chapter 5

[5.1] Natalie Giroux and Sudhakar Ganti. *Quality of Service in ATM Networks.* Upper Saddle River, NJ: Prentice Hall, 1998.

[5.2] R. Jain. "Congestion Control and Traffic Management in ATM Networks; Recent Advances and a Survey." *Computer Networks and ISDN Systems, 28,* 13, October 1996, pp. 1723–1780.

[5.3] R. Onvural. *ATM Networks: Performance Issues.* Norwood, MA: Artech House, 1994.

[5.4] C. Partridge. *Gigabit Networking.* Reading, MA: Addison Wesley, 1994.

[5.5] W. Stallings. *ISDN and B-ISDN with Frame Relay and ATM,* 3rd ed. Upper Saddle River, NJ: Prentice Hall, 1995.

[5.6] L. Kleinrock. *Queuing Systems.* New York: Wiley, 1976.

[5.7] ITU-T Recommendation I.371, "Traffic Control and Congestion Control in B-ISDN," 2000.

[5.8] ATM Forum, af-tm-0121.000, "Traffic Management Specification" Version 4.1, 1996.

[5.9] ITU-T Recommendation I.356, "B-ISDN ATM Layer Cell Transfer Performance," 2000.

[5.10] ATM Forum, af-uni-0010.002, "User Network Interface Specification," Version 3.1, 1994.

References for Chapter 6

[6.1] ATM Forum, af-bici-0013.003, "B-ISDN Inter Carrier Interface (B-ICI) Specification," Version 2.0, 1995.

[6.2] ATM Forum, af-sig-0061.000, "User Network Interface Signaling Specification," Version 4.0, 1996.

[6.3] ITU-T Recommendation Q.2761, "Functional Description of the B-ISDN User Part (B-ISUP) of Signaling System No. 7," 1999.

[6.4] ITU-T Recommendation Q.2762, "General Functions of Messages and Signals of the B-ISDN User Part (B-ISUP) of Signaling System No. 7," 1999.

[6.5] ITU-T Recommendation Q.2763, "Signaling System No. 7 B-ISDN User Part (B-ISUP)—Formats and Codes," 1999.

[6.6] ITU-T Recommendation Q.2764, "Signaling System No. 7 B-ISDN User Part (B-ISUP)—Basic Call Procedures," 1999.

[6.7] ISO/IEC 8348, "Information Technology—Open Systems Interconnection—Network Service Definition," Ed. 2, 1996.

[6.8] ITU-T Recommendation Q.2931, "Digital Subscriber Signaling System Nr. 2—User Network Interface (UNI) Layer 3 Specification for Basic Call/Connection Control," 1995.

[6.9] ITU-T Recommendation Q.931, "ISDN User–Network Interface Layer 3 Specification for Basic Call Control," 1998.

[6.10] ITU-T Recommendation I.430, "Basic User–Network Interface—Layer 1 Specification," 1995.

[6.11] ITU-T Recommendation I.431, "Primary Rate User–Network Interface—Layer 1 Specification," 1993.

[6.12] ITU-T Recommendation Q.2650, "Interworking between Signaling System No. 7 Broadband ISDN User Part (B-ISUP) and Digital Subscriber Signaling System No. 2 (DSS 2)," 1999.

[6.13] ITU-T Recommendation Q.2120, "B-ISDN Meta-Signaling Protocol," 1995.

[6.14] ITU-T Recommendation E.164, "The International Public Telecommunication Numbering Plan," 1997.

[6.16] ITU-T Recommendation E.191, "B-ISDN Addressing," 2000.

[6.17] ISO 3166-1 to 3, "Codes for the Representation of Names of Countries and Their Subdivisions"—Parts 1 to 3, ed., 1999.

References for Chapter 7

[7.1] ATM Forum, af-cs-0125.000, "ATM Inter Network Interface (AINI) Specification," 1999.

[7.2] ATM Forum, af-pnni-0026.000, "Interim Inter Switch Signaling Protocol (IISP)," Phase 0, 1994.

[7.3] ATM Forum, af-pnni-0055.000, "PNNI Specification," Phase 1, 1996.

[7.4] Radia Perlman. *Interconnections,* 2nd ed. Reading, MA: Addison Wesley, 2000.

References for Chapter 8

[8.1] ITU-T Recommendation M.3010, "Principles for a Telecommunications Management Network," 2000.

[8.2] Lakshmi Raman. *Fundamentals of Telecommunications Network Management.* New York: IEEE Press, 1999.

[8.3] ITU-T Recommendation I.751, "Asynchronous Transfer Mode Management of the Network Element View," 1996.

[8.4] Walter J. Goralski. *Sonet,* 2nd ed. New York: McGraw-Hill, 2000.

[8.5] ITU-T Recommendation I.610, "B-ISDN Operations and Maintenance Principles and Functions," 1999.

[8.6] ATM Forum, af-sec-0100.000, "ATM Security Specification," Version 1.0, 1999.

[8.7] ITU-T Recommendation I.630, "ATM Protection Switching," 1999.

[8.8] ATM Forum, af-nm-0020.000, "M4 I/F Requirements and Logical MIB: ATM NE View," Version 1.0, 1994.

[8.9] ITU-T Recommendation O.191, "Equipment to Measure the Cell Transfer Performance of ATM Connections," 2000.

References for Chapter 9

[9.1] ITU-T Recommendation I.731, "Types and General Characteristics of ATM Equipment," 1996.
[9.2] ITU-T Recommendation I.732, "Functional Characteristics of ATM Equipment," 1996.
[9.3] ITU-T Recommendation I.751, "Asynchronous Transfer Mode Management of the Network Element View," 1996.
[9.4] ITU-T Recommendation I.356, "B-ISDN ATM Layer Cell Transfer Performance," 2000.
[9.5] Bellcore GR-1110-CORE, "Broadband Switching System (BSS) Generic Requirements," Issue 4, 2000.
[9.6] Bellcore GR-1248-CORE, "Generic Requirements for Operations of ATM Network Elements (NEs)," Issue 4, 1998.
[9.7] ITU-T Recommendation I.311, "B-ISDN General Network Aspects," 1996.
[9.8] ITU-T Recommendation G.826, "Error Performance Parameters and Objectives for International, Constant Bit Rate Digital Paths at or Above the Primary Rate," 1999.

References for Chapter 10

[10.1] ITU-T Recommendation I.363, "B-ISDN ATM Adaptation Layer (AAL) Specification Series," 1996–2000.
[10.2] ITU-T Recommendation I.363.1, "B-ISDN ATM Adaptation Layer (AAL) Specification, Type 1 AAL," 1996.
[10.3] ITU-T Recommendation I.363.2, "B-ISDN ATM Adaptation Layer (AAL) Specification, Type 2 AAL," 2000.
[10.4] ITU-T Recommendation I.363.3, "B-ISDN ATM Adaptation Layer (AAL) Specification, Type 3/4 AAL," 1996.
[10.5] ITU-T Recommendation I.363.5, "B-ISDN ATM Adaptation Layer (AAL) Specification, Type 5 AAL," 1996.
[10.6] ITU-T Recommendation I.365.1, "B-ISDN ATM Adaptation Layer Sublayers: Frame Relaying Service Specific Convergence Sublayer (FR-SSCS)," 1993.
[10.7] ITU-T Recommendation I.365.2, "B-ISDN ATM Adaptation Layer Sublayers: Service-Specific Coordination Function to Provide the Connection-Oriented Network Service," 1995.
[10.8] ITU-T Recommendation Q.2110, "B-ISDN ATM Adaptation Layer—Service Specific Connection Oriented Protocol (SSCOP)," 1994.
[10.9] ITU-T Recommendation I.365.3, "B-ISDN ATM Adaptation Layer Sublayers: Service-Specific Coordination Function to Provide the Connection-Oriented Transport Service, 1995.
[10.10] ITU-T Recommendation I.364, "Support of the Broadband Connectionless Data Bearer Service by the B-ISDN," 1999.
[10.11] ITU-T Recommendation I.510, "Definitions and General Principles for ISDN Interworking," 1993.
[10.12] ITU-T Recommendation I.520, "General Arrangements for Network Interworking between ISDNs," 1993.
[10.13] ITU-T Recommendation I.580, "General Arrangements for Interworking between B-ISDN and 64 kbit/s Based ISDN," 1995.
[10.14] ITU-T Recommendation I.581, "General Arrangements for B-ISDN Interworking," 1997.
[10.15] ITU-T Recommendation G.823, "The Control of Jitter and Wander within Digital Networks which are Based on the 2048 kbit/s Hierarchy," 2000.

[10.16] ITU-T Recommendation G.824, "The Control of Jitter and Wander within Digital Networks which are Based on the 1544 kbit/s Hierarchy," 2000.

[10.17] ITU-T Recommendation Q.2630.1, "AAL Type 2 Signaling Protocol (Capability Set 1)," 1999.

References for Chapter 11

[11.1] W. Stallings, *ISDN and B-ISDN with Frame Relay and ATM,* 3rd ed. Upper Saddle River, NJ: Prentice Hall, 1995.

[11.2] ITU-T Recommendation Q.922, "ISDN Data Link Layer Specification for Frame Mode Bearer Services," 1992.

[11.3] ITU-T Recommendation I.122, "Framework for Frame Mode Bearer Services," 1993.

[11.4] ITU-T Recommendation I.620, "Frame Relay Operation and Maintenance Principles and Functions," 1996.

[11.5] ITU-T Recommendation I.555, "Frame Relaying Bearer Service Interworking," 1997.

[11.6] Frame Relay Forum, FRF 5, "Frame Relay/ATM PVC Network Interworking Implementation Agreement," 1995.

[11.7] Frame Relay Forum, FRF 8, "Frame Relay/ATM PVC Service Interworking Implementation Agreement," 1995.

[11.8] ITU-T Recommendation Q.933, "Digital Subscriber Signaling system No.1 (DSS 1)—Signaling Specifications for Frame Mode Switched and Permanent Virtual Connection Control and Status Monitoring," 1995.

[11.9] ITU-T Recommendation I.365.1, "B-ISDN ATM Adaptation Layer Sublayers: Frame Relaying Service Specific Convergence Sublayer (FR-SSCS)," 1993.

[11.10] ITU-T Recommendation X.36, "Interface between Data Terminal Equipment (DTE) and Data Circuit-Terminating Equipment (DCE) for Public Data Networks Providing Frame Relay Data Transmission Service by Dedicated Circuit," 2000.

[11.11] ITU-T Recommendation X.76, "Network-to-Network Interface between Public Networks Providing PVC and/or SVC Frame Relay Data Transmission Service," 2000.

[11.12] ITU-T Recommendation I.233.1, "ISDN Frame Relaying Bearer Service," 1991.

[11.13] IETF RFC 2427 (formerly RFC 1490), "Multiprotocol Interconnect over Frame Relay," September 1998.

[11.14] IETF RFC 2684 (formerly RFC 1483), "Multiprotocol Encapsulation over ATM Adaptation Layer 5," September 1999.

[11.15] ITU-T Recommendation Q.2933, "Digital Subscriber Signaling System No. 2—Signaling Specification for Frame Relay Service," 1996.

[11.16] ITU-T Recommendation Q.2727, "B-ISDN User Part—Support of Frame Relay," 1996.

[11.17] ITU-T Recommendation Q.921, "ISDN User–Network Interface—Data Link Layer Specification," 1997.

References for Chapter 12

[12.1] Radia Perlman. *Interconnections,* 2nd ed. Reading, MA: Addison-Wesley, 2000.

[12.2] W.R. Stevens. *TCP/IP Illustrated,* Vols. 1 to 3. Reading, MA: Addison-Wesley, 1998.

[12.3] RFC 791, "Internet Protocol," September 1981; and RFC 793, "Transmission Control Protocol," September 1981.

[12.4] IEEE 802.2, "Local Area Networks: Logical Link Control," 1989.

[12.5] IEEE 802.3, "Local Area Networks: Carrier Sense Multiple Access with Collision Detection (CSMA/CD)—Ethernet," 1985.

[12.6] IEEE 802.4, "Standard for Token-Passing Bus Access Method and Physical Layer Specifications," 1990.

[12.7] IEEE 802.5, "IEEE Standard for Local Area Networks: Token Ring Access Method and Physical Layer Specifications," 1997.

[12.8] RFC 2225 (formerly RFC 1577), "Classical IP and ARP over ATM," April 1998.

[12.9] L. Zhang et al., "RSVP: A New Resource Reservation Protocol," *IEEE Network,* IEEE, September 1993.

[12.10] RFC 2205, "Resource ReSerVation Protocol (RSVP)—Version 1 Functional Specification," September 1997.

[12.11] RFC 2475, "An Architecture for Differentiated Service," December 1998.

[12.12] RFC 2474 (formerly RFC 1455), "Definition of the Differentiated Services Field (DS Field) in the IPv4 and IPv6 Headers," December 1998.

[12.13] RFC 1122, "Requirements for Internet Hosts—Communication Layers," October 1989.

[12.14] RFC 1812, "Requirements for IP Version 4 Routers," June 1995.

[12.15] RFC 2598, "An Expedited Forwarding PHB," June 1999.

[12.16] RFC 2597, "Assured Forwarding PHB Group," June 1999.

References for Chapter 13

[13.1] ATM Forum, af-lane-0084-000, "LAN Emulation v.2 LUNI Specification," 1997.

[13.2] P. Newman et al., "IP Switching: The Inteligence of Routing, the Performance of Switching." Article published by Ipsilon Corporation (www.ipsilon.com).

[13.3] P. Newman et al., "IP Switching and Gigabit Routers." *IEEE Communications Magazine,* January 1997.

[13.4] RFC 3031, "Multiprotocol Label Switching Architecture," January 2001.

[13.5] RFC 3032, "MPLS Label Stack Encoding," January 2001.

[13.6] RFC 3035, "MPLS using LDP and ATM VC Switching," January 2001.

[13.7] RFC 3036, "LDP Specification," January 2001.

[13.8] IETF, draft-ietf-mpls-cr-ldp-05.txt, "Constraint Based LSP Setup Using LDP," 2000.

[13.9] ATM Forum, af-mpoa-0114.000, "MPOA Specification, Version 1.1," 1999.

[13.10] ITU-T Recommendation Y.1310, "Transport of IP over ATM in Public Networks," 2000.

[13.11] RFC 1548 (formerly RFC 1331), "The Point-to-Point Protocol (PPP)," December 1993.

[13.12] RFC 2332, "NMBA Next Hop Resolution Protocol (NHRP)," April 1998.

[13.13] RFC 1771 (former RFC 1654), "A Border Gateway Protocol 4 (BGP-4)," March 1995.

[13.14] IETF, draft-ietf-gsmp-08.txt, "General Switch Management Protocol v3," 2000.

Index

About the Author

Dr. Khalid Ahmad has been involved at the forefront of the development of ATM, Frame Relay, and MPLS protocols and their interworking from the inception of these technologies. He has contributed in all key areas of the technology including ATM traffic management, OAM, signaling, and adaptation layer and interworking protocols, authoring or co-authoring numerous technical proposals, reports, and studies in ATM-related fields. As a senior research engineering manager at Nortel Networks, a leading global manufacturer of ATM, optical fiber, and FR network elements, Dr. Ahmad has been involved in the practical aspects of telecommunications network planning and design based on ATM, IP, and SDH technologies.

Dr. Ahmad is an active participant in both the International Telecommunications Union (ITU, formerly the CCITT), and the ATM Forum. In the ITU he has chaired technical committees engaged in developing technical specifications for ATM equipment functional characteristics and ATM–FR Interworking, and is currently Rapporteur for ATM and AALs. In the last 5 years, he has been manager responsible for development of ATM, FR, and Interworking standards.

Prior to his research in networking technologies, Dr. Ahmad was involved in R&D work on semiconductor and optoelectronic device physics and technology for both telecommunications and space applications. He has published numerous papers and reports in this field.

Khalid Ahmad holds a BSc. with honors in Electrical and Electronic Engineering and a MSc. in Solid State Electronics from the University of Manchester Institute of Science and Technology (UMIST). He obtained a Ph.D in solid state physics from the Cavendish Laboratory, Cambridge University, U.K.